VOLUME FIVE HUNDRED AND NINETY NINE

METHODS IN ENZYMOLOGY

Fe-S Cluster Enzymes Part B

METHODS IN ENZYMOLOGY

Editors-in-Chief

ANNA MARIE PYLE
Departments of Molecular, Cellular and Developmental Biology and Department of Chemistry
Investigator, Howard Hughes Medical Institute
Yale University

DAVID W. CHRISTIANSON
Roy and Diana Vagelos Laboratories
Department of Chemistry
University of Pennsylvania
Philadelphia, PA

Founding Editors

SIDNEY P. COLOWICK and NATHAN O. KAPLAN

VOLUME FIVE HUNDRED AND NINETY NINE

METHODS IN ENZYMOLOGY

Fe-S Cluster Enzymes Part B

Edited by

SHEILA S. DAVID

University of California, Davis,
Davis, CA, United States

ELSEVIER

ACADEMIC PRESS

An imprint of Elsevier

Academic Press is an imprint of Elsevier
50 Hampshire Street, 5th Floor, Cambridge, MA 02139, United States
525 B Street, Suite 1800, San Diego, CA 92101–4495, United States
The Boulevard, Langford Lane, Kidlington, Oxford OX5 1GB, United Kingdom
125 London Wall, London, EC2Y 5AS, United Kingdom

First edition 2018

ISBN: 978-0-12-814717-7
ISSN: 0076-6879

For information on all Academic Press publications
visit our website at https://www.elsevier.com/books-and-journals

Publisher: Zoe Kruze
Acquisition Editor: Zoe Kruze
Editorial Project Manager: Alina Cleju
Production Project Manager: Divya Krishna Kumar
Cover Designer: Mark Rogers

Typeset by SPi Global, India

CONTENTS

CONTRIBUTORS

Angela-Nadia Albetel
University of South Carolina, Columbia, SC, United States

Fraser A. Armstrong
Inorganic Chemistry Laboratory, Oxford, United Kingdom

Andrey G. Baranovskiy
Eppley Institute for Research in Cancer, Fred & Pamela Buffett Cancer Center, University of Nebraska Medical Center, Omaha, NE, United States

Somsuvro Basu
Institut für Zytobiologie, Philipps-Universität Marburg, Marburg, Germany

Janina Bergmann
Institut für Zytobiologie, Philipps-Universität Marburg, Marburg, Germany

Eckhard Bill
Max-Planck-Institut für Chemische Energiekonversion, Mülheim an der Ruhr, Germany

Eric J. Camire
Boston University, Boston, MA, United States

Stephen P. Cramer
University of California, Davis, Davis, CA, United States

Sheila S. David
University of California, Davis, Davis, CA, United States

Serena DeBeer
Max Planck Institute for Chemical Energy Conversion, Mülheim an der Ruhr, Germany

Rhiannon M. Evans
Inorganic Chemistry Laboratory, Oxford, United Kingdom

Sven-Andreas Freibert
Institut für Zytobiologie, Philipps-Universität, Marburg, Germany

Leland B. Gee
University of California, Davis, Davis, CA, United States

Manik C. Ghosh
Molecular Medicine Branch, *Eunice Kennedy Shriver* National Institute of Child Health and Human Development, Bethesda, MD, United States

John D. Grossman
Boston University, Boston, MA, United States

Gregory P. Holmes-Hampton
Molecular Medicine Branch, *Eunice Kennedy Shriver* National Institute of Child Health and Human Development, Bethesda, MD, United States

Cindy Khuu
University of California, Davis, Davis, CA, United States

Kori T. Lay
University of California, Davis, CA, United States

Roland Lill
Institut für Zytobiologie, Philipps-Universität, Marburg; LOEWE Zentrum für Synthetische Mikrobiologie SynMikro, Marburg, Germany

Chandrima Majumdar
University of California, Davis, Davis, CA, United States

Clare F. Megarity
Inorganic Chemistry Laboratory, Oxford, United Kingdom

Andrew Melber
University of Utah Health Sciences Center, Salt Lake City, Utah, United States

Sarah L.J. Michel
School of Pharmacy, University of Maryland, Baltimore, MD, United States

Ulrich Mühlenhoff
Institut für Zytobiologie, Philipps-Universität, Marburg, Germany

Nicole N. Nuñez
University of California, Davis, Davis, CA, United States

Caryn E. Outten
University of South Carolina, Columbia, SC, United States

Cedric P. Owens
Schmid College of Science and Technology, Chapman University, Orange, CA, United States

Viktoria D. Paul
Institut für Zytobiologie, Philipps-Universität Marburg, Marburg, Germany

Youri I. Pavlov
Eppley Institute for Research in Cancer, Fred & Pamela Buffett Cancer Center, University of Nebraska Medical Center, Omaha, NE, United States

Deborah L. Perlstein
Boston University, Boston, MA, United States

Antonio J. Pierik
Chemistry and Biochemistry, University of Kaiserlautern, Kaiserlautern, Germany

Jordan D. Pritts
School of Pharmacy, University of Maryland, Baltimore, MD, United States

Alan G. Raetz
University of California, Davis, Davis, CA, United States

Tracey A. Rouault
Molecular Medicine Branch, *Eunice Kennedy Shriver* National Institute of Child Health and Human Development, Bethesda, MD, United States

Geoffrey D. Shimberg
School of Pharmacy, University of Maryland, Baltimore, MD, United States

Hollie M. Siebler
Creighton University, Omaha, NE, United States

Oliver Stehling
Institut für Zytobiologie, Philipps-Universität Marburg, Marburg, Germany

Tahir H. Tahirov
Eppley Institute for Research in Cancer, Fred & Pamela Buffett Cancer Center, University of Nebraska Medical Center, Omaha, NE, United States

John A. Tainer
The University of Texas M. D. Anderson Cancer Center, Houston, TX; Molecular Biophysics and Integrated Bioimaging, Lawrence Berkeley National Laboratory, Berkeley, CA, United States

Faik A. Tezcan
University of California, San Diego, La Jolla, CA, United States

Chi-Lin Tsai
The University of Texas M. D. Anderson Cancer Center, Houston, TX, United States

Hongxin Wang
University of California, Davis, Davis, CA, United States

Benjamin D. Weiler
Institut für Zytobiologie, Philipps-Universität, Marburg, Germany

Dennis R. Winge
University of Utah Health Sciences Center, Salt Lake City, Utah, United States

PREFACE

Iron–sulfur clusters in proteins are ancient cofactors with links to the very beginnings of life. Nature has adeptly used iron–sulfur cluster cofactors, like its heme cousins, for a wide range of purposes. The most common Fe–S clusters contain Fe (II/III) and sulfide to make the common rhombic [2Fe–2S] or cuboidal [4Fe–4S] and [3Fe–4S] cores that are coordinated within the protein via primarily Cys ligands, but also His, Asp, or Ser, or with an open coordination site accessible to water. Many Fe–S proteins contain multiple Fe–S clusters, as well as more complex clusters of higher nuclearity and mixed metal clusters containing other transition metals such as Mo, V, or Ni. The ability of Fe–S cluster-containing proteins to access different oxidation states, and tune the redox potential by the cluster structure and protein environment has made these clusters ideally suited to participate in electron-transfer reactions. Early in the history of Fe–S cofactors it was thought that the functions were primarily related to their redox chemistry. However, the diversity of roles in biology for Fe–S clusters continues to grow, and nonredox roles for Fe–S clusters have been revealed. The rise of genomics has allowed for new Fe–S proteins to be identified and the membership of Fe–S families to grow. Major advances have also been made in deciphering the complexity of players involved in the in vivo synthesis of Fe–S clusters and their insertion into their protein homes. Indeed, the critical functions of Fe–S cofactors are also underscored by the growing number of human diseases associated with genetic mutations in Fe–S cluster biogenesis genes. The scope of topics that interface with Fe–S clusters continues to expand.

In two volumes of *Methods in Enzymology* (Volumes 595 and 599), leading investigators working on a diverse range of Fe–S proteins present important and insightful methods ranging from biochemistry, genetics, cell biology, and spectroscopy to study these fascinating systems. Though the topics are diverse, many common themes emerge such as special considerations for purification for Fe–S cofactor-containing proteins, and spectroscopic methods to probe the properties of Fe–S cluster cofactors in proteins. Due to the diversity of topics and techniques, the volumes are loosely organized into specific topics, but there are also topics that are covered from different angles in both volumes. As one might expect, some topics that could be covered in this volume have been or will be covered in other volumes organized by biological function or chemistry. In particular, there is an upcoming volume on radical *S*-adenosylmethionine (SAM) enzymes, and therefore, this important, diverse, and growing class of Fe–S cofactor-containing enzymes is only briefly covered in the two volumes dedicated to Fe–S enzymes.

Volume 599 (Fe–S cluster enzymes, Part B) picks up where Volume 595 ended with the emerging and fascinating topic of roles of Fe–S clusters in nucleic acid-processing enzymes. In Chapter 1, Tahirov and coworkers discuss the studies to date addressing the roles of Fe–S clusters in DNA polymerases and primases. The chapter has an emphasis on purification and structural characterization of these complex enzymes. Chapters 2 and 3 (by David and coworkers) focus on methods used to study the $[4Fe–4S]^{2+}$-containing bacterial base excision (BER) enzyme MutY and its human homolog MUTYH. These two chapters discuss purification and in vitro kinetics and DNA-binding methods (Chapter 2) and cellular repair assays (Chapter 3) used to understand the roles these enzymes play in preventing deleterious DNA mutations and how the biological roles in repair are intricately linked to its Fe–S cofactor. The focus of Chapter 4 by Shimberg, Pritts, and Michel turns to a relatively new topic of the finding of proteins classified as Zinc Finger proteins based on sequence that contains Fe–S sites instead or in addition to Zn. In this chapter, the focus is on a cleavage and polyadenylation specificity factor 30 (CPSF30) and discerning the identity of the different metal cofactor sites and the impact on RNA binding. The fifth chapter by Holmes-Hampton, Ghosh, and Rouault continues the RNA theme with a focus on iron regulatory proteins 1 and 2 (IRP1 and IRP2); these cytosolic enzymes regulate genes involved in iron metabolism by binding of the apo-form, but not Fe–S cluster-containing form, to iron-responsive elements in mRNA. Chapter 6 by Tsai and Tainer turns to the broad topic of optimization of production of Fe–S proteins for structural characterization with a specific emphasis on molybdopterin-guanine-dinucleotide enzymes. The topics in this volume then turn to a very important and "hot" topic of Fe–S cluster assembly, and further elaborate aspects of this topic discussed in Volume 595. Lill and coworkers provide in two chapters detailed biochemical methods used to elucidate Fe–S cluster assembly. In Chapter 7, the focus is on in vitro reconstitution of iron–sulfur cluster assembly systems. Chapter 8 provides a comprehensive strategy for biochemical analysis of the Fe–S cluster assembly process in mitochondria, cytosol, and the nucleus. Melber and Winge detail the mitochondrial Fe/S biogenesis (ISC) pathway and Fe–S cluster transfer to mature Fe–S proteins in Chapter 9. This chapter highlights various experimental approaches used to define these pathways, with a special emphasis on the use of genetic methods. An important, yet poorly understood, aspect of metallocluster assembly is the role of nucleotide hydrolysis; this is the subject of Chapter 10 by Grossman, Camire, and Perlstein. In this chapter, methods using fluorescent nucleotides to interrogate the role of nucleotide hydrolysis by the cytosolic iron–sulfur cluster assembly scaffold Nbp35–Cfd1 are described. Albetel and Outten describe methods to elucidate the roles of glutaredoxin Fe–S cluster-binding proteins in Fe–S cluster trafficking in Chapter 11. Specifically this chapter outlines the use of UV–visible circular dichroism spectroscopy as a fast and simple method to investigate glutaredoxin Fe–S cluster-dependent interactions. Chapter 12 by Tezcan and Owens moves to the topic of nitrogenase, which was also an important part of Volume 595. In this chapter, the focus is on methods for purifying and characterizing nitrogenase from a new model organism, *Gluconacetobacter diazotrophicus*; in addition, the authors describe new insights gained into nitrogenase function from structural and spectroscopic studies facilitated by the use of this new enzyme source. The last three chapters in this volume focus on methods for studying iron–sulfur proteins. Armstrong, Evans, and Megarity describe protein film electrochemistry methods with applications to the study of Fe–S enzymes in Chapter 13. Chapters 14 and 15 describe advanced synchrotron-based techniques and applications to date; these techniques are likely to be amenable in the future to the study of more complex Fe–S proteins described in these two volumes. Gee, Wang, and Cramer describe the basic background and methods of nuclear resonance vibrational spectroscopy (NRVS), and applications of NRVS to the study of Fe proteins, including the simple Fe–S site in rubredoxin. The volume concludes with an insightful chapter by DeBeer on new advanced X-ray spectroscopic methods for studying Fe–S-containing proteins and model complexes.

Sheila S. David
University of California, Davis
Davis CA

CHAPTER ONE

Iron–Sulfur Clusters in DNA Polymerases and Primases of Eukaryotes

Andrey G. Baranovskiy*, Hollie M. Siebler†, Youri I. Pavlov*,[1], Tahir H. Tahirov*,[1]

*Eppley Institute for Research in Cancer, Fred & Pamela Buffett Cancer Center, University of Nebraska Medical Center, Omaha, NE, United States
†Creighton University, Omaha, NE, United States
[1]Corresponding authors: e-mail address: ypavlov@unmc.edu; ttahirov@unmc.edu

Contents

Abstract

Research during the past decade witnessed the discovery of [4Fe–4S] clusters in several members of the eukaryotic DNA replication machinery. The presence of clusters was confirmed by UV–visible absorption, electron paramagnetic resonance spectroscopy, and metal analysis for primase and the B-family DNA polymerases δ and ζ. The crystal structure of primase revealed that the [4Fe–4S] cluster is buried inside the protein and fulfills a structural role. Although [4Fe–4S] clusters are firmly established in the C-terminal domains of catalytic subunits of DNA polymerases δ and ζ, no structures are currently available and their precise roles have not been ascertained.

Methods in Enzymology, Volume 599
ISSN 0076-6879
https://doi.org/10.1016/bs.mie.2017.09.003

The [4Fe–4S] clusters in the polymerases and primase play a structural role ensuring proper protein folding and stability. In DNA polymerases δ and ζ, they can potentially play regulatory role by sensing hurdles during DNA replication and assisting with DNA polymerase switches by oscillation between oxidized-reduced states.

ABBREVIATIONS

CTD C-terminal domain of Pol catalytic subunits
CTD–B complex between CTD and B-subunit
MBS metal-binding site
Pol polymerase
TLS translesion synthesis
Ts temperature sensitive

1. INTRODUCTION

The story of the discovery of iron–sulfur clusters in eukaryotic DNA polymerases (Pols) begins with a genetic study of *pol3–13* mutations involving mutations of one cysteine (Cys) of the second metal-binding motif of yeast Polδ to an alanine. The *pol3–13* mutation was found to be synthetically lethal with mutations in the genes *MMS19, NBP35*, *TAH18*, and *DRE2* that encode for proteins enigmatic at that time (Chanet & Heude, 2003). We now know these are components of the cytosolic iron–sulfur protein assembly machinery (Netz, Mascarenhas, Stehling, Pierik, & Lill, 2014). The connection of iron–sulfur clusters with yeast and human DNA Pols was made 8 years later (Baranovskiy et al., 2012; Netz et al., 2011). During this time, the large subunit of yeast and human DNA primase was excitingly discovered to possess a [4Fe–4S] cluster (Klinge, Hirst, Maman, Krude, & Pellegrini, 2007; Weiner et al., 2007). Now iron–sulfur clusters have been documented or suspected at various locations characteristic for each B-family Pol (Fig. 1A). These clusters play vital, though not completely understood, roles in Pol transactions at the replication fork (Burgers & Kunkel, 2017). We review here what is known, what inferences can be made, and provide protocols for Pol purification and iron content analysis.

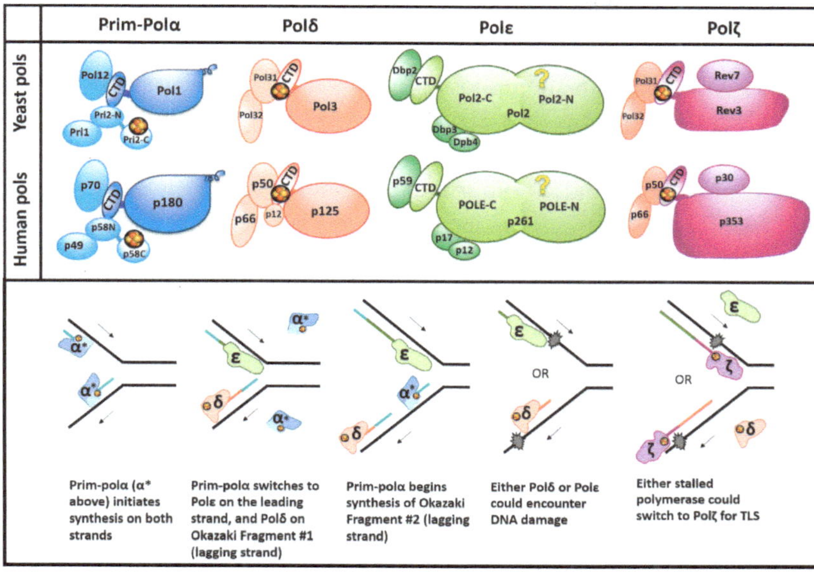

Fig. 1 B-family DNA polymerases at the replication fork. (A) Subunit structure of yeast and human B-family DNA polymerases. Groups of *yellow* and *orange spheres* represent [4Fe–4S] clusters. Only those clusters supported by multiple studies (see Burgers & Kunkel, 2017) are depicted here; others have been proposed but have not been verified yet. These include clusters in CTDs of Polα and Polε. Names of subunits for human Pols corresponding to their molecular mass are shown; the names of the subunits are as follows (from larger to smaller): Polα: POLA1, POLA2, PRIM2, PRIM1; Polδ: POLD1–POLD4; Polε: POLE, POLE2, POLE3, POLE4; Polζ: REV3L, hREV7 (or MAD2L2), POLD3, POLD4. See detailed nomenclature in (Baranovskiy, Gu, et al., 2017). (B) DNA replication under normal conditions and in the presence of DNA damage. Iron–sulfur clusters are shown as in (A). *Arrows* indicate the direction of DNA synthesis on the respective strand (leading strand *top*, lagging strand *bottom*). Newly synthesized DNA is color coded to match the Pol that synthesized that stretch. *Blue* represents both the primer laid by primase and the DNA laid by Polα. Pol colors match panel (A). For concision, the word "pol" has been omitted to leave only the *greek letters* to denote Pols, and α^* refers to pol α (*dark blue portion*) plus primase (*light blue portion*). Recent evidence has suggested that Polδ may also play a role in leading strand synthesis under certain conditions (Burgers & Kunkel, 2017), but this is not shown here.

2. IRON–SULFUR CLUSTERS IN DNA POLYMERASES AND PRIMASE

2.1 DNA Polymerases at the Replication Fork

The basis of heredity is accurate copying of genomic information. Because all DNA Pols synthesize DNA in $5' \rightarrow 3'$ direction, the mechanics of replication of the two DNA strands differ. The leading strand is synthesized continuously in the direction of the moving replication fork, and the lagging strand is synthesized in the opposite direction in ~150-base pair segments called Okazaki fragments (Fig. 1B) (Smith & Whitehouse, 2012). Bulk chromosomal replication is accomplished primarily by the B-family Pols. In eukaryotes, this family is comprised of multisubunit Pols α, δ, ε, and ζ (Fig. 1A). These Pols harbor divergent catalytic domains and assortments of accessory subunits, making each Pol specialized in its role in DNA replication. The C-terminal domains of Pol catalytic subunits (CTDs) contain two metal-binding sites (MBS1 and MBS2) previously thought to contain exclusively Zn^{2+}, as suggested by bioinformatics analysis of B-family Pols (Tahirov, Makarova, Rogozin, Pavlov, & Koonin, 2009) and a crystal structure of the CTD–B-subunit complex of yeast Polα (Klinge, Nunez-Ramirez, Llorca, & Pellegrini, 2009). In 2011, an exciting discovery suggested that MBS2, which participates in interaction with the conserved B-subunit, contains a [4Fe–4S] cluster (Netz et al., 2011). The presence of iron–sulfur clusters in MBS2 of Polα and Polε (Netz et al., 2011) has not been independently confirmed but rather opposed (Baranovskiy, Gu, et al., 2017; Baranovskiy et al., 2012; Klinge et al., 2009; Suwa et al., 2015), as discussed later.

Polα is present in eukaryotic cells in a tight complex with primase (Prim-Polα or primosome, Fig. 1A). Prim-Polα initiates DNA synthesis by synthesizing a primer at the start of the leading strand and each Okazaki fragment (Fig. 1B). Primase synthesizes a 9-mer RNA primer (Baranovskiy, Babayeva, et al., 2016), and then Polα extends it with dNTPs, making a suitable site for loading Polδ or Polε. Polα synthesis is only fairly accurate, as it possesses no internal proofreading capability. Polα is comprised of two subunits (Fig. 1A): catalytic Pol1, possessing $5' \rightarrow 3'$ Pol activity, and accessory B-subunit, Pol12 (yeast subunit names are used in this section; human nomenclature is shown in Fig. 1A and its legend). Primase is also comprised of two subunits: smaller catalytic Pri1 and larger accessory Pri2. The C-terminal domains of Pri2 and p58 (human) contain [4Fe–4S] clusters (Klinge et al., 2007; Weiner et al., 2007) and play important roles in substrate binding and replication initiation (Baranovskiy & Tahirov, 2017).

Once Prim-Polα has dissociated, synthesis is continued by either Polδ or Polε. Polε is required for early steps of DNA replication and is tightly associated with the CMG (Cdc45, Mcm2–7, and GINS) helicase, which travels along the leading strand and unwinds DNA ahead of the fork. Polε participates in DNA synthesis on the same strand (Burgers & Kunkel, 2017;

Georgescu et al., 2017; Yeeles, Janska, Early, & Diffley, 2017). Polε synthesis is highly processive due to its unique P-domain (Hogg et al., 2014) and is highly accurate (Ganai & Johansson, 2016; Pavlov, Shcherbakova, & Rogozin, 2006). Its catalytic subunit possesses both $5' \rightarrow 3'$ Pol activity and $3' \rightarrow 5'$ proofreading exonuclease activity. Polε is comprised of four subunits: Pol2, Dpb2, Dpb3, and Dpb4 (Fig. 1A). Catalytic Pol2 is a fusion of two different ancestral B-family Pols and has three domains: two copies of the polymerase/exonuclease module and the CTD (Tahirov et al., 2009). The N-terminal polymerase/exonuclease domain is catalytically active, while the second one is inactive and plays a vital structural role. It has been proposed that the active polymerase/exonuclease domain of Pol2 contains a [4Fe–4S] cluster at the base of the P-domain that enhances template:primer binding (Jain et al., 2014) (question mark in Fig. 1A). However, the crystal structure shows Zn^{2+} in that position (Hogg et al., 2014). The CTD of Pol2 binds Dpb2, which tethers Polε to the GINS complex of the replisome (Dmowski, Rudzka, Campbell, Jonczyk, & Fijalkowska, 2017; Zhou et al., 2017). Dpb3 and Dpb4 bind the central part of Pol2 and may participate in DNA binding and chromatin remodeling (Bermudez, Farina, Raghavan, Tappin, & Hurwitz, 2011; Hogg & Johansson, 2012).

Polδ takes over synthesis from Polα to complete each Okazaki fragment (Fig. 1B). It is well accepted that Polδ synthesizes primarily the lagging strand, but increasing evidence supports that it also contributes to leading strand synthesis (Pavlov & Shcherbakova, 2010; Stillman, 2015; Waga & Stillman, 1998; Yeeles et al., 2017; Zhou et al., 2017). Polδ is processive when bound to the accessory factor PCNA. It is also very accurate, possessing $3' \rightarrow 5'$ exonuclease activity. Polδ is comprised of three subunits in budding yeast, but four in most organisms (Fig. 1A). MBS2 of the catalytic subunit Pol3 contains a $[4Fe\text{–}4S]^{2+}$ cluster necessary for binding to Pol31 (Baranovskiy et al., 2012; Netz et al., 2011). Pol31 binds the other accessory subunit, Pol32, which can interact with many replication proteins including PCNA.

The fourth B-family member, Polζ, aids in replication when DNA is damaged or when replisome components are defective (Gan, Wittschieben, Wittschieben, & Wood, 2008; Waisertreiger et al., 2012) (Fig. 1B). DNA damage can stall the replication fork because the active sites of accurate Pols ε and δ cannot accommodate the aberrant, bulky structures. When this occurs, one response is translesion synthesis (TLS). During TLS, a A-, B-, or Y-family Pol incorporates a nucleotide across from the damaged site, leaving an aberrant primer terminus that is difficult to extend. Polζ can insert nucleotides across from some lesions, but is most specialized in extending aberrant termini. In this way, Polζ allows for continued DNA replication, but at the risk of mutation due to promiscuous synthesis by the inserter Pol, coupled with Polζ's low fidelity and lack of proofreading activity. Polζ is responsible for nearly all mutations caused by DNA-damaging agents or by replication stress (Pavlov et al., 2006). Fully functional Polζ is comprised of four subunits (Fig. 1A); accessory subunits include Rev7 and Pol31/Pol32 (latter pair is shared with Polδ) (Baranovskiy et al., 2012; Makarova & Burgers, 2015). Catalytic Rev3 possesses $5' \rightarrow 3'$ Pol activity, and, like Polδ, coordinates a critical [4Fe–4S] cluster in MBS2 (Baranovskiy et al., 2012; Netz et al., 2011).

Pol functions at the fork are summarized in the lower panel of Fig. 1B. This is a simplistic view; many more proteins are involved in vivo. Pols are sophisticated multisubunit and multidomain complexes whose intrinsic and extrinsic acrobatics ensure a variety of transactions during replication. Iron–sulfur clusters play indispensable, though still enigmatic, roles in the dynamics of Pols at the fork. In addition to structural roles discussed later, it has been speculated that they may play roles in Pol switching (Baranovskiy et al., 2012; Stepchenkova, Tarakhovskaya, Siebler, & Pavlov, 2017; Waisertreiger et al., 2012).

2.2 Iron–Sulfur Cluster in the Large Subunit of DNA Primase

Crystal structures have been reported for full-length human primase alone (Baranovskiy et al., 2015) and for primase within the primosome complex (Baranovskiy, Babayeva, et al., 2016). In the primase structure, the molecule has an extended cashew-like shape. The p58 subunit of primase is folded into two separate domains (Fig. 2): the N-terminal ($p58_N$, residues 1–252) and the C-terminal ($p58_C$, 271–509), connected by a flexible 18-residue linker (253–270). Primase contains three MBS: the Zn^{2+}-binding site of p49 with a proposed structural role, the catalytic site of p49 with two Mg^{2+} (or Mn^{2+}) ions, and the [4Fe–4S]-binding site of $p58_C$. The latter domain is all-helical and coordinates the iron–sulfur cluster with four conserved cysteines: Cys287, Cys367, Cys384, and Cys424 (Fig. 2). The cluster is fully buried and is in a stable $[4Fe–4S]^{2+}$ state (Klinge et al., 2007; Weiner et al., 2007). Mutations of iron-coordinating cysteines resulted in disruption of $p58_C$ and $Pri2_C$ folding, indicating that the cluster plays an indispensable structural role (Klinge et al., 2007; Liu & Huang, 2015; Weiner et al., 2007).

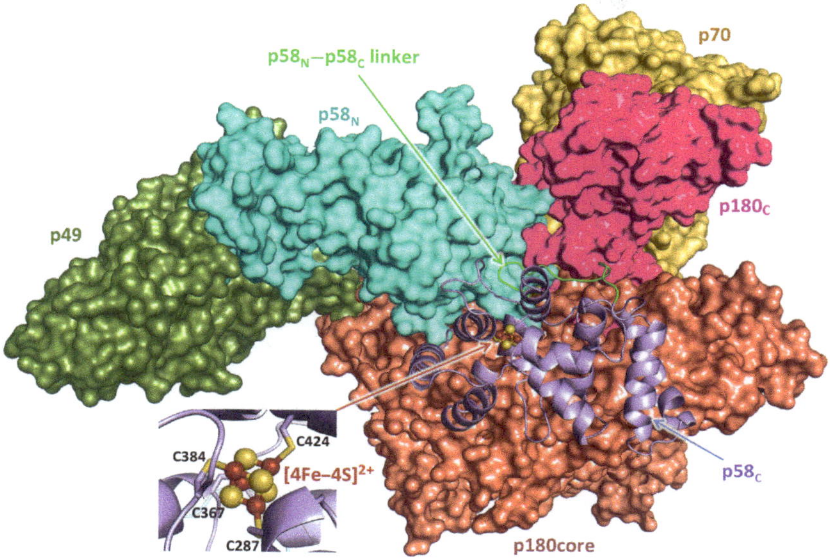

Fig. 2 Overall view of the human primosome. $p58_C$ with an adjacent linker and the rest of the molecule are represented as cartoon and surface, respectively. [4Fe–4S] is represented as *ball-sticks*, with iron and sulfur colored *red* and *yellow*, respectively. The *inset* shows close-up view of $[4Fe–4S]^{2+}$ coordination by four conservative Cys of $p58_C$. Coordinates of the human primosome (PDB ID 5EXR) were used to draw this structure. All structural figures are prepared using the PyMOL Molecular Graphics System (version 1.8, Schrödinger, LLC).

Biochemical studies of substrate binding by human primase revealed that $p58_C$ is responsible for the bulk of primase interactions with the template: primer (Baranovskiy, Zhang, et al., 2016). In particular, the 5′-triphosphate of the RNA primer, which is retained from the initiating nucleotide, and the 3′-overhang of the DNA template interact with $p58_C$ during all stages of primer synthesis. During primer synthesis, $p58_C$ stays bound to the junction at the 5′-end of the primer, while p49 moves with the growing 3′-end. Confirmation for such a model of primase–substrate interactions was provided by X-ray crystallography (Fig. 3) (Baranovskiy, Babayeva, et al., 2016). The iron–sulfur cluster is located 10 Å from the single-stranded DNA-binding surface and 18 Å from the double-stranded DNA-binding surface. This structure, together with structures of human primosome and primase with incoming NTP, provided the basis for models of the primosome during initiation, elongation, and termination of primer synthesis and the switch of the template:primer to Polα (Baranovskiy, Babayeva, et al., 2016;

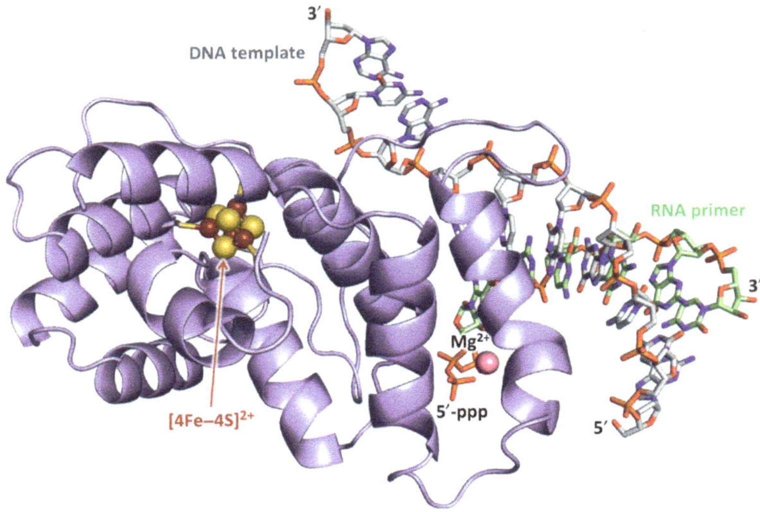

Fig. 3 Structure of $p58_C$–DNA:RNA complex. DNA template and RNA primer with 5′-triphosphate are shown as sticks with carbons colored *gray* and *green*, respectively. Oxygens, nitrogens, and phosphorus are colored *red*, *blue*, and *orange*, respectively. Mg^{2+} ion coordinated by 5′-triphosphate is shown as *pink sphere* with scaling 0.5. The figure was drawn using the coordinates of the $p58_C$/DNA:RNA complex (PDB ID 5F0Q).

Baranovskiy & Tahirov, 2017; Kilkenny, Longo, Perera, & Pellegrini, 2013). The formation of the tight $p58_C$/template:primer complex ensures: (a) the synthesis of a predominantly 9-mer RNA primer by primase; (b) additional verification of primer length after the switch from p49 to Polα, preventing dNTP addition to RNA primers shorter than 9-mer; and (c) coordination of primase and Polα catalytic activities, since primase is unable to initiate new primer synthesis until dissociation of $p58_C$ from the template:primer. Recent work proposed the role of redox switching and charge transfer in initiation and termination of RNA synthesis by primase (O'Brien et al., 2017). This idea still awaits rigorous testing because the proposed charge transfer channel in $p58_C$ may not exist in the natural protein (Baranovskiy, Babayeva, et al., 2017; Pellegrini, 2017). It is also unclear how the charge transfer is involved in initiation of RNA synthesis on single-stranded DNA.

2.3 Iron–Sulfur Clusters in B-Family DNA Polymerases

Critical analysis of current biochemical and structural data indicates that only two B-family DNA Pols, δ and ζ, contain the [4Fe–4S] cluster in MBS2

(Baranovskiy et al., 2012; Baranovskiy & Tahirov, 2017; Netz et al., 2011). Iron content in the human CTD–B-subunit (CTD–B) complexes of Polδ and Polζ was about three ions/protein (Baranovskiy et al., 2012), while in purified yeast Polδ it was approximately four (Netz et al., 2011). Variation could be due to different estimates of protein concentration and/or to higher stability of the cluster in intact Pol complexes. Several factors may lead to reduced iron content: oxidation of the $[4Fe–4S]^{2+}$ cluster to $[3Fe–4S]^{1+}$, complete disruption of the cluster in a fraction of molecules, and/or incomplete extraction of iron during measurement. Stoichiometry of iron binding below four is quite common after purification of proteins containing $[4Fe–4S]^{2+}$ clusters; for example, the iron/protein ratio was around three for both $p58_C$ and $Pri2_C$, where the iron–sulfur cluster is considered to be stable at aerobic conditions (Klinge et al., 2007; Weiner et al., 2007). Only a small fraction of iron–sulfur clusters in the yeast Polδ sample (<10%) was oxidized upon treatment with ferricyanide (Netz et al., 2011), which might be due to low accessibility of the cluster, its low sensitivity to this reagent, or its high stability, as seen for other cluster-containing enzymes (Fuss, Tsai, Ishida, & Tainer, 2015; Lukianova & David, 2005).

MBS2 is important for interactions between the catalytic and B-subunits of all B-family Pols (Johansson & Macneill, 2010). The crystal structures of the CTD–B complexes of human Pols α and ε (Baranovskiy, Gu, et al., 2017; Suwa et al., 2015) show that Zn^{2+}-MBS2 snugly fits a docking site on the B-subunit (Fig. 4A and B). Sulfur atoms of four zinc-coordinating cysteines form the tetrahedron with length of edges from 3.5 to 3.9 Å. In comparison, the edges of $[4Fe–4S]^{2+}$-coordinating tetrahedron in primase vary from 6.5 to 7.0 Å. Therefore, coordination of the [4Fe–4S] by MBS2 of Pols α and ε would require the cysteines to move apart by ~3 Å, resulting in significant conformational changes affecting the CTD–B interaction. Consistent with that, iron–sulfur clusters were not detected in pure and fully functional samples of human Polα and Polε, nor in their CTD–B complexes (Baranovskiy, Babayeva, et al., 2016; Baranovskiy et al., 2012; Zahurancik, Baranovskiy, Tahirov, & Suo, 2015; Zhang, Baranovskiy, Tahirov, & Pavlov, 2014). Earlier detection of iron–sulfur clusters in partially purified yeast Polα and Polε (Netz et al., 2011) can be explained by the presence of free catalytic subunits with misincorporated clusters in their CTDs, cluster binding by the Polε catalytic core (Jain et al., 2014), and the presence of primase in the Polα sample. Unfortunately, there is no high resolution structural information for Polδ and Polζ CTDs or their CTD–B complexes. Their [4Fe–4S]-binding modules may interact

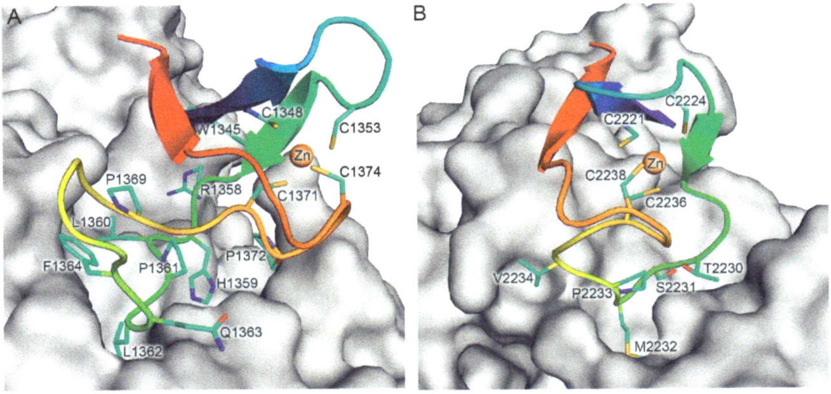

Fig. 4 Interaction interface of MBS2 of human Polα (A) and Polε (B) with the corresponding B-subunits. MBS2 and the B-subunit are represented as *cartoon* and *surface*, respectively. Main chain atoms of MBS2 are colored by a *gradient* from *blue* (N-terminus) to *red* (C-terminus). Side chains of cysteines coordinating Zn^{2+} as well as the residues participating in interaction with the B-subunit (except those located on β-strands) are shown as *sticks* and colored *cyan* for carbons, *red* for oxygens, and *blue* for nitrogens. Zn^{2+} ions are shown as *orange spheres* with scaling 0.5. The figure was drawn using the coordinates of human Polα and Polε CTD–B complexes (PDB IDs 4Y97 and 5VBN, respectively).

with the B-subunit differently than the Zn^{2+}-MBS2 in Polα and Polε. In support of this, protein fold predictions suggest the absence of β-strands in the [4Fe–4S]-binding module of Polδ and Polζ, and that two cysteines coordinating the cluster are located on the α-helix (Baranovskiy et al., 2012).

It is worth emphasizing the fixed relative position of two cysteines in MBS2 of Polα: Cys1348 is located on the β-sheet and Cys1353 is located on the short structured loop between the two β-strands (Fig. 4A). This makes [4Fe–4S] cluster coordination difficult and explains why CTD of Polα purified anaerobically was almost iron free, while CTDs of other B-family Pols contained significant amounts of iron and acid-labile sulfide (Netz et al., 2011). In MBS2 of Polε, the relative position of these two Cys is not so well fixed as in Polα (Fig. 4B): there is no proline in the loop between Cys2221 and Cys2224, the β-strand following Cys2224 is shorter, and its residues make four less hydrogen bonds with other residues of MBS2. This structural peculiarity explains the significant level of iron present in Polε CTD samples after purification under aerobic or anaerobic conditions (Baranovskiy et al., 2012; Netz et al., 2011). However, the Polε catalytic subunit (or its CTD) with a misincorporated [4Fe–4S] cluster at MBS2 cannot form a stable complex with the B-subunit, and is efficiently separated during appropriately designed purification (Baranovskiy et al., 2012; Zahurancik et al., 2015) (see Section 4). MBS1 is well structured in both Polα and Polε, precluding [4Fe–4S] coordination.

Intriguingly, coordination of [4Fe–4S] clusters by the CTDs of Polδ and Polζ could contribute to their ability to share the same B-subunit. According to fold predictions, CTDs of human and yeast Pols δ and ζ have very similar secondary structures, especially in the [4Fe–4S]-binding modules (Baranovskiy et al., 2012; Johnson, Prakash, & Prakash, 2012). It is possible that the iron–sulfur cluster integrity and/or changes in its oxidation state affect CTD–B complex stability and determine the mechanism of catalytic subunit switching between Polδ and Polζ during TLS (Baranovskiy et al., 2012; Waisertreiger et al., 2012). For example, oxidation to the $[3Fe–4S]^{1+}$ state would result in disruption of one of the four bonds between the cluster and the protein, with potential effects on MBS2 folding and its interaction with a B-subunit. In support of this idea, the fourth subunit of human Polδ (p12), proposed to stabilize the complex between the catalytic (p125) and B (p50) subunits (Li et al., 2006), undergoes degradation in response to DNA damage induced by UV, alkylating agents, oxidative, and replication stresses (Lee et al., 2012). The stabilizing role of p12 could be in protecting the iron–sulfur cluster from oxidation. Dissociation and degradation of p12 upon Polδ stalling would trigger dissociation of the catalytic subunit from p50/p66-PCNA/DNA and its replacement by two-subunit Polζ (p353/p30 in humans or Rev3/Rev7 in yeast) (Baranovskiy et al., 2012). In this scenario, the catalytic subunit of Polζ makes a temporary complex with Polδ accessory subunits p50/p66 only during lesion bypass. The alternative model of Polδ to Polζ switching during TLS, where the entire Polδ dissociates from PCNA and is replaced by four-subunit Polζ (p353/p30-p50/p66), is based on successful purification of four-subunit yeast and human Polζ after overexpression (Johnson et al., 2012; Lee, Gregory, & Yang, 2014; Makarova, Stodola, & Burgers, 2012). This model of Pol switch is consistent with the fact that p12, along with p66 and maybe p125, participates in interaction with PCNA (Li et al., 2006); therefore, dissociation of p12 could weaken the Polδ–PCNA complex. The additional role of four-subunit Polζ might be in replacing stalled Polε on the leading strand (Kraszewska, Garbacz, Jonczyk, Fijalkowska, & Jaszczur, 2012) or being a repository of the stabilized form of p353/p30 (with the iron–sulfur cluster protected from oxidation by p50), which replaces p125 on the lagging strand.

It is possible that different mechanisms of Polδ to Polζ switching operate in yeast (Polδ ↔ Polζ) and humans (p125 ↔ p353/p30). Yeast tetrameric Polζ and trimeric Polδ (naturally missing the fourth subunit) are shown to be stable; therefore, their dissociation during Pol switching might be difficult (Johnson et al., 2012; Makarova et al., 2012). In contrast, our studies with human proteins indicate that the CTD–B complexes of Polδ and Polζ are less stable than complexes of Polα and Polε (unpublished data). In line with that, p50 and p66 were underrepresented in purified human Polζ, indicating that the interaction between p353 and p50 is not strong (Lee et al., 2014). Reduced stability might be an intrinsic property of the CTD–B complexes of human Polδ and Polζ.

3. GENETIC EVIDENCE FOR THE IMPORTANT ROLES OF IRON–SULFUR CLUSTERS IN DNA REPLICATION IN VIVO

The [4Fe–4S] clusters in primase and the CTDs of Polδ and Polζ are coordinated by four Cys residues (Cys1–4, simplified naming according to their order in the polypeptide). However, mutational analysis suggests that not all of them are equally important for the role of any given cluster in vivo.

The [4Fe–4S] cluster in Pri2 has been proposed to contribute to the stability of the Prim-Polα complex and its recruitment to replication start sites (Liu & Huang, 2015). Single mutations affecting Cys1, Cys2, and Cys4 had no effect on the growth of yeast cells, but a strain with C434A substitution (Cys3) was temperature sensitive (Ts), displaying severe growth defects at 30°C and 37°C (normal/permissive is 25–30°C). Notably, substitution of Cys3 with Tyr (C434Y) was lethal to cell growth at 37°C, possibly due to steric interference between the bulky tyrosine and the neighboring residues. Various double mutant combinations (excluding Cys3) also showed growth defects at 30°C and 37°C. The triple mutant (C336A/C417A/C474A) showed severe growth defects and cells permanently arrested in G1 phase, unable to replicate DNA. Mutations leading to C434A, C336A/C417A, and the triple mutant also resulted in decreased Pri2 protein levels at 23°C. Genetic data are generally in agreement with biochemical data with mutant proteins; mutations that cause the Pri2 protein to lose the most iron (signifying degradation/destabilization of the cluster) lead to the most severe phenotypes in vivo in yeast (Klinge et al., 2007).

Single mutations affecting Cys that coordinate the [4Fe–4S] cluster of Pol3 (Polδ) led to a variety of phenotypes (Giot, Chanet, Simon,

Facca, & Faye, 1997). The C1059S (Cys2) substitution had no effect on yeast growth; C1069S (Cys3) was Ts, while C1069N was lethal. The C1074S (Cys4) mutant (*pol3–13,* mentioned first in Section 1) has been more thoroughly characterized. This mutant is Ts at 37°C, reflecting defects in DNA replication. These cells also show hypersensitivity to DNA damaging agents, severely decreased levels of induced mutagenesis (Giot et al., 1997), and, surprisingly, high levels of Polζ-dependent spontaneous mutations (Stepchenkova et al., 2017). This highlights the role of Polδ and its cluster in DNA repair and damage-avoidance mechanisms, potentially through its direct participation or through Pol switches required during TLS.

When the catalytic activity of Polζ is disrupted by mutations affecting the critical aspartate in the active site, almost all induced mutations in yeast vanish (Johnson et al., 2012; Siebler, Lada, Baranovskiy, Tahirov, & Pavlov, 2014). Strains with substitutions in all four cysteines coordinating the [4Fe–4S] of Rev3 behave similarly to the catalytically inactive mutant, showing a drastic decrease in survival after UV and almost complete loss of mutagenesis (Baranovskiy et al., 2012). The C1468S (Cys3) single substitution results in milder responses to UV; no effect was seen in the *CAN1* forward mutation assay, but there was a decrease in reversion mutations induced by UV (Johnson et al., 2012). This difference may be because the reversion assay tests for specific base pair substitutions at a particular site, whereas the forward assay tests for a variety of different mutations across the whole *CAN1* gene arising by diverse mechanisms. A Cys3/Cys4 double mutant (C1449S/C1473S) showed no binding to Pol31 and an almost complete loss of mutagenesis (Makarova et al., 2012). The complexity of transactions dependent on the iron–sulfur cluster in Polζ is illustrated by a mutant where the whole CTD is deleted, but induced mutagenesis is partially retained (Siebler et al., 2014). Overall, changes of single-Cys residues responsible for [4Fe–4S] binding have varying effects on Pol function, and substitution of more cysteines causes progressively more severe phenotypes.

4. PURIFICATION OF EUKARYOTIC B-FAMILY DNA POLYMERASES

The enzymatic and metal-binding properties of catalytic subunits may significantly differ from those of entire Pol complexes. Therefore, it is important to reduce the level of free catalytic subunits in purified Pol samples to the minimum. This task can be achieved only by placing the affinity tag on the accessory subunit, because it is difficult to remove the excess of catalytic subunit from Pol sample by the ion-exchange or size-exclusion chromatography. Here, we provide the core of a purification scheme applicable to all B-family Pols (yeast and human). Immobilized metal affinity chromatography is used as the first step; a 6 × histidine tag (His6) is placed on the N-terminus of the B-subunit. The next important step is chromatography on Heparin HiTrap HP column (GE Healthcare Life Sciences), which allows for the removal of aggregated molecules and excess of the B-subunit. Heparin imitates the sugar-phosphate backbone of DNA substrate and provides tight binding for Pols. The combination of these two chromatography steps usually yields Pol samples with decent purity in 1 day. Higher purity can be achieved by applying additional purification steps, for example, chromatography on MonoS or MonoQ columns (GE Healthcare Life Sciences). Chromatography on the hydroxyapatite column usually does not provide significant improvement in purity, but allows direct loading of protein samples with high salt concentration. Size-exclusion chromatography is preferable as a final polishing step that yields the protein in a storage buffer. Expression of Pol accessory subunits and CTD–B complexes is usually performed in *Escherichia coli* cell culture at 20°C (the growth media is supplemented with 50 μ*M* $FeCl_3$). Obtaining substantial levels of human Pols or their subcomplexes containing the large catalytic subunits require the employment of more expensive and time-consuming expression in insect cells, which are infected with a recombinant *Autographa californica* baculovirus. We prefer to use Sf21 cells because of fast duplication, as well as a separate baculovirus for each Pol subunit with the cDNA placed under control of the polyhedrin promoter.

4.1 Procedure

All steps are performed on ice or at 4°C unless otherwise stated. The frozen cell pellet (stored at −80°C in 3–6 g aliquots) is quickly thawed by incubation in room temperature water bath for 5 min and placed on ice. Cells are resuspended in nine volumes of buffer N (bN) containing 20 m*M* Tris–HCl, pH 7.9, 200 m*M* NaCl, 10 m*M* K-phosphate, pH 7.9, 3% glycerol, 4 m*M* β-mercaptoethanol, and supplemented with protease inhibitors: 0.5 m*M* phenylmethylsulfonyl fluoride (Sigma-Aldrich) and 1 μg/mL leupeptin (Acros Organics). After cell disruption by three passes through the EmulsiFlex-C3 homogenizer (Avestin Inc., Canada), the extract is cleared by centrifugation at 45,000 × *g* for 15 min and loaded to Profinity IMAC resin (Bio-Rad, USA) charged with Ni^{2+} ions. In the case of robust protein expression, 7 mL of resin is optimal for 14 g of cell pellet. The resin is washed with the following buffers: three column volumes (CV) of bN; five CV of bN containing 400 m*M* NaCl and 40 m*M* K-phosphate, pH 7.9; and one CV of bN. The bound material is eluted by 10 CV gradient (0–0.2 *M*) of imidazole–HCl, pH 7.9, in bN. After analysis by electrophoresis in 12% SDS-PAGE, the purest fractions are combined (approximately two CV) and loaded to 5 mL Heparin HiTrap HP column (GE Healthcare Life Sciences). It is imperative to estimate protein amount and load no more than 3 mg of protein per mL of resin. After washing the column with two CV of bN, Pols are eluted by 10 CV gradient bN_{200}–bN_{900} (subscript indicates NaCl concentration in m*M*). After purity analysis by SDS-PAGE, the purest fractions are combined (approximately one CV), dialyzed 3 h against 50 volumes of storage buffer: 10 m*M* Tris–HCl, pH 7.9, 200 m*M* NaCl (or KCl), 3% glycerol, and 1 m*M* dithiothreitol, and flash frozen in small aliquots.

4.2 Notes

1. Purity analysis by electrophoresis may be skipped if the elution place of Pol peak is well known.
2. According to structural and biochemical data, His-tag on the N-terminus of the B-subunit does not affect the formation of the Pol complex or its activity, but it can potentially affect interaction with other replication factors. In this case, a cleavable affinity tag is preferable, which can be implemented by adding the short recognition sequence for highly specific protease, like human rhinovirus 3C protease (common name of precision protease) or nuclear inclusion protease from tobacco etch virus. Tag cleavage usually requires overnight digestion, which prolongs purification and increases possibility of sample degradation and oxidation.
3. We found that most of the expressed Pol complexes accumulate in the cytoplasm of insect cells and can be easily extracted by cell pellet resuspension in bN after one freeze–thaw cycle.
4. Buffer exchange is carried out by placing the Pol sample into the Spectra/Pore 7 dialysis membrane with molecular weight cut off 25 kDa. This membrane is made of regenerated cellulose and chemically treated to minimize the heavy metal and sulfur content. Dialysis is conducted with continuous buffer mix by placing the beaker on a stirrer; beaker contains a stir bar of adequate size and must be covered by Saran to prevent the buffer and sample aeration.

5. ANALYSIS OF IRON CONTENT IN PROTEIN SAMPLES

Protein samples with [4Fe–4S] clusters have a yellow-brown color, which turns darker upon higher concentration. Iron–sulfur clusters show a wide absorbance peak around 400 nm, which is masked by a peak at 280 nm in large proteins. A detailed review of iron–sulfur cluster detection and analysis methods is provided in (Fuss et al., 2015). Most of these methods require expensive equipment and a significant amount of pure protein. Here, we provide a simple protocol for nonheme iron detection, routinely used in our laboratory to quickly analyze freshly purified protein samples. This calorimetric method is based on the reaction between ferrous ions (Fe^{2+}) and the chromogen ferrozine, resulting in a violet-colored complex whose intensity can be measured spectrophotometrically at 560 nm. Ferrozine shows higher sensitivity in comparison to other chromogens used in iron determination, including thiocyanate *o*-phenanthroline, bathophenanthroline, and 2,4,6-Tris(2-pyridyl)-*s*-triazine (TPTZ). Ferrozine does not react with ferric ions (Fe^{3+}), so the reducing agent hydroxylamine hydrochloride should be present to have all iron in Fe^{2+} form. The method described here is based on the protocol provided with the proprietary reagent set for measurement of the total iron-binding capacity of serum (Pointe Scientific, Inc.). Treatment at 37°C in acidic buffer containing surfactant allows for protein denaturation and quantitative iron extraction. Unlike serum, pure DNA Pol samples do not have detectable absorbance at 560 nm, so the original protocol was simplified by excluding absorbance reading before adding the ferrozine to a protein sample treated with an acidic buffer. The color intensity is proportional to the iron concentration in the range of 0–90 μ*M*.

5.1 Procedure

The reaction is set up at room temperature in a 0.6-mL tube; 10 μL of protein sample is diluted with 51 μL of premixed reagent solution containing 50 μL of buffer (220 m*M* hydroxylamine hydrochloride in acetate buffer, pH 4.5, with surfactant) and 1 μL of iron color reagent (16.7 m*M* ferrozine in hydroxylamine hydrochloride). After 10 min at 37°C, samples were spun for 3 min at 15,000 × *g* at room temperature to remove denatured protein. Supernatant (60 μL) is used to record the absorbance at 560 nm against a blank solution, using trUView cuvette (Bio-Rad; allows for measuring absorbance in volumes as low as 50 μL) and Nanodrop 2000c

(Thermo Scientific). Blank solution and iron standard are prepared and incubated in a similar way, except 10 μL of the iron-free buffer and 90 μ*M* $FeCl_2$ in hydroxylamine hydrochloride, respectively, are added instead of protein. Iron concentration in the protein sample is calculated using the following formula: A560(sample)/A560(standard) × 90 μ*M*. Finally, the ratio of iron and protein concentration gives the iron content/protein molecule.

5.2 Notes

1. For accurate iron determination, protein samples should contain 0.1–0.9 nmol of iron, which corresponds to 25–225 pmol of protein with [4Fe–4S] cluster (corresponding to 1.25–11.25 μg of 50 kDa protein).
2. Protein concentration determined by different methods can vary. We prefer measuring absorbance at 280 nm and using extinction coefficients calculated with the ProtParam software.
3. The absorbance reading of the iron standard should be close to 0.4.
4. Increasing reaction time to 1 h and temperature to 45°C might be useful for iron extraction from highly stable proteins.
5. The readings are stable for several hours at room temperature after sample preparation.

ACKNOWLEDGMENT

This work was supported by the National Institute of General Medical Sciences (NIGMS) Grant GM101167 to T.H.T. and Eppley Institute Pilot grant to Y.I.P. The Eppley Institute's X-ray Crystallography Core Facility is supported by the Cancer Center Support Grant P30CA036727.
The authors declare no competing financial interests.

REFERENCES

Baranovskiy, A. G., Babayeva, N. D., Zhang, Y., Blanco, L., Pavlov, Y. I., & Tahirov, T. H. (2017). Comment on "The [4Fe4S] cluster of human DNA primase functions as a redox switch using DNA charge transport" *Science*, *357*(6348), eaan2396. https://doi.org/10.1126/science.aan2396.

Baranovskiy, A. G., Babayeva, N. D., Zhang, Y., Gu, J., Suwa, Y., Pavlov, Y. I., et al. (2016). Mechanism of concerted RNA-DNA primer synthesis by the human primosome. *The Journal of Biological Chemistry*, *291*(19), 10006–10020. https://doi.org/10.1074/jbc.M116.717405.

Baranovskiy, A. G., Gu, J., Babayeva, N. D., Kurinov, I., Pavlov, Y. I., & Tahirov, T. H. (2017). Crystal structure of human Pol epsilon B subunit in complex with C-terminal domain of catalytic subunit. *The Journal of Biological Chemistry*, *292*(38), 15717–15730. https://doi.org/10.1074/jbc.M117.792705.

Baranovskiy, A. G., Lada, A. G., Siebler, H. M., Zhang, Y., Pavlov, Y. I., & Tahirov, T. H. (2012). DNA polymerase delta and zeta switch by sharing accessory subunits of DNA polymerase delta. *The Journal of Biological Chemistry*, *287*(21), 17281–17287. https://doi.org/10.1074/jbc.M112.351122.

Baranovskiy, A. G., & Tahirov, T. H. (2017). Elaborated action of the human primosome. *Genes (Basel)*, *8*(2), E62. https://doi.org/10.3390/genes8020062.

Baranovskiy, A. G., Zhang, Y., Suwa, Y., Babayeva, N. D., Gu, J., Pavlov, Y. I., et al. (2015). Crystal structure of the human primase. *The Journal of Biological Chemistry*, *290*(9), 5635–5646. https://doi.org/10.1074/jbc.M114.624742.

Baranovskiy, A. G., Zhang, Y., Suwa, Y., Gu, J., Babayeva, N. D., Pavlov, Y. I., et al. (2016). Insight into the human DNA primase interaction with template-primer. *The Journal of Biological Chemistry*, *291*(9), 4793–4802. https://doi.org/10.1074/jbc.M115.704064.

Bermudez, V. P., Farina, A., Raghavan, V., Tappin, I., & Hurwitz, J. (2011). Studies on human DNA polymerase epsilon and GINS complex and their role in DNA replication. *The Journal of Biological Chemistry*, *286*(33), 28963–28977. https://doi.org/10.1074/jbc.M111.256289.

Burgers, P. M. J., & Kunkel, T. A. (2017). Eukaryotic DNA replication fork. *Annual Review of Biochemistry*, *86*, 417–438. https://doi.org/10.1146/annurev-biochem-061516-044709.

Chanet, R., & Heude, M. (2003). Characterization of mutations that are synthetic lethal with pol3-13, a mutated allele of DNA polymerase delta in *Saccharomyces cerevisiae*. *Current Genetics*, *43*(5), 337–350. https://doi.org/10.1007/s00294-003-0407-2.

Dmowski, M., Rudzka, J., Campbell, J. L., Jonczyk, P., & Fijalkowska, I. J. (2017). Mutations in the non-catalytic subunit Dpb2 of DNA polymerase epsilon affect the Nrm1 branch of the DNA replication checkpoint. *PLoS Genetics*, *13*(1), e1006572. https://doi.org/10.1371/journal.pgen.1006572.

Fuss, J. O., Tsai, C. L., Ishida, J. P., & Tainer, J. A. (2015). Emerging critical roles of Fe-S clusters in DNA replication and repair. *Biochimica et Biophysica Acta*, *1853*(6), 1253–1271. https://doi.org/10.1016/j.bbamcr.2015.01.018.

Gan, G. N., Wittschieben, J. P., Wittschieben, B. O., & Wood, R. D. (2008). DNA polymerase zeta (pol zeta) in higher eukaryotes. *Cell Research*, *18*(1), 174–183. https://doi.org/10.1038/cr.2007.117.

Ganai, R. A., & Johansson, E. (2016). DNA replication—A matter of fidelity. *Molecular Cell*, *62*(5), 745–755. https://doi.org/10.1016/j.molcel.2016.05.003.

Georgescu, R., Yuan, Z., Bai, L., de Luna Almeida Santos, R., Sun, J., Zhang, D., et al. (2017). Structure of eukaryotic CMG helicase at a replication fork and implications to replisome architecture and origin initiation. *Proceedings of the National Academy of Sciences of the United States of America*, *114*(5), E697–E706. https://doi.org/10.1073/pnas.1620500114.

Giot, L., Chanet, R., Simon, M., Facca, C., & Faye, G. (1997). Involvement of the yeast DNA polymerase delta in DNA repair in vivo. *Genetics*, *146*(4), 1239–1251.

Hogg, M., & Johansson, E. (2012). DNA polymerase epsilon. *Sub-Cellular Biochemistry*, *62*, 237–257. https://doi.org/10.1007/978-94-007-4572-8_13.

Hogg, M., Osterman, P., Bylund, G. O., Ganai, R. A., Lundstrom, E. B., Sauer-Eriksson, A. E., et al. (2014). Structural basis for processive DNA synthesis by yeast DNA polymerase epsilon. *Nature Structural & Molecular Biology*, *21*(1), 49–55. https://doi.org/10.1038/nsmb.2712.

Jain, R., Vanamee, E. S., Dzikovski, B. G., Buku, A., Johnson, R. E., Prakash, L., et al. (2014). An iron-sulfur cluster in the polymerase domain of yeast DNA polymerase epsilon. *Journal of Molecular Biology*, *426*(2), 301–308. https://doi.org/10.1016/j.jmb.2013.10.015.

Johansson, E., & Macneill, S. A. (2010). The eukaryotic replicative DNA polymerases take shape. *Trends in Biochemical Sciences*, *35*(6), 339–347. https://doi.org/10.1016/j.tibs.2010.01.004.

Johnson, R. E., Prakash, L., & Prakash, S. (2012). Pol31 and Pol32 subunits of yeast DNA polymerase delta are also essential subunits of DNA polymerase zeta. *Proceedings of the National Academy of Sciences of the United States of America, 109*(31), 12455–12460. https://doi.org/10.1073/pnas.1206052109.

Kilkenny, M. L., Longo, M. A., Perera, R. L., & Pellegrini, L. (2013). Structures of human primase reveal design of nucleotide elongation site and mode of Pol alpha tethering. *Proceedings of the National Academy of Sciences of the United States of America, 110*(40), 15961–15966. https://doi.org/10.1073/pnas.1311185110.

Klinge, S., Hirst, J., Maman, J. D., Krude, T., & Pellegrini, L. (2007). An iron-sulfur domain of the eukaryotic primase is essential for RNA primer synthesis. *Nature Structural & Molecular Biology, 14*(9), 875–877. https://doi.org/10.1038/nsmb1288.

Klinge, S., Nunez-Ramirez, R., Llorca, O., & Pellegrini, L. (2009). 3D architecture of DNA Pol alpha reveals the functional core of multi-subunit replicative polymerases. *The EMBO Journal, 28*(13), 1978–1987. https://doi.org/10.1038/emboj.2009.150.

Kraszewska, J., Garbacz, M., Jonczyk, P., Fijalkowska, I. J., & Jaszczur, M. (2012). Defect of Dpb2p, a noncatalytic subunit of DNA polymerase epsilon, promotes error prone replication of undamaged chromosomal DNA in Saccharomyces cerevisiae. *Mutation Research, 737*(1–2), 34–42. https://doi.org/10.1016/j.mrfmmm.2012.06.002.

Lee, Y. S., Gregory, M. T., & Yang, W. (2014). Human Pol zeta purified with accessory subunits is active in translesion DNA synthesis and complements Pol eta in cisplatin bypass. *Proceedings of the National Academy of Sciences of the United States of America, 111*(8), 2954–2959. https://doi.org/10.1073/pnas.1324001111.

Lee, M. Y., Zhang, S., Lin, S. H., Chea, J., Wang, X., LeRoy, C., et al. (2012). Regulation of human DNA polymerase delta in the cellular responses to DNA damage. *Environmental and Molecular Mutagenesis, 53*(9), 683–698. https://doi.org/10.1002/em.21743.

Li, H., Xie, B., Zhou, Y., Rahmeh, A., Trusa, S., Zhang, S., et al. (2006). Functional roles of p12, the fourth subunit of human DNA polymerase delta. *The Journal of Biological Chemistry, 281*(21), 14748–14755. https://doi.org/10.1074/jbc.M600322200.

Liu, L., & Huang, M. (2015). Essential role of the iron-sulfur cluster binding domain of the primase regulatory subunit Pri2 in DNA replication initiation. *Protein & Cell, 6*(3), 194–210. https://doi.org/10.1007/s13238-015-0134-8.

Lukianova, O. A., & David, S. S. (2005). A role for iron-sulfur clusters in DNA repair. *Current Opinion in Chemical Biology, 9*(2), 145–151. https://doi.org/10.1016/j.cbpa.2005.02.006.

Makarova, A. V., & Burgers, P. M. (2015). Eukaryotic DNA polymerase zeta. *DNA Repair (Amst), 29*, 47–55. https://doi.org/10.1016/j.dnarep.2015.02.012.

Makarova, A. V., Stodola, J. L., & Burgers, P. M. (2012). A four-subunit DNA polymerase zeta complex containing Pol delta accessory subunits is essential for PCNA-mediated mutagenesis. *Nucleic Acids Research, 40*(22), 11618–11626. https://doi.org/10.1093/nar/gks948.

Netz, D. J., Mascarenhas, J., Stehling, O., Pierik, A. J., & Lill, R. (2014). Maturation of cytosolic and nuclear iron-sulfur proteins. *Trends in Cell Biology, 24*(5), 303–312. https://doi.org/10.1016/j.tcb.2013.11.005.

Netz, D. J., Stith, C. M., Stumpfig, M., Kopf, G., Vogel, D., Genau, H. M., et al. (2011). Eukaryotic DNA polymerases require an iron-sulfur cluster for the formation of active complexes. *Nature Chemical Biology, 8*(1), 125–132. https://doi.org/10.1038/nchembio.721.

O'Brien, E., Holt, M. E., Thompson, M. K., Salay, L. E., Ehlinger, A. C., Chazin, W. J., et al. (2017). The [4Fe4S] cluster of human DNA primase functions as a redox switch using DNA charge transport. *Science, 355*(6327), eaag1789. https://doi.org/10.1126/science.aag1789.

Pavlov, Y. I., & Shcherbakova, P. V. (2010). DNA polymerases at the eukaryotic fork—20 years later. *Mutation Research, 685*(1–2), 45–53. https://doi.org/10.1016/j.mrfmmm.2009.08.002.

Pavlov, Y. I., Shcherbakova, P. V., & Rogozin, I. B. (2006). Roles of DNA polymerases in replication, repair, and recombination in eukaryotes. *International Review of Cytology, 255*, 41–132. https://doi.org/10.1016/S0074-7696(06)55002-8.

Pellegrini, L. (2017). Comment on "The [4Fe4S] cluster of human DNA primase functions as a redox switch using DNA charge transport" *Science, 357*(6348), eaan2954. https://doi.org/10.1126/science.aan2954.

Siebler, H. M., Lada, A. G., Baranovskiy, A. G., Tahirov, T. H., & Pavlov, Y. I. (2014). A novel variant of DNA polymerase zeta, Rev3DeltaC, highlights differential regulation of Pol32 as a subunit of polymerase delta versus zeta in Saccharomyces cerevisiae. *DNA Repair (Amst), 24*, 138–149. https://doi.org/10.1016/j.dnarep.2014.04.013.

Smith, D. J., & Whitehouse, I. (2012). Intrinsic coupling of lagging-strand synthesis to chromatin assembly. *Nature, 483*(7390), 434–438. https://doi.org/10.1038/nature10895.

Stepchenkova, E. I., Tarakhovskaya, E. R., Siebler, H. M., & Pavlov, Y. I. (2017). Defect of Fe-S cluster binding by DNA polymerase delta in yeast suppresses UV-induced mutagenesis, but enhances DNA polymerase zeta-dependent spontaneous mutagenesis. *DNA Repair (Amst), 49*, 60–69. https://doi.org/10.1016/j.dnarep.2016.11.004.

Stillman, B. (2015). Reconsidering DNA polymerases at the replication fork in eukaryotes. *Molecular Cell, 59*(2), 139–141. https://doi.org/10.1016/j.molcel.2015.07.004.

Suwa, Y., Gu, J., Baranovskiy, A. G., Babayeva, N. D., Pavlov, Y. I., & Tahirov, T. H. (2015). Crystal structure of the human Pol alpha B subunit in complex with the C-terminal domain of the catalytic subunit. *The Journal of Biological Chemistry, 290*(23), 14328–14337. https://doi.org/10.1074/jbc.M115.649954.

Tahirov, T. H., Makarova, K. S., Rogozin, I. B., Pavlov, Y. I., & Koonin, E. V. (2009). Evolution of DNA polymerases: An inactivated polymerase-exonuclease module in Pol epsilon and a chimeric origin of eukaryotic polymerases from two classes of archaeal ancestors. *Biology Direct, 4*, 11. https://doi.org/10.1186/1745-6150-4-11.

Waga, S., & Stillman, B. (1998). The DNA replication fork in eukaryotic cells. *Annual Review of Biochemistry, 67*, 721–751. https://doi.org/10.1146/annurev.biochem.67.1.721.

Waisertreiger, I. S., Liston, V. G., Menezes, M. R., Kim, H. M., Lobachev, K. S., Stepchenkova, E. I., et al. (2012). Modulation of mutagenesis in eukaryotes by DNA replication fork dynamics and quality of nucleotide pools. *Environmental and Molecular Mutagenesis, 53*(9), 699–724. https://doi.org/10.1002/em.21735.

Weiner, B. E., Huang, H., Dattilo, B. M., Nilges, M. J., Fanning, E., & Chazin, W. J. (2007). An iron-sulfur cluster in the C-terminal domain of the p58 subunit of human DNA primase. *The Journal of Biological Chemistry, 282*(46), 33444–33451. https://doi.org/10.1074/jbc.M705826200.

Yeeles, J. T., Janska, A., Early, A., & Diffley, J. F. (2017). How the eukaryotic replisome achieves rapid and efficient DNA replication. *Molecular Cell, 65*(1), 105–116. https://doi.org/10.1016/j.molcel.2016.11.017.

Zahurancik, W. J., Baranovskiy, A. G., Tahirov, T. H., & Suo, Z. (2015). Comparison of the kinetic parameters of the truncated catalytic subunit and holoenzyme of human DNA polymerase varepsilon. *DNA Repair (Amst), 29*, 16–22. https://doi.org/10.1016/j.dnarep.2015.01.008.

Zhang, Y., Baranovskiy, A. G., Tahirov, T. H., & Pavlov, Y. I. (2014). The C-terminal domain of the DNA polymerase catalytic subunit regulates the primase and polymerase activities of the human DNA polymerase alpha-primase complex. *The Journal of Biological Chemistry, 289*(32), 22021–22034. https://doi.org/10.1074/jbc.M114.570333.

Zhou, J. C., Janska, A., Goswami, P., Renault, L., Abid Ali, F., Kotecha, A., et al. (2017). CMG-Pol epsilon dynamics suggests a mechanism for the establishment of leading-strand synthesis in the eukaryotic replisome. *Proceedings of the National Academy of Sciences of the United States of America, 114*(16), 4141–4146. https://doi.org/10.1073/pnas.1700530114.

CHAPTER TWO

Fe–S Clusters and MutY Base Excision Repair Glycosylases: Purification, Kinetics, and DNA Affinity Measurements

Nicole N. Nuñez, Chandrima Majumdar, Kori T. Lay, Sheila S. David[1]
University of California, Davis, CA, United States
[1]Corresponding author: e-mail address: ssdavid@ucdavis.edu

Contents

Abstract

A growing number of iron–sulfur (Fe–S) cluster cofactors have been identified in DNA repair proteins. MutY and its homologs are base excision repair (BER) glycosylases that prevent mutations associated with the common oxidation product of guanine (G),

Methods in Enzymology, Volume 599
ISSN 0076-6879
https://doi.org/10.1016/bs.mie.2017.11.035

8-oxo-7,8-dihydroguanine (OG) by catalyzing adenine (A) base excision from inappropriately formed OG:A mispairs. The finding of an $[4Fe–4S]^{2+}$ cluster cofactor in MutY, Endonuclease III, and structurally similar BER enzymes was surprising and initially thought to represent an example of a purely structural role for the cofactor. However, in the two decades subsequent to the initial discovery, purification and in vitro analysis of bacterial MutYs and mammalian homologs, such as human MUTYH and mouse Mutyh, have demonstrated that proper Fe–S cluster coordination is required for OG: A substrate recognition and adenine excision. In addition, the Fe–S cluster in MutY has been shown to be capable of redox chemistry in the presence of DNA. The work in our laboratory aimed at addressing the importance of the MutY Fe–S cluster has involved a battery of approaches, with the overarching hypothesis that understanding the role(s) of the Fe–S cluster is intimately associated with understanding the biological and chemical properties of MutY and its unique damaged DNA substrate as a whole. In this chapter, we focus on methods of enzyme expression and purification, detailed enzyme kinetics, and DNA affinity assays. The methods described herein have not only been leveraged to provide insight into the roles of the MutY Fe–S cluster but have also been provided crucial information needed to delineate the impact of inherited variants of the human homolog MUTYH associated with a colorectal cancer syndrome known as MUTYH-associated polyposis or MAP. Notably, many MAP-associated variants have been found adjacent to the Fe–S cluster further underscoring the intimate relationship between the cofactor, MUTYH-mediated DNA repair, and disease.

1. INTRODUCTION

One of the earliest DNA repair enzymes to be discovered containing a $[4Fe–4S]^{2+}$ cluster was the MutY glycosylase, based on its sequence homology to Endonuclease III (Endo III) (Michaels, Pham, Nghiem, Cruz, & Miller, 1990). MutY and Endo III are DNA glycosylases that initiate base excision repair (BER) by hydrolyzing the glycosidic bond of a damaged or mispaired base from the sugar (David & Wiliams, 1998; Michaels et al., 1990). Subsequent action of other BER pathway enzymes provides for resection of the sugar fragment and replacement with an appropriate undamaged nucleotide (Manlove, Nuñez, & David, 2016). MutY was initially discovered through a combination of genetics and biochemistry (Au, Cabrera, Miller, & Modrich, 1988; Banda, Nuñez, Burnside, Bradshaw, & David, 2017); these studies showed that mutations in the *mutY* gene resulted in increased levels of G:C to T:A transversion mutations and identified an activity in cell lysates capable of mediating repair of G:A mismatches back to G:C base pairs (bps). Purification of the protein product from *Escherichia coli* (*E. coli*, or *Ec*) and analysis on mismatch-containing substrates showed that MutY was an adenine glycosylase that removes A from G:A mismatches, allowing for downstream BER restoration of a G:C base pair (Au, Clark, Miller, & Modrich, 1989). Subsequent work showed that MutY also exhibited in vitro activity on OG:A mismatches, where OG is 8-oxo-7,8-dihydroguanine (Michaels & Miller, 1992; Porello, Leyes, & David, 1998). In addition, mutation frequency measurements indicated exceptionally high levels of G:C to T:A transversions upon inactivation of the gene for MutY along with the OG glycosylase Fpg (MutM) in *Ec* (Michaels, Cruz, Grollman, & Miller, 1992; Michaels & Miller, 1992; Michaels, Tchou, Grollman, & Miller, 1992; Miller & Michaels, 1996), leading to the suggestion that OG:A bps are the most important biological substrate for MutY (Banda et al., 2017). The importance of prevention of OG-induced mutations in humans is underscored by the correlation between the inheritance of biallelic mutations within the gene encoding the human MutY homolog (MUTYH) and colorectal polyposis and adenomas (Al-Tassan et al., 2002; David, O'Shea, & Kundu, 2007). This correlation provides for a high probability of colorectal cancer and is referred to as MUTYH-associated polyposis or MAP (Al-Tassan et al., 2002).

Nearly all MutY homologs, excluding a few distinct bacterial lineages, possess an Fe–S cluster of the $[4Fe\text{–}4S]^{2+}$ cluster subtype with a unique contiguous spacing of the Cys ligands (Cys-X_6-Cys-X_2-Cys-X_5-Cys) (Fig. 1; Lukianova & David, 2005; Trasvina-Arenas, Lopez-Castillo, Sanchez-Sandoval, & Brieba, 2016). The Fe–S cluster in Endo III was originally thought to play a purely structural role in these enzymes due to the fact that the known base excision chemistry would not require a redox cofactor, and the Fe–S cluster was observed to be resistant to reduction or oxidation (Lukianova & David, 2005). Early work in our laboratory involving refolding experiments of MutY performed in the presence or absence of added Fe (II) and sulfide indicated that the cluster is not needed for overall MutY folding or protein thermal stability (Porello, Cannon, & David, 1998). However, only forms harboring the Fe–S cluster exhibited affinity for substrate DNA and adenine base excision activity. Notably, these studies were done prior to the development of detailed protocols for analyzing MutY active fraction and kinetics of base excision (methods described in Section 3). Although enzyme activity in the Fe–S cluster containing refolded forms was detected, this method is not a robust approach for obtaining significant amounts of highly active, cluster-loaded MutY. Subsequent

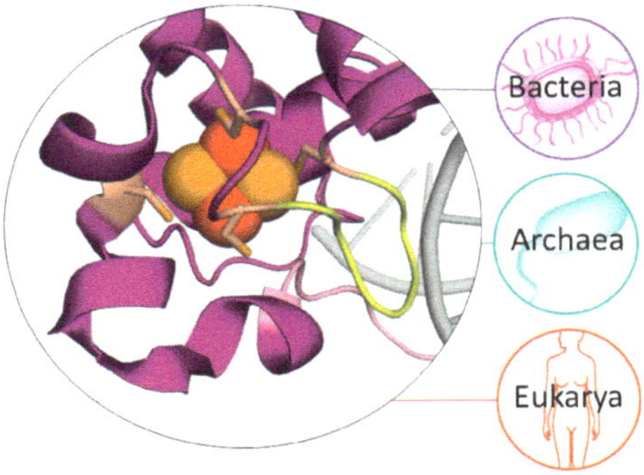

Fig. 1 MutY Fe–S cluster. As seen in the crystal structure of *Geobacillus stearothermophilus* MutY (PDB ID: 5DPK) (*purple*) (Woods et al., 2016), the $[4Fe–4S]^{2+}$ cluster (*orange* and *yellow spheres*) is coordinated by four Cys ligands (*beige-colored sticks*), and homolog lineages bearing the iconic Fe–S cluster span throughout all three domains of life. Notably, the Fe–S cluster is important for providing key contacts with the DNA (*gray*) via positioning of the FCL (*green*) generated by the first two Cys in the Cys-X_6-Cys-X_2-Cys-X_5-Cys conserved motif and is located adjacent to the start of the IDC (*light pink*) (Guan et al., 1998).

experiments using site-directed mutagenesis (SDM) to individually replace Fe–S cluster Cys ligands of *Ec* MutY with Ser, His, and Ala demonstrated that a fully coordinated cluster is needed for robust levels of overexpression in bacteria (Golinelli, Chmiel, & David, 1999; Lukianova & David, 2005). Methods for obtaining cluster-loaded MutY and mammalian homologs such as human MUTYH are also discussed in detail (Section 2).

The first crystal structures of Endo III and the N-terminal domain of MutY suggested that though the Fe–S cluster is located remotely from the active site, it would be critical for substrate engagement (Guan et al., 1998; Lee & Verdine, 2009; Luncsford et al., 2010; Woods et al., 2016). Tainer and coworkers reported the first crystal structure of *Ec* MutY that comprised the catalytic N-terminal domain alone and harbored the adenine base (Guan et al., 1998). Verdine and coworkers reported a subsequent structure of a catalytically inactive variant of the thermophilic MutY homolog, Asp144Asn *Geobacillus stearothermophilus* (*Gs*) MutY, bound to its natural substrate (an OG:A duplex) in what is known as the lesion recognition complex (LRC) (Fromme, Banerjee, Huang, & Verdine, 2004). Employment of the cleavage-resistant substrate analog *arabino*-2′-fluoro-2′-deoxyadenosine (FA) provided a structure with WT *Gs* MutY enzyme, referred to as the fluorinated lesion recognition complex (FLRC) (Lee & Verdine, 2009).

Notably, both the LRC and FLRC structures utilized disulfide cross-linking between the protein and DNA, presumably to stabilize the complex and facilitate crystallization. More recently, we reported the structure of *Gs* MutY bound to an azaribose-based transition state (TS) analog-containing duplex which we refer to as the transition state analog complex (TSAC) (Woods et al., 2016; Fig. 2A). In this case, disulfide cross-linking was not used and was presumably not required due to the high affinity of *Gs* MutY for the TS mimic duplex.

The TSAC structure, along with kinetics experiments, resulted in a revised S_N1 mechanism for MutY-catalyzed base excision (Fig. 2B; Woods et al., 2016). Though the reaction catalyzed by MutY appears deceptively simple on paper, MutY must locate rare OG:A bps within the context of the DNA polymer in cells containing a vast excess of structurally similar T:A bps; on average, one OG is present per 10^6–10^7 G bases, with only a fraction of those being OG:A bps. In addition, the adenine base excision process mediated by MutY produces toxic abasic sites opposite the damaged base, OG. Due to the potential for causing single-strand breaks, the abasic site product is potentially more detrimental to cells than the initial damaged and mismatched OG:A bp. These complexities require that the process of MutY/MUTYH-mediated base excision be carefully orchestrated with downstream BER enzymes and other DNA-dependent processes in cells.

A conspicuous feature of the MutY structures is a solvent-exposed loop, referred to as the Fe–S cluster loop (FCL) motif, that consists of positively charged residues positioned in between two of the Cys residues that coordinate the Fe–S cluster (Guan et al., 1998). The FCL motif is positioned for interactions with the phosphodiester backbone of DNA and appears to be stabilizing the bent DNA conformation of the bound substrate DNA (Bruner, Norman, & Verdine, 2000; Fromme et al., 2004). Indeed, key DNA interactions occur via positively charged amino acids within the FCL motif (Brinkmeyer & David, 2015; Chepanoske, Golinelli, Williams, & David, 2000; Guan et al., 1998). Binding analyses (methods described in Section 4) on mutated forms of *Ec* MutY, where the positively charged residues were replaced with alanine, illustrated the importance of these positively charged residues in mediating high affinity for the damaged DNA substrate. Many MAP variants are noted to be positioned in regions localized near the critically bound Fe–S cluster and the FCL (Manlove et al., 2016). For example, MAP variant Pro281Leu is located in the FCL, and this

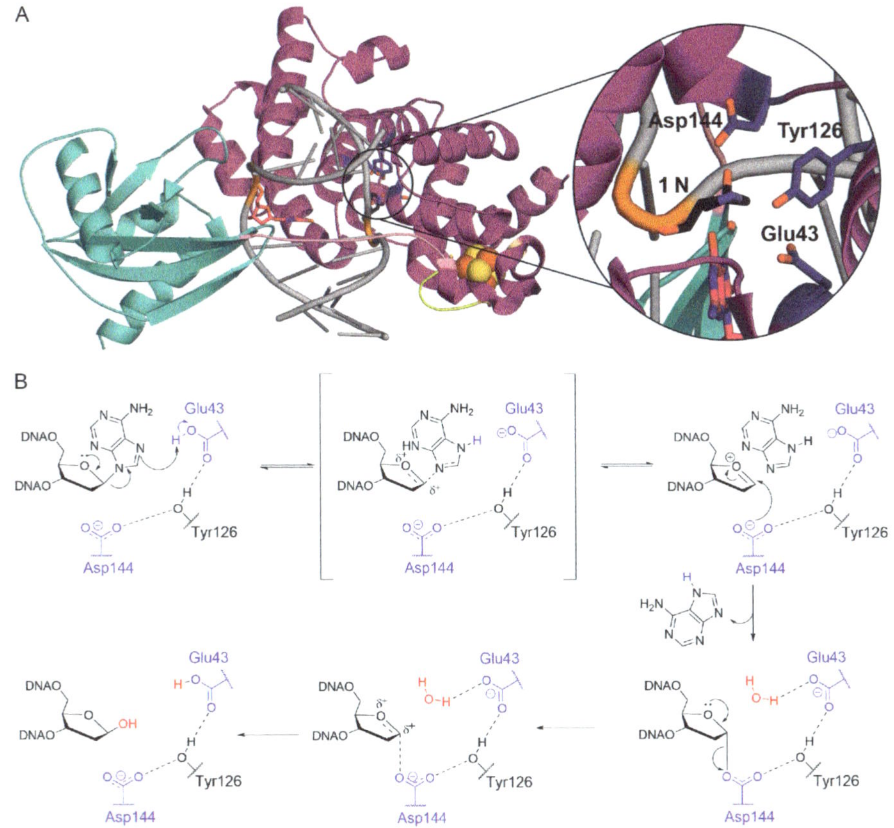

Fig. 2 Structure and updated mechanism for MutY. (A) Crystal structure of *Gs* MutY (PDB ID: 5DPK) bound to DNA (*gray*) containing OG (*hot pink*) opposite the pyrrolidine transition state mimic, 1 N (*black*) (Woods et al., 2016). MutY has several structural features, including the N-terminal domain (*purple*) that houses the Fe–S cluster (*orange* and *yellow spheres*) and FCL (*green*), which is connected to the C-terminal domain (*teal*), via the IDC (*light pink*). (B) Mechanism of adenine removal by MutY and homologs where active site residues are marked in *blue* and *numbered* according to their positions in *Gs* MutY. The adenine base is first protonated by Glu43 to form a good leaving group and dissociates from the sugar via an S_N1-like mechanism. The resulting oxocarbenium ion is proposed to be stabilized though the formation of a covalent intermediate with Asp144, which dissociates upon nucleophilic attack by an activated water molecule (*red*) to form the abasic site product.

mutation dramatically hampers substrate DNA recognition and adenine excision (Brinkmeyer & David, 2015).

In the process of studying MAP variants localized in the interdomain connector (IDC) region of MUTYH that connects the catalytic N-terminal domain with the C-terminal OG recognition domain, we noticed the presence of three highly conserved Cys residues (Cys-X_6-Cys-X_2-Cys) with spacing surprisingly similar to the Cys spacing of the N-terminal Fe–S cluster (Engstrom et al., 2014). This region of the IDC is not present in bacterial homologs and appears disordered in the N-terminal fragment crystal structure of human MUTYH whose sequence ends abruptly just after these Cys residues (Luncsford et al., 2010). This led us to suspect the presence of an additional metal site due the potential for these residues to serve as an appropriate ligand type and spacing for metal chelation. Metal analysis of *Mus musculus* (*Mm*) Mutyh showed the presence of one zinc ion per four iron atoms (from the N-terminal Fe–S cluster). SDM of the proposed Cys ligands resulted in diminished levels of Zn^{2+}, but also demonstrated poor Fe–S cluster loading. Moreover, low levels of Zn^{2+} within the IDC correlated with low levels of active enzyme and ability to suppress DNA mutations. This led us to dub this region as the "Zn^{2+} linchpin motif" to convey its role in properly engaging the OG-binding and catalytic domains on the OG:A mismatch to facilitate adenine base excision (Engstrom et al., 2014).

The role of the FCL and the region around the Fe–S cluster, in mediating DNA binding, suggested that the redox properties of the cluster might be influenced by the presence of DNA. Electrochemical experiments using DNA-modified electrodes have shown that the Fe–S cluster in *Ec* MutY is able to redox cycle with a midpoint potential of +90 mV vs NHE, similar to that observed for high-potential (HIPIP) Fe–S cluster proteins (Boon, Livingston, Chmiel, David, & Barton, 2003). More recent work using thin film voltammetry on pyrolytic graphite edge electrodes allowed for measurements of both free and bound MutY, and showed a large negative shift in redox potential upon binding to DNA (Bartels et al., 2017). This is consistent with the idea that DNA stabilizes the oxidized form of the cluster (Ha et al., 2017). In addition, studies of MutY, Endo III, and a thermophilic uracil-DNA glycosylase (*Archaeoglobus fulgidus* UDG) have revealed similar potentials for these BER enzymes, and that the electrochemical signal is attenuated if the DNA contains an abasic site or a mismatch (Boal et al., 2005). These studies have led to an intriguing hypothesis that DNA-mediated charge transport (DCT) between Fe–S cluster proteins may aid in the location of DNA damage by sensing the integrity of the DNA helix between two distally located Fe–S cluster-containing repair enzymes (Boal et al., 2009; Boal, Yavin, & Barton, 2007). These ideas are supported by additional studies with Fe–S cluster-containing enzymes participating in DNA repair and replication, such as DNA primase (O'Brien et al., 2017).

Delineating the importance of the MutY Fe–S cluster has relied on a variety of enzyme expression and purification procedures, detailed enzyme kinetics, and DNA-binding affinity measurements. In addition, these same strategies were pivotal in establishing the connection between reduced repair of OG:A mismatches due to inheritance of specific MUTYH variants and colorectal cancer (i.e., MAP). This chapter describes protocols that we have developed to study MutY enzymes. We also include examples of how these approaches were used in revealing important features of MutY. Of note, electrochemical approaches used to study MutY, and other Fe–S cluster-containing repair and replication proteins, have recently been reported in *Methods in Enzymology* (Barton, Bartels, Deng, & O'Brien, 2017). In addition, the chapter "Cellular assays for studying the Fe–S cluster containing base excision repair glycosylase MUTYH and homologs" by Majumdar et al. following this one (in this volume) focuses on cellular assays used to study MutY enzymes. Since its inception, our laboratory has been focused on bacterial and mammalian MutY glycosylases and the potential role(s) for the Fe–S cluster with the overarching hypothesis that revealing the role(s) of the Fe–S cluster cofactor is intimately related to understanding the chemical properties of MutY and MUTYH, as well as deciphering their complex biological functions.

2. OVER EXPRESSION AND PURIFICATION OF MutY HOMOLOGS

2.1 Considerations for Isolating MutY Homologs

The production of Fe–S cluster proteins for in vitro analyses can be particularly difficult and warrants thorough consideration of the chosen gene, expression conditions, and subsequent purification scheme. Most biochemical studies to date have used *Ec* MutY; however, the thermophilic bacterial homolog *Gs* MutY facilitates X-ray crystallography studies bound to DNA, presumably due to its higher stability (Fromme et al., 2004; Lee & Verdine, 2009; Wang, Chakravarthy, & Verdine, 2017; Wang, Lee, & Verdine, 2015; Woods et al., 2016). The mammalian homologs (i.e., MUTYH) have high homology to their bacterial cousins, and therefore, the bacterial protein can often be a useful surrogate for the human protein. Indeed, the defective activity of MAP variants was initially inferred by the kinetic analysis of the corresponding variants in the *Ec* MutY (Al-Tassan et al., 2002; Banda et al., 2017). However, some regions and positions of MAP-associated variants are only found in the mammalian enzymes requiring use of the mammalian protein (Brinkmeyer & David, 2015; Kundu, Brinkmeyer, Eigenheer, & David, 2010; Kundu, Brinkmeyer, Livingston, & David, 2009). Levels of overexpression are generally higher using the *Mm* Mutyh compared to *Hs* MUTYH. Depending on the type of experiment to be performed with the metalloprotein, specific considerations will be needed to ensure for accurate enzyme characterization. For example, X-ray absorption spectroscopy fine structure experiments required the use of a truncated form of *Mm* Mutyh in order to obtain the high concentrations necessary for the study (Engstrom et al., 2014).

Use of an affinity tag in production of Fe–S proteins may aid in overexpression and purification. For example, the use of an N-terminal maltose-binding protein (MBP) tag for MutY resulted in significantly higher yields of soluble protein (Boon et al., 2003; Ha et al., 2017). As is the case with MutY, additional constructs such as the *pRKISC* plasmid composed of Fe–S cluster assembly proteins that aid in cellular cluster assembly may be deemed necessary (Tokumoto & Takahashi, 2001). There are commercial sources of MutY (MyBioSource); however, our laboratory's experience is that MutY enzymes are highly sensitive to the storage conditions and do not tolerate repeated freeze–thaw cycles. These considerations prompt us to suggest that the commercial enzyme only be used for qualitative applications.

During purification of Fe–S cluster-containing enzymes, consideration of the type of environment, such as aerobic vs anaerobic conditions, the presence of reducing agents, degassed buffers, and length of time following cell rupture, may influence the quality of protein produced (Crack, Green, Thomson, & Le Brun, 2014; Py & Barras, 2010). In addition, supplementation of various media components may also be needed for improved cluster assembly. It is important to bear in mind that perturbations to the Fe–S cluster via introduced amino acid substitutions, especially of the Cys ligands, may result in low yields of protein, resulting in difficult protein characterization (Engstrom et al., 2014; Golinelli et al., 1999).

2.2 Bacteria as an Overexpression Host

Bacterial overexpression of proteins has the advantage of the ease of growth of bacteria and subsequent purification; however, these hosts lack the ability to facilitate posttranslational modifications, and may not have optimal codon usage (Kundu et al., 2009). These general procedures were followed for the bacterial host expression of both prokaryotic and mammalian homologs of MutY. In all cases, after the gene is confirmed through sequencing, the cell line (Table 1) is transformed, either through heat shock or through electroporation, with the appropriate expression vector. Transformed cells are then streaked onto LB agar plates with appropriate antibiotic selection (Table 1) and incubated at 37°C overnight.

The next day, a single colony is selected and used to inoculate LB starter media (Table 1) which is incubated overnight at 37°C and 220 rpm in a shaker (Table 3). The following morning, 50 mL of LB starter media is added to 2 L of LB growth media in a 6-L flask that has the same final media components as the starter media. Each flask is incubated at 37°C and 150 rpm until an optimal OD_{600nm} is reached for induction (Table 1). This is followed by a reduction in temperature to 30°C, and induction of overexpression by adding IPTG and supplementing with an Fe (II) source (Table 1). After the overexpression period, growth media is centrifuged at 7500 rpm for 15 min at 4°C, and, following decanting of the broth, the pellet is suspended in resuspension buffer (Table 1) and stored at −80°C. Pelleted cells are thawed in an incubator at 37°C and 150 rpm for 30 min prior to being sonicated on ice for 6 min in 30 s cycles at 70% power (Table 3). The lysate is centrifuged at 12,000 rpm for 15 min, at 4°C (Table 3). All the steps following this should be carried out at 4°C, and all buffers used should be prechilled. The supernatant is decanted into a centrifuge tube on ice, and, if desired, the pellet can be resuspended and undergo a repeat of sonification and centrifugation.

Troubleshooting tips: Due to the presence of the Fe–S cluster in MutY and homologs, it is important to use reductants when applicable, such as 1 m*M* DTT. Keep in mind that reductants may not be suitable for certain types of affinity tag separation procedures or tag cleavage enzyme incubations. See manufacturer instructions for appropriate reductant concentrations for the type of affinity tag and tag cleavage enzyme being used.

2.3 Purification of *E. coli* MutY

Cloning of the *mutY* gene allowed for sequencing and placement into DNA plasmids for various purposes (Michaels et al., 1990). The first expression plasmid used for MutY was pKKYEco that contained the *mutY* gene under the control on an IPTG inducible promotor (Porello, Williams, Kuhn,

Table 1 Summary of Components and Reagents Necessary for the Overexpression of MutY and Homologs

Enzyme Homolog	Expression Vector	Expression Host	Agar Growth Media	Culture Growth Media	Induction Conditions	Resuspension Buffer
Ec MutY	pKK223-3 (no tag)	JM101 *muty*$^{-}$ cells	LB media 15 μg/mL tetracycline 100 μg/mL ampicillin 50 m*M* agar	LB media 15 μg/mL tetracycline 100 μg/mL ampicillin	1 m*M* IPTG 30°C for 3 h 150 rpm $OD_{600nm} = 0.9$	50 m*M* Tris–HCl pH 7.5 2 m*M* EDTA 5 m*M* DTT 250 m*M* NaCl Protease inhibitor 5% glycerol
Gs MutY	*pET28* (cleavable 6 × His tag)	BL21 DE3 Rosetta 2 cells coexpressing *pRKISC*	LB media 15 μg/mL tetracycline 34 μg/mL kanamycin 50 m*M* agar	LB media 15 μg/mL tetracycline 34 μg/mL kanamycin	1 m*M* IPTG 30°C for 6 h 150 rpm 1 m*M* $FeCl_2$ $OD_{600nm} = 0.6$	20 m*M* Tris–HCl pH 8.0 300 m*M* NaCl 20 m*M* imidazole 5 m*M* β-ME Protease inhibitor
MBP-tagged *Ec* MutY	*pMAL-c2x* (MBP tag)	BL21 DE3 cells coexpressing *pRKISC*	LB media 15 μg/mL tetracycline 100 μg/mL ampicillin 50 m*M* agar	LB media 15 μg/mL tetracycline 100 μg/mL ampicillin 10 m*M* glucose	1 m*M* IPTG 30°C for 4 h 150 rpm 1 m*M* $FeCl_2$ $OD_{600nm} = 0.5$	20 m*M* sod phos. pH 7.5 1 m*M* EDTA 1 m*M* DTT 200 m*M* NaCl 1 m*M* PMSF 10% glycerol

Continued

Table 1 Summary of Components and Reagents Necessary for the Overexpression of MutY and Homologs—cont'd

Enzyme Homolog	Expression Vector	Expression Host	Agar Growth Media	Culture Growth Media	Induction Conditions	Resuspension Buffer
Mouse Mutyh	*pET28* (cleavable 6 × His tag)	BL21 DE3 cells coexpressing *pRKISC*	LB media 15 μg/mL tetracycline 34 μg/mL kanamycin 50 m*M* agar	LB media 15 μg/mL tetracycline 34 μg/mL kanamycin	1 m*M* IPTG 30°C for 6 h 150 rpm 1 m*M* $FeCl_2$ $OD_{600nm}=0.8$	20 m*M* sod. phos. pH 7.5 1 m*M* PMSF 10% glycerol
MBP-tagged human MUTYH	*pFastBacDual* (MBP tag)	Sf9 insect cells Baculovirus	N/A	Hyclone SfX media 25 μg/mL gentamycin	N/A	20 m*M* Tris–HCl pH 7.5 1 m*M* EDTA 1 m*M* DTT 200 m*M* NaCl Protease inhibitor

Michaels, & David, 1996), and this plasmid has been provided to many research groups (Manuel, Czerwinski, & Lloyd, 1996). Much of the biochemical information on MutY has been generated using this untagged *Ec* version of the protein. *Ec* MutY Δ226-350 (lacking the C-terminal domain) has been used for structural studies (Guan et al., 1998) and studies that revealed the OG recognition role of the C-terminal domain (Chmiel, Golinelli, Francis, & David, 2001). In another example, the X-ray structure of Cys199His MutY Δ226-350 demonstrated that the His199 side chain was coordinated to the Fe–S cluster; however, there was a reduction in iron occupancy at the site of His coordination giving rise to a larger amount of $[3Fe4S]^{1+}$ cluster (Messick et al., 2002).

The purification of MutY relies on the combination of several biochemical separation methods, including the precipitation of host cell DNA with streptomycin sulfate following cell lysis, and protein precipitation using ammonium sulfate. This step is followed by desalting and anion exchange columns, and final polishing with a Hi-Trap Heparin column (Chmiel et al., 2001; Porello et al., 1996). The details regarding the purification of the *Ec* MutY homolog are detailed in this section.

Once the cells expressing the protein of interest have been lysed via sonication (Table 3), the supernatant is transferred to a sterile beaker, or centrifuge bottle, and enough 25% (w/v) solution of streptomycin sulfate is added to a final concentration of 19%. The solution is stirred for 40 min at 4°C, and then centrifuged (Table 3) at 5000 rpm for 10 min to precipitate and pellet out nucleic acids. The supernatant, which appears yellow and viscous, is then decanted into a fresh, sterile beaker, or centrifuge bottle, and solid ammonium sulfate is slowly added until its concentration is 40% (w/v). The solution is stirred again for 1 h, to precipitate out the proteins, and then centrifuged at 8000 rpm for 1 h. The watery supernatant hence obtained is discarded, and the pellet is resuspended in Heparin Buffer A (Table 2) and filtered sequentially though a 1-μm and then a 0.45-μm filter prior to running on the FPLC (Table 3).

The preceding precipitation steps lead to a high salt concentration of the crude enzyme, and therefore necessitate the use of a desalting column (HiPrep 26/10 Desalting, GE Healthcare). After equilibration of the column with five-column volumes of buffer, the protein is eluted through an isocratic flow of Heparin Buffer A (Table 2), and fraction collection is cut off at a conductivity of 15–20 mS/cm^2. These fractions are then pooled together, immediately followed by a cation exchange column that is used to separate negatively charged and neutral proteins and residual nucleic acids. This type of separation is possible since the theoretical pI of MutY is 8.3,

Table 2 Summary of Components and Reagents Necessary for the Purification of MutY and Homologs

Enzyme Homolog	Tag	Separation Method	Separation Buffer A	Separation Buffer B	Heparin Buffer A	Heparin Buffer B	Storage Buffer Before Freezing
Ec MutY	N/A	Salt precipitation	40% (w/v) of ammonium sulfate	N/A	20 m*M* sod. phos. pH 7.5 1 m*M* EDTA 1 m*M* DTT 5% glycerol Filter/degas	20 m*M* sod. phos. pH 7.5 1 m*M* EDTA 1 m*M* DTT 5% glycerol 1 *M* NaCl Filter/degas	20 m*M* sod. phos. pH 7.5 1 m*M* EDTA 1 m*M* DTT 100 m*M* NaCl
Gs MutY	6 × His	Ni^{2+}-NTA column	20 m*M* Tris–HCl pH 8.0 300 m*M* NaCl 20 m*M* imidazole 5 m*M* β-ME	20 m*M* Tris–HCl pH 8.0 300 m*M* NaCl 250 m*M* imidazole 5 m*M* β-ME	N/A	N/A	20 m*M* Tris–HCl pH 8.0 150 m*M* NaCl 5 m*M* β-ME
Ec MutY	MBP	Amylose column	20 m*M* sod. phos. pH 7.5 1 m*M* EDTA 1 m*M* DTT 5% glycerol 200 m*M* NaCl	20 m*M* sod. phos. pH 7.5 1 m*M* EDTA 1 m*M* DTT 5% glycerol 200 m*M* NaCl 10 m*M* maltose	20 m*M* sod. phos. pH 7.5 1 m*M* EDTA 1 m*M* DTT 5% glycerol Filter/degas	20 m*M* sod. phos. pH 7.5 1 m*M* EDTA 1 m*M* DTT 5% glycerol 1 *M* NaCl Filter/degas	20 m*M* sod. phos. pH 7.5 1 m*M* EDTA 1 m*M* DTT 200 m*M* NaCl

Mutyh	6 × His	Ni^{2+}-NTA column	20 m*M* sod. phos. pH 7.5 10% glycerol 1 *M* NaCl 20 m*M* imidazole	20 m*M* sod. phos. pH 7.5 10% glycerol 300 m*M* NaCl 500 m*M* imidazole	20 m*M* sod. phos. pH 7.5 1 m*M* EDTA 5% glycerol Filter/degas	20 m*M* sod. phos. pH 7.5 1 m*M* EDTA 5% glycerol 1 *M* NaCl Filter/degas	20 m*M* sod. phos. pH 7.5 1 m*M* EDTA 200 m*M* NaCl
MUTYH	MBP	Amylose column	20 m*M* Tris–HCl pH 7.5 200 m*M* NaCl 1 m*M* EDTA 1 m*M* DTT Protease inhibitor	20 m*M* Tris–HCl pH 7.5 200 m*M* NaCl 1 m*M* EDTA 1 m*M* DTT Protease inhibitor 10 m*M* maltose	N/A	N/A	20 m*M* Tris–HCl pH 7.5 200 m*M* NaCl 1 m*M* EDTA 1 m*M* DTT

Table 3 Summary of Equipment Needed for the Over Expression and Purification of MutY Homologs and Variants

Equipment	Notable Specifications
Pipette tips/microcentrifuge tubes/ conical tubes	DNase/RNase/Pyrogen free and sterile
Vortex/mini centrifuge	Table top
Thermocycler	Applied Biosystems GeneAmp PCR System 2400
Syringe filters	0.2, 1, and 0.45 μm sterile
Gravity flow column	PD10 columns with 1 μm filter disc
Incubator/shaker	Temperature/speed control
Stir/hot plate	Temperature control
Centrifuge	Sorvall ultracentrifuge
Sonifier	Branson sonifier 250 at 70% power
FPLC	AKTApurifier FPLC system
Concentrator	Amicon ultrafiltration cell/spin filter, 10K MWCO
0.2 μm Spin filter	Corning Costar Spin-X
UV/vis	HP 8453 with OLISWorks
Western blot/SDS gel equipment	Bio-Rad
Agarose gel equipment	Bio-Rad mini sub gel GT
Cold room	Maintained at 4°C

allowing it to remain positively charged at the pH of the buffers used in this protocol. We use a Hi-Trap SP HP (Cation) (GE Healthcare) column with a linear gradient of 10%–70% Heparin Buffer B (Table 2) to elute the protein.

The eluted fractions are then combined and diluted fivefold with Heparin Buffer A, filtered with a 0.2 μm filter, and added to a 5-mL Pharmacia Hi-trap Heparin column. MutY is eluted using a linear gradient of 100% Heparin Buffer A to 100% Heparin Buffer B over 20 column volumes (Table 2). Fractions corresponding to MutY can be identified by the signature absorbances measured at 280 and 410 nm indicative a protein bearing an Fe–S cluster (Fig. 3). Pure MutY fractions are pooled and diluted in Heparin Buffer A. The sample is then concentrated to the mid μ*M* range (Table 3), and an equal volume of 50% glycerol is added to it, before storing at −80°C (short-term storage of 1–2 months) or in liquid nitrogen.

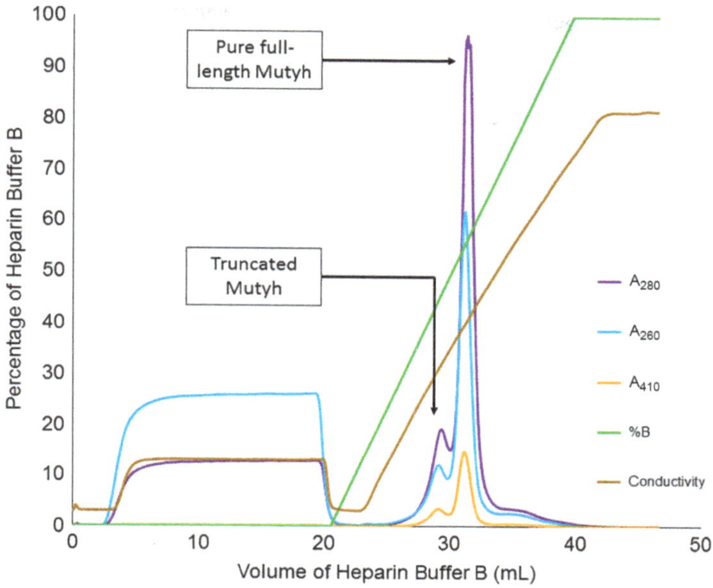

Fig. 3 Representative chromatogram of Mutyh during purification through a Hi-Trip Heparin column. The *purple*, *blue*, and *yellow lines* refer to the absorbance at 280, 260, and 410nm wavelengths respectively, while the *green* and *brown lines* refer to the percentage of Heparin Buffer B and the conductivity of the sample. Mutyh and its homologs have similar chromatograms and typically elute at a NaCl concentration of 450m*M* as shown by the peak above. The peak eluted just prior to the main protein peak is truncated protein and is collected in separate fractions to prevent contamination of the full-length enzyme (Pope & David, 2005).

Troubleshooting tips: Prior to use of the cation exchange column, it is important to ensure that the pH of the buffers being used is 7.5, as the MutY will not bind to the column if it is not positively charged. It is also important to keep in mind the binding capacity of each of the columns being used and ensure that limit is not exceeded. This might necessitate dividing the protein sample into multiple batches and purifying over several runs. It is recommended that liquid nitrogen is used for long-term storage of all MutY enzymes.

2.4 Purification of a Thermophilic Homolog, *G. stearothermophilus* MutY

The *Gs* MutY gene was identified via BLAST search of the genome (Fromme et al., 2004), and the gene was cloned into the *pET28a* expression vector (EMD Biosciences). This construct consists of an N-terminal hexa-His tag, followed by a thrombin protease cleavage site prior to the *Gs* MutY gene. In addition, this construct consists of two changes, Phe347Ser and Lys357Glu, to enhance expression yet still retain enzyme activity (Fromme et al., 2004). The protein is overexpressed in BL21 (DE3) Rosetta 2 cells (Novagen) along with the *pRKISC* plasmid that codes for the Fe–S cluster assembly proteins (Table 1; Tokumoto & Takahashi, 2001).

Following cell lysis, imidazole and NaCl are added to the supernatant to make a final concentration of 20 and 300 m*M*, respectively. The supernatant is batch bound to Ni^{2+}-NTA resin (Qiagen) for 1 h, poured over a PD10 column, and allowed to flow through by gravity. Following completion of flow through, the column is washed with 10 mL of separation wash buffer and eluted in 3 mL separation elutant buffer (Table 2) and is then concentrated with stirring (Table 3). The protein is filtered with a 0.2-μm filter then purified on an FPLC (Table 3) with an anion exchange MonoQ column (where the column volume, CV = 1 mL) that is preequilibrated with MonoQ start buffer (20 m*M* Tris–HCl pH 8.0, 5 m*M* β-mercaptoethanol) running at a flow rate of 0.5 mL/min. The sample is eluted over 20 CV of a linear gradient of 0%–100% MonoQ elution buffer (MonoQ start buffer with 1 *M* NaCl). *Gs* MutY should elute at approximately 400 m*M* NaCl, and these fractions are then combined and buffer exchanged into storage buffer (Table 2).

To cleave the His-tag from MutY, the solution is diluted and 2 U thrombin (Novagen) per mg enzyme is added to the solution and incubated for 16–20 h at 4°C. The thrombin reaction is subsequently inhibited by adding the protease inhibitor PMSF (final concentration of 1 m*M*) and incubated for 2 h at 4°C. To the supernatant, imidazole and NaCl are added to make a final concentration of 20 and 300 m*M*, respectively. The supernatant is again batch bound to fresh Ni^{2+}-NTA resin for 1 h, poured over a PD10 column, washed with an additional 2 mL separation wash buffer, which is added to the flow through containing the cleaved MutY protein.

The protein is then further purified by gel filtration on a Superdex 200 column (GE Healthcare). Prior to loading on the column, the protein sample is concentrated (Table 3) to approximately 300 μL and filtered through a 0.2-μm spin filter (Table 3). The column is preequilibrated with five CV of storage buffer (pH 8.0) at a flow rate of 0.5 mL/min, following which the protein sample is loaded onto it. Pure *Gs* MutY fractions are pooled and can be concentrated (Table 3) and buffer exchanged into the appropriate storage buffer (Table 2), and finally mixed with glycerol to 25% final concentration and stored at −80°C for future use.

Troubleshooting tips: Since the final purification step involves gel filtration, it is important to concentrate the protein sample to as small a volume as feasible, while avoiding aggregation. This can be achieved through the use of centrifugal concentration units, with regular mixing in-between concentrations or an ultrafiltration concentrator.

2.5 MBP-MutY for Higher Yields and Solubility

Use of a MBP tag fused to the N-terminus of MutY aids in purification and increases protein solubility. The MBP-MutY construct was created by cloning the *Ec MutY* gene into the *pMAL-c2x* expression vector (NEB). The resulting MBP-MutY protein overexpresses better than the untagged enzyme, with the overall yield and quality of protein produced being significantly higher (Boon et al., 2003; Chmiel et al., 2001). The MBP-MutY was useful for obtaining the concentrations needed for the first electrochemistry experiments (Boon et al., 2003), as well as more recent electrochemistry and XAS studies (Bartels et al., 2017; Ha et al., 2017).

Overexpression and growth conditions for expressing MBP-MutY that should be followed are outlined in Table 1, with a key feature of inclusion of glucose in the media. After sonication of cells expressing MBP-MutY, the combined supernatant is batch bound to amylose resin (NEB) for 1 h, poured over a PD10 column allowing flow through to complete, and washed with separation buffer A and eluted in separation buffer B (Table 2) followed by concentrating (Table 3). MBP-MutY is then diluted 10-fold in Heparin Buffer A (Table 2) to lower the salt concentration and is filtered with a 0.2-μm syringe filter.

Directly following, the sample can be added to a 5-mL Pharmacia Hi-trap Heparin column on an FPLC (Table 3) and eluted using a linear gradient in Heparin Buffer A to 100% Heparin Buffer B (Table 2) over 20 CV. Fractions corresponding to MBP-MutY can be identified by the absorbance measured at 280 nm (protein) and 410 nm (Fe–S cluster) (Fig. 3). Pure fractions can be combined and concentrated (Table 3) to mid to low μ*M* protein concentrations. The enzyme should be buffer exchanged into MutY storage buffer (Table 2) and concentrated to mid to low μ*M* concentrations, and stored as single-use aliquots at −80°C or liquid nitrogen. Purity of MBP-MutY samples can be confirmed via 12% SDS PAGE stained with SYPRO Ruby stain according to manufacturer's procedure.

2.6 Bacterial Expression of *M. musculus* Mutyh

Overexpression and purification of human MUTYH are notoriously finicky, producing low yields and quality of protein (Engstrom et al., 2014; Golinelli et al., 1999). However, high sequence homology between human MUTYH and other model organism homologs, such as mouse Mutyh, provides alternative routes to overcome the challenges of working with the human enzyme (Parker & Eshleman, 2003). An expression vector (*pQE30)* encoding Mutyh lacking 28 N-terminal amino acids (Yang et al., 2001) produces active mouse Mutyh; however, the *pQE30* vector does not allow for removal of the hexa-His tag (Pope & David, 2005). To utilize this platform but allow for the ability of tag cleavage, the *Mutyh* gene was removed using restriction digestion and cloned into a *pET28a* vector (EMD Biosciences) equipped with an N-terminal His_6 tag and thrombin cleavage site to allow for preparation of the untagged Mutyh enzyme (Engstrom et al., 2014).

Following cell lysis steps, imidazole and NaCl are added to the supernatant to make a final concentration of 20 m*M* and 1 *M*, respectively. The supernatant is batch bound to Ni^{2+}-NTA resin (Qiagen) for 1 h, poured over a PD10 column allowing flow through by gravity. Following completion of flow through, wash with 10 mL Mutyh separation wash buffer and elute in 3 mL Mutyh separation elutant buffer (Table 2).

The protein solution should then be diluted 10-fold with Heparin Buffer A (Table 2) to lower the salt and imidazole concentration. To cleave the His-tag from Mutyh, 2 U thrombin (Novagen) per mg Mutyh is added to the solution and incubated for 16–20 h. Subsequent incubation for 2 h with PMSF (final concentration of 1 m*M*). To the supernatant, imidazole and NaCl are added to make a final concentration of 20 m*M* and 1 *M*, respectively. The supernatant is again batch bound to clean Ni^{2+}-NTA resin for 1 h, poured over a PD10 column, and washed with an additional 2 mL Mutyh separation wash buffer, which is added to the flow through containing the cleaved Mutyh protein.

Mutyh should be diluted 10-fold in Heparin Buffer A (Table 2) to lower salt concentration, filtered with a 0.2-μm filter, and added to a 5-mL Pharmacia Hi-trap Heparin column on an FPLC (Table 3), and eluted using a linear gradient of 100% Heparin Buffer A to 100% Heparin Buffer B over 20 CV. Fractions corresponding to Mutyh can be combined and concentrated (Table 3), to mid to low μ*M* concentrations. The enzyme is buffer exchanged into Mutyh storage buffer (Table 2) and concentrated to mid to low μ*M* concentrations and stored as single-use aliquots at −80°C or in liquid nitrogen. Purity of Mutyh samples is confirmed via 12% SDS PAGE stained with SYPRO Ruby stain as according to manufacturers procedure.

2.7 Eukaryotic Expression System for Production of *Homo sapiens* MUTYH

Mammalian cells and their respective enzymes are inherently complex, with gene products experiencing an assortment of pre- and posttranslation modifications. The Bac-to-Bac Baculovirus Expression System (Invitrogen) employs an insect cell-based expression system that incorporates an improved baculovirus transfection procedure to aid in production of eukaryotic proteins. This expression system was found to be particularly useful in producing active human MUTYH amenable to quantitative kinetic analysis (Kundu et al., 2010). To accomplish this, the *MUTYH* 535 amino acid gene isoform and the *malE* gene that codes for an N-terminal MBP tag were previously cloned into the *MCS II* site in the *pFastBacDual* vector, and Arg codons were optimized for expression in the Bac-to-Bac system (Kundu et al., 2010).

The recombinant bacmid is created via transformation into chemically competent DH10Bac (Invitrogen) cells. Uptake colonies can be screened for the recombinant bacmid by performing a blue-white colony-screening assay followed by sequence analysis of the resultant DNA product to check for proper insertion. The transfection procedure closely follows the Bac-to-Bac kit protocol where 1×10^6 Sf9 cells are added to each well of a 6-well cell growth plate and allowed to attach for 1 h at 25°C. Two transfection reactions are prepared: one with 2 (low) and one with 6 (high) μL of stock plasmid DNA, each of which is then added to 100 μL of MUTYH culture growth media (Table 1), supplemented with 6 μL of Cellfectin, and incubated for 30 min. The media can be carefully decanted and separated from the attached cells, followed by addition of the transfection mixture along with 800 μL of additional MUTYH culture growth media. The cells are incubated at 28°C for 5 h, followed by removal of the current transfection mixture and addition of 2 mL fresh MUTYH culture growth media, and subsequently incubated for 72 h at 28°C to propagate the virus. Following incubation, the media can be centrifuged at 500 × *g* for 5 min to pellet cells and debris.

The supernatants are transferred to centrifuge tubes and stored at 4°C, and a small aliquot of supernatant should be tested via dot blot assay. This will be in comparison to a standard baculovirus of known high titer to determine the ratio of infectious virus particles to cells (or multiplicity of infection, MOI). This dot blot assay only requires 2 μL of the standard and viral MUTYH containing supernatants to be detected via dotting on a nitrocellulose membrane, followed by blocking in 5% milk in PBST for 15 min at 25°C with shaking. The blot can then be visualized with mouse anti-gp64 antibody as according to standard western blot protocol. To generate enough protein for characterization, 600 mL Sf9 cells exponentially growing at >95% viability are seeded at a concentration of 6×10^5 cells/mL in MUTYH culture growth media and then infected with an empirically determined amount of human WT MUTYH baculovirus where the MOI is approximately 3. Following postinfection incubation for 65 h, the cells can be centrifuged at 600 × *g* for 30 min, followed by suspension in 10 mL separation wash buffer.

The cells are to be lysed using a freeze/thaw method of three cycles at 37°C alternated with three cycles at −80°C, each lasting for 5 min. All of the following steps are then carried out at 4°C or on ice unless otherwise noted, and all buffers used are prechilled. The cell lysate should be further sonicated (Table 3) two times for 10 s each with a 10-s rest period in between. After sonication, an 18% PEG solution containing 1 *M* NaCl and 10% glycerol is added dropwise to the cell lysate with continuous stirring for 30 min to precipitate out the genomic DNA. The cell lysate is then centrifuged at 15000 × *g* for 30 min to pellet cell debris, and 1 mL amylose resin is added to the supernatant and batch bound for 1 h with end-over-end rotation. The slurry is added to a PD10 column allowing flow through to complete via gravity and washed with 10 column volumes of separation wash buffer (Table 2). Following completion of the wash, 5 mL of separation elution buffer (Table 2) is added to the column and incubated for 30 min. The eluted fractions are combined and concentrated (Table 3) to low μ*M* concentrations and stored as single-use aliquots at −80°C. Purity of MBP-MUTYH samples is confirmed via 12% SDS PAGE stained with SYPRO Ruby stain as according to manufacture procedure.

3. GEL-BASED ADENINE GLYCOSYLASE ASSAYS AND MEASUREMENTS OF KINETIC PARAMETERS

3.1 General Setup and Execution of the Glycosylase Assay

A general scheme for the gel-based adenine glycosylase assay is depicted in Fig. 4, where MutY (or homolog) is allowed to react with 30 bp radiolabeled

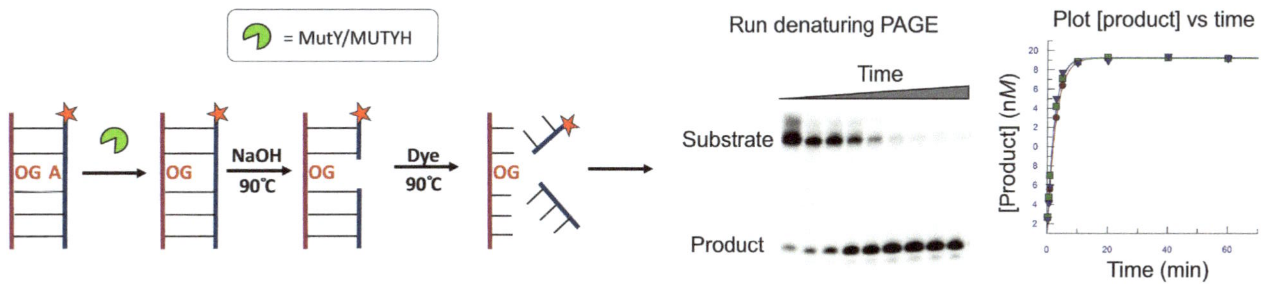

Fig. 4 MutY glycosylase assays. Schematic depiction of the glycosylase assay and PAGE for determination of kinetic parameters and active fractions of MutY/MUTYH.

DNA substrate containing a central OG:A mispair (Porello, Leyes, et al., 1998). The reaction is quenched with base at different time points to sever the AP site-containing DNA (resulting from adenine removal) forming a shorter radiolabeled oligonucleotide (nt) product. The 14 nt radiolabeled fragment derived from the product is separated from the 30 nt radiolabeled fragment derived from the duplex substrate by electrophoresis in a 15% denaturing polyacrylamide gel and visualized using storage phosphor autoradiography. Plots of product concentration as a function of time are obtained and used to determine the rate constants of the enzymatic reaction.

We were initially prompted to more thoroughly investigate the kinetic behavior of MutY in order to explain conflicting data in the literature as to whether OG:A vs G:A bp duplexes are the preferred substrates (Porello, Leyes, et al., 1998). Our kinetic analysis provided insight into reasons for the apparent contradictions in the literature; a major issue was the lack of appreciation for the much higher affinity of MutY to its OG:AP (abasic) site product duplex (and long half-life ~3 h) compared to the G:AP site product, and the consequences of this on the analysis of MutY activity. Indeed, the intrinsic rate of adenine excision for MutY with OG:A substrates is faster than G:A substrates when analyzed under single turnover (STO) conditions; however, the faster turnover with G:A substrates than OG:A substrates results in more product formed under multiple turnover (MTO) conditions with G:A substrates after long reaction times, and depending on amount of enzyme used. These features make it important to accurately determine the concentration of *active* MutY when performing kinetic experiments so that the experiments can be performed under the proper conditions (STO or MTO). These features also make measuring Michaelis–Menten parameters k_{cat} and K_m with MutY enzymes impractical due to minimal substrate turnover (especially with OG:A substrates) in a reasonable time frame. Based on our analysis of MutY kinetic behavior, we have developed a kinetic framework for analyzing the processing of DNA substrates by MutY enzymes that balances rigor and simplicity (Fig. 5). The behavior of MutY enzymes can be characterized into three basic steps: "binding" of MutY to the DNA and OG:A mispair (including base flipping), *N*-glycosidic bond cleavage (base excision) of the A, and release of the abasic site DNA product (Fromme et al., 2004; McCann & Berti, 2008; Michelson et al., 2012; Porello, Leyes, et al., 1998; Woods et al., 2016). In terms of "binding" we also consider that this involves at least two phases: an initial association with the DNA and encounter with the target bp, followed by flipping of the A from the helix into the enzyme active site (Fig. 5).

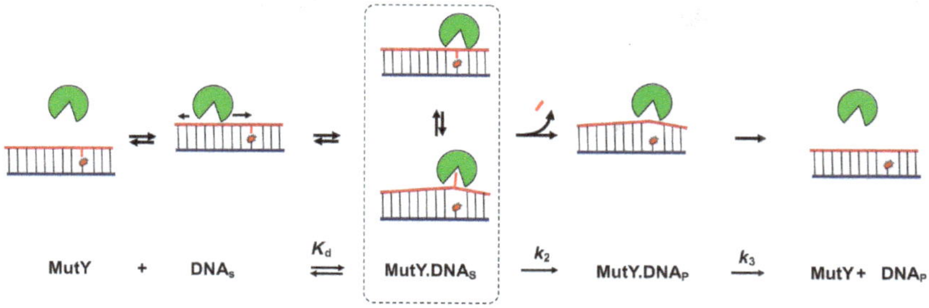

Fig. 5 Schematic representation of damage processing and minimal kinetics scheme for MutY and homologs. The free enzyme is known to nonspecifically engage with the DNA strand and use a processive search to detect the OG:A lesion. Upon encountering the lesion, it flips the adenine into its active site pocket. The binding affinity of the enzyme–substrate complex is determined by the dissociation constant K_d. The enzyme then catalyzes base excision (measured by the rate constant k_2) to form an abasic site product, which it then slowly releases (measured by rate constant k_3).

By performing the gel-based adenine glycosylase assay under different conditions (multiple turnover or single turnover) specific rate constants can be measured to determine the intrinsic rate of base excision catalysis (k_2) and rate of release of the AP site-containing DNA product (k_3) (Porello, Leyes, et al., 1998). Additionally, this assay can be used to determine the active fraction of MutY in a purified sample relative to the total protein concentration determined by absorbance at $A_{280\text{nm}}$ (Supplementary Video 1 in the online version at https://doi.org/10.1016/bs.mie.2017.11.035).

3.1.1 Radiolabeling the DNA Substrate

A variety of DNA sequences could be used for these assays; however, we have typically used 30 bp duplexes in order to ensure complete duplex formation at the concentrations used in these assays. Most routine assays of MutY enzymes use duplexes containing a central OG:A bp; however, analysis of substrate analogs has often been useful (Chmiel et al., 2001; Chu, Fettinger, & David, 2011; Manlove et al., 2017). We have also found that catalytic deficiencies in MutY enzymes are more often revealed using the nonoptimal G:A substrate (Chepanoske, Lukianova, Lombard, Golinelli-Cohen, & David, 2004; Chmiel, Livingston, & David, 2003; Pope, Porello, & David, 2002). The typical 30-nucleotide long DNA oligomers that we have used have the following sequences:

5′-d(CGA TCA TGG AGC CAC **OG**AG CTC CCG TTA CAG)-3′ (University of Utah Health Science Center Core Facilities, OG phosphoramidite purchased from Glen Research)

5′-d(CTG TAA CGG GAG CT**A** GTG GCT CCA TGA TCG)-3′ (Integrated DNA Technologies, IDT)

These oligonucleotides are HPLC purified and desalted using a Dionex DNAPac PA-100 anion exchange column. Oligonucleotides should be stored dried, and only dissolved in solution for quantitation and subsequent labeling. OG-containing oligonucleotides are particularly susceptible to degradation if stored in solution. The 5′-end-labeled 30 bp long DNA duplex with a centrally placed OG:A mispair is generated by incubating 2.5 pmol of the A-containing strand with 25 μCuries of [γ-^{32}P] ATP (Perkin-Elmer) in the presence of 10 units of T4 polynucleotide kinase (NEB) and the provided kinase buffer, at 37°C for 30 min. The reaction is terminated by incubating at 90°C for 5 min. The labeled product is purified using a G-50 microspin column (GE Healthcare), and then freeze dried in a speed-vac under the low heat setting. The dry oligonucleotide is suspended in 2 × annealing buffer (Table 4) and combined with an additional 47.5 pmol of unlabeled A-containing strand, followed by 60.0 pmol of complementary OG carrying strand to yield a solution of 5% labeled DNA. The final concentration of DNA duplex is brought to 100 n*M* with MilliQ H_2O, and the solution is incubated for 5 min at 90°C on a dry bath incubator, transferred to a Styrofoam box, and allowed to cool slowly to 4°C for 12–16 h to ensure complete annealing.

3.1.2 Preparation of a Denaturing Polyacrylamide Gel

The 15% acrylamide/bisacrylamide gel mixture is prepared according to the composition in Table 4. Degassing the solution for 30 min removes dissolved oxygen after which the solution can be stored in a dark bottle until further use. The gel sandwich for this assay is prepared in a manner similar to that used for DNA sequencing where polymerization of the gel mixture is induced by addition of catalytic amounts of TEMED and ammonium persulfate, and the gel is allowed to set for at least 2 h before use.

3.1.3 General Assay Setup

The steps in the glycosylase reaction are illustrated for an active site titration (AST) experiment (Fig. 6A) but are similar for STO experiments. Glycosylase assay reactions are carried out in 100 μL volumes total following enzyme addition to the reaction mix (Table 4). The positive and negative controls (excess enzyme and no enzyme, respectively) can be carried out in as little as 20 μL reaction volumes. While the reaction tubes are allowed to equilibrate at 37°C, the enzyme stock solution is thawed on ice and

Table 4 Composition of Solutions for Characterization Assays

Assay	2× Annealing Buffer	Gel Composition	Acrylamide: Bisacrylamide	10× Assay Buffer	MutY/MUTYH Dilution Buffer	Reaction Mix Composition
Glycosylase	40 m*M* Tris–HCl pH 7.6 20 m*M* EDTA 300 m*M* NaCl	15% acrylamide 7 *M* urea 1 × TBE buffer	40% 19:1	200 m*M* Tris–HCl pH 7.6 100 m*M* EDTA	20 m*M* Tris–HCl pH 7.6 10 m*M* EDTA 20% glycerol	1 × assay buffer 20 n*M* DNA* 30 m*M* NaCl 0.1 mg/mL BSA
EMSA	40 n*M* Tris–HCl pH 7.6 20 m*M* EDTA 300 m*M* NaCl	6% acrylamide 0.5 × TBE buffer	40% 29:1	200 m*M* Tris–HCl pH 7.6 10 m*M* EDTA 1.0 *M* NaCl	20 m*M* Tris–HCl pH 7.6 10 m*M* EDTA 20% glycerol	2 × assay buffer 10 p*M* DNA* 0.2 mg/mL BSA 2 m*M* DTT 20% glycerol

DNA* refers to the radiolabeled DNA which is 5% labeled in the glycosylase assay and 100% labeled in EMSA.

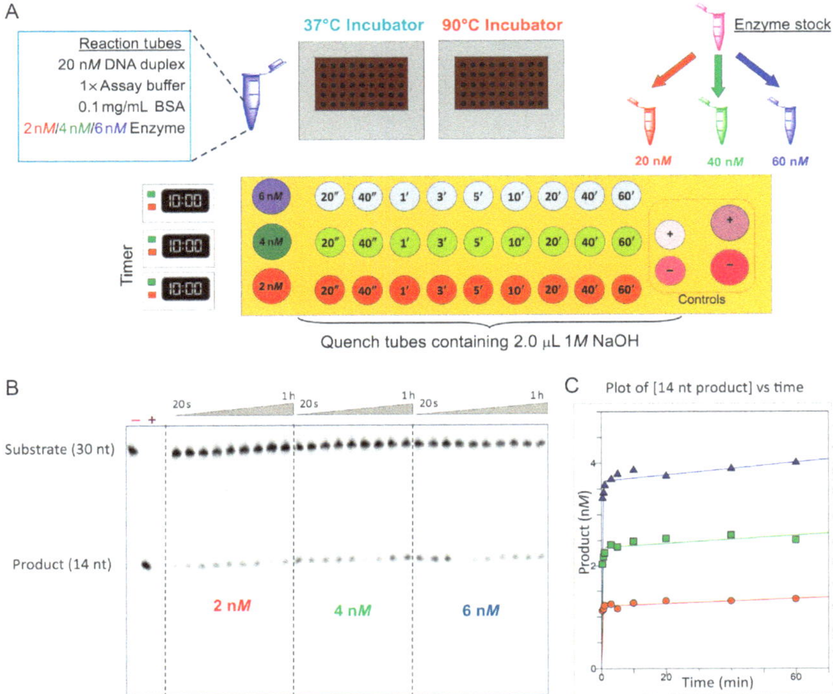

Fig. 6 General setup for glycosylase assays. (A) Setup for a multiple-turnover/active site titration experiment showing composition of the reaction and quench tubes. (B) Representative gel from a typical active site experiment. (C) Characteristic plot of the concentration of product produced over time. Note that single-turnover experiments would be performed similarly by adjusting the enzyme concentration to be in excess.

diluted using ice-cold MutY/MUTYH dilution buffer (Table 4) to give desired enzyme concentrations, in our case 20, 40, and 60 n*M* solutions. These solutions are mixed well before each subsequent tube transfer. The reaction is started by the addition of 10 μL of the 20 n*M* enzyme solution to the 2 n*M* reaction tube while simultaneously starting the timer. The contents of the tube are mixed by gentle and thorough pipetting, and, at each time point, 8 μL of the reaction mixture is quenched with 2 μL of 1.0 *M* NaOH, immediately incubated at 90°C for 5 min, and then placed on ice. This process is repeated for the 4 and 6 n*M* reactions using the 40 and 60 n*M* enzyme solutions, respectively. The positive and negative controls are incubated at 37°C for the hour-long duration of the experiment, after which they are also quenched with NaOH.

Troubleshooting tips: In order to avoid losing enzyme activity, mixing the enzyme must be done slowly with a pipette. It is also important to limit the number of freeze–thaw cycles the enzyme undergoes as this will also result in a reduction of activity. Negative controls (no enzyme) should indicate quality of labeled DNA. The positive control (excess enzyme) lane should show a single band corresponding to complete conversion to product; if not, this usually indicates a problem with the DNA, which can sometimes be remedied by reannealing the DNA or may require use of a new OG (or A) containing oligonucleotide strand.

3.1.4 Visualization and Quantitation of Results

Prior to loading the samples on the gel (Fig. 6B), 10 μL of formamide loading dye (80% formamide, 0.025% xylene cyanol, 0.025% bromophenol blue) is added to each sample, followed by incubation at 90°C for 10 min, and immediate cooling on ice. The samples are loaded on the premade gel with both wells filled with 1 × TBE and run at 1200 V for 2 h. The gel is then exposed to a phosphor imager screen for 12–15 h, and then scanned (Table 5). The substrate and product band intensities resulting from storage phosphor autoradiogram are quantitated and the data plotted using and appropriate graphing program (Table 5).

3.1.5 Salt Concentration in Assay Buffer

A very important consideration in carrying out these experiments is the type and final concentration of salt present in the reaction mixtures. In our laboratory, the glycosylase assay buffer contains NaCl (Table 4), and its concentration is optimized based on the effects of specific and nonspecific DNA interactions under low (30 m*M*) to high (150 m*M*) salt conditions (Livingston, Kundu, Pozzi, Anderson, & David, 2005; Raetz et al., 2012). The formation of the DNA–enzyme complex is facilitated through hydrogen bonding and electrostatic interactions between the DNA backbone and the amino acid side chains, and it is possible to modulate the extent of these interactions by altering the salt concentrations. At high salt concentrations, the Na^+ cations compete with the enzyme for phosphate-binding sites, resulting in lower nonspecific interactions between the enzyme and DNA substrate (Record, Ha, & Fisher, 1991) allowing for the design of experiments to gauge the contribution of these interactions.

Indeed, when the *Ec* homolog of the MAP variant, Tyr82Cys MutY was tested under low and high salt conditions, it was demonstrated to be sensitive to an increase in NaCl concentration, displaying up to a 25-fold reduction in

Table 5 Summary of Equipment Needed for the Glycosylase Assay in Order to Determine the Percent Active Fraction and Intrinsic Rate of Glycosidic Bond Cleavage of MutY Homologs and Variants

Equipment	Notable Specifications
Pipettes/tips/microcentrifuge tubes/conical tubes	DNase/RNase/Pyrogen free and sterile
HPLC	Shimadzu prominence LC-2AT
Speed-vac centrifuge	Savant speed vac SC110
Rapid Quench Flow Instrument	Kintek Rapid Quench Flow (RQF-3)
Timer	Recording time points throughout reaction
Vortex/mini centrifuge	Table top
Ice bucket	Experiment carried out on ice
UV/vis	HP 8453 with OLISWorks
Glass plates, spacers, and comb	0.35 mm thick spacers and a 32-well comb
Dry bath incubator	Fits 1.6 mL microcentrifuge tubes, temperature controlled
Phosphorimager screen	Molecular Dynamics phosphor imager screen
Phosphorimager	Typhoon 9400 scanner, GE Healthcare
Software	ImageQuaNT V.5.2. and GraFit V 5.0.2

the rate of adenine removal at 150 m*M* NaCl compared to 30 m*M* NaCl, and a 60-fold reduction compared to WT MutY under the same conditions (Livingston et al., 2005). This salt concentration sensitivity has also been observed with mouse Mutyh and human MUTYH Tyr165Cys disease variants. Under buffer conditions of physiological salt (150 m*M* NaCl) Tyr165Cys displays no OG:A activity yet at lower salt (100 m*M* NaCl), adenine removal is detected albeit with lower extents of reaction completion (Raetz et al., 2012). Substitution of the DNA-intercalating residue Tyr at position 82 with Cys at this position may attenuate the possibility for stabilization of the distorted DNA through intercalation and force the enzyme to bind through electrostatic interactions and hydrogen bonding. Binding of this kind would be disrupted by high salt conditions, and therefore, the enzyme is unable to enter its catalytically competent state at 150 m*M* salt, resulting in lower adenine excision (Livingston et al., 2005). Herein, we describe the glycosylase assay, where we carry out this experiment under low salt (i.e., 30 m*M* final NaCl) conditions.

3.2 Correcting the Enzyme Concentration for the Percent Active Fraction

Active fraction measurements of enzymes are important since proteins over-expressed and purified in the laboratory may be improperly folded or truncated, and thus fractions of the enzyme population are inherently, catalytically incompetent (Fersht, 1999; Porello, Leyes, et al., 1998). The active fraction of different preparations of the same enzyme can widely vary, and these differences must be considered when determining efficiency of catalysis and substrate binding. For example, initial studies concluded that the MAP variants Tyr165Cys and Gly382Asp are completely devoid of activity due to lack of consideration of active fraction (Wooden, Bassett, Wood, & McCullough, 2004). However, due to poor bacterial cell-based expression of human MUTYH, the enzyme activity must be corrected for the percent active fraction to isolate specific defects in catalysis (Kundu et al., 2009; Pope & David, 2005). Indeed, even WT *Ec* MutY displays variations in percent active fraction between different preparations. These differences may be further exacerbated when comparing MUTYH variants, such as those indicated in MAP, where the amino acid variation may influence stability, folding, and ability to actively engage the DNA substrate.

Under conditions of MTO, the reactions of MutY enzymes are characterized by biphasic kinetics, in which an exponential "burst" of product, formed by *N*-glycosidic bond cleavage, is followed by a steady-state phase of product formation. This kinetic behavior indicates a slow step after the chemistry step, which, in this case, is the release of the enzyme from the AP site-containing DNA product. A convenient aspect of such kinetic behavior is that it allows us to relate the amplitude of the "burst" phase to the active concentration of the enzyme in an AST experiment (Fersht, 1999; Porello, Leyes, et al., 1998). This percent active enzyme in a given sample of MutY can only be determined if the concentration of the enzyme is greater than K_d. Eq. (1) displays the change in product formation over time, where k_{obs} is the observed rate constant for the burst and k_{ss} is the slope of the linear phase. A_0 is the amplitude of the "burst" and is equal to the concentration of active MUTYH. This value is obtained by extrapolating the linear portion of the plot to determine the y intercept.

$$[P]_t = A_0\{1 - \exp(-k_{\text{obs}}t)\} + k_{\text{ss}}t \tag{1}$$

Each assay is comprised of three trials using a final enzyme concentration of 2, 4, and 6 nM (concentration calculated from the $A_{280\text{nm}}$ of the protein stock solution). The reaction is quenched at time points 20 s, 40 s, 1 s, 3 min, 5 min, 10 min, 20 min, 40 min, and 1 h using 2.0 μL of 1 M NaOH using a setup similar to that shown in (Fig. 6A). The data are fitted to Eq. (1) to determine the A_0, k_{obs}, and k_{ss}. The active enzyme concentration is based on the percentage of active protein out of the total protein concentration based on the $A_{280\text{nm}}$.

3.3 Determining the Rate of Product Release

The rate of release of the AP site-containing DNA product from the MutY active site is denoted as the rate constant k_3 (Porello, Leyes, et al., 1998). Due to the cytotoxicity of AP sites generated by MutY, enzyme variants that result in perturbations of the rate of product release may disrupt the delicate hand off between the glycosylase and downstream repair enzymes (Kundu et al., 2009). Under MTO conditions, the rate constant k_3 is determined from the slope of the linear portion of the product vs time plots. It is important to perform an AST of samples incubated in the reaction buffer for various times spanning the planned MTO to ensure the enzyme does not lose activity over the course of the reaction (Kundu et al., 2009). This ensures that the rate of product turnover is not reflective of a decrease in MutY activity, but a steady change in product formation. The rate constant, k_3, is described by Eq. (2) below (Porello, Leyes, et al., 1998).

$$k_3 = k_{\text{ss}}/A_0 \tag{2}$$

3.4 Assessing the Rate of Glycosidic Bond Cleavage

The intrinsic rate on N-glycosidic bond cleavage (k_2) of MutY is most conveniently measured using glycosylase assays performed under STO conditions (Porello, Leyes, et al., 1998), wherein the concentration of active enzyme is higher than that of the substrate DNA (also $[E] > K_d$). Using these types of experiments has allowed for delineation of the specific effects of amino acid variations in MutY and changes in the mismatch substrate on base excision. The observed rate of product formation can determined from fitting of the production curve to a single exponential using expression (3):

$$[P]_t = A_0\{1 - \exp(-k_{\text{obs}}t)\} \tag{3}$$

Assuming enzyme–substrate binding is in rapid equilibrium ($k_{\text{off}} \gg k_2$), k_{obs} is relate to k_2 by:

$$k_{\text{obs}} = [E]_0 k_2 / (K_d + [E]_0) \tag{4}$$

where $[E]_0$ refers to the initial enzyme concentration, and K_d is the dissociation constant. Under the conditions of STO, and when the concentration of enzyme used is well above K_d, this equation reduces to $k_{\text{obs}} = k_2$. Thus, the rate constant obtained by plotting the product concentration as a function of time is a measure of the intrinsic rate of adenine base excision by the enzyme (Porello, Leyes, et al., 1998). The creation of the 5′-end-labeled DNA substrate and the preparation of reaction and quench tubes is identical to that discussed in Section 2.2. For this assay, the substrate DNA concentration remains 20 n*M*, whereas the enzyme concentration (corrected for active fraction) is at least twofold higher than that of the substrate DNA. In case of the mammalian homologs of MutY, it may be necessary to increase the enzyme concentration to as much as fivefold that of the substrate DNA concentration. In order to ensure that conditions have been obtained to allow for the $k_{\text{obs}} = k_2$ simplification to be made, several concentrations of enzyme in excess should be tested such that k_{obs} is maximized. The product formation is monitored by using the same 15% PAGE described earlier, and the product concentration is fitted to Eq. (3) using a graphing program to determine the rate of glycosidic bond cleavage.

In the case of enzymatic reactions that have a high k_2, such as in the case of WT *Ec* MutY with an OG:A substrate, the rate of the reaction cannot be accurately estimated using the manual kinetic assay described in Section 3.3. In this scenario, an automated rapid quench flow instrument may be employed to measure the extent of product formation at time points as short as 0.1 s. In setting up for this experiment, enzyme and DNA reaction mixtures are prepared in 500 μL reaction volumes under conditions consistent with the manual experiments, with the exception that the enzyme is initially incubated with unlabeled nonspecific DNA duplex (the 30 bp duplex described in Section 3.1.1 with a centrally located G:C base pair) to stabilize it. Typically, 80 n*M* active enzyme is incubated on ice in MutY dilution buffer (Table 4) containing 25 n*M* nonspecific duplex supplemented with

NaCl for a final concentration of 30 mM salt. This enzyme mixture is added to an equal volume of the 40 nM labeled DNA substrate to make the final DNA concentration 20 nM and final enzyme concentration 40 nM. The reaction mixtures are quenched with 0.5 M NaOH at time points 0.2, 0.5, 1, 3, 7, 10, 30, and 60 s and subsequently incubated at 90°C for 5 min. This is followed by electrophoresis on a 15% denaturing polyacrylamide gel, imaging the gel, and fitting the change in product concentration over time to Eq. (3) using a graphing program to determine the rate of glycosidic bond cleavage.

Troubleshooting tips: In order to minimize risk of radioactive contamination while using the rapid quench instrument, tubes used to collect quenched reaction should be covered with parafilm beforehand to avoid splash back during ejection. Using rounded tubes, such as 2 mL SealRite rounded microcentrifuge tubes, reduces the risk of a radioactive spill while loading DNA into the syringe used on the rapid quench by eliminating the need to use a needle. To keep the enzyme cold during the experiment, the syringe containing the enzyme should be surrounded by ice using an ice bucket full of ice to cover the bottom and a glove filled with ice to cover the top.

4. GEL-BASED ASSAYS FOR DETERMINING MutY-DNA AFFINITY

4.1 General Features and Considerations of Gel-Based Binding Assays With MutY

Previous investigations by our group regarding the roles of the MutY Fe–S cluster demonstrated the importance of the region containing the Fe–S cluster for specific binding to the OG:A-containing DNA substrate, via mutations of the ligands to Fe–S cluster, and residues near the Fe–S cluster or part of the FCL motif (Chepanoske et al., 2000, 2004; Porello, Cannon, et al., 1998). Electrophoretic mobility shift assays (EMSA) are useful for measuring the equilibrium dissociation constant (K_d) for MutY with various DNA duplexes. Though other types of binding assays may also be used, the high affinity of MutY for substrate, product, and transition state analogs makes use of the EMSA gel-based assay advantageous due to low concentrations of the DNA duplex that can be used, in comparison to other common DNA-binding methods, such as fluorescence anisotropy. In these experiments, use of catalytically inactive enzyme or a noncleavable 2′-deoxyadenosine mimic allows for isolation of substrate binding by preventing catalysis.

For example, catalytically inactive Glu37Ser *Ec* MutY has previously been used to study binding interactions with the natural substrate OG:A (Livingston et al., 2005). In order to study mutated MutY enzymes and MAP variants, it is convenient to use a noncleavable analog of A, such as 2′-fluoro-2′-deoxyadenosine (FA). Measurements of affinity to product analogs can also provide a means to evaluate specific aspects of the MutY mechanism. The AP site product formed by MutY adenine excision can be mimicked via incorporation of a product analog such as tetrahydrofuran (THF) into DNA. This analog lacks the hydroxyl group present in the natural AP site; however, its stability and ability to be incorporated into DNA via solid-phase DNA synthesis makes it an attractive substitute for measuring AP site affinity (Chu et al., 2011; Woods et al., 2016). Implementation of EMSAs for determining the dissociation constant of MutY with various DNA contexts is explained in Section 4.2.

Transition state analogs are expected to be tight-binding inhibitors of enzymes due to the high-affinity of enzymes to their transition states; this suggests that the tighter the binding, the more closely the analog resembles the transition state for the reaction (Pauling, 1948). Kinetic isotope effect studies on MutY have suggested that the reaction mechanism has two distinct oxocarbenium ion-like transition states (McCann & Berti, 2008). In analogy to related nucleosidases and glycosylases, we have developed transition state analogs that feature pyrrolidine nucleotides (Chu et al., 2011). An exceedingly useful general transition state mimic for BER glycosylases is (3*R*,4*R*)-(hydroxymethyl)pyrrolidin-3-ol (1 N), which has a nitrogen in the corresponding 1′-position in the deoxyribose sugar (Woods et al., 2016). At physiological pH, the protonated amine in 1 N mimics the positive charge buildup in the transition state.

Although a powerful tool, the use of tight binding, transition state analogs can result in difficult evaluation and accurate comparison of the dissociation constant, K_d. This is due to limits of detection of the radiolabeled substrate for measuring subnanomolar dissociation constants, which are dependent on the assumption that [enzyme] ≫ [substrate]. To overcome these limitations, another perspective on binding between the enzyme and DNA can be gained by measuring k_{off}, the dissociation rate constant of the enzyme–substrate complex resulting in free DNA and MutY. Previous work with WT *Gs* MutY and transition state analog, 1 N, base paired across from G indicated a K_d below the detection limit, precluding comparing affinity with Asp144Asn MutY (Chu et al., 2011). However, k_{off} for WT MutY could be accurately calculated using the off rates assay and it showed a

127-fold increased k_{off} for Asp144Asn compared to WT MutY (Chu, 2011). This variation of EMSA for determining enzyme affinity for DNA is described in Section 4.3.

4.1.1 Radiolabeling the DNA Substrate

MutY/MUTYH enzymes have high affinity for their preferred OG:A substrate, and therefore, these assays should be performed using DNA concentrations well below the K_d (at least 10-fold lower, ideally 100-fold). Typically, 5 pmol of the A (or FA) containing oligonucleotide is 5′ end labeled. The radiolabeled strand is annealed to a 20% excess of the complementary OG strand, and the final concentration of the mixture is brought to, typically, 5 n*M* by the addition of MilliQ H_2O, although higher concentrations can be used so long as the DNA concentration is 10-fold lower than K_d. Annealing is done by incubating the solution for 5 min at 90°C on a dry bath incubator, transferring the solution to a Styrofoam box, and slowly cooling to 4°C for 12–16 h. Note, in considering DNA concentration, we assume that no labeled DNA is lost during the postlabeling procedures, so the DNA concentration is likely an overestimate; however, this ensures that the DNA concentration is kept below both the K_d and the enzyme concentration.

Troubleshooting tips: Due to the low concentrations of DNA used in EMSAs, it is best to use freshly labeled DNA (no more than 1 week old) in order to facilitate band quantitation via storage phosphor autoradiography.

4.1.2 Preparation of a Nondenaturing Polyacrylamide Gel

A 6% nondenaturing polyacrylamide gel (Table 4) is prepared for pouring in the assembled glass plates (Table 6). Polymerization of the gel mixture is induced by addition of catalytic amounts of TEMED and ammonium persulfate, and the gel is allowed to set for at least 1 h. Gels are cooled to 4°C and prerun at 120 V for 30 min prior to loading.

4.1.3 Running an EMSA

After running the experiment as described in Section 4.2, the nondenaturing polyacrylamide gel is loaded with reaction mixtures in a cold room at 4°C to prevent gel heating. To ensure efficient penetration of the samples into the gel, the running voltage of the gel is increased to 250 V for 10 min prior to and during loading. After the last sample (negative control) has been loaded,

Table 6 Summary of the Equipment Needed for an EMSA in Order to Determine the Affinity of MutY Homologs and Variants to DNA

Equipment	Any Notable Specifications
Pipette tips/microcentrifuge tubes/ conical tubes	DNase/RNase/Pyrogen free and sterile
HPLC	Shimadzu prominence LC-2AT
Speed-vac centrifuge	Savant speed vac SC110
Timer	Recording time points throughout reaction
Vortex/mini centrifuge	Table top
Ice bucket	Experiment carried out on ice
UV/vis	HP 8453 with OLISWorks
Glass plates, spacers, and comb	3 mm thick spacers and a 14-well comb
Dry bath incubator	Fits 1.6 mL microcentrifuge tubes, temperature controlled
Gel dryer	Bio-Rad Model 583 Gel Dryer
Phosphorimager screen	Molecular Dynamics phosphor imager screen
Phosphorimager	Typhoon 9400 scanner, GE Healthcare
Software	ImageQuaNT V.5.2. and GraFit V 5.0.2

the gel is run at 250 V for an additional 10 min. Following this, the voltage is dropped to 120 V and the gel is run for 1.5–2 h or until the lower band of the dye front is about two-thirds of the way down the gel.

Troubleshooting tips: To minimize temperature variation when working with enzyme, allow all pipette tips, gels, and buffers to cool to 4°C before running the EMSA. Running the gel at higher voltage after loading minimizes dissociation of complexes before entering gel, but may also result in heating, thus the 10 min time frame is a guideline, and the voltage can be turned down once all of the samples have entered the gel. Loading dye can be added to samples to monitor migration through gel; however, in some cases we have observed that loading dye can disrupt weak enzyme–DNA complexes. In this case it is best to just add the loading dye to the control lane (no enzyme). We have found the *Methods in Enzymology* article by J. Carey an excellent resource for tips on running EMSAs of protein–DNA complexes (Carey, 1991).

4.1.4 Visualization and Quantitation of Results

Once the run is complete, the gel is removed from the glass plates and placed in between blotting paper and cellophane (Bio-Rad). The gel is dried for 2 h using a gel dryer (Table 6). The dried gel is developed for 3 days on a phosphor imager screen, and the resulting autoradiogram is scanned and quantitated. The percent DNA bound is plotted against log[enzyme] and fitted to a one-site-binding isotherm using a graphing program to determine K_d.

4.2 Measurements of MutY-DNA Dissociation Constants

Solutions of varying concentrations of the enzyme are titrated against a labeled DNA duplex and allowed to equilibrate at a specific temperature, following which native PAGE is employed to separate the DNA–enzyme complex from free DNA substrate (Fig. 7). For a one-site-binding model, the K_d of the enzyme–substrate complex is expressed as:

$$K_d = [\text{Enzyme}][\text{DNA}]/[\text{EnzymeDNA}] \quad (5)$$

In using EMSAs to determine dissociation constants, a constant low concentration of duplex DNA (in the case of *Ec* MutY, a final DNA concentration of 5 p*M*) is equilibrated with a range of enzyme concentrations to determine the enzyme concentration at which half of the DNA is bound. By visual inspection, the lane having roughly the same intensities of bound and free DNA corresponds roughly to the K_d; however, a more accurate value is obtained by fitting to a one-site-binding isotherm. Notably, relating enzyme concentration to K_d is only accurate if the DNA concentration is negligible (at least 10- to 100-fold lower than the [*E*]). Thus, the [DNA] should be below the lowest enzyme concentration and below the expected K_d. Prior to starting the experiment, a reaction mix with the composition detailed in Table 4 is prepared. For a single experiment, 20 μL of the reaction mix is aliquoted into each of 14 microcentrifuge tubes and allowed to equilibrate at 25°C. Meanwhile, serial dilutions of the enzyme (based on active concentration) are carried out in MutY/MUTYH dilution buffer (Table 4)

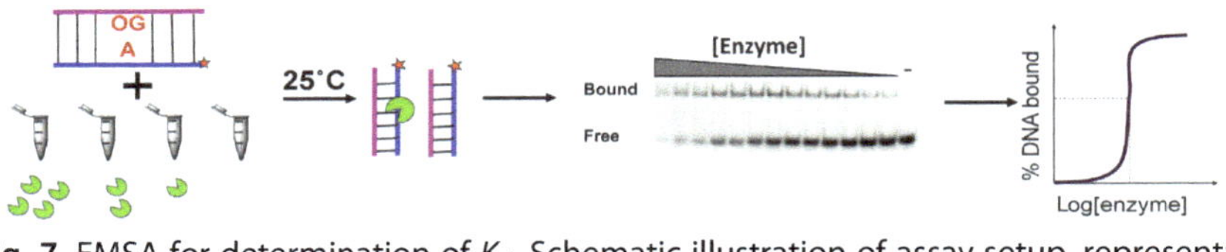

Fig. 7 EMSA for determination of K_d. Schematic illustration of assay setup, representative gel image, and expected plot of gel quantitation of K_d curve.

to give 13 different concentrations. These concentrations should range from at least 10-fold greater than the DNA concentration, up to several concentrations that will provide 100% bound DNA. To initiate the experiment, 20 μL of the highest enzyme concentration is added to 20 μL of DNA solution, mixed gently, and then incubated at 25°C for 30 min. This is repeated for each of the enzyme concentrations as well as the negative control. In our assays, the K_d is reported as the average of at least four separate experiments with enzyme concentrations approaching low μ*M* to reach the maximum concentrations needed for saturated substrate binding.

Troubleshooting tips: If the binding titration curves do not fit well to a one-site-binding isotherm, this often indicates that the DNA concentration is too high. Binding curves that plateau below 100% bound often indicate a problem with the DNA duplex. This may be due to incomplete annealing or degraded OG-containing oligonucleotide. In addition, homologs or variants that have a lower affinity for the DNA substrate may require the use of DNA concentrations as high as 50–100 p*M* in the reaction mixture.

4.3 Measurement of DNA Dissociation Rate (k_{off})

It can be difficult to measure an accurate K_d between MutY and tight-binding substrate, transition state and product analogs due to subnanomolar affinity. In addition, use of subnanomolar concentrations of enzyme prevents the use of the assumption that [enzyme] ≫ [substrate], and therefore, the model mentioned in Section 4.2 for calculating K_d fails. To circumvent this issue, another perspective on binding between the enzyme and DNA can be gained by calculating k_{off}, which allows us to measure the rate of release of the DNA and MutY from the bound substrate–enzyme complex. This is related to the dissociation constant as shown:

$$K_d = k_{off}/k_{on} \tag{6}$$

Through Eq. (6), it can be deduced that a larger k_{off} indicates a larger K_d value, suggesting a looser binding interaction. Similar to using an EMSA for calculating the dissociation constant, solutions of enzyme and DNA duplex are allowed to equilibrate at a specific temperature, followed by running a native PAGE to separate the DNA–enzyme complex from free DNA substrate. What differs between the two experiments is the variable is not the concentration of enzyme but rather the time of incubation of the enzyme with DNA. In this procedure, a constant concentration of radiolabeled duplex DNA is equilibrated with excess enzyme, then excess

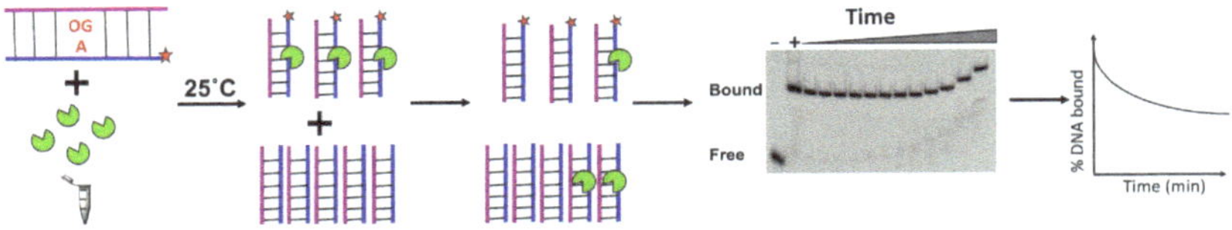

Fig. 8 EMSA determination of k_{off}. Schematic of assay setup, representative gel of separation of enzyme bound and unbound DNA, and graphing of the image results. The first lane in gel is the negative control, containing no enzyme, second lane is positive control with no unlabeled DNA added.

unlabeled duplex DNA is added and incubated for varying lengths of time (Fig. 8).

To begin the experiment, the radiolabeled DNA (Section 4.1.1) is incubated with enough enzyme for 30 min at 25°C to allow for 100% bound DNA. Following equilibration, a 1000-fold excess of unlabeled DNA is added to the reaction mixture. The unlabeled DNA mixture is prepared following the annealing procedure in Section 4.1.1. In the case of MutY, 220 μL of the DNA master mix is made with the appropriate amount of enzyme diluted in MutY/MUTYH dilution buffer (Table 4) and 20 pM labeled DNA. This is allowed to equilibrate at 25°C for 30 min. 20 μL of the solution, after 30 min, is loaded onto a gel, prepared by following Section 4.1.3, as the negative control. To initiate the experiment, the reaction volume is doubled by the addition of 200 μL of unlabeled duplex DNA to the reaction mixture, resulting in 10 nM unlabeled duplex DNA, and mixed gently using a pipette. By adding an excess of unlabeled DNA, it is assumed that any enzyme that releases labeled DNA will then bind to unlabeled DNA. 20 μL aliquots of the reaction mixture are loaded onto the gel at the appropriate time points for up to 1 h. The gel is run and quantitated following the procedures found in Sections 4.1.3 and 4.1.4, respectively. In the case of a tight-binding interaction, it is expected that the labeled DNA will remain mostly bound to the enzyme, resulting in a faint band of unbound DNA (Fig. 8). Alternatively, a loose binding interaction would produce a darker free DNA band, as more of the labeled DNA will be released. From these results, k_{off} can be calculated using the following equation:

$$[\text{DNA}]/[\text{EnzymeDNA}] = \text{span}^{-k_{off}t} + \text{span}_0^{-(k_{off})_0 t} + \text{plateau} \quad (7)$$

In Eq. (7), a linear transformation is used in the graphing software (Table 6) with the span equal to intercept/gradient, off rate equal to −1/gradient, and plateau is equal to the percent DNA bound when the slope is zero.

Troubleshooting tips: If the concentration of enzyme is too high, this could lead to a supershifted band in the gel, making quantification difficult. Reduce the enzyme concentration to address this issue.

5. APPLICATION OF METHODS TO REVEAL ROLES OF THE Fe–S CLUSTER COFACTOR IN MutY HOMOLOGS

To study the Fe–S cluster of MutY and homologs in vitro requires efficient overexpression and purification of the enzyme followed by assessment of repair activity via glycosylase and binding assays. Bacterial overexpression and purification of *Ec* MutY and Cys to His, Ser, and Ala variants demonstrated that cluster coordination is essential for producing viable glycosylases capable of recognizing and excising adenine (Golinelli et al., 1999; Lukianova & David, 2005). The utilization of the thermophilic homolog, *Gs* MutY, has aided in the development of several crystal structures of the protein in complex with DNA (Fromme et al., 2004; Woods et al., 2016). The ability to obtain highly concentrated samples of MBP-MutY allowed for XAS studies on MBP-MutY that provided insight into the influence of solvation and DNA on the Fe–S covalency of the MutY Fe–S cluster and its influence on its redox properties (Ha et al., 2017). Production of mammalian homologs, such as the bacterial expression of mouse Mutyh, resulted in identification of a novel Zn^{2+} linchpin motif in higher eukaryotic MUTYH homologs (Engstrom et al., 2014). An insect cell-based expression of human MUTYH allowed for detailed characterization of the functional consequences of several MAP variants (Kundu et al., 2009) and identification of a functionally relevant phosphorylation site (Kundu et al., 2010).

Following purification, it is important to establish the percent active fraction to account for variances between different protein preparations, which is particularly important with mammalian homologs (Brinkmeyer & David, 2015; Porello, Leyes, et al., 1998). The amplitude of the burst phase from adenine glycosylase assay under MTO conditions is directly related to the active enzyme concentration and provides a means to correct for the active fraction of a given aliquot of a purified MutY enzyme stock (Porello, Leyes, et al., 1998). Analysis of the active fraction can also be useful in determining the functional consequences of an amino acid variation and may be illuminating in terms of the origin of reduced activity for a given variant. Indeed, since MutY adenine excision activity requires proper engagement of MutY on the OG:A mismatch DNA substrate, a reduced active fraction in a mutated MutY/MUTYH enzyme may be due to its inability to bind to the DNA substrate in a productive manner that leads to catalysis.

Analysis of multiple preparations of two MAP variants Pro281Leu and Arg295Cys MUTYH (Fig. 9) indicated that both enzymes have significantly reduced active fractions of 0.1% and 5% relative to the WT enzyme, respectively (Brinkmeyer & David, 2015). This suggested reduced engagement with the DNA substrate; indeed, further analysis showed the Pro281Leu MUTYH was completely unable to bind to the OG:FA substrate analog duplex, while Arg295Cys was able to bind, but with reduced affinity (~10-fold). Notably, in the case of Arg295Cys, using correction for active fraction, the intrinsic rate of adenine glycosylase activity (k_2) was similar to WT, indicating that the mutation reduces the fraction of bound catalytically competent enzyme. In contrast, the reduced engagement of Pro281Leu could not simply be corrected by using higher concentrations of enzyme, and in this case, k_2 could not be determined. The results with Pro281Leu and Arg295Cys illustrate the key importance of the FCL motif in facilitating interactions between the positively charged residues of MUTYH and the

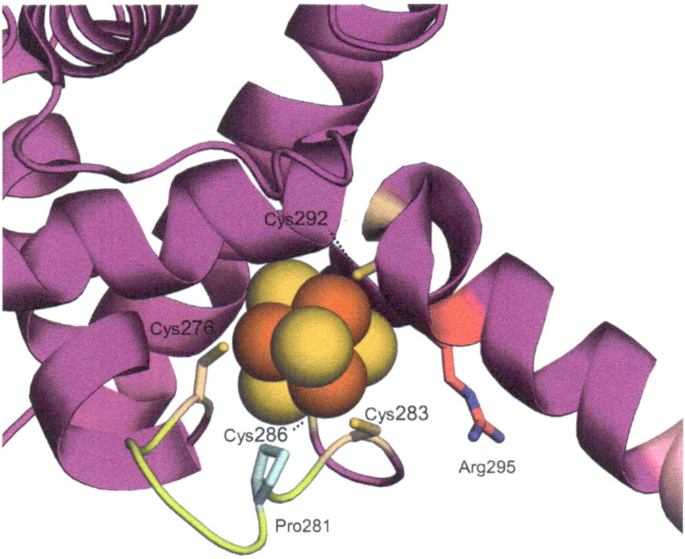

Fig. 9 MAP variants located at Pro281 (*cyan*) and Arg295 (*magenta*) are adjacent the Fe–S cluster. The MUTYH N-terminal (*purple*) fragment crystal structure of *Hs* MUTYH (PDB ID: 3N5N) (Luncsford et al., 2010) contains the FCL (*light green*), IDC (*light pink*), and Fe–S cluster (*orange* and *yellow spheres*) that is coordinated by four cysteine residues (*beige-colored sticks*).

Table 7 Kinetics and Binding Data for Pro281Leu, Arg295Cys, and WT MUTYH

	% Active Fraction[a]	**k_2 (min^{-1})**[b]	**K_d (nM)**[c]
WT	100	0.90 ± 0.02	1.9 ± 0.6
Arg295Cys	5	0.4 ± 0.1	>9
Pro281Leu	0.1	No cleavage	Not bound

[a]Representative data showing the percent active fraction of Brinkmeyer and David (2015).
[b]The intrinsic rate (k_2) of glycosidic bond cleavage determined using the glycosylase assay under STO conditions and 150 mM NaCl.
[c]The equilibrium dissociation constant, K_d, as determined by EMSA of the enzyme bound to a DNA duplex with central OG:FA base pair in the presence of 100 mM NaCl.

negatively charged DNA (Fig. 9, Table 7; Brinkmeyer & David, 2015). Of note, similar types of results were obtained in studying the Zn^{2+} linchpin mutants, showing the keen insight that can be gained by this type of careful analysis (Engstrom et al., 2014).

Detailed analysis of the kinetic parameters using the glycosylase assays and K_d measurements via EMSA has also revealed the relative importance of specific residues around the Fe–S cluster in mismatch recognition vs adenine excision (Banda et al., 2017; Lukianova & David, 2005). For example, the impact of replacement of three positively charged amino acids in the FCL of MutY (e.g., Arg194Ala, Lys196Ala, and Lys198Ala) was discerned by measuring the kinetic parameters of the adenine glycosylase activity (k_2, k_3) with both OG:A and G:A substrates (Chepanoske et al., 2000). In this case, the G:A substrate was particularly useful in revealing the catalytic defect and has the advantage of not requiring use of the rapid quench instrument for determination of k_2. Notably, all three variants were shown to have similarly decreased affinity (10-fold) for an OG:FA duplex; however, Lys198Ala was shown to have the most dramatic effect on the glycosylase activity measured under STO conditions, illustrating a direct role of Lys198 in mediating mismatch recognition and excision. Indeed, a simple "DNA-binding" defect would not be expected to impact k_2 since these experiments are performed under concentrations significantly above K_d. Thus, the reduced k_2 for Lys198Ala indicates that the "binding" interaction of Lys198 with the OG:A substrate is critical for catalysis. Notably, mutation of several Lys residues to Ala near the active site had no effect on the measured k_2, illustrating that, though seemingly far from the active site, the FCL and the region around the Fe–S cluster play a key role in mediating interactions with the damaged substrate, perhaps by helping to facilitate base flipping and proper placement of the adenine within the active site.

Similarly, Arg143 in MutY was identified to be close to DNA by photocrosslinking, and along with nearby Arg147 is localized within a highly conserved Arg(Val/Leu)XXArg motif in Fe–S cluster BER glycosylases (Chepanoske et al., 2004). These two Arg residues hydrogen-bond to the Fe–S cluster and are close to the DNA phosphodiester backbone. Both Arg143Ala and Arg147Ala MutY exhibit reduced OG:FA affinity and reduced catalysis with a G:A substrate. Notably, the MAP variant Arg227Trp has also been shown to have reduced mismatch affinity and glycosylase activity (Bai et al., 2005). Kisker and coworkers also noted that there is an Arg in the nucleotide excision repair enzyme XPD positioned similarly these two Arg residues in MutY, suggesting that a similar role of the XPD Fe–S cluster in positioning residues for specific interactions with DNA (Wolski et al., 2008).

Fe–S cluster cofactors provide a wealth of diverse biochemical possibilities within the cell, and their discovery in a variety of DNA-binding enzymes has significantly increased since the discovery of the cofactor in MutY and Endo III (Boon et al., 2003; O'Brien et al., 2017). The BER glycosylase MutY and MUTYH have served as notable examples, where the methods described herein aided in illustrating the importance of the highly conserved $[4Fe–4S]^{2+}$ cluster. The cofactor was demonstrated essential for catalytic domain structural stability and base lesion recognition (Lukianova & David, 2005). In addition to being essential for making intimate contacts with DNA, electrochemistry experiments on purified MutY demonstrated that the Fe–S cluster is also capable of performing redox chemistry well within cell potential ranges and participates in DCT processes (Bartels et al., 2017). The DCT abilities of MutY and other DNA-binding enzymes are hypothesized to aid these low copy number proteins in locating DNA damage within a biological timescale. The presence of the Fe–S cluster would help circumvent this issue, by allowing the protein to participate in DCT as a means to coordinate DNA replication and repair. These discoveries build upon the development and implementation of methods described in this chapter and the accompanying chapter "Cellular assays for studying the Fe–S cluster containing base excision repair glycosylase MUTYH and homologs" by Majumdar et al. in this volume. Indeed, a suite of methods and approaches are needed to reveal the complexity of roles for Fe–S cluster-bearing DNA-binding enzymes, such as MutY, in maintaining genomic integrity.

Supplementary data to this article can be found online at https://doi.org/10.1016/bs.mie.2017.11.035.

ACKNOWLEDGMENTS

We are grateful to all of the past and present members of the David laboratory who contributed to developing these methods. The research in the David laboratory on MutY/MUTYH has been supported by the National Cancer Institute of the National Institutes of Health (CA067985) and the National Science Foundation (CHE-1610721). N.N.N. has been supported by a GAANN fellowship, a UCD Humanities Graduate Research Award, and a Schwall Dissertation Fellowship. K.T.L. was supported by an NIH Chemical Biology training program (T32 GM113770).

REFERENCES

Al-Tassan, N., Chmiel, N. H., Maynard, J., Fleming, N., Livingston, A. L., Williams, G. T., et al. (2002). Inherited variants of MYH associated with somatic G: C→T: A mutations in colorectal tumors. *Nature Genetics*, *30*, 227–232.

Au, K. G., Cabrera, M., Miller, J. H., & Modrich, P. (1988). Escherichia coli mutY gene product is required for specific A-G—C.G mismatch correction. *Proceedings of the National Academy of Sciences of the United States of America*, *85*, 9163–9166.

Au, K. G., Clark, S., Miller, J. H., & Modrich, P. (1989). Escherichia coli mutY gene encodes an adenine glycosylase active on G-A mispairs. *Proceedings of the National Academy of Sciences of the United States of America*, *86*, 8877–8881.

Bai, H., Jones, S., Guan, X., Wilson, T. M., Sampson, J. R., Cheadle, J. P., et al. (2005). Functional characterization of two human MutY homolog (hMYH) missense mutations (R227W and V232F) that lie within the putative hMSH6 binding domain and are associated with hMYH polyposis. *Nucleic Acids Research*, *33*, 597–604.

Banda, D. M., Nuñez, N. N., Burnside, M. A., Bradshaw, K. M., & David, S. S. (2017). Repair of 8-oxoG:A mismatches by the MUTYH glycosylase: Mechanism, metals and medicine. *Free Radical Biology and Medicine*, *107*, 202–215.

Bartels, P. L., Zhou, A., Arnold, A. R., Nuñez, N. N., Crespilho, F. N., David, S. S., et al. (2017). Electrochemistry of the [4Fe4S] cluster in base excision repair proteins: Tuning the redox potential with DNA. *Langmuir*, *33*, 2523–2530.

Barton, J. K., Bartels, P. L., Deng, Y., & O'Brien, E. (2017). Electrical probes of DNA-binding proteins. *Methods in Enzymology*, *591*, 355–414.

Boal, A. K., Genereux, J. C., Sontz, P. A., Gralnick, J. A., Newman, D. K., & Barton, J. K. (2009). Redox signaling between DNA repair proteins for efficient lesion detection. *Proceedings of the National Academy of Sciences of the United States of America*, *106*, 15237–15242.

Boal, A. K., Yavin, E., & Barton, J. K. (2007). DNA repair glycosylases with a [4Fe-4S] cluster: A redox cofactor for DNA-mediated charge transport? *Journal of Inorganic Biochemistry*, *101*, 1913–1921.

Boal, A. K., Yavin, E., Lukianova, O. A., O'Shea, V. L., David, S. S., & Barton, J. K. (2005). DNA-bound redox activity of DNA repair glycosylases containing [4Fe-4S] clusters. *Biochemistry*, *44*, 8397–8407.

Boon, E. M., Livingston, A. L., Chmiel, N. H., David, S. S., & Barton, J. K. (2003). DNA-mediated charge transport for DNA repair. *Proceedings of the National Academy of Sciences of the United States of America*, *100*, 12543–12547.

Brinkmeyer, M. K., & David, S. S. (2015). Distinct functional consequences of MUTYH variants associated with colorectal cancer: Damaged DNA affinity, glycosylase activity and interaction with PCNA and Hus1. *DNA Repair (Amst)*, *34*, 39–51.

Bruner, S. D., Norman, D. P., & Verdine, G. L. (2000). Structural basis for recognition and repair of the endogenous mutagen 8-oxoguanine in DNA. *Nature*, *403*, 859–866.

Carey, J. (1991). Gel retardation. *Methods in Enzymology*, *208*, 103–117.

Chepanoske, C. L., Golinelli, M. P., Williams, S. D., & David, S. S. (2000). Positively charged residues within the iron-sulfur cluster loop of E-coli MutY participate in damage recognition and removal. *Archives of Biochemistry and Biophysics*, *380*, 11–19.

Chepanoske, C. L., Lukianova, O. A., Lombard, M., Golinelli-Cohen, M. P., & David, S. S. (2004). A residue in MutY important for catalysis identified by photocross-linking and mass spectrometry. *Biochemistry*, *43*, 651–662.

Chmiel, N. H., Golinelli, M. P., Francis, A. W., & David, S. S. (2001). Efficient recognition of substrates and substrate analogs by the adenine glycosylase MutY requires the C-terminal domain. *Nucleic Acids Research*, *29*, 553–564.

Chmiel, N. H., Livingston, A. L., & David, S. S. (2003). Insight into the functional consequences of inherited variants of the hMYH adenine glycosylase associated with colorectal cancer: Complementation assays with hMYH variants, and pre-steady-state kinetics of the corresponding mutated E-coli enzymes. *Journal of Molecular Biology*, *327*, 431–443.

Chu, A. M. (2011). *Synthesis and characterization of azaribose transition state analogs of base excision repair glycosylases*. Ann Arbor: University of California—Davis. Thesis Dissertation.

Chu, A. M., Fettinger, J. C., & David, S. S. (2011). Profiling base excision repair glycosylases with synthesized transition state analogs. *Bioorganic & Medicinal Chemistry Letters*, *21*, 4969–4972.

Crack, J. C., Green, J., Thomson, A. J., & Le Brun, N. E. (2014). Techniques for the production, isolation, and analysis of iron-sulfur proteins. *Methods in Molecular Biology*, *1122*, 33–48.

David, S. S., O'Shea, V. L., & Kundu, S. (2007). Base-excision repair of oxidative DNA damage. *Nature*, *447*, 941–950.

David, S. S., & Wiliams, S. D. (1998). Chemistry of glycosylases and endonucleases involved in base-excision repair. *Chemical Reviews*, *98*, 1221–1261.

Engstrom, L. M., Brinkmeyer, M. K., Ha, Y., Raetz, A. G., Hedman, B., Hodgson, K. O., et al. (2014). A zinc linchpin motif in the MUTYH glycosylase interdomain connector is required for efficient repair of DNA damage. *Journal of the American Chemical Society*, *136*, 7829–7832.

Fersht, A. (1999). *Structure and mechanism in protein science: A guide to enzyme catalysis and protein folding*. New York: W.H. Freeman.

Fromme, J. C., Banerjee, A., Huang, S. J., & Verdine, G. L. (2004). Structural basis for removal of adenine mispaired with 8-oxoguanine by MutY adenine DNA glycosylase. *Nature*, *427*, 652–656.

Golinelli, M. P., Chmiel, N. H., & David, S. S. (1999). Site-directed mutagenesis of the cysteine ligands to the [4Fe-4S] cluster of Escherichia coli MutY. *Biochemistry*, *38*, 6997–7007.

Guan, Y., Manuel, R. C., Arvai, A. S., Parikh, S. S., Mol, C. D., Miller, J. H., et al. (1998). MutY catalytic core, mutant and bound adenine structures define specificity for DNA repair enzyme superfamily. *Nature Structural Biology*, *5*, 1058–1064.

Ha, Y., Arnold, A. R., Nunez, N. N., Bartels, P. L., Zhou, A., David, S. S., et al. (2017). Sulfur K-edge XAS studies of the effect of DNA binding on the [Fe4S4] site in EndoIII and MutY. *Journal of the American Chemical Society*, *139*, 11434–11442.

Kundu, S., Brinkmeyer, M. K., Eigenheer, R. A., & David, S. S. (2010). Ser 524 is a phosphorylation site in MUTYH and Ser 524 mutations alter 8-oxoguanine (OG): A mismatch recognition. *DNA Repair*, *9*, 1026–1037.

Kundu, S., Brinkmeyer, M. K., Livingston, A. L., & David, S. S. (2009). Adenine removal activity and bacterial complementation with the human MutY homologue (MUTYH) and Y165C, G382D, P391L and Q324R variants associated with colorectal cancer. *DNA Repair (Amst)*, *8*, 1400–1410.

Lee, S., & Verdine, G. L. (2009). Atomic substitution reveals the structural basis for substrate adenine recognition and removal by adenine DNA glycosylase. *Proceedings of the National Academy of Sciences of the United States of America*, *106*, 18497–18502.

Livingston, A. L., Kundu, S., Pozzi, M. H., Anderson, D. W., & David, S. S. (2005). Insight into the roles of tyrosine 82 and glycine 253 in the Escherichia coli adenine glycosylase MutY. *Biochemistry*, *44*, 14179–14190.

Lukianova, O. A., & David, S. S. (2005). A role for iron-sulfur clusters in DNA repair. *Current Opinion in Chemical Biology*, *9*, 145–151.

Luncsford, P. J., Chang, D.-Y., Shi, G., Bernstein, J., Madabushi, A., Patterson, D. N., et al. (2010). A structural hinge in eukaryotic MutY homologues mediates catalytic activity and Rad9–Rad1–Hus1 checkpoint complex interactions. *Journal of Molecular Biology*, *403*, 351–370.

Manlove, A. H., McKibbin, P. L., Doyle, E. L., Majumdar, C., Hamm, M. L., & David, S. S. (2017). Structure-activity relationships reveal key features of 8-oxoguanine: A mismatch detection by the MutY glycosylase. *ACS Chemical Biology*, *12*, 2335–2344.

Manlove, A. H., Nuñez, N. N., & David, S. S. (2016). The GO repair pathway: OGG1 and MUTYH. In *The base excision repair pathway* (pp. 63–115). World Scientific.

Manuel, R. C., Czerwinski, E. W., & Lloyd, R. S. (1996). Identification of the structural and functional domains of MutY, an Escherichia coli DNA mismatch repair enzyme. *The Journal of Biological Chemistry*, *271*, 16218–16226.

McCann, J. A. B., & Berti, P. J. (2008). Transition-state analysis of the DNA repair enzyme MutY. *Journal of the American Chemical Society*, *130*, 5789–5797.

Messick, T. E., Chmiel, N. H., Golinelli, M. P., Langer, M. R., Joshua-Tor, L., & Davi, S. S. (2002). Noncysteinyl coordination to the $[4Fe\text{-}4S]^{2+}$ cluster of the DNA repair adenine glycosylase MutY introduced via site-directed mutagenesis. Structural characterization of an unusual histidinyl-coordinated cluster. *Biochemistry*, *41*, 3931–3942.

Michaels, M. L., Cruz, C., Grollman, A. P., & Miller, J. H. (1992). Evidence that MutY and MutM combine to prevent mutations by an oxidatively damaged form of guanine in DNA. *Proceedings of the National Academy of Sciences of the United States of America*, *89*, 7022–7025.

Michaels, M. L., & Miller, J. H. (1992). The GO system protects organisms from the mutagenic effect of the spontaneous lesion 8-hydroxyguanine (7,8-dihydro-8-oxoguanine). *Journal of Bacteriology*, *174*, 6321–6325.

Michaels, M. L., Pham, L., Nghiem, Y., Cruz, C., & Miller, J. H. (1990). MutY, an adenine glycosylase active on G-A mispairs, has homology to endonuclease III. *Nucleic Acids Research*, *18*, 3841–3845.

Michaels, M. L., Tchou, J., Grollman, A. P., & Miller, J. H. (1992). A repair system for 8-oxo-7,8-dihydrodeoxyguanine. *Biochemistry*, *31*, 10964–10968.

Michelson, A. Z., Rozenberg, A., Tian, Y., Sun, X. J., Davis, J., Francis, A. W., et al. (2012). Gas-phase studies of substrates for the DNA mismatch repair enzyme MutY. *Journal of the American Chemical Society*, *134*, 19839–19850.

Miller, J. H., & Michaels, M. (1996). Finding new mutator strains of Escherichia coli—A review. *Gene*, *179*, 129–132.

O'Brien, E., Holt, M. E., Thompson, M. K., Salay, L. E., Ehlinger, A. C., Chazin, W. J., et al. (2017). The [4Fe4S] cluster of human DNA primase functions as a redox switch using DNA charge transport. *Science*, *355*, 6327–6338.

Parker, A. R., & Eshleman, J. R. (2003). Human MutY: Gene structure, protein functions and interactions, and role in carcinogenesis. *Cellular Molecular Life Science*, *60*, 2064–2083.

Pauling, L. (1948). Nature of forces between large molecules of biological interest. *Nature*, *161*, 707–709.

Pope, M. A., & David, S. S. (2005). DNA damage recognition and repair by the murine MutY homologue. *DNA Repair*, *4*(1), 91–102.

Pope, M. A., Porello, S. L., & David, S. S. (2002). Escherichia coli apurinic-apyrimidinic endonucleases enhance the turnover of the adenine glycosylase MutY with G: A substrates. *The Journal of Biological Chemistry*, *277*, 22605–22615.

Porello, S. L., Cannon, M. J., & David, S. S. (1998). A substrate recognition role for the [4Fe-4S](2+) cluster of the DNA repair glycosylase MutY. *Biochemistry*, *37*, 6465–6475.
Porello, S. L., Leyes, A. E., & David, S. S. (1998). Single-turnover and pre-steady-state kinetics of the reaction of the adenine glycosylase MutY with mismatch-containing DNA substrates. *Biochemistry*, *37*, 14756–14764.
Porello, S. L., Williams, S. D., Kuhn, H., Michaels, M. L., & David, S. S. (1996). Specific recognition of substrate analogs by the DNA mismatch repair enzyme MutY. *Journal of the American Chemical Society*, *118*, 10684–10692.
Py, B., & Barras, F. (2010). Building Fe–S proteins: Bacterial strategies. *Nature Reviews. Microbiology*, *8*, 436–446.
Raetz, A. G., Xie, Y. L., Kundu, S., Brinkmeyer, M. K., Chang, C., & David, S. S. (2012). Cancer-associated variants and a common polymorphism of MUTYH exhibit reduced repair of oxidative DNA damage using a GFP-based assay in mammalian cells. *Carcinogenesis*, *33*, 2301–2309.
Record, M. T., Ha, J.-H., & Fisher, M. A. (1991). [16] Analysis of equilibrium and kinetic measurements to determine thermodynamic origins of stability and specificity and mechanism of formation of site-specific complexes between proteins and helical DNA. *Methods in Enzymology*, *208*, 291–343.
Tokumoto, U., & Takahashi, Y. (2001). Genetic analysis of the isc operon in Escherichia coli involved in the biogenesis of cellular iron-sulfur proteins. *Journal of Biochemistry*, *130*, 63–71.
Trasvina-Arenas, C. H., Lopez-Castillo, L. M., Sanchez-Sandoval, E., & Brieba, L. G. (2016). Dispensability of the [4Fe-4S] cluster in novel homologues of adenine glycosylase MutY. *The FEBS Journal*, *283*, 521–540.
Wang, L., Chakravarthy, S., & Verdine, G. L. (2017). Structural basis for the lesion-scanning mechanism of the MutY DNA glycosylase. *The Journal of Biological Chemistry*, *292*, 5007–5017.
Wang, L., Lee, S. J., & Verdine, G. L. (2015). Structural basis for avoidance of promutagenic DNA repair by MutY adenine DNA glycosylase. *The Journal of Biological Chemistry*, *290*, 17096–17105.
Wolski, S. C., Kuper, J., Hanzelmann, P., Truglio, J. J., Croteau, D. L., Van Houten, B., et al. (2008). Crystal structure of the FeS cluster-containing nucleotide excision repair helicase XPD. *PLoS Biology*, *6*, e149.
Wooden, S. H., Bassett, H. M., Wood, T. G., & McCullough, A. K. (2004). Identification of critical residues required for the mutation avoidance function of human MutY (hMYH) and implications in colorectal cancer. *Cancer Letters*, *205*, 89–95.
Woods, R. D., O'Shea, V. L., Chu, A., Cao, S., Richards, J. L., Horvath, M. P., et al. (2016). Structure and stereochemistry of the base excision repair glycosylase MutY reveal a mechanism similar to retaining glycosidases. *Nucleic Acids Research*, *44*, 801–810.
Yang, H., Clendenin, W. M., Wong, D., Demple, B., Slupska, M. M., Chiang, J. H., et al. (2001). Enhanced activity of adenine-DNA glycosylase (Myh) by apurinic/apyrimidinic endonuclease (Ape1) in mammalian base excision repair of an A/GO mismatch. *Nucleic Acids Research*, *29*(3), 743–752.

CHAPTER THREE

Cellular Assays for Studying the Fe–S Cluster Containing Base Excision Repair Glycosylase MUTYH and Homologs

Chandrima Majumdar, Nicole N. Nuñez, Alan G. Raetz, Cindy Khuu, Sheila S. David[1]
University of California, Davis, Davis, CA, United States
[1]Corresponding author: e-mail address: ssdavid@ucdavis.edu

Contents

Methods in Enzymology, Volume 599
ISSN 0076-6879
https://doi.org/10.1016/bs.mie.2017.12.006

Abstract

Many DNA repair enzymes, including the human adenine glycosylase MUTYH, require iron–sulfur (Fe–S) cluster cofactors for DNA damage recognition and subsequent repair. MUTYH prokaryotic and eukaryotic homologs are a family of adenine (A) glycosylases that cleave A when mispaired with the oxidatively damaged guanine lesion, 8-oxo-7,8-dihydroguanine (OG). Faulty OG:A repair has been linked to the inheritance of missense mutations in the *MUTYH* gene. These inherited mutations can result in the onset of a familial colorectal cancer disorder known as MUTYH-associated polyposis (MAP). While in vitro studies can be exceptional at unraveling how MutY interacts with its OG:A substrate, cell-based assays are needed to provide a cellular context to these studies. In addition, strategic comparison of in vitro and in vivo studies can provide exquisite insight into the search, selection, excision process, and the coordination with protein partners, required to mediate full repair of the lesion. A commonly used assay is the rifampicin resistance assay that provides an indirect evaluation of the intrinsic mutation rate in *Escherichia coli* (*E. coli* or *Ec*), read out as antibiotic-resistant cell growth. Our laboratory has also developed a bacterial plasmid-based assay that allows for direct evaluation of repair of a defined OG:A mispair. This assay provides a means to assess the impact of catalytic defects in affinity and excision on overall repair. Finally, a mammalian GFP-based reporter assay has been developed that more accurately models features of mammalian cells. Taken together, these assays provide a cellular context to the repair activity of MUTYH and its homologs that illuminates the role these enzymes play in preventing mutations and disease.

1. INTRODUCTION

Bacterial MutY and its mammalian MUTYH homologs are glycosylases that catalyze the excision of adenine mispaired opposite 8-oxo-7,8-dihydroguanine (OG) as the first step in base excision repair (BER) to prevent OG-mediated G:C → T:A transversion mutations (David, O'Shea, & Kundu, 2007). These glycosylases carry a $[4Fe–4S]^{2+}$ (Fe–S) cluster coordinated by four Cys residues, two of which anchor a solvent-exposed Fe–S cluster loop (FCL) containing positively charged amino acid residues that have been implicated in DNA binding and enabling subsequent catalysis (Chepanoske, Golinelli, Williams, & David, 2000). Compromised activity of inherited variants of MUTYH has been linked to a familial colorectal cancer (CRC) syndrome called MUTYH-associated polyposis (MAP) (Al-Tassan et al., 2002). Individuals with MAP exhibit colorectal polyposis and have a high likelihood of eventual CRC onset (Farrington et al., 2005; Jenkins et al., 2006) due to an accumulation of

G:C to T:A transversion mutations in the *adenomatous polyposis coli* tumor suppressor gene (Al-Tassan et al., 2002; Sieber et al., 2003; Viel et al., 2017).

The discovery of MAP spurred interest in understanding how specific MAP variants alter the protein structure and associated enzymatic activity of MUTYH. Interestingly, many MAP variants are located in regions near the Fe–S cluster (Fig. 1) and have the potential to disrupt metal coordination and proper positioning of the FCL residues for mismatch engagement (Banda, Nuñez, Burnside, Bradshaw, & David, 2017). In our laboratory, the functional properties of the first identified MAP variants (Tyr165Cys and Gly382Asp in human MUTYH) were assessed using kinetic and binding assays. These experiments were initially performed using the corresponding variants in *Escherichia coli* (*E. coli* or *Ec*) MutY (Tyr82Cys and Gly253Asp) (Al-Tassan et al., 2002) and revealed a compromised adenine glycosylase activity on G:A and OG:A mispair-containing substrates. Our laboratory also showed that both variants have a compromised affinity for an OG:FA-containing substrate analog duplex (where FA = 2′-fluoro-2′-deoxyadenosine) relative to that of the WT enzyme (Chmiel, Livingston, & David, 2003). Subsequent analysis of the corresponding variants in the mouse homolog Mutyh (expressed in bacteria) and the insect-cell-expressed human MBP-MUTYH protein indicated similarly reduced levels of activity (Brinkmeyer & David, 2015; Engstrom et al., 2014; Kundu, Brinkmeyer, Livingston, & David, 2009; Pope & David, 2005). In vitro kinetics and binding studies of *Ec* MutY have also been useful in establishing the

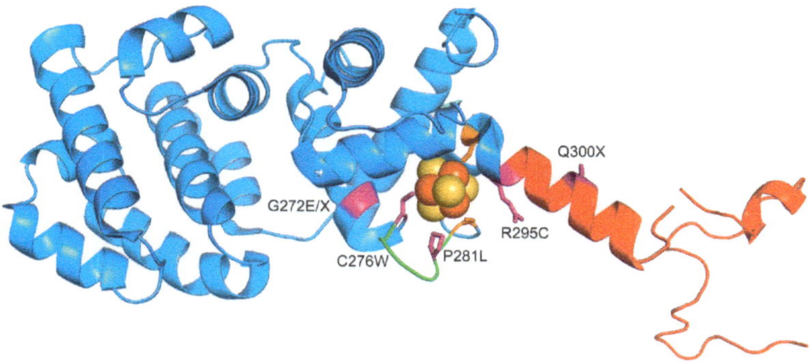

Fig. 1 Fe–S cluster in MUTYH. N-terminal fragment crystal structure of *Homo sapiens* (*Hs*) MUTYH (PDB 3N5N) depicting the position of MAP variants adjacent to the Fe–S cluster (Luncsford et al., 2010). The N-terminal domain is in *blue*, the FCL in *green*, IDC in *red*, the Fe–S cluster coordinating Cys ligands in *orange*, and reported MAP variants in *purple*, where X denotes a nonsense mutation (Out et al., 2010).

importance of the Fe–S cluster domain in substrate recognition and excision by MutY (Chepanoske et al., 2000). An accompanying MutY chapter "Fe–S clusters and MutY base excision repair glycosylases: Purification, kinetics, and DNA affinity measurements" by Nuñez et al. in this volume provides additional details regarding in vitro experiments.

In vitro analyses were a key element in cementing the link between inheritance of catalytically compromised MUTYH variants and CRC (Al-Tassan et al., 2002; David et al., 2007). A notable feature of the in vitro studies of Tyr165Cys and Gly382Asp, as well as the corresponding mutated versions in the bacterial and murine enzymes, was that the adenine excision activity of the Tyr-to-Cys variants was consistently more compromised than that of the Gly-to-Asp variants. This difference could be interpreted as implicating a higher CRC predisposition for individuals with Tyr165Cys over Gly382Asp; however, we also considered that the activity of Gly382Asp MUTYH may be more compromised in a cellular context. Indeed, in vitro experiments are limited in accurately mimicking the activity of MUTYH in a cellular environment. For example, within mammalian cells, MUTYH is subject to transcriptional regulation (Fry et al., 2008) and undergoes posttranslational protein modifications such as phosphorylation and ubiquitination (Dorn, Ferrari, Imhof, Ziegler, & Hubscher, 2014; Kundu, Brinkmeyer, Eigenheer, & David, 2010). Furthermore, cellular DNA is packaged within chromatin structure and access to genomic DNA for repair is regulated by epigenetic factors (Amouroux, Campalans, Epe, & Radicella, 2010; Menoni, Di Mascio, Cadet, Dimitrov, & Angelov, 2017).

The cellular environment also requires that MUTYH searches for OG:A mispairs within the vast excess of undamaged genomic DNA; this complicating feature is clearly not accounted for in in vitro assays, which use short 30 base pair (bp) duplexes. Moreover, the search process inside cells is likely highly coordinated with other processes in the cell, such as replication and transcription, and requires precise interactions with other protein partners. Several MUTYH-protein partner interactions have been identified such as interactions with the downstream AP site processor protein, AP endonuclease 1, and the proliferating cellular nuclear antigen sliding DNA clamp (Parker et al., 2001). In addition, MUTYH also interacts with proteins from other DNA repair pathways outside of BER, such as the mismatch repair (MMR) protein MSH6α (Gu et al., 2002) and the epigenetic regulator SIRT6 (Hwang et al., 2015). Intriguingly, MUTYH also interacts with the Hus1 protein of the Rad9-Rad1-Hus1 (911) DNA damage-sensing complex, suggesting a role in mediating signaling cascades during the response to oxidative stress (Shi et al., 2006; Turco et al., 2013).

Structural and functional studies have elucidated features of MutY and its substrates that are important for recognition and catalysis (Banda et al., 2017; Manlove, Nuñez, & David, 2016). However, the relative importance of specific features of MMR and adenine excision in overall cellular repair remains unclear. To address these issues, we have developed bacterial cell assays to evaluate the impact of specific modifications of the OG or A in the OG:A substrate (Livingston, O'Shea, Kim, Kool, & David, 2008; Manlove et al., 2017), or to evaluate specific amino acid variations in MutY, on the overall repair in cells (Brinkmeyer, Pope, & David, 2012). These studies have revealed that features that aid in recognition of OG within the OG:A mispair are the most important for mediating high levels of overall repair. Indeed, we have found that "binding" defects measured in vitro are magnified in a cellular context, consistent with the stringent requirements for base pair location and discrimination. For example, we recently showed using a battery of OG analogs that initial recognition of an OG:A mispair and discrimination from a T:A bp may be attributed to a single interaction with the 2-amino group of OG localized in the major groove (Manlove et al., 2017). These studies with substrate analogs illustrate the power of such cellular assays to reveal unique molecular details that are most impactful to cellular repair.

Using cell-based assays, it will be possible to more fully evaluate the roles of the MutY/MUTYH Fe–S cluster. In addition, these assays will allow researchers to consider the implications of Fe–S cluster-associated variants in the development of MAP. In some cases, analysis of variations in the MUTYH protein in vitro may be difficult due to challenges with expressing and purifying mammalian homologs. This obstacle is also further highlighted by the observation that mutations to residues surrounding the Fe–S cluster can result in disruption of protein folding and consequently poor protein yields (Chepanoske, Porello, Fujiwara, Sugiyama, & David, 1999; Golinelli, Chmiel, & David, 1999). As a result of these limitations, the development of cellular assays is crucial for understanding MUTYH function on many levels and will be especially useful for understanding the consequences of mutations to the Fe–S cluster coordinating ligands and surrounding amino acids, and delineating the effects of adjacent MAP variants (Brinkmeyer et al., 2012; Golinelli et al., 1999; Raetz et al., 2012).

In this chapter, we describe the cellular assays that have been developed in our laboratory to probe the features of MUTYH-mediated repair.

We elaborate on studies using rifampicin reversion assays to indirectly monitor deficiencies in MUTYH activity by measuring the mutation frequency of cells expressing a MUTYH variant. In addition, we detail the protocols of bacterial and mammalian cell assays that employ a reporter plasmid to quantify substrate-specific repair by MutY/MUTYH variants. These techniques in combination with in vitro characterization of MutY enzymes, described in the accompanying chapter "Fe–S clusters and MutY base excision repair glycosylases: Purification, kinetics, and DNA affinity measurements" by Nuñez et al., make it possible to holistically understand the role(s) of the Fe–S cluster within MutY/MUTYH, and the intimately related roles of MUTYH in diseases, such as MAP.

2. MUTATION SUPPRESSION ACTIVITY MEASURED IN RIFAMPICIN RESISTANCE ASSAYS

2.1 Overview of the Rifampicin Resistance Assay

The *mutY* gene was initially discovered as a locus that resulted in increased G:C to T:A transversion mutations when disrupted (David et al., 2007; David & Wiliams, 1998; Miller & Michaels, 1996). A convenient method for analyzing mutation frequency in *Ec* uses the ability of the antibiotic rifampicin to block bacterial RNA polymerases (Wehrli, Knüsel, Schmid, & Staehelin, 1968). *Ec* strains lacking *mutY* and *mutM* (MutM removes OG from OG:C bps) exhibit extremely high levels of G:C to T:A mutations in the *rpoβ* gene that results in modification of the rifampicin binding pocket and, consequently, an increase in the frequency of surviving colonies on rifampicin selection plates (Fig. 2) (Garibyan et al., 2003). The mutation frequency is reduced, as inferred from reduced number of colonies in the presence of rifampicin, by expression of MutY or MUTYH on a plasmid vector. The supplemented gene is able to complement for the lack of the bacterial MutY, providing a useful means to analyze the activity of a MutY or MUTYH variant in these bacterial cell assays (Chmiel et al., 2003; Golinelli et al., 1999; Slupska, Luther, Chiang, Yang, & Miller, 1999).

Site-directed mutagenesis of individual *Ec* MutY Fe–S cluster Cys ligands to Ser, His, and Ala followed by evaluation in the rifampicin resistance assays demonstrated varying abilities to suppress mutation frequency depending on the Cys ligand replaced (Golinelli et al., 1999). For example, expression of *Ec* MutY where the Fe–S cluster ligand Cys199 was replaced with Ala, Ser, or His resulted in similar low numbers of rifampicin-resistant (Rif^R) colonies as WT, suggesting flexibility in amino acid replacements at position 199 of

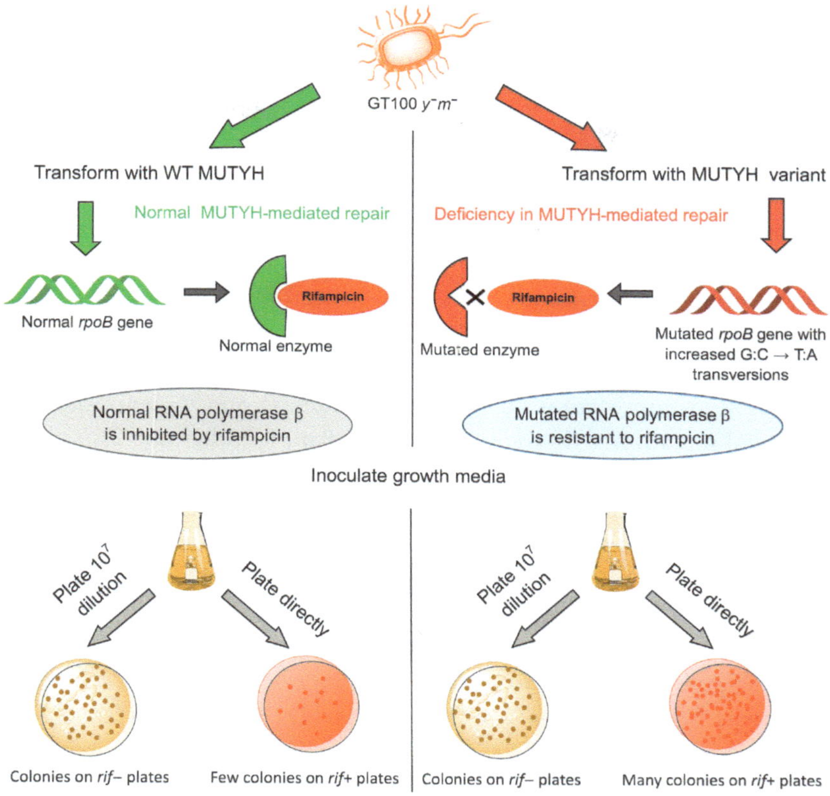

Fig. 2 Schematic representation of the rifampicin resistance assay. This assay is used to establish the cellular mutation frequency and correlate these mutagenesis events to deficiencies in MUTYH-mediated repair. This assay is accomplished through counting the number of colonies resistant to rifampicin (depicted as *red plates*) in comparison to samples grown on a nonrifampicin containing plate (depicted in *brown*). The ratio of the median number of resistant colonies to the average number of viable colonies from the corresponding cultures is used to determine the mutation frequency (Garibyan et al., 2003).

MutY. Notably, in the case of Cys199Ala, the polypeptide chain could not be detected in overexpression tests, suggesting that loss of this Fe-coordinating ligand makes the enzyme exceedingly unstable. This result indicates that within these cellular assays, very low levels of active protein are needed to study the effects of the mutants. A distinctly different trend was observed at position 192, where relative levels of overexpression (Cys192His > Cys192Ser > Cys192Ala) were found to not correlate with relative ability to suppress mutations (Cys192Ser > Cys192Ala > Cys192His)

(Golinelli et al., 1999). In this case, modification at Cys192 not only affects stability, but also the ability to properly mediate repair of OG:A mismatches and prevent mutations.

2.2 Transformation and Growth of Cells

This rifampicin resistance assay follows closely to a previously published protocol for analyzing mutated versions of MutY with the following added specifications for analyzing MUTYH (Golinelli et al., 1999). The *Ec* strain used in the rifampicin resistance assay is the GT100 $y^- m^-$ cell line that lacks functional MutY and OG glycosylase MutM. Previous work has shown that only in the absence of both glycosylases are the levels of G:C to T:A transversions high, consistent with both enzymes acting on OG lesion substrates to prevent mutations (Michaels, Cruz, Grollman, & Miller, 1992). Transformation of a *pMAL-c2x* expression plasmid bearing the WT *MBP-MUTYH* gene into the GT100 $y^- m^-$ cells significantly reduces the number of rifampicin revertant colonies, resulting in a lower mutation frequency compared to GT100 $y^- m^-$ cells alone and GT100 $y^- m^-$ cells carrying the *pMAL-c2x* empty vector (Kundu et al., 2009).

The *pMAL-c2x* (NEB) expression plasmid containing the WT *MBP-MUTYH* gene, for example, is transformed via heat shock into chemically competent *Ec* GT100 $y^- m^-$ and plated onto LB agar plates (Table 1) and incubated overnight at 37°C (Table 3). Each trial is comprised of $n = 15$ for WT as well as for each associated variant being tested to account for the variability in mutation onset during cell growth. Single colonies are selected to inoculate liquid cultures (1 mL each, Table 1) and allowed to grow for 16–18 h at 37°C and 220 rpm (Table 3). Colonies are picked based on morphology and isolation to ensure the analysis of the mutation rate of a single starter *Ec* colony. An assumption made when selecting colonies is that a single colony likely represents over 10^7 cells (Lodish et al., 2000), and that these colonies are progeny of a mutant conferring rifampicin resistance.

Table 1 Agar and Media Composition for the Rifampicin Resistance Assay

Culture Agar	Culture Media	Rifampicin Agar
LB media	LB media	LB media
15 μg/mL Tetracycline	15 μg/mL Tetracycline	15 μg/mL Tetracycline
100 μg/mL Ampicillin	100 μg/mL Ampicillin	100 μg/mL Ampicillin
50 m*M* Agar		100 μg/mL Rifampicin
		50 m*M* Agar

2.3 Determining the Mutation Frequency

The following day, culture agar and rifampicin agar are warmed at 37°C for 1 h prior to plating. 100 μL of a 10^{-7} dilution is plated on culture agar plates to determine the number of viable cells, while 100 μL of the undiluted culture is plated onto the rifampicin selection plates (Table 1) to determine the number of resistant colonies. The plates are incubated at 37°C for 16 h, after which the number of colonies on each plate is counted. The mutation frequency (f, $\times 10^{-8}$) is calculated by Eq. (1) for each trial:

$$f = \frac{\text{median number of resistant colonies}}{\text{average number of viable colonies}} \quad (1)$$

It is important to note that the mutation frequency is the proportion of cells in a population that are mutated, but this cannot be directly equated with the population's mutation rate (Foster, 2006; Rosche & Foster, 2000). The majority of colonies likely arise from cells that developed a mutation during the incubation process, and consequently, these cells and their progeny will confer rifampicin resistance. Thus, a single mutation early in the growth period yields an exponential number of mutant colonies as its progeny (Rosche & Foster, 2000). A particular culture may have had a low mutation rate, but due to random chance, it can generate assay results with high mutant counts. As a result, the relationship between cellular mutation rate and number of mutants at the endpoint is a complex function.

Fortunately, continual improvement to mathematical models has recently converged on a set of guidelines that are both statistically rigorous and easily accessible to experimentalists (Zheng, 2015). Notably, the Ma-Sandri-Sarkar (MSS) maximum likelihood estimator (Sarkar, Ma, & Sandri, 1992) has been shown to be a robust estimator of mutation rate (Couce & Blazquez, 2011). Recent web-based bioinformatic programs simplify maximum likelihood analysis (Gillet-Markowska, Louvel, & Fischer, 2015). In the following section, we will discuss an application of one such method to compare the mutation rate of WT and a mutant MUTYH.

2.3.1 Troubleshooting Tips

Due to the large variability in cellular assays, it is imperative to ensure a proper sample size when using the rifampicin resistance assay. As evidenced in Table 2, the number of colonies spans a wide range, and therefore, each trial should contain a minimum of 15 colonies due to the variability in cell growth, as fewer cultures can result in a misrepresentation of the mutation frequency of the enzyme. Furthermore, we recommend running a wild-type enzyme (or other control) alongside every set of mutants for proper comparison of colonies. In the case of consistently high numbers of mutant colonies, especially in the wild type, rifampicin selection plates should be poured again and used within a week to ensure efficacy of the antibiotic. Plates should be dark red in color; if plates become lighter this indicates degradation of rifampicin and the ability to inhibit RNA polymerase.

2.4 Case Study: Use of Rifampicin Resistance Assay to Characterize Zn-Linchpin Motif in MUTYH

In addition to the Fe–S cluster, the mammalian homologs, Mutyh and MUTYH, have been found to possess a Zn^{2+} ion within the interdomain connecter (IDC) that connects the catalytic N-terminal and OG-recognition C-terminal domains. The critical importance of the Zn^{2+} ion within the IDC, resulting in its designation as the "zinc-linchpin" motif, was determined using metal analysis and glycosylase assays of Cys to Ser mutations in the mouse Mutyh homolog, along with rifampicin resistance assays of the analogous versions of MUTYH (Engstrom et al., 2014). Replacement of two Cys ligands (Cys325/307 and Cys328/310 in human/mouse enzyme) with Ser resulted in reduced levels of Zn^{2+} and Fe–S cluster (Engstrom et al., 2014). Evaluation of these same mutations in human MUTYH showed a substantial increase in mutation frequency compared to WT when evaluated using the rifampicin resistance assay (Engstrom et al., 2014).

In Table 2, we have shown representative data for one trial of the rifampicin resistance assay for the Cys325Ser MUTYH relative to WT MUTYH. In this experiment, each trial is comprised of at least 15 individual cultures of the same genotype. Each data set is entered into the *bzRates* calculator (http://www.lcqb.upmc.fr/bzrates) to get a mutation rate per cell division shown in the last two rows of Table 2, with additional correction for the 10^7 dilution factor. In our case, relative fitness of the mutants to wild type is not known and 100 μL of 1 mL culture was used for plating efficiency. Our laboratory and others (Wolff, Kim, Hu, Yang, & Miller, 2004) have reported the median number of rifampicin revertant colonies divided by the average number of cells in an untreated control as the mutation frequency (Brinkmeyer et al., 2012; Chmiel et al., 2003; Engstrom et al., 2014; Golinelli et al., 1999; Kundu et al., 2009). When the Cys325Ser MUTYH variant was transformed into GT100 $y^- m^-$ cells, the mutation frequency of cells containing this variant was greater than 11-fold higher than WT (Engstrom et al., 2014), consistent with the analysis in Table 2.

Table 2 Rifampicin Resistance Assay Example Data for Calculating Mutation Rate per Cell Division, Corresponding to the Median Number of Colonies for Number of Mutants (Rifampicin Resistant) or Average Number of Total Cells (Viable)

	Wild-Type MUTYH		Cys325Ser MUTYH	
Culture Number	**Number of Mutants**	**Total Cells (10^7)**	**Number of Mutants**	**Total Cells (10^7)**
1	14	562	114	462
2	38	478	158	842
3	16	493	120	637
4	26	526	121	615
5	6	483	105	350
6	25	501	133	450
7	104	633	182	500
8	24	373	127	850
9	12	540	136	510
10	6	370	162	648
11	50	650	104	350
12	21	471	142	596
13	10	672	203	806
14	6	406	307	850
15	34	487	131	528
16	N/A	N/A	106	674
Trial totals	21	510	132	604
Mutation rate per cell division	6×10^{-10}		5×10^{-9}	
95% Confidence interval	4×10^{-10} to 8×10^{-10}		3×10^{-9} to 6×10^{-9}	

3. ANALYSIS OF MutY-MEDIATED REPAIR OF DEFINED PLASMID SUBSTRATES IN *E. COLI*

3.1 Designing a Plasmid-Based Bacterial Cell Assay

In the accompanying chapter "Fe–S clusters and MutY base excision repair glycosylases: Purification, kinetics, and DNA affinity measurements" by

Nuñez et al. in this volume, we have described various in vitro assays utilized by our laboratory (Brinkmeyer & David, 2015; Chepanoske et al., 1999; Porello, Leyes, & David, 1998) to measure kinetic parameters of MutY enzymes. These assays typically use a substrate mispair within a defined 30 bp duplex. These in vitro assays allow for quantitative determination of substrate binding affinities and rates of the base excision reaction. However, the relationship of the enzymatic parameters to overall repair in a more complex biological context is important to delineate. Indeed, the task required of MutY in cells to recognize a single OG:A mispair among the large excess of similar bps would be expected to be more challenging than within a short DNA duplex. To address these issues, we have developed a plasmid-based assay that allows for evaluating MutY (WT or variant)-mediated repair of a defined substrate within plasmid DNA in bacteria. By comparing enzymatic and binding parameters to the extent of repair in the plasmid-based cellular assays, we have revealed distinct structural features of the OG:A mispair and MutY that are important for various steps in the detection, engagement, and excision process of BER (Brinkmeyer et al., 2012; Livingston et al., 2008; Manlove et al., 2017). The important features of the OG:A substrate have been revealed by evaluating the MutY-mediated repair of synthetically derived modified substrates, such as OI:A (where OI = 8-oxoinosine) (Livingston et al., 2008; Manlove et al., 2017). We have also evaluated the repair of variants of catalytic residues in *Ec* MutY (such as Asp138Cys MutY) on OG:A repair (Brinkmeyer et al., 2012). Ongoing efforts in our laboratory are aimed at using these assays to perform similar studies to address the roles of the Fe–S cluster and the FCL motif in MutY.

This assay measures the cellular repair activity of MutY in terms of conversion of a strategically placed OG:A lesion to G:C in *Ec* cells; full repair requires the initial MutY-initiated removal of A, and then subsequent BER, and action of MutM to provide the G:C bp (Fig. 3). The key feature of this assay is placement of the OG:A bp at a potential *Bmt*I restriction site such that MutY-mediated repair leads to creation of a restriction site that can be monitored by restriction digestion. Lack of MutY-mediated repair of OG:A leads to significant levels of T:A at the lesion site. The %G:C bps in cells possessing MutY are compared to those lacking MutY to determine the extent of repair. To study the effect of the MutY variants, MutY-deficient cells are first transformed with the appropriate MutY-expressing plasmids (*pKK223 MutY*). This is followed by transformation of the substrate containing reporter plasmid.

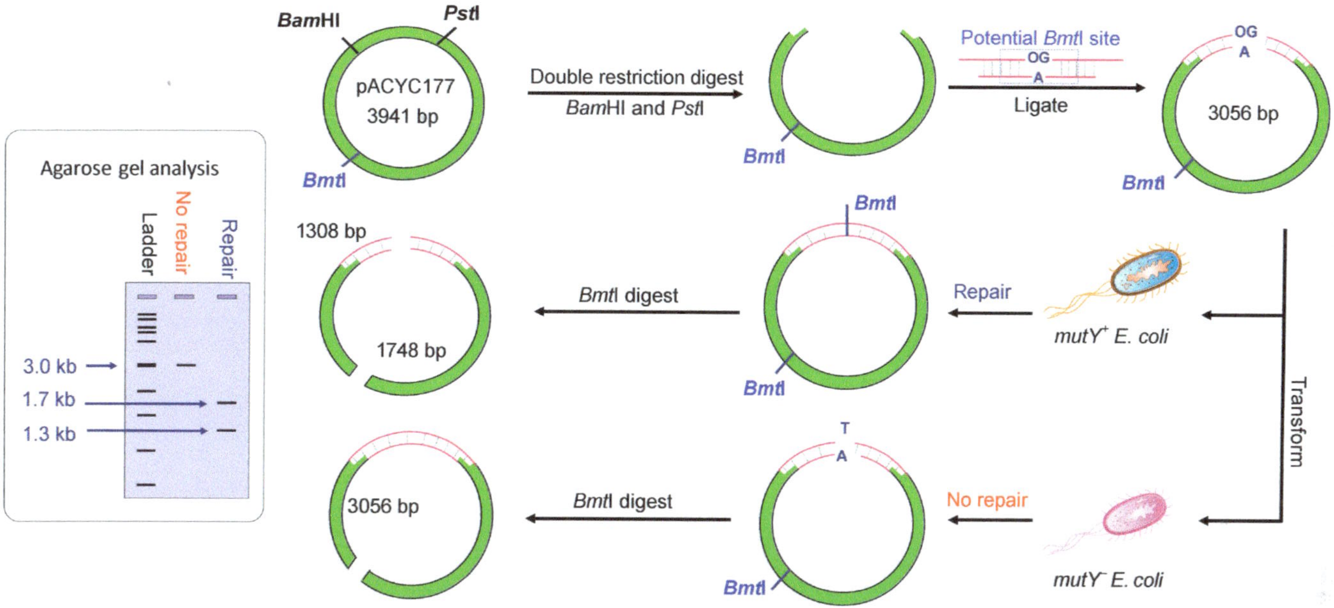

Fig. 3 Schematic representation of the plasmid-based bacterial cell assay to assay MutY-mediated OG:A repair. The restriction enzyme sites for *Bam*HI, *Pst*I, and *Bmt*I are indicated on the *pACYC177* (*green*) plasmid, and the insert duplex carrying the OG:A mispair is shown in *pink*. The representative agarose gel on the *left* shows the expected bands formed after *Bmt*I digestion of the recovered plasmids. To analyze MutY variants, *muty*$^-$ *Ec* are transformed with the appropriate vector expressing the mutant enzyme and compared to the results from cells transformed with plasmids expressing the WT enzyme (Brinkmeyer et al., 2012).

3.2 Selection of Cloning Vector and Design of Insertion Sequence

The *Ec* strain used for this assay, JM101(*mutY::Tet*^{R}), and the *pkk223.3* expression vector that contains the *Ec muty* gene (e.g., *pkkWTMutY*) were kindly provided to us by Dr. M. Michaels and Dr. J.H. Miller (Michaels, Pham, Nghiem, Cruz, & Miller, 1990; Miller & Michaels, 1996). Variants of interest are generated via site-directed mutagenesis and confirmation of the resultant gene mutation by sequencing. The *pACYC177* reporter plasmid is isolated from ER2424 cells (NEB) using a Promega Miniprep DNA purification system, following the manufacturer's protocol. The plasmid is amplified using *dam*$^-$ GM2929 cells (provided by Dr. M. Marinus (Barras & Marinus, 1989)) to ensure the isolated plasmid is not methylated and does not activate MMR proteins that would interfere with MutY-mediated repair (Pukkila, Peterson, Herman, Modrich, & Meselson, 1983). 10 μg of the isolated plasmid is digested, with 100 U each of *BamH*I and *Pst*I (NEB) in the appropriate buffer (NEB 3.1) to generate sticky ends. The double digested plasmid (*pACYC*), purified using a 0.8% agarose gel (prepared in TAE—40 m*M* Tris pH 7.6, 20 m*M* acetic acid, and 1 m*M* EDTA pH 8.0) (Table 3) and extracted using Qiagen gel extraction kit, is ligated to 30 pmol of the insert duplex containing OG:A using 15 U of T4 ligase (Promega) followed by purification using a PCR clean-up kit (Qiagen).

The insert duplex, containing the OG:A mispair in a potential *Bmt*I recognition site (underlined), has the following sequence:

5′-d(GAT CCG ATC ATG GAG <u>GCT A**OG**C</u> GCT CCC GTT ACA GCT GCA)-3′

5′-d(GC TAG TAC CTC <u>CGA T **A** G</u> CGA GGG CAA TGT CG)-3′

This duplex is created by individually phosphorylating 300 pmol of each strand with 10 m*M* ATP using T4 polynucleotide kinase (PNK) (NEB) in the provided buffer. The strands are then combined and annealed in solution with a final duplex concentration of 1 μ*M*. A positive control is also prepared by phosphorylating and annealing a duplex with a centrally located G:C base pair in place of the OG:A mispair.

3.3 Electrocompetency and Electroporation Protocol

To ensure maximum transformation efficiency of the plasmids, cells are made freshly electrocompetent. Briefly, a single colony of JM101 from a fresh agar plate is inoculated into 10 mL LB medium dosed with tetracycline

Table 3 Summary of Equipment Needed for MutY/MUTYH Cell-Based Assays

Equipment	Notable Specifications
Pipette tips/microcentrifuge tubes/conical tubes	DNase/RNase/Pyrogen free and sterile
Vortex/mini centrifuge	Table top
Thermocycler	Applied Biosystems GeneAmp PCR System 2400
Incubator/shaker	Temperature/speed control
Stir/hot plate	Temperature control
Centrifuge	Sorvall Ultracentrifuge
Sonifier	Branson sonifier 250 at 70% power
UV/Vis	HP 8453 with OLISWorks
Electroporator and cuvettes	Bio-Rad MicroPulser Electroporator and Bio-Rad Gene Pulser Cuvettes, 0.1 cm
Midiprep kit	Promega Wizard midiprep kit
Western blot/SDS gel equipment	Bio-Rad
Agarose gel equipment	Bio-Rad Mini-Sub Gel GT
Flow cytometer	Becton Dickinson FACScan Flow cytometer
Cold room	Maintained at 4°C

(15 mg/mL). The culture is incubated overnight at 37°C and 220 rpm (Table 3). 1.5 mL of this culture, pelleted at 8000 rpm for 10 min (Table 3), is used to inoculate 50 mL of 2× YT (3.2 g yeast extract, 2 g tryptone, and 1 g NaCl in 200 mL H_2O) dosed with tetracycline and allowed to incubate with shaking. When the culture OD_{600nm} reaches 0.6, it is divided equally into three prechilled 15-mL centrifuge tubes and cooled in an ice bath for 30 min. Meanwhile, all solutions, tubes, and pipettes are prechilled to 4°C. The cells are pelleted by centrifugation at 3000 × *g* for 15 min at 4°C (Table 3), and the supernatant is decanted. The pellet is resuspended sequentially in 14 mL Milli-Q H_2O, 10 mL Milli-Q H_2O, 4 mL 10% glycerol, and finally in 300 μL of 10% glycerol and pelleted after each resuspension. 50 μL of the final glycerol solution is aliquoted into prechilled 0.6 mL microcentrifuge tubes kept ready in an ice tray. The electrocompetent cells hence prepared are best used immediately, but may be stored at −80°C for up to 2 weeks.

The MutY expression vector is added to each tube and mixed by pipetting. The cell/DNA mix is then transferred to an electroporation cuvette (Table 3) and pulsed with 1.80 kV (Table 3). SOC medium (300 μL, Invitrogen) is immediately added to the cuvette and used to resuspend and transfer the cells to a 1.6-mL microcentrifuge tube kept on ice. The cells are then incubated at 37°C with shaking at 220 rpm for 1 h, following which 150 μL of each transformation is plated on LB agar plates containing the appropriate antibiotics. Single colonies from this transformation are used to inoculate 10 mL LB dosed with the appropriate antibiotics. The electroporation described above is repeated with the ligated reporter plasmid (Kan^R), with one modification: after the hour-long incubation of the cells in SOC medium, equal volumes (150 μL) are used to inoculate agar plates and 10.0 mL LB growth media that are incubated overnight at 37°C and 220 rpm (Table 3).

3.4 Amplification and Plasmid Extraction

For transformations that produce greater than 100 colonies, the corresponding overnight cultures are used to inoculate 100 mL LB media dosed with the appropriate antibiotics. The cells are allowed to grow until their OD_{600nm} reaches 0.6, after which 100 μL of chloramphenicol is added to each culture and incubation is continued overnight at 37°C and 220 rpm. The cells are then harvested by centrifugation at 10,000 × *g* for 10 min at 4°C (Table 3) and plasmid DNA is isolated using a midiprep kit (Table 3) following the manufacturer's protocol. The concentrations of the recovered plasmids are determined by measuring the A_{260nm} of the samples (Table 3).

3.5 Agarose Gel and Confirmation by Sequencing

To test for MutY-mediated repair of the OG:A base pair in the substrate plasmid to a *Bmt*I site, 1500 ng of the isolated plasmids are incubated overnight with 20 U of *Bmt*I-HF (NEB) and its corresponding buffer at 37°C. The reactions are inhibited by the addition of the provided agarose loading dye, and the fragments are resolved and analyzed by agarose gel electrophoresis using ethidium bromide (2% agarose gel prepared in TAE—40 m*M* Tris pH 7.6, 20 m*M* acetic acid, and 1 m*M* EDTA pH 8.0). The bands are visualized by detecting the fluorescence of the DNA intercalator EtBr.

Typically, bands corresponding to both the MutY-expressing plasmid and the reporter plasmid are observed. However, the high percentage of the agarose gel serves to resolve them adequately, and molecular weight bands higher than 3 kb may be disregarded during further analysis. The percent conversion to G:C at the lesion site is determined using Eq. (2):

$$\%\mathrm{G:C} = \frac{I_{1748} + I_{1308}}{I_{3056} + I_{1748} + I_{1308}} \times 100 \quad (2)$$

In the above equation, *I* represents the intensity of the indicated molecular weight band. The extent of digestion is assessed by comparing to the digestion of a plasmid carrying a G:C at the lesion site. The percent repair by the each MutY/MUTYH variant is reported as an average of six independent transformations ± the standard deviation of these trials. The identity of the base incorporated in the lesion site (G:C or T:A) is confirmed by Sanger sequencing.

3.5.1 Troubleshooting Tips

Proper ligation of the insert duplex to the double digest plasmid is important for the successful implementation of this assay. We have found that a 1:1 or 1:3 ligation ratio of double digest to insert DNA results in the most successful transformations. However, if fewer than 100 colonies are routinely obtained after electroporation, it may be worthwhile to test different ligation ratios, such as 1:2 or 1:5. Additionally, ligations carried out at 16°C for 12–14 h have been found to be more efficient than those carried out at room temperature. Another consideration is that the JM101 cell line is $endA^{+}$ and therefore the midiprep to recover the plasmids must be done as rapidly as possible to prevent degradation of genomic DNA (which is seen as dark smears near the wells of the agarose gel).

3.6 Case Study: Use of Bacterial OG:A Repair Assay to Assess Relative Impacts of Catalytic vs Binding Defects in MutY

Using the bacterial cell assay, we previously evaluated variants of MutY at two key catalytic residues (Asp138, Glu37) and a truncated form of MutY lacking the C-terminal domain (MutY Δ226–350) (Brinkmeyer et al., 2012). In the S_N1 mechanism for adenine excision catalyzed by MutY, the deprotonated form of Asp138 has been proposed to stabilize the oxocarbenium ion intermediate during the initial adenine departure step as well as mediate formation of a transient covalent acetal intermediate (Woods et al., 2016). The protonated form of Glu37 has been proposed to mediate protonation of N7 of adenine to enhance its lability and, then subsequently in its deprotonated state, act as a general base to activate the water molecule in the final hydrolysis step to form the abasic site product

(Woods et al., 2016). The cellular OG:A repair activity of Asp138Glu, Glu37Asp, and Asp138Cys allowed for evaluation of the extent that catalytic defects due to amino acid changes translate into overall repair within a cellular context. Importantly, these three mutated enzymes exhibited similar level of overexpression and stability, allowing differences in repair to be related to rates of adenine excision (k_2) measured in vitro. Asp138Glu MutY exhibited similar % G:C at the OG:A lesion as the WT enzyme (100%), consistent with its WT catalytic activity. However, Glu37Asp and Asp138Cys exhibited G:C levels of 68% ± 2% and 47% ± 3% compared to 39% ± 2% G:C in the *muty*$^{-}$ control. The observation of significant levels of OG:A repair mediated by Glu37Asp and Asp138Cys was surprising given that the adenine excision rates were found to be 200- and 600-fold reduced, respectively, compared to the WT enzyme. In contrast, the presence of the C-terminal OG-recognition domain (Chmiel, Golinelli, Francis, & David, 2001; Wang, Chakravarthy, & Verdine, 2017) was essential for OG:A-mediated repair in these assays. The percent G:C observed at the OG:A lesion site in cells expressing MutY Δ226–350 was 27% ± 3%, a value similar to that observed in the absence of MutY, despite the fact that the adenine excision rate (k_2) is only reduced 30-fold for MutY Δ226–350 compared to WT MutY. Notably, mutated forms of catalytic residues (e.g., Glu37Asp) retain WT-like high affinity (K_d) for the mismatch as judged by electrophoretic mobility shift assays (EMSA) with an OG:FA substrate analog-containing duplex (Chepanoske et al., 1999). In contrast, the OG:FA-duplex affinity of MutY Δ226–350 is approximately 200-fold less than WT MutY. These studies illustrated that high affinity for the OG:A substrate is the dominant feature that impacts repair in a cellular context. This is consistent with the arduous task of MutY to identify the OG:A bp within the large excess of undamaged DNA and other DNA-binding proteins in cells.

4. A MAMMALIAN CELL-BASED GFP REPORTER ASSAY

4.1 Evaluating DNA Repair in a Mammalian Cell-Based System

The use of purified enzymes and substrates in vitro allows for rapid interrogation of amino acid changes that can alter enzyme function. However, cellular assays have the potential to reveal subtle as well as complex protein–substrate and protein–protein interactions. It is somewhat self-evident that studying mammalian enzymes in bacterial cells is not ideal, especially for MUTYH, which has numerous mammalian protein partners (Gu et al.,

2002; Hwang et al., 2015; Parker et al., 2001; Shi et al., 2006; Turco et al., 2013) and undergoes posttranslational modifications not present in bacteria (Dorn et al., 2014; Kundu et al., 2010). Following the discovery of MAP in 2002 (Al-Tassan et al., 2002), further work to study MUTYH variants in mammalian cells hinged on developing MUTYH-deficient mice and their resulting mouse embryonic fibroblast (MEF) cell lines (Hirano et al., 2003; Russo et al., 2004; Xie et al., 2004). The mouse Mutyh enzyme has high sequence homology to the human enzyme, and this was further complimented by utilization of mature gene knockout technology in mice (Lee & Threadgill, 2004). However, methods to determine MUTYH activity in mammalian cells were still under development. For example, Molatore et al. measured the levels of OG in genomic DNA as a method to evaluate MUTYH-mediated repair (Molatore et al., 2010); however, it is well established that MUTYH does not directly remove OG lesions and therefore this assay, at best, is only indirectly reporting on MUTYH activity (McGoldrick, Yeh, Solomon, Essigmann, & Lu, 1995; Pope & David, 2005). A more appropriate assay to directly measure adenine excision in mammalian cells was clearly needed. Here, we detail the approach and methods employed to directly measure OG:A repair in mammalian cells. These methods were used to measure repair in *Mutyh*$^{-/-}$ MEFs stably expressing MUTYH variants (Raetz et al., 2012).

4.2 Design of Mammalian Plasmid Reporter Constructs

The ability to measure the repair of a single lesion type in a quantitative manner in cells necessitates the import of synthetic DNA lesions and ideally a simple way to quantify their repair. A common approach is to introduce a synthetic base lesion into cells within an exogenous reporter construct. There are numerous creative methods developed where a single damaged base placed within a gene region can modulate a reporter gene signal. For example, when placed in the transcribed strand, many DNA lesions block or reduce transcriptional efficiency, which can be quantified as a change in the expression of a measurable gene product such as GFP (green fluorescent protein) (You & Wang, 2015). Others have shown that transcriptional mutagenesis of a strategically placed damaged base results in miscoded RNA products (Doetsch, 2002), which can either activate or inactivate the reporter gene product (Nagel et al., 2014). Within the context of controlled experiments these signals can be compared against undamaged positive and negative controls to quantify cellular repair.

Here, we focus on the measurement of OG:A repair by MUTYH using GFP as a reporter (Raetz et al., 2012). This general approach can be applied to other lesions, BER glycosylases, and reporter genes. For example, fluorescent and luminescent gene products have been used as reporters in DNA repair assays (Burger, Matt, Kieser, Gebhard, & Bergemann, 2010; Masaoka, Horton, Beard, & Wilson, 2009). As depicted in Fig. 4, this mammalian cell-based assay is able to reveal the level of MUTYH-mediated repair of OG:A mispairs located within plasmid DNA.

The A to C repair of the GFP gene is amplified by the strong cytomegalovirus promoter located upstream of the coding region. Transfection status of individual cells is quantified by the constitutive expression of destabilized red fluorescent protein (dsRed) translated from an internal ribosome entry site (IRES) on the same reporter plasmid (Clonetech). The repair status of individual cells is detected as green fluorescence by flow cytometry. Although endonuclease, polymerase, and ligase activity is required to complete repair on the transcribed strand, this scheme should not require the repair of the OG lesion on the nontemplate strand. One concern was that cellular pathways that enforce nonsense-mediated mRNA destruction would inhibit red fluorescence since the transcript has the stop codon in the GFP cDNA (Lykke-Andersen & Jensen, 2015). However, we found that the same level of red fluorescence was measured both with and without the GFP stop codon, and these constructs were used as GFP-negative and GFP-positive controls in all experiments.

4.3 Production of Plasmid Reporter Constructs

To produce a plasmid DNA construct containing a single site-specific lesion, the first step is to produce a pure closed-circular single-stranded DNA (ssDNA) template. The second step is to complete a primer extension reaction that produces a double-stranded DNA that contains the synthetic OG:A mispair. ssDNA is produced by infecting *Ec* carrying the pGFP-off (GFP^-/$dsRed^+$) plasmid with M13K07 phage and grown overnight at 37°C and 220 rpm (Table 3). This culture is then centrifuged at 10,000 × *g* for 30 min (Table 3). The supernatant is combined with 0.2 volumes of precipitation buffer (2.5 *M* NaCl, 20% w/v PEG 8000 (Teknova)) and incubated 1 h on ice at 4°C. This mixture is centrifuged at 6000 × *g* for 30 min at 4°C (Table 3), and the supernatant decanted. The remaining film formed by the phage is gently resuspended in 10 mL cold TE buffer (10 m*M* Tris pH 7.5, 1 m*M* EDTA pH 8.0) and briefly centrifuged at room temperature, the

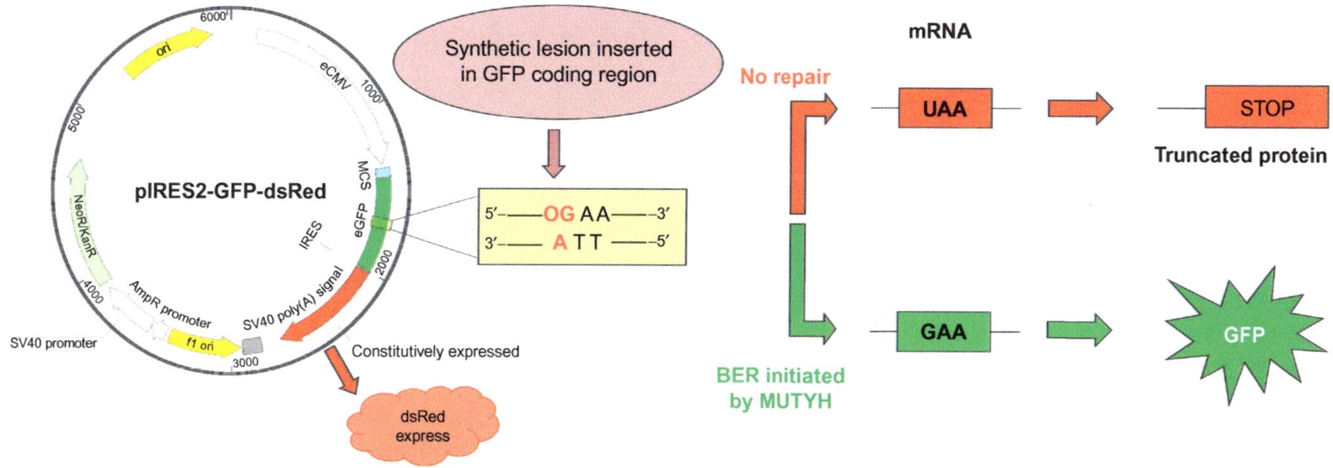

Fig. 4 Schematic representation of the GFP reporter assay. Repair of the OG:A base pair initiated by MUTYH results in GFP expression. Here, a chemically synthesized OG containing nucleotide is placed opposite A within the coding region of the GFP reporter construct. By default, full GFP expression is interrupted with a stop codon, leading to a 33 amino acid-truncated protein product that does not contain the chromophore core needed for fluorescence (Tsien, 1998). However, excision of A and subsequent repair to C causes the complementary mRNA codon to change from UAA to GAA, and translation of this change restores the wild-type amino acid, Glu, allowing expression of full-length wild-type green fluorescent protein.

soluble fraction is decanted and reprecipitated with 0.2 volumes of precipitation buffer, and the pellet is resuspended in cold TE buffer. This process is repeated for a total of three PEG precipitations.

ssDNA is extracted from the final phage product by 15-min phenol incubation followed by two phenol/chloroform extractions using Phase Lock gel tubes (5-Prime), and ethanol precipitation. The ssDNA is then resuspended in 10 m*M* Tris pH 7.5 in a volume of 1 mL per 200 mL starting culture, typically yielding ssDNA concentrations between 300 and 800 ng/μL. The amount and purity of the ssDNA is determined by 0.9% agarose gel electrophoresis (Table 3) in TAE buffer.

Second strand synthesis via primer extension begins with phosphorylation and annealing of the OG-containing oligonucleotide to the ssDNA. The oligonucleotide is 5′ phosphorylated using T4 PNK per manufacturer's protocol (NEB), typically with 10 pmol DNA in a total reaction volume of 50 μL. 2–10 μg of single-stranded pIRES2-dsRed-GFP DNA is combined with threefold molar excess of the phosphorylated oligonucleotide in a final buffer concentration of 28 m*M* Tris–HCl, 5 m*M* $MgCl_2$, 10 m*M* NaCl, and 1 m*M* DTT. To maximize annealing efficiency, the above solution is denatured in 100°C water, immediately removed from the heat source, and slowly cooled to room temperature. Second strand synthesis and ligation are carried out with the addition of 6% w/v PEG 8000, 0.6 μg/L bovine serum albumin, 1.5 m*M* fresh DTT, 1 m*M* ATP, 0.5 m*M* dNTPs, 4 U/μL of T4 DNA ligase, and 0.5 U/μL of T7 DNA polymerase. This reaction mixture is incubated at 16°C for 30 min, 37°C for 1 h, 30°C for 30 min, 16°C for 6 h, and then held at 4°C overnight (Table 3). The product must be immediately purified by phenol/chloroform extraction with ethanol precipitation, and the product is verified by agarose gel electrophoresis (Table 3) vs single-stranded and double-stranded vector controls.

Importantly, the double-stranded DNA starting material used to produce the ssDNA contains the premature GFP stop codon; thus, it cannot be a source of extraneous GFP expression. We have observed that the use of second strand synthesis on a circular single-stranded substrate was hampered by inconsistent yield of the final double-stranded DNA product needed for the assay. The use of nicking enzymes with a double-stranded plasmid substrate with strategically placed restriction sites in the same strand allows for the direct ligation of a short oligonucleotide into a plasmid reporter, and appears to an easier and faster way to produce reporter vectors (You & Wang, 2015).

4.3.1 Troubleshooting Tips

The sequence recognized by the nicking enzymes must not be present anywhere else on the plasmid to prevent off-target strand scission. In addition, we have found that producing sufficient quantities of the plasmid reporter for mammalian cell culture transfections is the rate-limiting step in performing these cellular repair activity assays. Thus, we highly recommend exploring the use of nicking enzymes and other methods to produce these substrates more efficiently.

4.4 Quantification of Expression With Competitive RT-PCR

To understand how specific amino acid changes in the MUTYH protein alter OG:A DNA repair, we created stable cell lines containing variants using the *Mutyh*$^{-/-}$ MEFs as the parental line. A critical variable was the level of MUTYH expression, since the rate and amount of OG:A repair at any time point is dependent on enzyme expression. Thus, modulating or normalizing for MUTYH expression was a critical aspect of this cell-based assay approach. We found that quantification of western blot results was time intensive and not highly consistent and thus turned to measuring MUTYH transcript levels via competitive reverse transcriptase PCR (RT-PCR).

To accurately measure transcription of MUTYH in stably transfected cell lines, competitive RT-PCR is used with a synthetic RNA standard. This standard is created by blunt end ligation of two PCR products, where the products correspond to the 5′ and 3′ ends of the MUTYH transcript. The goal is to create an RNA fragment that will be amplified by the RT-PCR but will have a shorter length. Since this RNA can be carefully quantified before being added to the reaction, it provides a relative reference between band density and the amount of RNA. For MUTYH, a central 337-nucleotide region was removed (MUTYH Δ337), and the PCR product inserted into the *pET15b* expression vector (Novagen) and transcribed in vitro with T7 RNA polymerase using standard protocols (Green, Sambrook, & Sambrook, 2012; Sambrook, Fritsch, & Maniatis, 1989). This RNA standard was quantified by spectrophotometry (Table 3) and stored in small aliquots at −80°C.

For RT-PCR, Trizol extraction of total RNA from a single well of a six-well cell culture plate (approximately 1×10^6 cells) is done based on the manufacturer's protocol (Invitrogen). RT-PCR is performed using the Qiagen One Step RT-PCR kit using 200 ng of total RNA typically with

1 pg of the MUTYH Δ337 RNA standard per manufacturer's protocol. The truncated RNA product is reverse-transcribed and PCR amplified (Table 3) together with any cellular MUTYH mRNA using the same two primers. The resulting PCR will have two products, with the endogenous MUTYH appearing as the full-length product (approximately 800 bp based on our primer locations), whereas the truncated synthetic RNA creates a DNA product of approximately 500 bp. Since both the DNA substrates compete for primers in the same reaction, the relative intensity of these two bands allows for the quantification of MUTYH mRNA to a known quantity of MUTYH RNA. These methods yielded quantification of MUTYH expression that was highly consistent under repeated experiments with the same cell line, allowing for reliable baseline expression level quantification of MUTYH in multiple cell lines.

4.5 Measuring Repair With a GFP Reporter

To measure OG:A repair of the reporter, MEFs are grown to 100% confluency at the time of cell harvest. In our case, we found GFP repair signal was maximized after three nights' posttransfection of the reporter plasmid. Typically, 2 μg of plasmid per six-well plate is transfected, in parallel with both positive (GFP enabled, no OG insert) and negative (GFP stop codon insert present) conditions. Comparison of GFP signals between cell lines without MUTYH (*Mutyh*$^{-/-}$ MEFs) vs cell lines with normal wild-type MUTYH (WT MEFs) allowed these to be used as comparative baseline measurements of negative and positive controls (respectively) vs cell lines stably transfected with MUTYH variants. MEFs are not known to have high transfection efficiency and transfection reagents can be highly toxic, so careful titration experiments must be performed to determine optimal transfection conditions. Although we typically saw transfection efficiency of only 1% (i.e., 1000 fluorescent events in 100,000 cells processed), the sample of 1000 events was abundant enough to collect meaningful data. It is typical to be able to process over 100,000 cells for each six-well plate condition.

Cells are harvested for flow cytometry using 0.5% trypsin and washed three times with ice-cold sterile phosphate-buffered saline (PBS, pH 7.2), then suspended in ice-cold sterile PBS (pH 7.2) containing 1% paraformaldehyde, and stored at 4°C. Flow cytometry must be completed within 3 h. Events are gated by forward vs side scatter to include only single cells, and the influence in overlap between red and green fluorescence signals is compensated for using a 6%–8% compensation level. Three control transfection wells (*pUC18* transfection control, *pGFP-off* (GFP$^-$/dsRed$^+$), and *pGFP-on* (GFP$^+$/dsRed$^+$)) should be transfected, harvested, and analyzed in parallel with the OG:A lesion-containing construct in all experiments.

The data obtained from these experiments can be analyzed using FlowJo Software Version 7.0 (Ashland, OR). We used a simple four-quadrant analysis method to categorize cells with regard to their red and green fluorescence. With green fluorescence measured on the horizontal scale and red fluorescence on the vertical scale, events (single cells) are categorized as being GFP positive or GFP negative. This characterization is based on a vertical quadrant boundary determined by minimizing the number of GFP-negative events in the GFP-positive (*pGFP-on*) transfection, while also symmetrically minimizing the number of GFP-positive events in the GFP-negative (*pGFP-off*) transfection. Additionally, these boundaries are adjusted to maximize exclusion of untransfected cells transfected with a nonfluorescent *pUC18* construct. Nonfluorescent cells are minimized by the position of the horizontal quadrant boundary, while trying to simultaneously maximize legitimate events in the *pGFP-off* (dsRed positive) and *pGFP-on* (dsRed and GFP positive) transfection controls. Lesion repair is calculated as the percent of GFP-positive cells in the dsRed (transfection) positive cell population. This calculation is then normalized within each experiment so that the double-positive GFPon/dsRed transfection condition is considered the maximum possible GFP signal. Differences in the ratio of GFP-positive cells to GFP-negative cells between cell lines are tested for significance using Fisher's exact test with a two-tailed P value. Finally, the standard deviation is calculated from the differences in OG:A repair between multiple experiments.

4.5.1 Troubleshooting Tips

OG-containing oligomers are subject to further oxidation or degradation, and thus are best preserved by storing frozen lyophilized aliquots at −20°C. Both the optimum condition for transfection of plasmid DNA and the harvesting of cells for flow cytometry will greatly depend on the particular cell line under evaluation.

4.6 Case Study: Evaluating MAP Variants in Mammalian Cells

Cancer-associated variants of MUTYH tested with the cellular OG:A repair assay described above led to some key insights. Our results using this assay have shown that the Gly382Asp variant, which was relatively mildly impaired in binding affinity and excision rate (Al-Tassan et al., 2002), was more significantly impaired in the cellular GFP assay, suggesting that minor differences with in vitro assays can be diagnostic for more severe deficiencies in cells. In addition, we showed that even strong overexpression of cancer-associated variant Tyr165Cys at the mRNA level did not compensate for its lack of activity.

Furthermore, we quantified the repair of MEFs stably expressing the common MUTYH polymorphism Gln324His (Raetz et al., 2012). Although the high allele frequencies of this polymorphism in some ethnic groups suggest that it would be unlikely to significantly alter cellular repair, there is some clinical evidence associating it with higher cancer risk (Kasahara et al., 2008; Miyaishi et al., 2009; Picelli et al., 2010). Indeed, we found the Gln324His variant was significantly less capable of repairing the OG:A reporter plasmid as measured by flow cytometry, evidence of the sensitivity of this assay given the subtle phenotype in humans. Binding and glycosylase assays with purified Gln324His under low salt (60 and 100 m*M* NaCl) showed adenine excision is similar to WT MUTYH (Brinkmeyer & David, 2015). Interestingly, physiological salt concentrations (150 m*M* NaCl) with Gln324His show reduced adenine excision (Brinkmeyer & David, 2015), which is similar to the MAP-associated Gly382Asp variant (Raetz et al., 2012).

5. SUMMARY AND CONCLUSION

Our laboratory has developed and utilized cellular assays to study MutY and MUTYH to better elucidate their roles in facilitating repair of oxidative lesions. The rifampicin resistance assay offers an indirect measurement of DNA repair via the intrinsic mutation rate of *Ec* conferred by antibiotic resistance due to mutations in RNA polymerase. The plasmid-based bacterial repair assay is a more direct measure of repair. It can employ WT MutY, and its variants, as well as synthetic analogs of the natural OG:A substrate, and allows for measurement of the efficiency of cellular lesion search and repair process. Both the rifampicin resistance and bacterial plasmid assays evaluate repair within a bacterial host, which may not possess the same post-translational modifications and protein partners encountered by mammalian homologs. To address this, we have developed a quantitative mammalian cell GFP-based assay.

Without the cellular assays, our understanding of MutY enzyme function would be limited. For example, in vitro assays only minimally replicate the search process of MutY/MUTYH locating DNA damage, and therefore cellular assays more accurately report on these processes by embedding the lesion within an excess of undamaged DNA. Taken together, in vitro and cellular assays for studying MUTYH provide powerful means to unravel the complex roles played by its unique Fe–S cluster cofactor. Indeed, using such approaches new insights into key molecular features of MutY enzymes, as well as important clinical information on MAP variants have already been revealed, as described in this chapter. Future applications of these approaches for MUTYH and related enzymes will further delineate the complex roles of Fe–S clusters in DNA repair, and the important ways that DNA repair impacts disease.

ACKNOWLEDGMENTS

We are grateful to all members of the David laboratory past and present who contributed to developing these methods. Research in the David laboratory on MutY/MUTYH has been supported by the National Cancer Institute of the National Institutes of Health (CA067985). N.N.N. gratefully acknowledges support from a GAANN fellowship, and a UCD Humanities Graduate Research Award and the Schwall Dissertation Fellowship. A.G.R. was supported by a National Institute of Environmental Health Sciences training program in Environmental Health Sciences (T32 ES007058).

REFERENCES

Al-Tassan, N., Chmiel, N. H., Maynard, J., Fleming, N., Livingston, A. L., Williams, G. T., et al. (2002). Inherited variants of MYH associated with somatic G: C -> T: A mutations in colorectal tumors. *Nature Genetics*, *30*(2), 227–232.

Amouroux, R., Campalans, A., Epe, B., & Radicella, J. P. (2010). Oxidative stress triggers the preferential assembly of base excision repair complexes on open chromatin regions. *Nucleic Acids Research*, *38*(9), 2878–2890.

Banda, D. M., Nuñez, N. N., Burnside, M. A., Bradshaw, K. M., & David, S. S. (2017). Repair of 8-oxoG:A mismatches by the MUTYH glycosylase: Mechanism, metals and medicine. *Free Radical Biology and Medicine*, *107*, 202–215.

Barras, F., & Marinus, M. G. (1989). The great GATC: DNA methylation in *E. coli*. *Trends in Genetics*, *5*, 139–143.

Brinkmeyer, M. K., & David, S. S. (2015). Distinct functional consequences of MUTYH variants associated with colorectal cancer: Damaged DNA affinity, glycosylase activity and interaction with PCNA and Hus1. *DNA Repair (Amst)*, *34*, 39–51.

Brinkmeyer, M. K., Pope, M. A., & David, S. S. (2012). Catalytic contributions of key residues in the adenine glycosylase MutY revealed by pH-dependent kinetics and cellular repair assays. *Chemistry & Biology*, *19*(2), 276–286.

Burger, K., Matt, K., Kieser, N., Gebhard, D., & Bergemann, J. (2010). A modified fluorimetric host cell reactivation assay to determine the repair capacity of primary keratinocytes, melanocytes and fibroblasts. *BMC Biotechnology*, *10*, 46.

Chepanoske, C. L., Golinelli, M. P., Williams, S. D., & David, S. S. (2000). Positively charged residues within the iron-sulfur cluster loop of E-coli MutY participate in damage recognition and removal. *Archives of Biochemistry and Biophysics*, *380*(1), 11–19.

Chepanoske, C. L., Porello, S. L., Fujiwara, T., Sugiyama, H., & David, S. S. (1999). Substrate recognition by Escherichia coli MutY using substrate analogs. *Nucleic Acids Research*, *27*(15), 3197–3204.

Chmiel, N. H., Golinelli, M. P., Francis, A. W., & David, S. S. (2001). Efficient recognition of substrates and substrate analogs by the adenine glycosylase MutY requires the C-terminal domain. *Nucleic Acids Research*, *29*(2), 553–564.

Chmiel, N. H., Livingston, A. L., & David, S. S. (2003). Insight into the functional consequences of inherited variants of the hMYH adenine glycosylase associated with colorectal cancer: Complementation assays with hMYH variants and pre-steady-state kinetics of the corresponding mutated E. coli enzymes. *Journal of Molecular Biology*, *327*(2), 431–443.

Couce, A., & Blazquez, J. (2011). Estimating mutation rates in low-replication experiments. *Mutation Research*, *714*(1–2), 26–32.

David, S. S., O'Shea, V. L., & Kundu, S. (2007). Base-excision repair of oxidative DNA damage. *Nature*, *447*(7147), 941–950.

David, S. S., & Wiliams, S. D. (1998). Chemistry of glycosylases and endonucleases involved in base-excision repair. *Chemical Reviews*, *98*(3), 1221–1261.

Doetsch, P. W. (2002). Translesion synthesis by RNA polymerases: Occurrence and biological implications for transcriptional mutagenesis. *Mutation Research*, *510*(1–2), 131–140.

Dorn, J., Ferrari, E., Imhof, R., Ziegler, N., & Hubscher, U. (2014). Regulation of human MutYH DNA glycosylase by the E3 ubiquitin ligase mule. *The Journal of Biological Chemistry*, *289*(10), 7049–7058.

Engstrom, L. M., Brinkmeyer, M. K., Ha, Y., Raetz, A. G., Hedman, B., Hodgson, K. O., et al. (2014). A zinc linchpin motif in the MUTYH glycosylase interdomain connector is required for efficient repair of DNA damage. *Journal of the American Chemical Society*, *136*(22), 7829–7832.

Farrington, S. M., Tenesa, A., Barnetson, R., Wiltshire, A., Prendergast, J., Porteous, M., et al. (2005). Germline susceptibility to colorectal cancer due to base-excision repair gene defects. *American Journal of Human Genetics*, 77(1), 112–119.

Foster, P. L. (2006). Methods for determining spontaneous mutation rates. *Methods in Enzymology*, *409*, 195–213.

Fry, R. C., Svensson, J. P., Valiathan, C., Wang, E., Hogan, B. J., Bhattacharya, S., et al. (2008). Genomic predictors of interindividual differences in response to DNA damaging agents. *Genes & Development*, *22*(19), 2621–2626.

Garibyan, L., Huang, T., Kim, M., Wolff, E., Nguyen, A., Nguyen, T., et al. (2003). Use of the rpoB gene to determine the specificity of base substitution mutations on the Escherichia coli chromosome. *DNA Repair (Amst)*, *2*(5), 593–608.

Gillet-Markowska, A., Louvel, G., & Fischer, G. (2015). bz-rates: A web tool to estimate mutation rates from fluctuation analysis. *G3: Genes|Genomes|Genetics*, *5*(11), 2323–2327.

Golinelli, M. P., Chmiel, N. H., & David, S. S. (1999). Site-directed mutagenesis of the cysteine ligands to the [4Fe-4S] cluster of Escherichia coli MutY. *Biochemistry*, *38*(22), 6997–7007.

Green, M. R., Sambrook, J., & Sambrook, J. (2012). *Molecular cloning: A laboratory manual* (4th ed.). Cold Spring Harbor, NY: Cold Spring Harbor Laboratory Press.

Gu, Y., Parker, A., Wilson, T. M., Bai, H., Chang, D. Y., & Lu, A. L. (2002). Human MutY homolog, a DNA glycosylase involved in base excision repair, physically and functionally interacts with mismatch repair proteins human MutS homolog 2/human MutS homolog 6. *The Journal of Biological Chemistry*, *277*(13), 11135–11142.

Hirano, S., Tominaga, Y., Ichinoe, A., Ushijima, Y., Tsuchimoto, D., Honda-Ohnishi, Y., et al. (2003). Mutator phenotype of MUTYH-null mouse embryonic stem cells. *The Journal of Biological Chemistry*, *278*(40), 38121–38124.

Hwang, B. J., Jin, J., Gao, Y., Shi, G., Madabushi, A., Yan, A., et al. (2015). SIRT6 protein deacetylase interacts with MYH DNA glycosylase, APE1 endonuclease, and Rad9-Rad1-Hus1 checkpoint clamp. *BMC Molecular Biology*, *16*, 12.
Jenkins, M. A., Baglietto, L., Dowty, J. G., Van Vliet, C. M., Smith, L., Mead, L. J., et al. (2006). Cancer risks for mismatch repair gene mutation carriers: A population-based early onset case-family study. *Clinical Gastroenterology and Hepatology*, *4*(4), 489–498.
Kasahara, M., Osawa, K., Yoshida, K., Miyaishi, A., Osawa, Y., Inoue, N., et al. (2008). Association of MUTYH Gln324His and APEX1 Asp148Glu with colorectal cancer and smoking in a Japanese population. *Journal of Experimental & Clinical Cancer Research: CR*, *27*, 49.
Kundu, S., Brinkmeyer, M. K., Eigenheer, R. A., & David, S. S. (2010). Ser 524 is a phosphorylation site in MUTYH and Ser 524 mutations alter 8-oxoguanine (OG): A mismatch recognition. *DNA Repair*, *9*(10), 1026–1037.
Kundu, S., Brinkmeyer, M. K., Livingston, A. L., & David, S. S. (2009). Adenine removal activity and bacterial complementation with the human MutY homologue (MUTYH) and Y165C, G382D, P391L and Q324R variants associated with colorectal cancer. *DNA Repair (Amst)*, *8*(12), 1400–1410.
Lee, D., & Threadgill, D. W. (2004). Investigating gene function using mouse models. *Current Opinion in Genetics & Development*, *14*(3), 246–252.
Livingston, A. L., O'Shea, V. L., Kim, T., Kool, E. T., & David, S. S. (2008). Unnatural substrates reveal the importance of 8-oxoguanine for in vivo mismatch repair by MutY. *Nature Chemical Biology*, *4*(1), 51–58.
Lodish, H., Berk, A., Zipursky, S. L., Matsudaira, P., Baltimore, D., & Darnell, J. (2000). *Molecular cell biology* (4th ed.). National Center for Biotechnology InformationÕs Bookshelf.
Luncsford, P. J., Chang, D. Y., Shi, G., Bernstein, J., Madabushi, A., Patterson, D. N., et al. (2010). A structural hinge in eukaryotic MutY homologues mediates catalytic activity and Rad9-Rad1-Hus1 checkpoint complex interactions. *Journal of Molecular Biology*, *403*(3), 351–370.
Lykke-Andersen, S., & Jensen, T. H. (2015). Nonsense-mediated mRNA decay: An intricate machinery that shapes transcriptomes. *Nature Reviews. Molecular Cell Biology*, *16*(11), 665–677.
Manlove, A. H., McKibbin, P. L., Doyle, E. L., Majumdar, C., Hamm, M. L., & David, S. S. (2017). Structure–activity relationships reveal key features of 8-Oxoguanine: A mismatch detection by the MutY glycosylase. *ACS Chemical Biology*, *12*, 2335–2344.
Manlove, A. H., Nuñez, N. N., & David, S. S. (2016). The GO repair pathway: OGG1 and MUTYH. In *The base excision repair pathway* (pp. 63–115). World Scientific.
Masaoka, A., Horton, J. K., Beard, W. A., & Wilson, S. H. (2009). DNA polymerase beta and PARP activities in base excision repair in living cells. *DNA Repair (Amst)*, *8*(11), 1290–1299.
McGoldrick, J. P., Yeh, Y. C., Solomon, M., Essigmann, J. M., & Lu, A. L. (1995). Characterization of a mammalian homolog of the Escherichia coli MutY mismatch repair protein. *Molecular and Cellular Biology*, *15*(2), 989–996.
Menoni, H., Di Mascio, P., Cadet, J., Dimitrov, S., & Angelov, D. (2017). Chromatin associated mechanisms in base excision repair—Nucleosome remodeling and DNA transcription, two key players. *Free Radical Biology & Medicine*, *107*, 159–169.
Michaels, M. L., Cruz, C., Grollman, A. P., & Miller, J. H. (1992). Evidence that MutY and MutM combine to prevent mutations by an oxidatively damaged form of guanine in DNA. *Proceedings of the National Academy of Sciences of the United States of America*, *89*(15), 7022–7025.

Michaels, M. L., Pham, L., Nghiem, Y., Cruz, C., & Miller, J. H. (1990). MutY, an adenine glycosylase active on G-A mispairs, has homology to endonuclease III. *Nucleic Acids Research*, *18*(13), 3841–3845.

Miller, J. H., & Michaels, M. (1996). Finding new mutator strains of Escherichia coli—A review. *Gene*, *179*(1), 129–132.

Miyaishi, A., Osawa, K., Osawa, Y., Inoue, N., Yoshida, K., Kasahara, M., et al. (2009). MUTYH Gln324His gene polymorphism and genetic susceptibility for lung cancer in a Japanese population. *Journal of Experimental & Clinical Cancer Research: CR*, *28*, 10.

Molatore, S., Russo, M. T., D'Agostino, V. G., Barone, F., Matsumoto, Y., Albertini, A. M., et al. (2010). MUTYH mutations associated with familial adenomatous polyposis: Functional characterization by a mammalian cell-based assay. *Human Mutation*, *31*(2), 159–166.

Nagel, Z. D., Margulies, C. M., Chaim, I. A., McRee, S. K., Mazzucato, P., Ahmad, A., et al. (2014). Multiplexed DNA repair assays for multiple lesions and multiple doses via transcription inhibition and transcriptional mutagenesis. *Proceedings of the National Academy of Sciences of the United States of America*, *111*(18), E1823–32.

Out, A. A., Tops, C. M., Nielsen, M., Weiss, M. M., van Minderhout, I. J., Fokkema, I. F., et al. (2010). Leiden open variation database of the MUTYH gene. *Human Mutation*, *31*(11), 1205–1215.

Parker, A., Gu, Y., Mahoney, W., Lee, S. H., Singh, K. K., & Lu, A. L. (2001). Human homolog of the MutY repair protein (hMYH) physically interacts with proteins involved in long patch DNA base excision repair. *The Journal of Biological Chemistry*, *276*(8), 5547–5555.

Picelli, S., Zajac, P., Zhou, X. L., Edler, D., Lenander, C., Dalen, J., et al. (2010). Common variants in human CRC genes as low-risk alleles. *European Journal of Cancer*, *46*(6), 1041–1048.

Pope, M. A., & David, S. S. (2005). DNA damage recognition and repair by the murine MutY homologue. *DNA Repair*, *4*(1), 91–102.

Porello, S. L., Leyes, A. E., & David, S. S. (1998). Single-turnover and pre-steady-state kinetics of the reaction of the adenine glycosylase MutY with mismatch-containing DNA substrates. *Biochemistry*, *37*(42), 14756–14764.

Pukkila, P. J., Peterson, J., Herman, G., Modrich, P., & Meselson, M. (1983). Effects of high levels of DNA adenine methylation on methyl-directed mismatch repair in *Escherichia coli*. *Genetics*, *104*(4), 571–582.

Raetz, A. G., Xie, Y., Kundu, S., Brinkmeyer, M. K., Chang, C., & David, S. S. (2012). Cancer-associated variants and a common polymorphism of MUTYH exhibit reduced repair of oxidative DNA damage using a GFP-based assay in mammalian cells. *Carcinogenesis*, *33*(11), 2301–2309.

Rosche, W. A., & Foster, P. L. (2000). Determining mutation rates in bacterial populations. *Methods*, *20*(1), 4–17.

Russo, M. T., De Luca, G., Degan, P., Parlanti, E., Dogliotti, E., Barnes, D. E., et al. (2004). Accumulation of the oxidative base lesion 8-hydroxyguanine in DNA of tumor-prone mice defective in both the Myh and Ogg1 DNA glycosylases. *Cancer Research*, *64*(13), 4411–4414.

Sambrook, J., Fritsch, E. F., & Maniatis, T. (1989). *Molecular cloning: A laboratory manual*. Cold Spring Harbor, NY: Cold Spring Harbor Laboratory Press.

Sarkar, S., Ma, W. T., & Sandri, G. H. (1992). On fluctuation analysis: A new, simple and efficient method for computing the expected number of mutants. *Genetica*, *85*(2), 173–179.

Shi, G., Chang, D. Y., Cheng, C. C., Guan, X., Venclovas, C., & Lu, A. L. (2006). Physical and functional interactions between MutY glycosylase homologue (MYH) and checkpoint proteins Rad9-Rad1-Hus1. *The Biochemical Journal*, *400*(1), 53–62.

Sieber, O. M., Lipton, L., Crabtree, M., Heinimann, K., Fidalgo, P., Phillips, R. K., et al. (2003). Multiple colorectal adenomas, classic adenomatous polyposis, and germ-line mutations in MYH. *The New England Journal of Medicine*, *348*(9), 791–799.

Slupska, M. M., Luther, W. M., Chiang, J. H., Yang, H., & Miller, J. H. (1999). Functional expression of hMYH, a human homolog of the Escherichia coli MutY protein. *Journal of Bacteriology*, *181*(19), 6210–6213.

Tsien, R. Y. (1998). The green fluorescent protein. *Annual Review of Biochemistry*, *67*, 509–544.

Turco, E., Ventura, I., Minoprio, A., Russo, M. T., Torreri, P., Degan, P., et al. (2013). Understanding the role of the Q338H MUTYH variant in oxidative damage repair. *Nucleic Acids Research*, *41*(7), 4093–4103.

Viel, A., Bruselles, A., Meccia, E., Fornasarig, M., Quaia, M., Canzonieri, V., et al. (2017). A specific mutational signature associated with DNA 8-oxoguanine persistence in MUTYH-defective colorectal cancer. *eBioMedicine*, *20*, 39–49.

Wang, L., Chakravarthy, S., & Verdine, G. L. (2017). Structural basis for the lesion-scanning mechanism of the MutY DNA glycosylase. *The Journal of Biological Chemistry*, *292*(12), 5007–5017.

Wehrli, W., Knüsel, F., Schmid, K., & Staehelin, M. (1968). Interaction of rifamycin with bacterial RNA polymerase. *Proceedings of the National Academy of Sciences of the United States of America*, *61*(2), 667–673.

Wolff, E., Kim, M., Hu, K., Yang, H., & Miller, J. H. (2004). Polymerases leave fingerprints: Analysis of the mutational spectrum in Escherichia coli rpoB to assess the role of polymerase IV in spontaneous mutation. *Journal of Bacteriology*, *186*(9), 2900–2905.

Woods, R. D., O'Shea, V. L., Chu, A., Cao, S., Richards, J. L., Horvath, M. P., et al. (2016). Structure and stereochemistry of the base excision repair glycosylase MutY reveal a mechanism similar to retaining glycosidases. *Nucleic Acids Research*, *44*(2), 801–810.

Xie, Y., Yang, H., Cunanan, C., Okamoto, K., Shibata, D., Pan, J., et al. (2004). Deficiencies in mouse Myh and Ogg1 result in tumor predisposition and G to T mutations in codon 12 of the K-ras oncogene in lung tumors. *Cancer Research*, *64*(9), 3096–3102.

You, C., & Wang, Y. (2015). Quantitative measurement of transcriptional inhibition and mutagenesis induced by site-specifically incorporated DNA lesions in vitro and in vivo. *Nature Protocols*, *10*(9), 1389–1406.

Zheng, Q. (2015). A new practical guide to the Luria–Delbrück protocol. *Mutation Research: Fundamental and Molecular Mechanisms of Mutagenesis*, *781*, 7–13.

CHAPTER FOUR

Iron–Sulfur Clusters in Zinc Finger Proteins

Geoffrey D. Shimberg[2], Jordan D. Pritts[2], Sarah L.J. Michel[1]
School of Pharmacy, University of Maryland, Baltimore, MD, United States
[1]Corresponding author: e-mail address: smichel@rx.umaryland.edu

Contents

Abstract

Zinc finger (ZF) proteins are proteins that use zinc as a structural cofactor. The common feature among all ZFs is that they contain repeats of four cysteine and/or histidine residues within their primary amino acid sequence. With the explosion of genome sequencing in the early 2000s, a large number of proteins were annotated as ZFs based solely upon amino acid sequence. As these proteins began to be characterized

[2] G.S. and J.P. contributed equally to this work.

Methods in Enzymology, Volume 599
ISSN 0076-6879
https://doi.org/10.1016/bs.mie.2017.09.005

experimentally, it was discovered that some of these proteins contain iron–sulfur sites either in place of or in addition to zinc. Here, we describe methods to isolate and characterize one such ZF protein, cleavage and polyadenylation specificity factor 30 (CPSF30) with respect to its metal-loading and RNA-binding activity.

1. INTRODUCTION

Zinc fingers (ZFs) are a large family of principally eukaryotic proteins that utilize zinc as a cofactor to fold and function (Jantz, Amann, Gatto, & Berg, 2004; Lee & Michel, 2014; Maret, 2012; Michalek, Besold, & Michel, 2011). ZFs contain repeats of four invariant cysteine and/or histidine residues within their primary amino acid sequences, and these residues serve as ligands for the Zn^{2+} ion (Lee & Michel, 2014; Michalek et al., 2011). Although first identified in the 1980s, the ubiquity of ZFs was not fully appreciated until the late 1990s/early 2000s with the advent of whole-genome sequencing (Andreini, Banci, Bertini, & Rosato, 2006; Berg, 1986, 1988; Bertini, Decaria, & Rosato, 2010; Decaria, Bertini, & Williams, 2010; Laity, Lee, & Wright, 2001). From these sequencing efforts, between 3% and 10% of all proteins were annotated as ZFs, based upon the presence of cysteine/histidine-rich sequences (Lee & Michel, 2014). These ZFs have been separated into different classes, with ZFs grouped based upon the number and spacing of cysteine and histidine residues within each ZF domain (Krishna, Majumdar, & Grishin, 2003; Lee & Michel, 2014; Magyar & Godwin, 2003; Matthews & Sunde, 2002; Michalek et al., 2011). Presently, at least 14 classes of ZFs have been defined.

Most of the ZFs that have been identified from genome sequencing have not yet been studied experimentally; therefore, in many cases whether these "zinc finger" proteins are bona fide zinc fingers is not known. In the last few years, several proteins annotated as "zinc fingers" have been found to harbor iron–sulfur clusters, in lieu of or in addition to zinc. The first of these proteins to be identified was mitoNEET, which contains a 3-cysteine, 1-histidine (CCCH) motif within its primary sequence (Baxter, Jennings, & Onuchic, 2012; Conlan et al., 2009; Tamir et al., 2015; Wiley, Murphy, Ross, van der Geer, & Dixon, 2007; Wiley, Paddock, et al., 2007). MitoNEET attracted initial interest because it was found to be a target of the antidiabetes drug pioglitazone (Colca et al., 2004). Subsequently, mitoNEET was found to play a key role in electron transport, and there is emerging evidence for roles in iron regulation, mitochondrial energy metabolism, and Fe–S cluster regulation

(Lipper et al., 2015; Wang, Landry, & Ding, 2017). Despite being annotated as a "zinc finger" because it contains a conserved CCCH domain that is typically related to zinc binding, mitoNEET turned red upon protein expression and purification and was found to contain a 2Fe–2S cluster, coordinated to the CCCH "zinc finger" ligands (Wiley, Paddock, et al., 2007) (Fig. 1). Subsequently, two close homologs of mitoNEET, Miner1 and Miner2, were also shown to contain 2Fe–2S clusters bound to their CCCH motifs (Conlan et al., 2009; Lin, Zhang, Lai, & Ye, 2011). In addition, 2Fe–2S clusters were identified in the *Escherichia coli* iron–sulfur cluster assembly proteins IscR and IscU and the yeast Grx3/4/Fra2-signaling proteins (Blanc, Gerez, & Ollagnier de Choudens, 2015; Li et al., 2009). These findings brought into question the dogma that proteins with conserved cysteine/histidine sequences are *always* ZFs; moreover, this underscores the need for experimental validation of the metal identity and coordination.

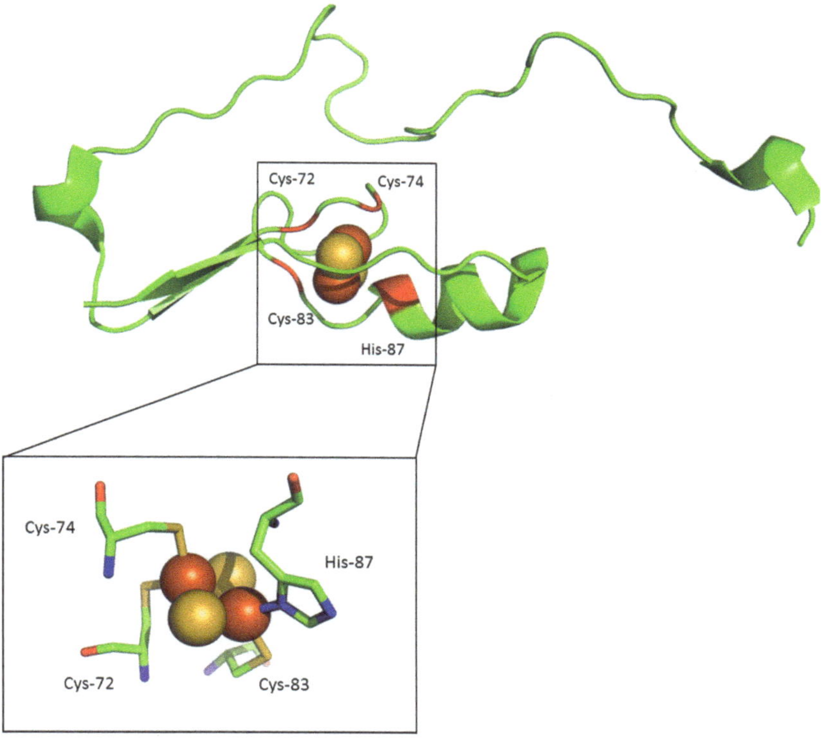

Fig. 1 Structure of mitoNEET with the 2Fe–2S cluster highlighted (PDB 2R13, figure made in PyMol).

More recently, our laboratory identified another intriguing eukaryotic protein annotated as a ZF that contains a 2Fe–2S cluster (as initially observed as a reddish colored protein) (Figs. 2 and 3A). This protein, cleavage and polyadenylation specificity factor 30 (CPSF30) contains five CCCH domains and is a hybrid of an iron–sulfur cluster/ZF protein (Shimberg et al., 2016). CPSF30 houses a 2Fe–2S cluster with one CCCH ligand set, analogous to mitoNEET, Miner1 and Miner2, and four zinc-loaded CCCH ZF sites, analogous to traditional ZFs (Shimberg et al., 2016). (Fig. 2) The full biological role of CPSF30 is not yet understood; however, it is known to be involved in pre-mRNA regulation. We have shown that CPSF30 binds the AU-rich hexamer of RNA present in the majority of pre-mRNA molecules (Yang & Doublie, 2011) by measuring protein/RNA binding with a synthetic RNA sequence that corresponds to α-synuclein pre-RNA. The interaction is sequence selective and requires that both iron and zinc sites be present (Shimberg et al., 2016). As this hexamer is conserved in approximately 90% of pre-mRNA molecules (Beaudoing, Freir,

ZF1	S	G	A	A	V	C	E	F	F	L	K	A	A	–	C	G	K	G	G	M	C	P	F	R	H	I	S	G
ZF2		K	T	V	V	C	K	H	W	L	R	G	L	–	C	K	K	G	D	Q	C	E	F	L	H	E	Y	D
ZF3	T	K	M	P	E	C	Y	F	Y	S	K	F	G	E	C	S	N	K	E		C	P	F	L	H	I	D	P
ZF4	S	K	I	K	D	C	P	W	Y	D	R	G	F	–	C	K	H	G	P	L	C	R	H	R	H	T	R	R
ZF5					I	C	V	N	Y	L	V	G	F	–	C	P	E	G	P	S	C	K	F	M	H	P	R	F

Fig. 2 Sequence alignment of the CPSF30 CCCH domains.

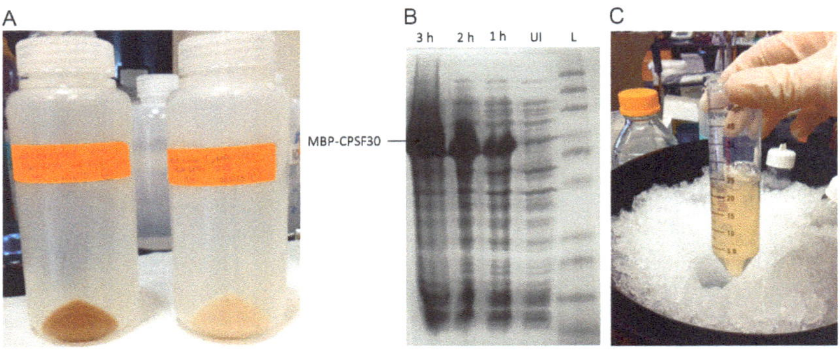

Fig. 3 The induction through sonication of CPSF30. (A) Comparison of CPSF30 (*left*) cell paste vs MBP (*right*), postoverexpression. CPSF30 results in a rust-colored protein sample indicative of the presence of an iron cofactor. (B) 15% SDS-PAGE of CPSF30 protein induction trial. From right to left is the Invitrogen BenchMark Protein Ladder (Thermo Fisher), uninduced pellet, and induced pellet after 1, 2, and 3 h. (C) Image of the soluble fraction of CPSF30 after sonication of the CPSF30 pellet. Note that the protein retains the *reddish hue*.

Wyatt, Claverie, & Gautheret, 2000), it can be inferred that CPSF30 has a broad application to bind various pre-mRNA sequences aiding poly-adenylation. In this chapter, we describe the methods we utilize to isolate metal-loaded CPSF30 and assess its RNA-binding activity. Our approach utilizes established methods used to isolate traditional ZF proteins combined with those for Fe–S proteins. This approach has the potential to be utilized for the isolation and analysis of other ZF proteins that have been identified from genome sequences, but not yet characterized experimentally.

2. APPROACHES TO CLONE ZINC FINGER/Fe–S CLUSTER GENES

2.1 Cloning Strategy

When choosing the expression system for a ZF and/or Fe–S protein, like most proteins, a wealth of options are commercially available. Choosing an appropriate vector is critical to ensure that proper protein production is successful. The expression of proteins containing ZFs as well as ZFs with Fe–S clusters is often achieved using commercial pET vectors (example: pET-28a, Novagen, Cat. No. 69864-3) containing either C- or N-terminal hexahistidine tags (his-tag). This is an effective method when the protein is inherently soluble (Woestenenk, Hammarström, van den Berg, Härd, & Berglund, 2004) and there are multiple examples of the use of a hexahistidine tag to purify a protein with an Fe–S cluster, with some recent examples in these references (Boal et al., 2005; Engstrom, Partington, & David, 2012; Poor et al., 2014; Tan et al., 2012) as well as for ZFs. We note that in some cases, if the metal site is labile, the use of this approach may disrupt native metal binding and therefore should be considered on a case-by-case basis (Zhao & Huang, 2016). As an example, the pET-28a expression system contains an N-terminal his-tag, thrombin cleavage site, T7 tag, and an optional C-terminal his-tag. Upon protein overexpression, purification is accomplished via immobilized metal affinity chromatography (IMAC), in which the hexahistidine tag binds to nickel (cobalt, copper, or iron)-loaded resin in the solid phase (Persikov & Singh, 2014). The protein of interest can then be eluted and separated from other cellular proteins using an imidazole buffer gradient. The T7 tag can be utilized for further purification of the protein of interest if needed. Incorporation of a thrombin cut site allows cleavage of the N-terminal his-tag yielding native protein postpurification. Determination of whether to utilize a C- vs N-terminal his-tag is accomplished empirically—proteins with either tag appended are produced and the resultant protein's stability, solubility, and activity are assessed. N-terminal his-tags are generally more common, as cloning design and gene insertion into the vector is more straightforward (see, for example, Boal et al., 2005). We note that for some iron–sulfur proteins, a C-terminal hexahistidine tag has proven more robust, as seen with RimO (Lee et al., 2009). If the protein of interest contains an Fe–S cluster near the far end of the N-terminus, a C-terminal tag may be preferred as there is concern that the hexahistidine residues may alter the Fe-binding properties at the Fe–S ligand site (Lanz et al., 2012). The pET-28a expression system also contains kanamycin resistance allowing bacterial selection containing the plasmid (Pattenden & Thomas, 2008). These expression systems are well suited for small ZFs with moderate to high solubility.

Where native protein solubility is a concern, particularly for larger ZF proteins, we and other laboratories have found that the utilization of a vector that encodes for a maltose-binding protein (MBP) tag [e.g., pMAL-c5E (discontinued), pMAL-c5X or pMAL-p5X from New England Biolabs] can often produce a more soluble protein (Cat. No. E8200S) (Boon, Livingston, Chmiel, David, & Barton, 2003; Chmiel, Golinelli, Francis, & David, 2001; Liu, Xia, & Case, 2002; Vandevenne et al., 2013; Wiley, Murphy, et al., 2007). The pMAL Protein Fusion and Purification System contains a cloning site downstream of a *malE* gene encoding the 42.5 kDa MBP and results in expression of an MBP-fusion protein that increases protein solubility, while maintaining proper folding and biological activity (Kapust & Waugh, 1999). In some cases MBP appears to act as a chaperone to induce protein folding and stability (Kapust & Waugh, 1999). After expression, the MBP-fusion protein allows for simple purification using amylose affinity chromatography. The construction of the vector also places a linker containing 10 Asn residues between the MBP and protein of interest domains to inhibit significant binding interactions between the two proteins. As a result, the protein of interest is maintained in a more native environment away from the MBP tag and ensures the tag will be able to bind the amylose column efficiently during purification. The pMAL vector also incorporates a Factor Xa or Enterokinase recognition site immediately before the cloning region to allow cleavage of the MBP tag postpurification yielding native protein (Pavletich & Pabo, 1991). Furthermore, different pMAL vectors can allow protein expression in the cytoplasm or periplasm as desired by the user (Pattenden & Thomas, 2008). For added versatility, another cleavage site can be engineered into the expressed protein during cloning. For example, the recognition sequence of the tobacco etch virus (TEV) protease can be incorporated in the forward primer during

PCR (Kurz et al., 2006; Phan et al., 2002). Our laboratory uses the pMAL vector to clone ZFs with Fe–S clusters, as described here. We note that we have also produced CPSF30 with either a glutathione *s*-transferase or a hexahistidine tag. We find that the MBP tag produces the highest yields of soluble and active protein while also increasing our long-term storage stability.

2.1.1 Equipment

- Thermocycler
- Agarose gel running apparatus
- SPD1 SpeedVac Concentrator (Thermo Fisher)
- Light box
- Scalpel
- Centrifuge

2.1.2 Buffers and Reagents

- 10 × *Taq* DNA polymerase PCR buffer (200 m*M* Tris–HCl, 500 m*M* KCl, pH 8.4) (Thermo Fisher, Cat. No. 18067017)
- dNTPs
- $MgCl_2$
- Platinum Taq polymerase
- DH5α *E. coli* cells
- LB agar (10 g/L casein peptone, 5 g/L yeast extract, 5 g/L sodium chloride, and 15 g/L agar) (Affymetrix, Cat. No. 75851)
- Ampicillin (Sigma Aldrich, Cat. No. A9518)
- Ethidium bromide (Fisher, Cat. No. BP 1302-10)
- Tris/acetic acid/EDTA (TAE) buffer (Bio-Rad, Cat. No. 1610743)
- Qiagen miniprep kit (Qiagen, Cat. No. 27106)
- Qiagen PCR purification kit (Qiagen, Cat. No. 28104)
- Buffer 4 (NEB, Cat. No. B7004S)
- BSA (NEB, Cat. No. B9001S)
- *Nde*1 (NEB, Cat. No. R0111)
- *BamH*1 (NEB, Cat. No. R0136S)
- 6 × Loading dye (1 × contains 2.5% Ficoll-400, 11 m*M* EDTA, 3.3 m*M* Tris–HCl, 0.017% SDS, 0.015% bromophenol blue, pH 8.0) (NEB, Cat. No. B7021S)
- Qiagen gel extraction kit (Qiagen, Cat. No. 28704)
- Invitrogen Super Optimal broth with Catabolite repression (S.O.C.) media (2% tryptone, 0.5% yeast extract, 10 m*M* NaCl, 2.5 m*M* KCl, 10 m*M* $MgCl_2$, 10 m*M* $MgSO_4$, 20 m*M* glucose) (Thermo Fisher, Cat. No. 15544034)

2.1.3 Protocol

Note: As a general guide, the pMAL Protein Fusion and Purification System Instructional Manual (New England Biolabs) is followed with a few exceptions.

Primer design

1. Design primers, as an example CPSF30 primers are indicated:
   Forward: Noncoding (6 nucleotides)/restriction site (6 nucleotides)/ TEV cleavage site (21 nucleotides)/CPSF30 DNA (24 nucleotides).
   5′-TTC TTC/CAT ATG/GAA AAT TTA TAT TTT CAA GGT/ATG GAC AAG TCA GGG GCT GCT GTC-3′
   Note: A start codon is not necessarily needed as the beginning of the MBP sequence contains a start codon.
   Reverse: Noncoding (6 nucleotides)/restriction site (6 nucleotides)/ stop codons ×2 (6 nucleotides)/CPSF30 DNA (24 nucleotides).
   Note: DNA and stop codon sequences are reverse compliments of coding DNA.
   5′-TTC TTC/GGA TCC/TTA CTA/TTC AAA TCG AGG GTG CAT GAA TTT-3′

PCR setup

2. The following reagents are combined for the PCR experiment, and these values are adapted from recommendation by New England Biolabs (Biolabs, 2015).

Reagents	**Volume (μL)**	**Final Concentration**
Milli-Q water	34.5	—
PCR buffer (10×)	5	1×
$MgCl_2$	1.5	1.5 m*M*
dNTPs (10 m*M*)	1.5	3 m*M*
Forward primer (5 μ*M*)	3	0.3 μ*M*
Reverse primer (5 μ*M*)	3	0.3 μ*M*
Template DNA	1	—
Platinum Taq polymerase (5 μ/μL)	0.5	2.5 μ
Total volume	50	—

Note: Reagents are added to the PCR tube in the order listed.

PCR cycle

3. The PCR conditions explained here typically work well for ZF gene amplification in our lab. With other constructs, some steps (e.g., primer design, denaturation and annealing temperatures, length of extension times, etc.) may need to be varied to improve desirable yields.
   a. Initial denaturation of template DNA at 94°C for 3 min
   b. Denature double-stranded DNA at 94°C for 1 min
   c. Anneal primers at 65°C for 1 min
   d. Elongation of DNA at 72°C for 1 min
   e. Repeat for 25 cycles
   f. Final elongation step at 72°C for 15 min
   g. Final hold at 4°C
4. Load the amplified PCR products onto a 1% agarose gel with ethidium bromide staining to ensure that the correct length of DNA was copied. The agarose gel should be prepared by adding 0.3 g of ultrapure agarose and 1 μL of ethidium bromide to 30 mL of 1 × TAE buffer and allowed to solidify in the gel apparatus.
5. Follow Qiagen's PCR purification kit and elute DNA using Milli-Q water and store the purified DNA at −20°C (Qiagen protocol, Cat. No. 28104).
6. Load and run a 1% agarose gel, as previously described (step 4) with all PCR products and extracted pMAL-c5E plasmid to ensure the products and vector have been amplified successfully.

pMAL-c5E plasmid prep

The pMAL-c5E vector can be used as received, or amplified and stored as described here.

7. Transform 1 μL of pMAL-c5E plasmid into 50 μL of DH5α *E. coli* chemically competent cells by pipetting 1 μL of plasmid into a 50-μL aliquot of DH5α cells in a 1.8-mL Eppendorf tube and gently vortex the tube to mix the reagents.
8. Incubate on ice for 5 min.
9. Heat shock the plasmid and cell mixture for 45 s at 45°C.
10. Incubate on ice for 2 min.
11. Add 500 μL of S.O.C. media and triturate the solution to ensure proper mixing.
12. Incubate at 37°C for 45 min.
13. Plate 100 μL of the solution onto a LB agar plate containing 100 μg/mL ampicillin. Allow to incubate overnight at 37°C.
14. Remove the plates and allow them to cool to room temperature.

15. Select a single colony to grow in an overnight flask of 50 mL containing 100 μg/mL of ampicillin at 37°C.
16. The 50 mL overnight cultures of pMAL-c5E vector should be centrifuged at 5250 × *g* at 4°C for 10 min.
17. Isolate the DNA via a Qiagen miniprep kit (Plasmid Miniprep Kit: QIAprep Spin Miniprep Kit-Cat. No. 27106).
18. Elute the DNA with Milli-Q water and store at −20°C.

Double digest using Nde*1 and* BamH*1*

19. Dry all samples for 1 h in vacuo in a SpeedVac concentrator.
20. Using New England Biolabs "double digest finder," find compatible working conditions for both digestion enzymes. For *Nde*1 and *BamH*1, Buffer 4 was used with BSA at 37°C for 2 h.
21. Resuspend the pMAL-c5E vector in 32 μL of Milli-Q water and PCR product in 15.8 μL so that each is at a concentration of about 25 ng/μL.
22. Add the following to each reaction.
    a. 2 μL of Buffer 4
    b. 0.5 μL of BSA
    c. 1 μL of *Nde*1
    d. 1 μL of *BamH*1
23. Incubate for 2 h at 37°C.

    Note: If the double-digestion step does not produce satisfactory yields of digested DNA, single digests of the plasmid with each restriction enzyme should be performed followed by analysis on a 1% agarose gel to verify that both restriction enzymes are active.
24. Add 2 μL of 6 × loading dye to each reaction mixture and load 10 μL per well on a 1% agarose gel. Run the gel at room temperature for 50 min, or until the loading dye is approximately 80% down the gel, at 100 V.

    Note: Voltage can typically be varied between 80 and 150 V depending on the concentration of agarose in the gel to prevent it from overheating.
25. Excise the bands from the gel using a light box and scalpel.
26. Follow Qiagen's gel extraction kit and elute DNA using 50 μL of Milli-Q water (Plasmid Miniprep Kit: QIAprep Spin Miniprep Kit-Cat. No. 28704)
27. Remove 1 μL aliquot of each reaction to verify purity by 1% agarose gel as previously described.
28. Dry samples in SpeedVac for 35 min.

Ligation

29. We find that ligation is best performed by testing several ratios of plasmid to insert and then selecting the best product for future studies.

	Reaction A (μL)	Reaction B (μL)	Reaction C (μL)	Reaction D (μL)
Ligase	1	1	1	1
10 × Ligase buffer	2	2	2	2
Insert DNA	1	3	4	5
Plasmid DNA	7	5	4	3
Water	9	9	9	9
Total volume	20	20	20	20

30. Incubate the ligation reactions for 10 min at room temperature and then transform as previously described (step 7) into DH5α *E. coli* cells with the following exception: 5 μL of ligation mixture into 50 μL of cells.
31. Plate 150 μL of transformed cells onto LB agar plates containing 100 μg/mL of ampicillin.
32. Incubate the LB agar plates overnight at 37°C.
33. Choose a single colony from the plate and grow overnight in 50 mL of sterilized LB containing 100 μg/mL of ampicillin at 37°C with shaking at 230 rpm.
34. Spin down overnight culture in a 50-mL conical tube at 5250 × *g* for 10 min at 4°C.
35. Follow the Qiagen mini prep kit (Cat. No. 27106) to extract the DNA; use 50 μL of Milli-Q water to elute the DNA.
36. Store at −20°C.
37. Aliquot 10 μL of each sample to verify ligation by 1% agarose gel.
38. Dilute sample with 5.8 μL of Milli-Q water and perform a single digestion of the ligated product using one of the restriction enzymes chosen during ligation.
39. Add 2 μL of DNA 6 × loading buffer and run a 1% agarose gel as previously described.

Note: A successful ligation should result in the presence of a single band on the agarose gel at the length of the pMAL-c5E plasmid plus the length of your gene of interest. If only the plasmid (no insert) is observed, a different colony should be selected.

40. Quantitate DNA concentration by ultraviolet–visible spectroscopy (UV–vis) at A_{260}. Dilute 3 µL of pure DNA in a cuvette containing 500 µL of Milli-Q water and run full spectrum analysis.

Concentration can be determined by:

$$[\mathrm{DNA}] = (50\mu g/\mathrm{mL}) \times (\text{dilution factor}) \times A_{260}$$

Example

$$[\mathrm{DNA}] = (50\mu g/\mathrm{mL}) \times (503\mu L/3\mu L) \times (0.023924)$$
$$[\mathrm{DNA}] = 200\,\mathrm{ng}/\mu L$$

Note: 50 µg/mL is used to quantify dsDNA. For ssDNA and RNA 33 µg/mL and 40 µg/mL can be used, respectively (Barbas, Burton, Scott, & Silverman, 2007).

41. The purity of the sample can be determined by a ratio of A_{260}/A_{280}. The sample should have a ratio of 1.7–2.0 as pure DNA is ~1.8 (Barbas et al., 2007).

42. Concentrate the sample in vacuo to the concentration desired by the sequencing group you are using. For example, we needed samples at approximately 300 ng/µL.

43. Submit samples for DNA sequencing and verify that the correct sequence has been inserted and that the MBP and insert are in the correct reading frame.

3. EXPRESSION OF ZF PROTEINS AND ADAPTATIONS FOR INCLUSION OF IRON–SULFUR CLUSTERS

After the protein of interest has been successfully incorporated into your chosen vector, it needs to be overexpressed, retrieved from within the cell, and maintained in a soluble aqueous fraction. In this section, we describe our expression protocols as well as how to utilize sonication to obtain a desired protein from a cell pellet postinduction. As a reference, we have included sample data from CPSF30 in Fig. 3.

3.1 General Protocol for Expression of Zinc Finger Proteins Containing Iron–Sulfur Clusters

Optimizing bacterial growth mediums and expression conditions play an essential role in successful protein overproduction. We express zinc finger proteins using BL21 (DE3) *E. coli* cells in Lennox Luria Bertani Broth (LB) BL21 (DE3) cells (Invitrogen/Thermo Fisher, Cat. No. C600003). LB is one of the most commonly used growth mediums for cultivating *E. coli* since it was first described by Bertani (1951) and Larentis et al. (2014). BL21 (DE3) cells have high transformation efficiency, express T7 polymerase during isopropyl β-D-1-thiogalactopyranoside (IPTG) induction, and are deficient of Lon and OmpT proteases making it suitable for overproduction of nontoxic proteins (Newton, Mackay, & Crossley, 2001; Rosano & Ceccarelli, 2014).

Once the bacterial strain and growth medium have been selected, the growth conditions need to be optimized. Varying induction temperature, IPTG concentrations, shaking speed, and media supplementations can have large effects on protein yield and quality (Larentis et al., 2014). By reducing the induction temperature or IPTG concentration, bacterial growth and protein expression can be slowed down (Larentis et al., 2014). This can be beneficial if the protein of interest is being expressed too quickly and yielding low-quality or misfolded protein (Larentis et al., 2014). We typically overexpress zinc finger proteins at 37°C with a final IPTG concentration of 1 m*M*; however, if expression levels are not satisfactory, we alter the temperature and/or IPTG concentration. Incubator shaking speeds affect culture aeration, and higher speeds can help accommodate higher cell densities in the culture and increase protein yields (Larentis et al., 2014). To maintain proper aeration, we utilize a production volume of 1 L in 4-L Erlenmeyer flasks and a shaker speed of 230 rpm. Some laboratories have shown success expressing zinc finger proteins in BL21 cells without the need of supplementing their media (e.g., metals, metal donors, scaffold donors, etc.) as seen with GATA-1 and MBNL (Newton et al., 2001; Warf & Berglund, 2007), while other labs have shown that additional supplementation to the growth media during induction is needed to increase protein quality and biological activity. This has been observed with HIV-1 NC and a zinc finger protein designed by Sangamo Biosciences (Lee, De Guzman, Turner, Tjandra, & Summers, 1998; Liu et al., 2002). For expression of ZF proteins in our laboratory, we supplement the media with $ZnCl_2$ to improve zinc incorporation yields and 0.2% glucose to ensure it is the primary carbon source. For supplementation of iron to aid Fe–S cluster assembly we have recently begun to supplement with $FeCl_3$; for a more in-depth study of supplementation and Fe–S cluster incorporation yields, we refer you to the work of Jaganaman, Pinto, Tarasev, and Ballou (2007). This helps facilitate increased incorporation of the zinc and Fe–S clusters into the protein thereby improving overall activity. Another approach to optimize Fe–S cluster loading is to utilize a minimal media/metal supplementation approach which provides the essential salts and sugars that are required for cell growth. Booker and Krebs describe this protocol for the expression of RlmN and AstB Radical SAM proteins

(Lanz et al., 2012). This allows the levels of metal ions that are present in the growth media to be controlled, in contrast to LB broth, which contains variable levels of metal ions (including iron and zinc). Our general expression protocol for CPSF30 is described later.

3.1.1 Equipment

- Incubating Orbital Shaker, Model 3500I (VWR, Cat. No. 12620-946)
- Innova Orbital Shaker
- 250-mL Erlenmeyer flasks
- 4-L Erlenmeyer flasks
- Avanti J-20 XPI centrifuge (Beckman Coulter, SKU# 8043-30-1171)
- J-LITE® JLA-10.500 Rotor Assembly with 500-mL bottles and aluminum canisters (Beckman Coulter, Cat. No. 369681)
- Table top microcentrifuge equipped for 1.5-mL Eppendorf tubes (Denville Scientific, Cat. No. C0260-24)
- Lambda 25 UV–vis Spectrometer (Perkin Elmer, Cat. No. L600000B)
- Autoclave
- Mini-PROTEAN® Tetra Vertical Electrophoresis Cell (Bio-Rad, Cat. No. 1658004)
- Oscillating platform shaker (Stovall Life Sciences, Cat. No. BDRAA115S)
- Fisher Scientific dry bath incubator (Boekel, Cat. No. 110011)

3.1.2 Buffers and Reagents for LB Overexpression

- Milli-Q water system
- Invitrogen S.O.C. media (Thermo Fisher, Cat. No. 15544034)
- Lennox LB broth (American Bioanalytical Inc., Cat. No. AB01198)
- 100 mg/mL ampicillin prepared from ampicillin sodium salt (Sigma Aldrich, Cat. No. A9518)
- LB broth containing 0.2% glucose
- 100 m*M* $ZnCl_2$ (Sigma Aldrich, Cat. No. Z0152)
- 1 *M* IPTG (Research Products International Corp through Fisher Scientific, Cat. No. 50-488-727)
- 2 × Laemmli Sample Buffer (Bio-Rad, Cat. No. 1610737)
- 1 × SDS-PAGE running buffer (25 m*M* Tris base, 19 m*M* glycine, and 3.5 m*M* SDS)
- Invitrogen BenchMark Protein Ladder (Thermo Fisher, Cat. No. 10747012)
- Coomassie blue stain (3 m*M* Coomassie blue in 50% methanol, 10% acetic acid, and 40% Milli-Q water by volume)
- Destain (25% methanol, 10% acetic acid, and 65% Milli-Q water by volume)

3.1.3 Protocol

Day 1

1. Prepare overnight culture medium by adding 1 g of LB media to 50 mL of Milli-Q water.

   Note: We typically set up these cultures in duplicate in case one of the overnight cultures fails to grow.
2. Cover flasks with tin foil and autoclave the media for 20 min at 250°C.
3. Pipette 2 μL of the plasmid into a 50-μL aliquot of BL21 (DE3) cells in a 1.5-mL Eppendorf tube and flick the bottom of the tube to mix.
4. Incubate on ice for 5 min.
5. Heat shock the plasmid and cell mixture for 45 s at 45°C.
6. Incubate on ice for 2 min.
7. Add 500 μL of S.O.C. media and triturate the solution to ensure proper mixing.
8. Incubate at 37°C for 45 min.
9. Remove the overnight media from autoclave and allow to cool to room temperature (~30 min).
10. Pipette 50 μL of 100 mg/mL ampicillin into the overnight flasks and swirl gently to mix achieving an effective concentration of 100 μg/mL.
11. Pipette 150 μL of transformed cells into the overnight flasks.

    Note: LB agar plates containing 100 μg/mL of ampicillin could be made as well to save colonies for future overnight flasks.
12. Incubate the transformed cells in an incubating orbital shaker at 37°C with shaking at 230 rpm overnight for 12–16 h.
13. Prepare 1 L of induction media containing 0.2% glucose by adding 20 g of LB and 2 g of glucose to 1 L of Milli-Q water.
14. Cover flasks with tin foil and autoclave for 20 min.
15. Allow induction media to cool to room temperature.

    Note: Induction media can be left at room temperature overnight if tin foil is not removed and handled aseptically after autoclaving.

Day 2

1. Inoculate 1 L induction media with 15 mL of overnight culture.
2. Incubate induction flask at 37°C with shaking at 230 rpm.
3. Acquire 1 mL aliquot and obtain a starting UV–visible measurement at 600 nm.
4. Monitor protein induction over time by taking 1 mL aliquots and measuring the absorbance at 600 nm.

5. When the flask reaches an absorbance at 600 nm of approximately 0.3, supplement the media with 1 mL of 100 m*M* $ZnCl_2$ to reach an effective concentration of 1 m*M* Zn.
6. When the flask reaches an absorbance of 0.5–0.6 at 600 nm take 1 mL aliquot to reserve for SDS-PAGE analysis and induce with 1 mL of 1 *M* IPTG achieving an effective concentration of 0.1 m*M*.
7. Collect and preserve 1 mL aliquots of cells every hour to be used for SDS-PAGE analysis.
8. At 3 h postinduction stop the protein expression and centrifuge induction media at 7800 $\times g$ for 20 min at 4°C.

   Note: It is recommended to perform inductions in batches of two or four to allow simple balancing of the centrifuge. If only one batch is prepared, the centrifuge can be balanced using a bottle containing only water of identical weight.
9. Remove supernatant and store cell pellets at −20°C.

SDS-PAGE analysis

An example of SDS-PAGE analysis of CPSF30 is shown in Fig. 3B.

1. Pellet 1 mL aliquots by microcentrifugation and remove supernatant.
2. Resuspend cell pellets in 100 μL of Milli-Q water and add an equal volume of 2 × Laemmli sample buffer.
3. Incubate resuspended cell pellets on a heat block at 90°C for 3–5 min.
4. Run 5–15 μL of each sample on a 15% SDS-PAGE until the ladder reaches about 2 cm from the bottom of the gel.
5. Carefully remove the gel from the glass plates and transfer it to a suitable container for staining.
6. Add enough Coomassie blue stain to cover the gel and agitate for 30–60 min using an oscillating rocker.
7. Remove Coomassie blue stain and add enough destaining solution to cover the gel and return to the oscillating rocker.

   Note: Destaining solution may need to be exchanged with fresh solution two or three times.
8. Destain until background Coomassie stain dissipates and the protein bands can be resolved.
9. Wash remaining destaining solution from the container and the gel can be stored in Milli-Q water.

   Note: It is advised that the gel be imaged immediately for record keeping. Although the gel can be stored for up to 1 week in water, microbial growth and gel swelling can begin to obfuscate the bands. Be careful not to allow the water to evaporate, as the gel will dry, shrink, and become brittle. For storage of up to several weeks, gels can be stored in 5% acetic acid at 2–8°C. If the gel is being retained for further analysis or for a laboratory notebook it can be stored dry as previously described or through commercially available gel drying instruments (Smith, 1994).

3.2 Cell Lysis

Once CPSF30 has been successfully overexpressed as evidenced by an enhanced band on SDS-PAGE, the protein must be extracted and purified from the expression cells. Unless your protein has been engineered to be excreted from the cell during expression, the plasma membrane must be ruptured to release the protein of interest. There are many different techniques available to disrupt the cell membrane and extract the protein of interest. These include sonication, freeze–thaw, enzymatic digestion, chemical digestion, and French press. Our laboratory finds that sonication is an effective strategy to lyse cells containing overexpressed ZFs. Sonication uses ultrasonic sound waves to produce liquid shear and cavitation that subsequently disintegrates the cell wall of the bacteria. Cavitation is the formation, growth, and collapse of vapor bubbles created by high intensity sound waves (Feliu, Cubarsi, & Villaverde, 1998; Ho et al., 2006). The collapse of these vapor filled bubbles causes intense local shock up at levels of thousands of atmospheres disrupting the cells around them (Ho et al., 2006). One of the main drawbacks of sonication is the heat generated (Feliu et al., 1998). To overcome the potential risk of denaturing the protein at high heat, it is advised that the samples are kept on ice and sonication methods be optimized to short durations of pulses with resting periods to allow the sample to cool (Feliu et al., 1998).

3.2.1 Equipment

- Fisher Scientific Sonic Dismembrator Model 100 (Fisher, Model No. XL2000-350R)
- Avanti J-20 XPI centrifuge (Beckman Coulter, SKU# 8043-30-1171)
- JA-25.50 Fixed-Angle Rotor equipped with 50-mL centrifuge tubes (Beckman Coulter, Cat. No. 363055)

3.2.2 Buffers and Reagents

- Lysis buffer (20 m*M* Tris, 200 m*M* NaCl, adjust pH to 7.5 using NaOH or HCl)
- Pierce™ Protease Inhibitor Tablets, EDTA free (Thermo Scientific, Cat. No. 88266)

3.2.3 General Sonication Protocol

1. Remove the cell pellet from −20°C storage and place on ice.
2. Add one protease inhibitor tablet and resuspend the pellet in 25 mL of lysis buffer.
3. Split the resuspended pellet into two equal fractions in 50-mL centrifuge tubes.
4. Sonicate one fraction at 22.5 kHz on level 6 of 10 for 20 s on ice.
   Note: Sonicator tip should be submerged approximately half way into the solution and gently moved around without touching the walls of the centrifuge tube.
5. Let the solution rest on ice for 40 s.
6. Repeat steps 4 and 5.
7. Sonicate the fraction at 22.5 kHz on level 7 of 10 for 20 s.
8. Let the solution rest on ice for 40 s.
9. Repeat steps 7 and 8.
10. Repeat steps 4–9 with the other fraction.
11. Centrifuge both fractions for 20 min at 12,100 rpm (20,000 × *g*) at 4°C.
12. Combine both supernatants to form the load for amylose column purification.
    Note: Retain aliquots of both the pellet and lysate for SDS-PAGE analysis.

If you have an Fe–S cluster present, the supernatant will remain reddish-brown (Fig. 3C).

4. PROTEIN PURIFICATION

After conditions have been optimized to keep the protein of interest stable and soluble after expression, the protein must be isolated from the rest of the host cell's proteins. In this section, we discuss how to purify CPSF30 using amylose column chromatography and a polishing step using size exclusion chromatography. As a guide, we have included sample data from CPSF30 in Fig. 4.

4.1 Amylose Column Chromatography

CPSF30 includes an MBP tag, which allows us to utilize amylose chromatography to purify. This approach works well for all MBP ZFs our laboratory has investigated. In the cell, MBP mediates various maltodextrin metabolism pathways recognizing any alpha-(1 → 4)-D-glucose polysaccharide over eight repeating units (Pattenden & Thomas, 2008). Amylose

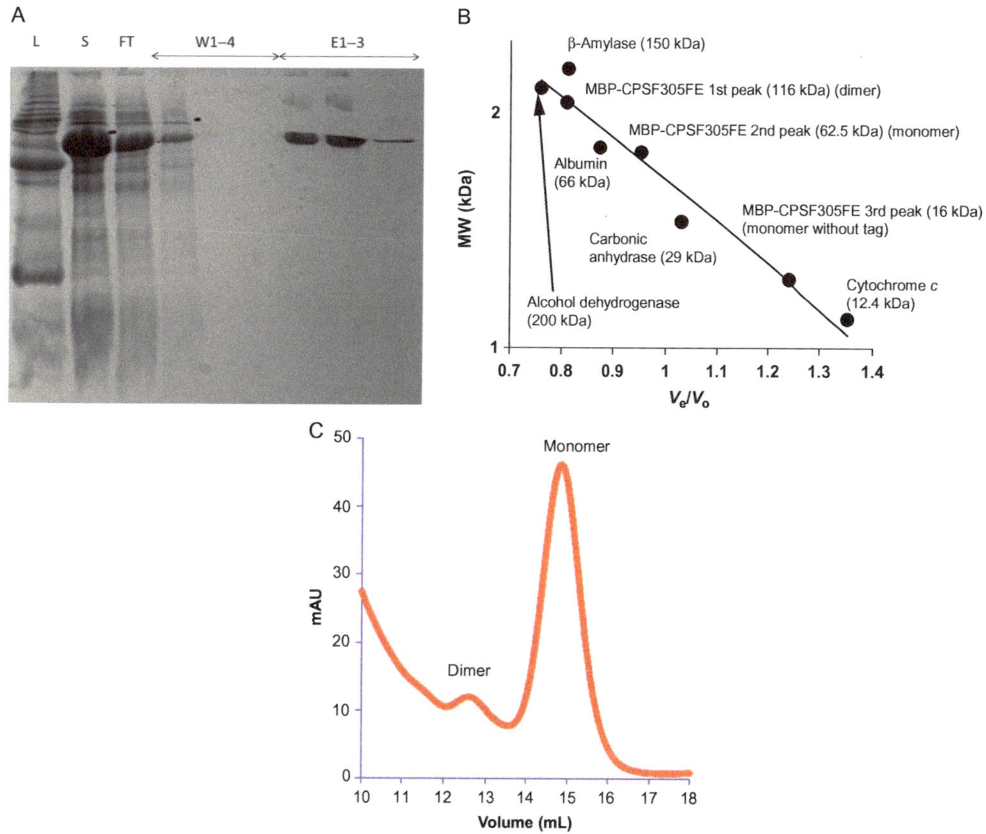

Fig. 4 The purification steps for CPSF30. (A) 15% SDS-PAGE of CPSF30 postamylose column chromatography. *E1–3* = elution 1–3, *FT* = flow-through, *L* = Invitrogen BenchMark Protein Ladder (Thermo Fisher), *S* = supernatant, *W1–4* = wash 1–4. (B) Calibration curve utilizing Sigma Aldrich gel filtration markers kit for protein molecular weights 12,000–200,000 Da used to determine the molecular weight of MBP-CPSF30. (C) UV–visible monitored chromatogram of MBP-CPSF30 during purification via Superdex 10/300. Both the dimer and monomer forms are shown.

affinity column chromatography takes advantage of this native-binding interaction by incorporating repeating maltose polymers with these alpha-$(1 \rightarrow 4)$ linkages covalently bonded to agarose beads in a stationary phase to work in a bind and elute purification procedure. This makes for a specific and effective purification method that can yield pure protein after just one purification step and we often obtain >95% purity of ZFs of interest. MBP's high activity in diverse environments increases its appeal as a purification method as it allows for a wide range of buffer conditions varying pH and ionic strength without sacrificing purification yields. The most common buffer conditions with high yields still remain around neutral pH at about 7.5–8.0 and salt concentrations in the range of 100–500 m*M* (Pattenden & Thomas, 2008). At high enough protein concentrations, MBP can affect pH as it has an acidic isoelectric point so higher buffering capacity is ideal, and it is generally recommended to have concentrations of at least 20 m*M* (Pattenden & Thomas, 2008). Even though purification of MBP using amylose column chromatography is highly robust, it does have some caveats. Nonionic detergents like Triton X-100, polysorbate 20, and other additives that inhibit hydrophobic interactions should be avoided during the purification process as they can reduce column loading capacity and result in lower purification yields (Pattenden & Thomas, 2008). Additionally, carbon sources in the induction media other than glucose should be avoided as they can upregulate expression of maltose scavenging proteins when maltodextrin concentrations are depleted. These scavenging proteins can make their way into the sonicated lysate and can bind, modify, or release maltose from the stationary phase of the column-decreasing loading capacity and allowing MBP loss in the flow-through (Pattenden & Thomas, 2008). To ensure glucose is the primary carbon source in our media, we supplement with an additional 0.2% glucose when expressing CPSF30 (Shimberg et al., 2016).

4.1.1 Equipment

- Glass Econo-Column® Columns 2.5 cm × 20 cm (Bio-Rad, Cat. No. 7374252)
- 50-mL conical tubes
- 3.5 kDa Dialysis tubing (VWR, Cat. No. 28170-166)
- Amicon Ultra-15 Centrifugal Filter Unit with Ultracel-30 membrane (Millipore, Cat. No. UFC903008)
- SDS-PAGE instrumentation as previously described

4.1.2 Buffers and Reagents

- Wash buffer (20 m*M* Tris, 200 m*M* NaCl, pH adjusted to 7.5)
- Elution buffer (20 m*M* Tris, 200 m*M* NaCl, 10 m*M* maltose, pH adjusted to 7.5)
- Amylose resin (New England Biolabs, Cat. No. E8021S)

  Note: The amylose resin has a binding capacity of 6–8 mg/mL of protein to column bed volume. 15 mL of amylose resin can be used for up to 100 mg of crude MBP-fusion protein at a time
- Dialysis buffer (20 m*M* Tris, 50 m*M* NaCl, pH adjusted to 7.0)
- Storage buffer (20% ethanol)
- SDS-PAGE buffers and reagents as previously described

  Note: Wash, elution, and dialysis buffers can vary depending on the protein of interest's stability in various pH, ionic strength, and buffering capacity conditions. These properties are usually determined empirically.

4.1.3 General Amylose Affinity Column Chromatography Purification Protocol

Note: Our purification procedure is performed at room temperature, but if protein stability is a concern, all steps may be performed in a cold room at 4°C.

1. Slurry the resin in 20% ethanol and add to column.
2. Allow the resin to form a gravity settled bed at the bottom of the column.
3. Equilibrate the column flowing through excess wash buffer (at least five column volumes).
4. Load cell lysis supernatant onto amylose column and seal the column.
5. Place the loaded column on an orbital rocker and rock for 15–20 min.
6. Allow the resin to form a gravity packed bed and open the bottom valve of the column capturing the flow-through until the supernatant is just above the resin bed.

   Note: Be careful to never let the column run dry at any time during the purification process.
7. Add 45 mL (three CVs) of wash buffer to the column and capture it in a new 50-mL conical tube. Place captured fractions on ice until purification procedure is completed.
8. Repeat step 7 three more times capturing each wash in a new conical tube.
9. Add 15 mL (one CV) of elution buffer and capture the flow-through.
10. Repeat step 9 two more times.
11. Retain 50 μL aliquots of all fractions for SDS-PAGE analysis.

12. Once purification has been verified by SDS-PAGE, combine all elution fractions containing pure protein and pipette protein solution into dialysis tubes for buffer exchange.

    Note: Dialysis tubing should be incubated in dialysis buffer for at least 5 min before filling with protein solution.
13. Place tubing containing protein in 4 L of dialysis buffer to exchange overnight at 2–8°C.
14. Concentrate protein samples to approximately 100–250 μ*M* using Amicon spin filters by centrifuging at 4000 × *g* for 25 min at 4°C.
15. Repeat concentration (step 14) by removing filtrate and adding elution fractions until all fractions are concentrated to approximately 1.5–5.0 mL.
16. Fill retentate side of spin filter with fresh dialysis buffer and centrifuge at 4000 × *g* for 25 min at 4°C.
17. Repeat step 16 two more times.
18. If pure, aliquot samples for storage at −80°C or store/prepare samples for size exclusion chromatography.
19. Add storage buffer to column and allow approximately three column volumes to flow-through. Store column in excess storage buffer at 4°C. The column can be regenerated by washing as follows: Milli-Q water (three CVs), 0.1% SDS (three CVs), Milli-Q water (one CV), and wash buffer (five CVs). The column is then ready to be loaded again.

    Note: Amylose resin can be reused up to five times.

A typical SDS-PAGE of purified CPSF30 (>95% purity) is shown in Fig. 4A.

4.2 Additional Polishing Step via Size Exclusion Chromatography

CPSF30 is typically obtained at >95% purity after one amylose column. If desired, CPSF30 can be further purified using size exclusion chromatography (also known as gel filtration chromatography). This method separates impurities based upon size. A standard curve for molecular weights should be constructed using a gel filtration markers kit for protein molecular weights 12,000–200,000 Da (Sigma Aldrich) on a Superdex 200 10/300 gel filtration column (GE) (Gel Filtration Molecular Weight Markers Kit for Molecular Weights 12,000–200,000 Da Technical Bulletin, n.d.). This kit uses six standard proteins to construct a calibration curve that can be used to determine molecular weights of proteins in a mixed sample. These standards are cytochrome *c* (12.4 kDa), carbonic anhydrase (29 kDa), bovine serum albumin (66 kDa), alcohol dehydrogenase (150 kDa), β-amylase

(200 kDa), and blue dextran (2000 kDa). Blue dextran is not used in the calibration curve, but is used to determine the void volume of the column as it does not interact with the bead's pores and elutes at the solvent front (Gel Filtration Molecular Weight Markers Kit for Molecular Weights 12,000–200,000 Da Technical Bulletin, n.d.). This allows normalization of the elution volume measurements of the other five calibration standards. The ratio of elution volume (V_e) to void volume (V_o) (V_e/V_o) can be determined for each standard allowing the calibration curve to estimate the molecular weight of unknown proteins (such as MBP-CPSF30 vs contaminants) (Gel Filtration Molecular Weight Markers Kit for Molecular Weights 12,000–200,000 Da Technical Bulletin). The calibration needs to be performed each time the size exclusion column is used as retention rates can shift slightly when the column is moved from one location to another changing packing factors or internal bead densities. Fig. 4B gives an example of CPSF30 that has been purified via size exclusion chromatography and referenced to the calibration curve. Prior to loading the column, it is important to centrifuge your sample on a table top centrifuge (Denville) at 14,000 rpm for 10–15 min to remove any aggregates in the protein sample that could damage the FPLC system or the column. The sample can then be loaded onto the Superdex 200 10/300 column (we use an AKTA Pure FPLC (GE) with Unicorn software) (AKTA pure: User Manual, 2012). The flow rate should be experimentally optimized to ensure proper separation of your protein of interest from other impurities; we use a flow rate of 0.4 mL/min (31 cm/h) to purify CPSF30. Fig. 4C shows the chromatogram for purification of CPSF30.

5. METHODS TO CHARACTERIZE ZF PROTEINS WITH Fe–S CLUSTERS

To characterize ZF proteins isolated with an Fe–S cluster, three methods are typically employed: optical spectroscopy to visualize thiolate-to-Fe(III) charge-transfer, ICP-MS (inductively coupled plasma mass spectrometry) to quantify iron and zinc, and XAS (X-ray absorption spectroscopy) spectroscopy to determine metal oxidation states, geometry at the metal center, and ligand identity. We perform spectroscopy and ICP-MS at Maryland, and our protocols are described later. XAS spectroscopy is performed by our collaborator, Tim Stemmler (Wayne State University), and we refer you to our published collaborative work (Shimberg et al., 2016).

5.1 Protein Characterization Using UV–vis

UV–visible spectroscopy can be utilized to determine the concentration, purity, and presence of any iron–sulfur cofactors in newly identified "zinc finger" proteins (Adrover, Howes, Iannuzzi, Smulevich, & Pastore, 2015; Mapolelo, Zhang, Naik, Huynh, & Johnson, 2012; Miller, McLachlan, & Klug, 1985). In a typical experiment, an optical spectrum from 200 to 800 nm is obtained. Absorbance peaks at 220 and 280 nm, which correspond to backbone amide residues and aromatic side chains (tyrosine, tryptophan, and phenylalanine) respectively are then identified. The protein concentration can be determined utilizing Beer's law, $A_{280} = \varepsilon bc$, where A is the absorbance of the sample in absorbance units, ε is the molar absorptivity in L mol^{-1} cm^{-1} (often measured at 280 nm), b is the path length of the sample in cm, and c is the concentration of the sample in mol L^{-1}. In some cases, zinc finger protein copurify with nucleic acids, and these are often detected by the presence of an additional absorbance peak at 260 nm from purine and pyrimidine residues (Miller et al., 1985). The presence of an iron–sulfur cluster in a "zinc finger" can also be detected via UV–visible spectroscopy as charge transfer bands between 220 and 600 nm (Adrover et al., 2015; Dailey, Finnegan, & Johnson, 1994; Mapolelo et al., 2012). Fig. 5 shows the UV–visible spectra of CPSF30 loaded with both zinc and the Fe–S site as well as the Fe–S only species. Absorbance bands at 340, 420, 456, and 583 nm are observed for CPF30.

5.1.1 Equipment

- UV–vis
- Quartz cuvette

5.1.2 Buffers and Reagents

- 20 mM Tris, pH 7.0, 50 mM NaCl

5.1.3 General UV–vis Characterization Protocol for Proteins Containing Zinc Finger and Iron–Sulfur Clusters

Note: Our method is performed under aerobic conditions, but if iron oxidation is a concern, UV–vis characterization should be done in the absence of an oxygen atmosphere as previously described (Adrover et al., 2015; Mapolelo et al., 2012).

1. Add approximately 500 μL of dialysis buffer to the cuvette and insert into spectrophotometer

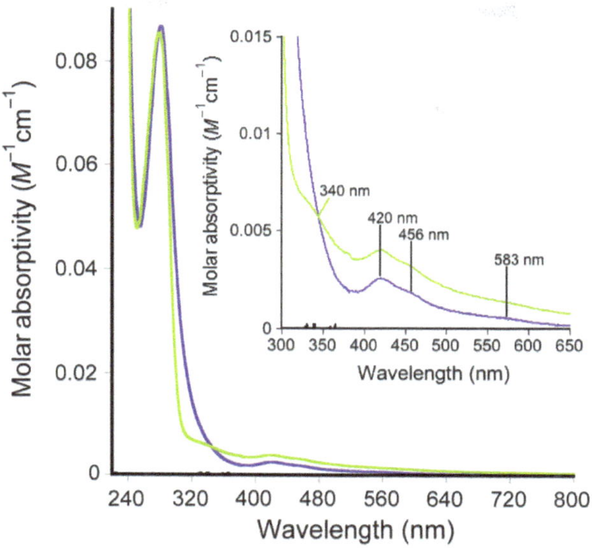

Fig. 5 Full UV–visible spectrum of CPSF30 protein in 20 m*M* Tris, 100 m*M* NaCl, pH 8, after purification. (*Inset*) Close up of 300–650 nm range denoting the Fe–S cluster charge transfer peaks. The *green band* shows the spectrum of isolated CPSF30 with both Fe and Zn bound; the *blue band* is the spectrum observed upon Zn chelation (Fe-only spectrum).

2. Blank the cuvette
3. Add 150 μL of pure protein solution and mix
4. Run a full UV–vis scan between 200 and 800 nm
5. Analyze spectrum for peaks around 280, 260, and between 300 and 600 nm.

   Note: If spectrum reaches maximum absorbance, a dilution may be necessary and the experiment will have to be repeated.
6. Calculate protein concentration using Beer's law ($A = \varepsilon bc$) and solve for concentration

   Note: If ε has not been determined empirically for your protein, a theoretical estimate can be obtained from http://web.expasy.org/protparam/. This estimate can vary slightly from actual molar absorptivity values in solution as protein folding, pH, and ionic strength can affect some aromatic residue's ability to absorb light (Simonian & Smith, 2006).
7. Look for peaks indicative of the presence of a $[2Fe—2S]^{2+}$ cluster or $[4Fe—4S]^{2+}$ cluster between 300 and 600 nm.

5.2 ICP-MS

ICP-MS is utilized to determine the metal content of proteins. ICP-MS is extremely sensitive—concentrations as low as one part per quadrillion can be theoretically measured (ppq) (Profrock & Prange, 2012). The instrument ionizes the sample with inductively coupled plasma and then the ionized sample is passed through a mass spectrometer to separate and quantify the metal/nonmetal ions (Zoorob, McKiernan, & Caruso, 1998). ICP-MS works with better speed, precision, and sensitivity to determine metal ions compared to inductively coupled plasma atomic emission spectroscopy (ICP-AES) which can also be used to determine metal content of proteins (Rommers & Boumans, 1996). The plasma used for ICP-MS is energized by heating argon gas with an electromagnetic coil, which generates electrically conductive argon ions that can interact with an aerosol sample to ionize elements for detection. The sample is converted into an aerosol by passing through a nebulizer to create consistent droplet sizes to interact with the charged argon gas. This is important to remove any large droplets from the sample and increase reproducibility of detection. An accurate calibration curve is integral to the instrument's ability to quantify the analytes of interest and should be conducted for each batch of samples. It is also important to utilize internal standards within the samples to monitor matrix effects. Matrix effects occur when a component of the sample, other than the analyte of interest, skews the reported values of detection, and enhance or suppresses the signal. Matrix effects can be monitored by quantifying a known element in a neat sample vs the same element spiked concentration in your matrix. Another caveat to ICP-MS when working with ^{56}Fe determination is interferences by ^{40}Ar^{16}O^{+} and ^{40}Ca^{16}O^{+} which have a molecular weight of 56 like iron and similar ionization states (Segura, Madrid, & Cámara, 2003). One way to overcome this interference is to use a helium (He) collision chamber before the detector. The He collision mode differentiates monoatomic elements vs polyatomic species by kinetic energy discrimination (McCurdy, Woods, & Potter, 2006). Since diatomic species have a larger cross-sectional area, they are prone to more collisions and move slower through the collision cell. This allows the monoatomic elements (^{56}Fe) with the same mass to pass to the detector with less interference and allows analysis down to concentrations as low as parts per billion (ppb). We routinely use a helium collision chamber for our analysis. Secondarily, ^{57}Fe can be analyzed with lower interference, but due to its lower abundance, sensitivity can be an issue.

ICP-MS protocol

1. Prepare 1 μ*M* CPSF30 in 5 mL 2% trace metal nitric acid (Fisher). 150 μL internal standard (100 μg/mL Bi, Ge, In, Li, Lu, Rh, Sc, and Tb; Agilent Technologies) is added to samples to ensure accuracy.
2. Zinc and iron calibration standards ranging from 0 to 500 ppb Zn/Fe are created using iron and zinc atomic absorption standard dilutions (Fluka Analytical).
3. Zinc and iron levels are detected on an Agilent 7700 × ICP-MS using an octopole reaction system in HE mode, an rf power of 1550 W, an argon carrier gas flow of 1.0 L/min, argon make-up gas flow of 0.1 L/min, helium gas flow of 4.5 mL/min, octopole rf of 160 V, QP bias of −15 V, and OctP bias of −18 V.
4. Data analysis was performed using the Agilent 7700 × ICP-MS instrument provided Mass Hunter software.

   Note: In a molar ratio, we usually see 0.5–1.7 equivalents of iron and 3.1–3.7 equivalents of zinc to CPSF30.

5.3 XAS Sample Preparation

To determine the geometry at the metal site, the ligands involved in coordination and metal oxidation state of ZF/Fe–S hybrid proteins, XAS is a common approach (Shimberg et al., 2016). Below we describe our protocol for sample preparation.

Protocol

1. CPSF30 samples are prepared in 20 m*M* Tris, 50 m*M* NaCl, pH 7 with 30% glycerol. Metal concentrations of CPSF30 are confirmed via ICP-MS analysis with metal concentrations greater than 0.5 m*M* of either Zn or Fe.
2. Samples are loaded into lucite XAS cells, prewrapped with kapton tape, flash-frozen in liquid nitrogen, and stored in liquid nitrogen until data collection.

6. ACTIVITY ASSAYS TO ASSESS DNA OR RNA BINDING FOR ZF/Fe–S HYBRID PROTEINS

Once isolated, the function of a ZF/Fe–S hybrid protein must be assessed. ZFs typically bind to other macromolecules (e.g., DNA or RNA) to promote transcription or translation (Brown, 2005). In addition, in recent years, Fe–S cofactored proteins have been found to also participate in DNA or RNA binding (Boal et al., 2005; Brown, 2005). Two common strategies to assess DNA or RNA binding are electrophoretic mobility shift assays (EMSA) and fluorescence anisotropy (FA). The application of these techniques for CPSF30/RNA binding is described later, and an example of these data are shown in Fig. 6.

6.1 Evaluation of CPSF30/RNA Binding via EMSA

In the EMSA assay, the RNA (or DNA) target is 5′ end labeled utilizing phosphate containing radioactively labeled ^{32}P, to allow determination of an interaction between protein and substrate (Fialcowitz-White et al., 2007; Hellman & Fried, 2007). In a typical experiment, the ^{32}P-RNA is incubated with increasing concentrations of protein, and the position of ^{32}P-RNA on the gel is shifted if binding occurs.

EMSA assays for CPSF30/RNA utilized α-synuclein pre-RNA; however, this assay can be adapted to examine any RNA sequence. An example of an EMSA assay for CPSF30 is shown in Fig. 6A.

Protocol

Radioactively labeling RNA

1. Quantitate RNA of interest as previously described (Section 2.1 step 40) and dilute to 5 pmol/μL using RNase-free water.
2. Add and mix the following reagents:

Reagent	**Volume (μL)**
5 μ*M* RNA	2
RNase-free water	4
10 × T4 PNK buffer (NEB Cat. No. B0201S)	1
[Υ^{32}P] ATP (6000 Ci/mmol stock)	2

3. Mix and then add 1 μL of T4 polynucleotide kinase to reach a final volume of 10 μL.
4. Incubate at 37°C for 10 min.
5. Add 50 μL of RNase-free water and 2 μL of 0.5 *M* EDTA.
6. Heat inactivate the reaction at 70°C for 15 min.
7. Extract once with 35 μL phenol: 35 μL of $CHCl_3$:IAA (chloroform: isoamyl alcohol) and mix by vortex.

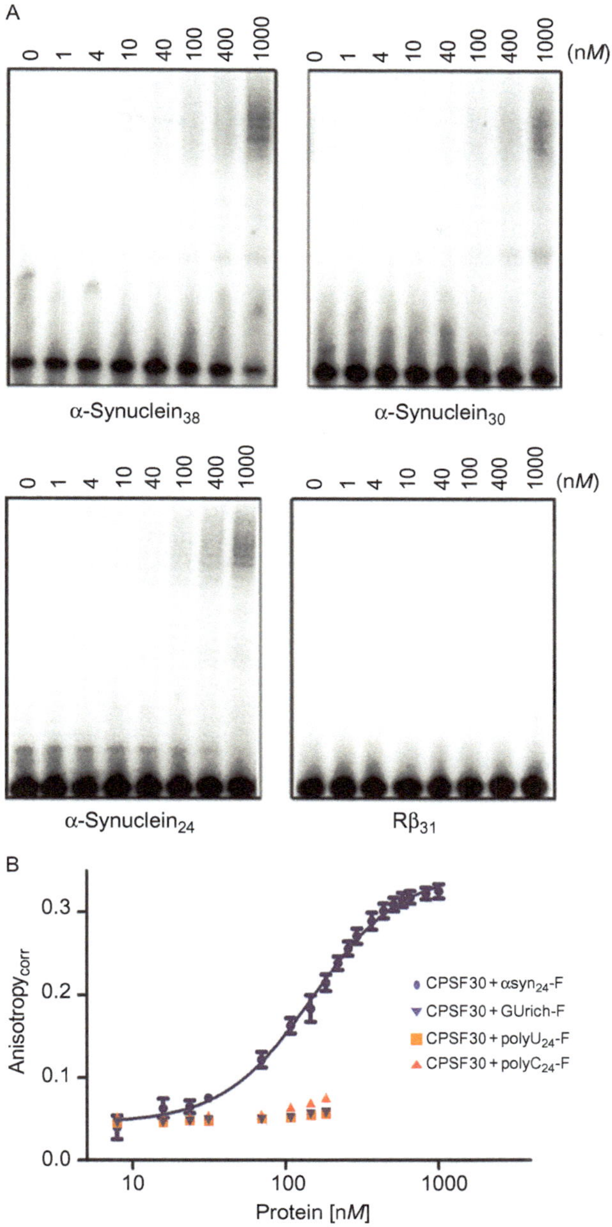

Fig. 6 Characterization of CPSF30/RNA binding. (A) EMSA data for CPSF30 with RNA (α-synuclein RNA sequence) at various sequence lengths, compared to a negative control with $R\beta_{31}$. (B) FA monitored titration of CPSF30 with α-synuclein RNA vs mutant RNA sequences. Binding is only observed with α-synuclein RNA, and these data are fit to a cooperative binding model with a $[P]_{1/2} = 143.8 \pm 3.8$ n*M* and a hill coefficient of 1.58 ± 0.07.

8. Remove unincorporated nucleotides by passage of the top aqueous layer through a G-25 spin column (Roche, Cat. No. 11273990001).
9. Quantify incorporation by liquid scintillation counting. Add 1 μL of RNA solution to 10 mL of scintillation fluid (Ecoscint H by National Diagnostics, Cat. No. LS-275) in a scintillation vial. This should yield between 2 and 3 K cpm/fmol.

   Note: The final concentration of RNA probes should be about 80 fmol/μL assuming 80% recovery in 100 μL of elution volume.

EMSA

Note: All water used must be RNase free and all reagents must be EDTA free.

1. Dilute the RNA probe to 2 n*M* using a 10-m*M* Tris, pH 8.0 buffer.
2. Prepare the following reaction mixture:

Reagent	**Volume (μL)**	**Final Concentration**
4 × LS-binding mix (400 m*M* Tris/800 m*M* KCl)	125	50 m*M* Tris pH 8.0 and 100 m*M* KCl
5 mg/mL poly-rC (Midland Certified Reagent Company, Cat. No. P-3002)	60	0.3 mg/mL
10 mg/mL acetylated BSA (Promega, Cat. No. R396D)	10	0.1 mg/mL
1.5 *M* DTT	1.6	2 m*M*
50% Glycerol	200	10% (v/v)
RNase-free water	590	N/A
10 m*M* $ZnCl_2$	10	100 μ*M*

3. Heat the RNA dilutions at 70°C for 5 min and then rest on ice for 3 min.
4. Incubate the ^{32}P-labeled RNA with increasing concentrations of CPSF30 (between 0 and 1000 n*M*) in the reaction mixture on ice for 15 min.

   Note: The final concentration of RNA probe in each assay should equal 0.2 n*M*.
5. Pipette the RNA or CPSF30/RNA mixture into lanes of a 5% (vol/vol) native polyacrylamide gel containing 10% (vol/vol) glycerol, with a 0.5 × (44.5 m*M*) Trisborate buffer (pH 8.0).

Note: The gel should be prerun at 150 V for 20 min before adding the RNA/CPSF30 mixture. During loading, run the gel at 80 V.

6. Run the gels at 200 V for approximately 3 h at 4°C.
7. Vacuum dry the gel for approximately 2 h (Bio-Rad model 583).
8. Expose the gel overnight using a phosphor screen.
9. Phosphorimage the gel next day (e.g., GE Typhoon FLA9500).

6.2 Quantification of ZF/RNA Binding via Fluorescence Anisotropy

Although EMSA can be utilized to quantify CPSF30/RNA binding (or other ZF/RNA binding), our laboratory uses FA. FA is a solution-based technique that measures the differences in polarization (anisotropy) that occur when a fluorescently labeled macromolecule binds to a non-fluorescently labeled macromolecule forming a complex (Lakowicz, 1999; Wilson, 2005). The FA measure is indirectly proportional to tumbling rate of the macromolecule, and when a complex forms, the tumbling rate decreases and an increase in FA is observed. A number of fluorophores can also be utilized. For the FA experiment with ZFs that bind RNA, we prefer to conjugate fluorescein to the 3′ end of our RNA, although 5′ end labeling can also be utilized. We excite at 495 n*M* and observe a maximum emission at 517 nm. An example of FA for CPSF30 with fluorescein-labeled RNA is shown in Fig. 6B and we describe our protocol below. The length of the target RNA sequence utilized should be optimized such that all regions of RNA involved in binding are present. For CPSF30, we find that RNA molecules between 24 and 38 nucleotides long are suitable for CPF30/RNA binding, and that our sequences do not appear to adopt any secondary structure (as measured by thermal denaturation). It should be noted that since FA readings are calculated from changes in apparent molecular size during binding, very large RNA sequences can attenuate measurements as the overall change in molecular size is lessened. FA is very versatile and can be adapted to any protein/DNA, protein/RNA, or protein/protein interaction of interest.

Protocol

1. *Fluorimeter schematics*. Experiments are conducted in the L format on an ISS PC-1 spectrofluorometer with polarizers. A full excitation/emission scan of F-labeled RNA to determine optimal excitation/emission wavelengths should be performed. We recommend an excitation wavelength/band pass of 495 nm/2 nm and an emission wavelength/bandpass of 517 nm/1 nm for fluorescein-labeled RNA.

2. Add 5 nM of fluorescently labeled RNA in 20 mM Tris, pH 7.0, 50 mM sodium chloride with 0.2 mg/mL BSA, and 0.4 mg/mL poly-rC to reach a final volume of 500 μL in a Spectrosil far UV quartz window fluorescence cuvette (Starna Cells). BSA serves to prevent protein adherence to the quartz cuvette walls and poly-rC is an internal negative control for nonspecific protein/RNA interactions.
3. Titrate the protein with the RNA and observe anisotropy changes until saturation. In a typical experiment, upon addition of the protein, the complex is incubated for 5 min. The protein is titrated in aliquots increasing concentrations beginning with 10 nM.

   Note: During the experiment, dilution of the sample should be taken into account and compensated for in a corrected anisotropy value that is calculated as follows:

$$\text{Corrected anisotropy} = \text{anisotropy} \times \frac{\text{starting reaction volume}}{\text{current volume}}$$

4. If some quenching of fluorescence is observed during the titration, the anisotropy (r) should be corrected for the change in quantum yield (Q, $f_{\text{free}}/f_{\text{bound}}$, protein-dependent change in fluorescence) using the following equation:

$$r_c = \frac{r_0(r_{\text{bound}} - r) + (r_{bound}Q(r - r_0))}{(r_{\text{bound}} - r + Q(r - r_0))}$$

   where r_c is the corrected anisotropy, r_0 is the anisotropy of the free fluorescein-labeled oligonucleotide, and r_{bound} is the anisotropy of the RNA–protein complex at saturation. Plot r_c against the concentration of protein.
5. Fit the data to an appropriate binding model. Our data were best fit to a cooperative binding model using nonlinear regression (GraphPad Prism 5):

$$n\text{P} + R \rightleftharpoons \text{P}_n\text{R}$$

$$\text{K} = \frac{[\text{P}_n\text{R}]}{[\text{P}]^n[\text{R}]}$$

$$r_{Tc} = r_0 + (r_{bound} - r_0)\left[\frac{\left(\frac{[P]}{[P]_{1/2}}\right)^h}{\left(1+\left(\frac{[P]}{[P]_{1/2}}\right)^h\right)}\right]$$

r_{Tc} is the total, corrected anisotropy, r_0 is the anisotropy of the free fluorescein-labeled oligonucleotide, r_{bound} is the anisotropy of the RNA–protein complex at saturation, [P] is the concentration of protein, $[P]_{1/2}$ is the concentration of protein at half-maximal saturation, and h is the Hill coefficient. (*Note*: it is always a best practice to fit data to several models, beginning with the simplest 1:1 binding).

7. CONCLUSIONS

A large number of proteins are annotated as ZF proteins in genome databases, but only a handful have been characterized experimentally. In Section 5, we described a general approach one can take to isolate a novel ZF protein, and evaluate which metal ions are present. We also present activity assays, EMSA and FA, that can be applied to identify whether the ZF binds to RNA or DNA.

ACKNOWLEDGMENTS

S.L.J.M. thanks the NSF for support of this work (CHE1708732, CHE1306208); G.D.S. has been partially supported by NIH training Grant T32GM066706-13. We thank Dr. Tim Stemmler (Wayne State) and Dr. Gerald Wilson (Maryland School of Medicine), for their fantastic collaborations.

REFERENCES

Adrover, M., Howes, B. D., Iannuzzi, C., Smulevich, G., & Pastore, A. (2015). Anatomy of an iron-sulfur cluster scaffold protein: Understanding the determinants of [2Fe-2S] cluster stability on IscU. *Biochimica et Biophysica Acta*, *1853*(6), 1448–1456. https://doi.org/10.1016/j.bbamcr.2014.10.023.

AKTA pure: User Manual. (2012). *General Electric*.

Andreini, C., Banci, L., Bertini, I., & Rosato, A. (2006). Zinc through the three domains of life. *Journal of Proteome Research*, *5*(11), 3173–3178. https://doi.org/10.1021/pr0603699.

Barbas, C. F., 3rd, Burton, D. R., Scott, J. K., & Silverman, G. J. (2007). Quantitation of DNA and RNA. *CSH Protocols* pdb ip47https://doi.org/10.1101/pdb.ip47.

Baxter, E. L., Jennings, P. A., & Onuchic, J. N. (2012). Strand swapping regulates the iron-sulfur cluster in the diabetes drug target mitoNEET. *Proceedings of the National Academy of Sciences of the United States of America*, *109*(6), 1955–1960. https://doi.org/10.1073/pnas.1116369109.

Beaudoing, E., Freir, S., Wyatt, J. R., Claverie, J. M., & Gautheret, D. (2000). Patterns of variant polyadnylation signal usage in human genes. *Genome Research*, *10*(7), 1001–1010.

Berg, J. M. (1986). Potential metal-binding domains in nucleic acid binding proteins. *Science*, *232*(4749), 485–487.

Berg, J. M. (1988). Proposed structure for the zinc-binding domains from transcription factor IIIA and related proteins. *Proceedings of the National Academy of Sciences of the United States of America*, *85*(1), 99–102.

Bertani, G. (1951). A method for detection of mutations, using streptomycin dependence in *Escherichia coli*. *Genetics*, *36*(6), 598–611.

Bertini, I., Decaria, L., & Rosato, A. (2010). The annotation of full zinc proteomes. *Journal of Biological Inorganic Chemistry*, *15*(7), 1071–1078. https://doi.org/10.1007/s00775-010-0666-6.

Biolabs, N. E. (2015). *PCR protocol for* Taq *DNA polymerase with standard Taq buffer (M0273)*.

Blanc, B., Gerez, C., & Ollagnier de Choudens, S. (2015). Assembly of Fe/S proteins in bacterial systems: Biochemistry of the bacterial ISC system. *Biochimica et Biophysica Acta*, *1853*(6), 1436–1447. https://doi.org/10.1016/j.bbamcr.2014.12.009.

Boal, A. K., Yavin, E., Lukianova, O. A., O'Shea, V. L., David, S. S., & Barton, J. K. (2005). DNA-bound redox activity of DNA repair glycosylases containing [4Fe-4S] clusters. *Biochemistry*, *44*(23), 8397–8407. https://doi.org/10.1021/bi047494n.

Boon, E. M., Livingston, A. L., Chmiel, N. H., David, S. S., & Barton, J. K. (2003). DNA-mediated charge transport for DNA repair. *Proceedings of the National Academy of Sciences of the United States of America*, *100*(22), 12543–12547. https://doi.org/10.1073/pnas.2035257100.

Brown, R. S. (2005). Zinc finger proteins: Getting a grip on RNA. *Current Opinion in Structural Biology*, *15*(1), 94–98. https://doi.org/10.1016/j.sbi.2005.01.006.

Chmiel, N. H., Golinelli, M. P., Francis, A. W., & David, S. S. (2001). Efficient recognition of substrates and substrate analogs by the adenine glycosylase MutY requires the C-terminal domain. *Nucleic Acids Research*, *29*(2), 553–564.

Colca, J. R., McDonald, W. G., Waldon, D. J., Leone, J. W., Lull, J. M., Bannow, C. A., et al. (2004). Identification of a novel mitochondrial protein ("mitoNEET") cross-linked specifically by a thiazolidinedione photoprobe. *American Journal of Physiology Endocrinology and Mebatolism*, *286*, E252–E260.

Conlan, A. R., Axelrod, H. L., Cohen, A. E., Abresch, E. C., Zuris, J., Yee, D., et al. (2009). Crystal structure of Miner1: The redox-active 2Fe-2S protein causative in Wolfram Syndrome 2. *Journal of Molecular Biology*, *392*(1), 143–153. https://doi.org/10.1016/j.jmb.2009.06.079.

Dailey, H. A., Finnegan, M. G., & Johnson, M. K. (1994). Human ferrochelatase is an iron-sulfur protein. *Biochemistry*, *33*(2), 403–407.

Decaria, L., Bertini, I., & Williams, R. J. (2010). Zinc proteomes, phylogenetics and evolution. *Metallomics*, *2*(10), 706–709. https://doi.org/10.1039/c0mt00024h.

Engstrom, L. M., Partington, O. A., & David, S. S. (2012). An iron-sulfur cluster loop motif in the Archaeoglobus fulgidus uracil-DNA glycosylase mediates efficient uracil recognition and removal. *Biochemistry*, *51*(25), 5187–5197. https://doi.org/10.1021/bi3000462.

Feliu, J. X., Cubarsi, R., & Villaverde, A. (1998). Optimized release of recombinant proteins by ultrasonication of E. coli cells. *Biotechnology and Bioengineering*, *58*(5), 536–540.

Fialcowitz-White, E. J., Brewer, B. Y., Ballin, J. D., Willis, C. D., Toth, E. A., & Wilson, G. M. (2007). Specific protein domains mediate cooperative assembly of HuR oligomers on AU-rich mRNA-destabilizing sequences. *The Journal of Biological Chemistry*, *282*(29), 20948–20959. https://doi.org/10.1074/jbc.M701751200.

Gel Filtration Molecular Weight Markers Kit for Molecular Weights 12,000–200,000 Da Technical Bulletin. n.d. Sigma-Aldrich.
Hellman, L. M., & Fried, M. G. (2007). Electrophoretic mobility shift assay (EMSA) for detecting protein-nucleic acid interactions. *Nature Protocols*, *2*(8), 1849–1861. https://doi.org/10.1038/nprot.2007.249.
Ho, C. W., Chew, T. K., Ling, T. C., Kamaruddin, S., Tan, W. S., & Tey, B. T. (2006). Efficient mechanical cell disruption of Escherichia coli by an ultrasonicator and recovery of intracellular hepatitis B core antigen. *Process Biochemistry*, *41*(8), 1829–1834. https://doi.org/10.1016/j.procbio.2006.03.043.
Jaganaman, S., Pinto, A., Tarasev, M., & Ballou, D. P. (2007). High levels of expression of the iron-sulfur proteins phthalate dioxygenase and phthalatedioxygenase reductase in Escherichia coli. *Protein Expression and Purification*, *52*(2), 273–279.
Jantz, D., Amann, B. T., Gatto, G. J., Jr., & Berg, J. M. (2004). The design of functional DNA-binding proteins based on zinc finger domains. *Chemical Reviews*, *104*(2), 789–799. https://doi.org/10.1021/cr020603o.
Kapust, R. B., & Waugh, D. S. (1999). Escherichia coli maltose-binding protein is uncommonly effective at promoting the solubility of polypeptides to which it is fused. *Protein Science*, *8*(8), 1668–1674. https://doi.org/10.1110/ps.8.8.1668.
Krishna, S. S., Majumdar, I., & Grishin, N. V. (2003). Structural classification of zinc fingers: Survey and summary. *Nucleic Acids Research*, *31*(2), 532–550.
Kurz, M., Cowieson, N. P., Robin, G., Hume, D. A., Martin, J. L., Kobe, B., et al. (2006). Incorporating a TEV cleavage site reduces the solubility of nine recombinant mouse proteins. *Protein Expression and Purification*, *50*(1), 68–73. https://doi.org/10.1016/j.pep.2006.05.006.
Laity, J. H., Lee, B. M., & Wright, P. E. (2001). Zinc finger proteins: New insights into structural and functional diversity. *Current Opinion in Structural Biology*, *11*(1), 39–46.
Lakowicz, J. R. (1999). Fluorescence anisotropy. In *Principles of fluorescence spectroscopy* (pp. 291–319). United States: Springer.
Lanz, N. D., Grove, T. L., Gogonea, C. B., Lee, K. H., Krebs, C., & Booker, S. J. (2012). RlmN and AtsB as models for the overproduction and characterization of radical SAM proteins. *Methods in Enzymology*, *516*, 125–152. https://doi.org/10.1016/B978-0-12-394291-3.00030-7.
Larentis, A. L., Nicolau, J. F., Esteves Gdos, S., Vareschini, D. T., de Almeida, F. V., dos Reis, M. G., et al. (2014). Evaluation of pre-induction temperature, cell growth at induction and IPTG concentration on the expression of a leptospiral protein in E. coli using shaking flasks and microbioreactor. *BMC Research Notes*, 7, 671. https://doi.org/10.1186/1756-0500-7-671.
Lee, B. M., De Guzman, R. N., Turner, B. G., Tjandra, N., & Summers, M. F. (1998). Dynamical behavior of the HIV-1 nucleocapsid protein. *Journal of Molecular Biology*, *279*(3), 633–649. https://doi.org/10.1006/jmbi.1998.1766.
Lee, S. J., & Michel, S. L. (2014). Structural metal sites in nonclassical zinc finger proteins involved in transcriptional and translational regulation. *Accounts of Chemical Research*, *47*(8), 2643–2650. https://doi.org/10.1021/ar500182d.
Lee, K.-H., Saleh, L., Anton, B. P., Madinger, C. L., Benner, J. S., Iwig, D. F., et al. (2009). Characterization of RimO, a new member of the methylthiotransferase subclass of the radical SAM superfamily. *Biochemistry*, *48*(42), 10162–10174. https://doi.org/10.1021/bi900939w.
Li, H., Mapolelo, D. T., Dingra, N. N., Naik, S. G., Lees, N. S., Hoffman, B. M., et al. (2009). The yeast iron regulatory proteins Grx3/4 and Fra2 form heterodimeric complexes containing a [2Fe-2S] cluster with cysteinyl and histidyl ligation. *Biochemistry*, *48*(40), 9569–9581. https://doi.org/10.1021/bi901182w.
Lin, J., Zhang, L., Lai, S., & Ye, K. (2011). Structure and molecular evolution of CDGSH iron-sulfur domains. *PLoS One*, *6*(9), e24790. https://doi.org/10.1371/journal.pone.0024790.

Lipper, C. H., Paddock, M. L., Onuchic, J. N., Mittler, R., Nechushtai, R., & Jennings, P. A. (2015). Cancer-related NEET porteins transfer 2Fe-2S clusters to Anamorsin, a protein required for cytosolic iron-sulfur cluster biogenesis. *PLoS One, 10*(10), e0139699, https://doi.org/10.1371/journal.pone.0139699.

Liu, Q., Xia, Z., & Case, C. C. (2002). Validated zinc finger protein designs for All 16 GNN DNA triplet targets. *The Journal of Biological Chemistry, 277*(6), 3850–3856. https://doi.org/10.1074/jbc.M110669200.

Magyar, J. S., & Godwin, H. A. (2003). Spectropotentiometric analysis of metal binding to structural zinc-binding sites: Accounting quantitatively for pH and metal ion buffering effects. *Analytical Biochemistry, 320*(1), 39–54.

Mapolelo, D. T., Zhang, B., Naik, S. G., Huynh, B. H., & Johnson, M. K. (2012). Spectroscopic and functional characterization of iron-sulfur cluster-bound forms of Azotobacter vinelandii (Nif)IscA. *Biochemistry, 51*(41), 8071–8084. https://doi.org/10.1021/bi3006658.

Maret, W. (2012). New perspectives of zinc coordination environments in proteins. *Journal of Inorganic Biochemistry, 111*, 110–116. https://doi.org/10.1016/j.jinorgbio.2011.11.018.

Matthews, J. M., & Sunde, M. (2002). Zinc fingers—Folds for many occasions. *IUBMB Life, 54*(6), 351–355. https://doi.org/10.1080/15216540216035.

McCurdy, E., Woods, G., & Potter, D. (2006). *Unmatched removal of spectral interferences in ICP-MS using the agilent octopole reaction system with helium collision mode.* Agilent Technologies.

Michalek, J. L., Besold, A. N., & Michel, S. L. (2011). Cysteine and histidine shuffling: Mixing and matching cysteine and histidine residues in zinc finger proteins to afford different folds and function. *Dalton Transactions, 40*(47), 12619–12632. https://doi.org/10.1039/c1dt11071c.

Miller, J., McLachlan, A. D., & Klug, A. (1985). Repetitive zinc-binding domains in the protein transcription factor IIIA from Xenopus oocytes. *EMBO Journal, 4*(6), 1609–1614.

Newton, A., Mackay, J., & Crossley, M. (2001). The N-terminal zinc finger of the erythroid transcription factor GATA-1 binds GATC motifs in DNA. *The Journal of Biological Chemistry, 276*(38), 35794–35801. https://doi.org/10.1074/jbc.M106256200.

Pattenden, L. K., & Thomas, W. G. (2008). Amylose affinity chromatography of maltose-binding protein: Purification by both native and novel matrix-assisted dialysis refolding methods. *Methods in Molecular Biology, 421*, 169–189.

Pavletich, N. P., & Pabo, C. O. (1991). Zinc finger-DNA recognition: Crystal structure of a Zif268-DNA complex at 2.1 Å. *Science, 252*(5007), 809–817.

Persikov, A. V., & Singh, M. (2014). De novo prediction of DNA-binding specificities for Cys2His2 zinc finger proteins. *Nucleic Acids Research, 42*(1), 97–108. https://doi.org/10.1093/nar/gkt890.

Phan, J., Zdanov, A., Evdokimov, A. G., Tropea, J. E., Peters, H. K., Kapust, R. B., et al. (2002). Structural basis for the substrate specificity of tobacco etch virus protease. *The Journal of Biological Chemistry, 277*(52), 50564–50572. https://doi.org/10.1074/jbc.M207224200.

Poor, C. B., Wegner, S. V., Li, H., Dlouhy, A. C., Schuermann, J. P., Sanishvili, R., et al. (2014). Molecular mechanism and structure of the Saccharomyces cerevisiae iron regulator Aft2. *Proceedings of the National Academy of Sciences of the United States of America, 111*(11), 4043–4048. https://doi.org/10.1073/pnas.1318869111.

Profrock, D., & Prange, A. (2012). Inductively coupled plasma-mass spectrometry (ICP-MS) for quantitative analysis in environmental and life sciences: A review of challenges, solutions, and trends. *Applied Spectroscopy, 66*(8), 843–868. https://doi.org/10.1366/12-06681.

Rommers, P., & Boumans, P. (1996). ICP-AES versus (LA-)IPC-MS: Competition or a happy marriage? A view supported by current data. *Fresenius Journal of Analytical Chemistry, 355*(7–8), 763–770.

Rosano, G. L., & Ceccarelli, E. A. (2014). Recombinant protein expression in microbial systems. *Frontiers in Microbiology, 5*, 341. https://doi.org/10.3389/fmicb.2014.00341.

Segura, M., Madrid, Y., & Cámara, C. (2003). Elimination of calcium and argon interferences in iron determination by ICP-MS using desferrioxamine chelating agent immobilized in sol–gel and cold plasma conditions. *Journal of Analytical Atomic Spectrometry, 18*(9), 1103–1108. https://doi.org/10.1039/B301719M.

Shimberg, G. D., Michalek, J. L., Oluyadi, A. A., Rodrigues, A. V., Zucconi, B. E., Neu, H. M., et al. (2016). Cleavage and polyadenylation specificity factor 30: An RNA-binding zinc-finger protein with an unexpected 2Fe-2S cluster. *Proceedings of the National Academy of Sciences of the United States of America, 113*(17), 4700–4705. https://doi.org/10.1073/pnas.1517620113.

Simonian, M. H., & Smith, J. A. (2006). Spectrophotometric and colorimetric determination of protein concentration. *Current Protocols in Molecular Biology* chapter 10, Unit 10 11A, 10.1.1–10.1A.9https://doi.org/10.1002/0471142727.mb1001as76.

Smith, B. (1994). Drying gels. In J. Walker (Ed.), *Basic protein and peptide protocols* (pp. 157–161). Humana Press.

Tamir, S., Paddock, M. L., Darash-Yahana-Baram, M., Holt, S. H., Sohn, Y. S., Agranat, L., et al. (2015). Structure-function analysis of NEET proteins uncovers their role as key regulators of iron and ROS homeostasis in health and disease. *Biochimica et Biophysica Acta, 1853*(6), 1294–1315. https://doi.org/10.1016/j.bbamcr.2014.10.014.

Tan, G., Landry, A. P., Dai, R., Wang, L., Lu, J., & Ding, H. (2012). Competition of zinc ion for the [2Fe-2S] cluster binding site in the diabetes drug target protein mitoNEET. *Biometals, 25*(6), 1177–1184. https://doi.org/10.1007/s10534-012-9580-4.

Vandevenne, M., Jacques, D. A., Artuz, C., Nguyen, C. D., Kwan, A. H., Segal, D. J., et al. (2013). New insights into DNA recognition by zinc fingers revealed by structural analysis of the oncoprotein ZNF217. *The Journal of Biological Chemistry, 288*(15), 10616–10627. https://doi.org/10.1074/jbc.M112.441451.

Wang, Y., Landry, A. P., & Ding, H. (2017). The mitochondrial outer membrane protein mitoNEET is a redox enzyme catalyzing electron transfer from FMNH2 to oxygen or ubiqunone. *The Journal of Biological Chemistry, 292*(24), 10061–10067. https://doi.org/10.1074/jbc.M117.789800Epub 2017 May 1.

Warf, M. B., & Berglund, J. A. (2007). MBNL binds similar RNA structures in the CUG repeats of myotonic dystrophy and its pre-mRNA substrate cardiac troponin T. *RNA, 13*(12), 2238–2251. https://doi.org/10.1261/rna.610607.

Wiley, S. E., Murphy, A. N., Ross, S. A., van der Geer, P., & Dixon, J. E. (2007). MitoNEET is an iron-containing outer mitochondrial membrane protein that regulates oxidative capacity. *Proceedings of the National Academy of Sciences of the United States of America, 104*(13), 5318–5323. https://doi.org/10.1073/pnas.0701078104.

Wiley, S. E., Paddock, M. L., Abresch, E. C., Gross, L., van der Geer, P., Nechushtai, R., et al. (2007). The outer mitochondrial membrane protein mitoNEET contains a novel redox-active 2Fe-2S cluster. *The Journal of Biological Chemistry, 282*(33), 23745–23749. https://doi.org/10.1074/jbc.C700107200.

Wilson, G. M. (2005). RNA folding and RNA-protein binding analyzed by fluorescence anisotropy and resonance engery transfer. In C. D. Geddes & J. R. Lakowicz (Eds.), *Reviews in fluorescence*, Vol. 2 (p. 20). New York: Springer Science + Business Media, Inc.

Woestenenk, E. A., Hammarström, M., van den Berg, S., Härd, T., & Berglund, H. (2004). His tag effect on solubility of human proteins produced in Escherichia coli: A comparison between four expression vectors. *Journal of Structural and Functional Genomics, 5*(3), 217–229. https://doi.org/10.1023/B:jsfg.0000031965.37625.0e.

Yang, Q., & Doublie, S. (2011). Structural biology of poly(A) site definition. *Wiley Interdisciplinary Reviews RNA, 2*(5), 732–747. https://doi.org/10.1002/wrna.88.

Zhao, D., & Huang, Z. (2016). Effect of His-Tag on expression, purification and structure of zinc finger protein, ZNF191(243-368). *Bioinorganic Chemistry and Applications 2016*, 8206854. https://doi.org/10.1155/2016/8206854.

Zoorob, G. K., McKiernan, J. W., & Caruso, J. A. (1998). ICP-MS for elemental speciation studies. *Mikrochimica Acta, 128*(3–4), 145–168. https://doi.org/10.1007/Bf01243044.

CHAPTER FIVE

Methods for Studying Iron Regulatory Protein 1: An Important Protein in Human Iron Metabolism

Gregory P. Holmes-Hampton, Manik C. Ghosh, Tracey A. Rouault[1]
Molecular Medicine Branch, *Eunice Kennedy Shriver* National Institute of Child Health and Human Development, Bethesda, MD, United States
[1]Corresponding author: e-mail address: rouault@mail.nih.gov

Contents

Abstract

Iron regulatory proteins 1 and 2 (IRP1 and IRP2) are two cytosolic proteins that maintain cellular iron homeostasis by regulating the expression of genes involved in iron metabolism. IRPs respond to cellular iron deficiency by binding to iron-responsive elements (IREs) found in the mRNAs of iron metabolism transcripts, enhancing iron import, and reducing iron storage, utilization, and export. IRP1, a bifunctional protein, exists in equilibrium between a $[Fe_4S_4]$ cluster containing cytosolic aconitase, and an apoprotein that binds to IREs. At high cellular iron levels, this equilibrium is shifted more toward iron–sulfur cluster containing aconitase, whereas IRP2 undergoes proteasomal degradation by an E3 ubiquitin ligase complex that contains an F-box protein, FBXL5. *Irp1*$^{-/-}$ mice develop polycythemia and pulmonary hypertension, whereas *Irp2*$^{-/-}$ mice develop microcytic anemia and progressive neurodegeneration, indicating that Irp1 has important functions in the erythropoietic and pulmonary systems, and Irp2 has essential roles in supporting erythropoiesis and nervous system functions. Mice

Methods in Enzymology, Volume 599
ISSN 0076-6879
https://doi.org/10.1016/bs.mie.2017.09.006

lacking both *Irp1* and *Irp2* die during embryogenesis, suggesting that functions of Irp1 and Irp2 are redundant. In this review, we will focus on the methods for studying IRP1 activities and function in cells and animals.

1. INTRODUCTION

Nearly five decades ago, it was shown that increased iron content in rat diet led to an increased ferritin protein without a corresponding increase in the iron to protein ratio (Linder, Munro, & Morris, 1970). This was one of the first examples of protein expression being linked to iron content. Years later, it was shown that this behavior was the result of an iron-responsive element (IRE) located in the 5′ untranslated region (5′UTR) of the mRNA for ferritin (Aziz & Munro, 1987; Hentze et al., 1987). Ultimately, it was shown that iron regulatory protein 1 (IRP1) binds to this IRE (Rouault, Stout, Kaptain, Harford, & Klausner, 1991) and is involved in numerous aspects of iron homeostasis in the cell (DeRusso et al., 1995; Klausner & Rouault,1993). IRP1 binds to IREs to modulate the expression of several proteins in response to the iron content of the cell. These IREs are stem-loop structures (Samaniego, Chin, Iwai, Rouault, & Klausner, 1994) in the UTR of various transcripts, and depending on whether they are in the 3′UTR or the 5′UTR, binding of IRP1 will increase or decrease expression of the protein encoded by the target transcript (Sanchez et al., 2011). Examples of proteins with IREs in the 3′UTR include transferrin receptor 1 (TFRC, also known as TfR1) (Casey, Koeller, Ramin, Klausner, & Harford, 1989) and the divalent metal transporter 1 (DMT1, also known as NRAMP2 or DCT1) (Gunshin et al., 1997), both of which are involved in iron uptake. Examples of proteins with IREs in the 5′UTR include L- and H-ferritin (Caughman, Hentze, Rouault, Harford, & Klausner, 1988; Rouault, Hentze, Caughman, Harford, & Klausner, 1988), erythroid 5-aminolevulinate synthase (ALAS2) (Duncan, Faggart, Roger, & Cornell, 1999), mitochondrial aconitase (ACO2) (Dandekar et al., 1991; Kim, LaVaute, Iwai, Klausner, & Rouault, 1996; Schalinske, Chen, & Eisenstein, 1998), and ferroportin 1 (FPN1, also known as IREG1 or SLC40A1) (Abboud & Haile, 2000; Donovan et al., 2000), proteins that are involved in the storage (L- and H-ferritin), utilization or export of iron. Essentially, when iron levels are low in the cell, IRP1 will function in its IRE-binding role to increase transcript levels of proteins responsible for import, and to decrease translation of proteins responsible for utilization, storage, and export of iron.

Another example of a sequence with an IRE is hypoxia-inducible factor 2 alpha, Hif2α (EPAS1), which harbors a conserved IRE-binding site in its 5′UTR (Sanchez, Galy, Muckenthaler, & Hentze, 2007). Hif2α is a protein that is involved in the response to changes in the oxygen level. The gene is active under low oxygen levels (i.e., hypoxia). As with other sequences that contain a 5′UTR IRE-binding site, binding of IRP1 to the Hif2α-IRE decreases new synthesis of the encoded protein.

IRP1 has also been shown to function as a cytosolic aconitase (Haile et al., 1992; Kennedy, Mende-Mueller, Blondin, & Beinert, 1992), a [Fe_4S_4] containing protein that catalyzes the conversion of citrate to isocitrate in the cytosol. In its IRE-binding form, the protein is devoid of this cluster (Pantopoulos, 2004; Rouault, 2006). Essentially, IRP1 acts as a register of the iron content of the cell. In iron-replete conditions, the protein is found in the holo (iron–sulfur cluster bound) form, whereas in iron-deficient conditions, the protein is found in its apo (noniron bound) form, and serves the cellular role of restoring proper iron homeostasis. As the cell begins to import more iron, the apo-IRE binding form switches more to the holo-cytosolic aconitase form. In this sense, the protein exists in equilibrium between the apo and the holo forms depending on the current iron status of the cell.

In addition to IRP1, there is another iron regulatory protein, IRP2, which also binds to the IREs (Guo, Yu, & Leibold, 1994; Samaniego et al., 1994). Unlike IRP1, IRP2 is not known to bind iron, and does not serve in the cytosolic aconitase role that has been established for IRP1. Mouse models lacking alleles for both *Irp1* and *Irp2* are not viable, whereas single complete knockouts of either *Irp1* or *Irp2* are viable (Smith, Ghosh, Ollivierre-Wilson, Hang Tong, & Rouault, 2006). These results suggest that there is a compensatory mechanism wherein Irp1 can fulfill the role of Irp2 and vice versa for the IRE-binding activity. In the case of Irp1 knockout, the cell will also be deficient in cytosolic aconitase. However, the mitochondrial aconitase will still be intact, and it is not known if the specific lack of cytosolic aconitase contributes to Irp1 knockout-specific phenotypes, or if phenotypes arise mainly from the lack of IRE-binding activity of Irp1, although the latter is thought to be the major contributing factor (Anderson et al., 2013; Ghosh et al., 2013; Wilkinson & Pantopoulos, 2013; Zhang, Ghosh, & Rouault, 2014). Irp1 and Irp2 also differ in their response to excess cellular iron; in the case of Irp1, the [Fe_4S_4] which ligates to cysteines in the active site cleft and prevents binding to IREs, whereas Irp2 is degraded under high iron conditions.

Although not all IREs have been studied for their affinity to IRP1 and IRP2, those that have been quantitatively studied for binding affinity have shown comparable affinity for both proteins (Allerson, Cazzola, & Rouault, 1999; Butt et al., 1996). Specifically, IRE containing genes for CDC14A (Sanchez et al., 2006), Hif2α (Ghosh, Zhang, & Rouault, 2015; Sanchez et al., 2007), and ACO2 (Kim et al., 1996) have shown affinity for both IRP1 and IRP2. Although previous reports have suggested differential roles for IRP1 and IRP2 (Butt et al., 1996; Henderson, Menotti, & Kuhn, 1996; Menotti, Henderson, & Kuhn, 1998), different phenotypes identified for *Irp1*$^{-/-}$ (polycythemia and pulmonary hypertension) and *Irp2*$^{-/-}$ (neurodegeneration, microcytic anemia, and erythropoietic protoporphyria) mice may be attributed to differential expression of Irp1 and Irp2 in individual cell types of various tissues (Ghosh et al., 2013, 2015).

2. ACONITASE ACTIVITY ASSAYS

Aconitase contains a [Fe_4S_4] cluster that includes three irons bound to cysteinyl ligands of the protein, whereas the fourth iron molecule is ligated by the sulfur of the [Fe_3S_4] cluster (Beinert, Kennedy, & Stout, 1996; Brown, Kennedy, Antholine, Eisenstein, & Walden, 2002; Kennedy, Emptage, Dreyer, & Beinert, 1983). This fourth iron is the site of substrate binding wherein citrate is converted to isocitrate through a *cis*-aconitate intermediate (Villafranca, 1974). Subsequently, isocitrate is converted to α-ketoglutarate by isocitrate dehydrogenase in a reaction that also converts $NADP^+$ to NADPH (Cherbavaz, Lee, Stroud, & Koshland, 2000). The formation of NADPH can be followed by UV–vis spectroscopy, or alternatively by coupling the formation of NADPH with the reduction of (3-(4,5-dimethylthiazol-2-yl)-2,5-diphenyl-2H-tetrazolium bromide) (MTT) to formazan in the presence of phenazine methosulfate, which produces a reaction product that can be quantitatively assessed via an in-gel assay. These two similar approaches will be discussed in the following sections and are outlined in Fig. 1. Each technique has advantages: the spectrophotometric technique can be performed much more quickly than the in-gel assay and can be performed on a 96-well plate, meaning that more samples can be analyzed simultaneously. The in-gel assay separates the two aconitases, enabling the experimentalist to distinguish the activity of the cytosolic aconitase from that of the mitochondrial aconitase. To distinguish these

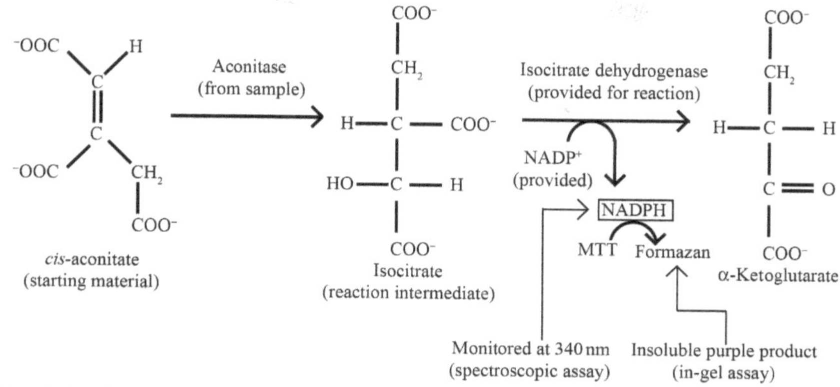

Fig. 1 A schematic representation of the aconitase assay. Both the spectroscopic and in-gel assays utilize *cis*-aconitate, $NADP^+$, and isocitrate dehydrogenase to assess the aconitase activity in the sample. The *cis*-aconitate is converted to isocitrate, which is in turn converted to α-ketoglutarate, and $NADP^+$ is concominantly converted to NADPH. In the spectroscopic assay, the formation of NADPH is monitored at 340 nm. For the in-gel assay, the NADPH is used to reduce MTT to formazan, which appears as a purple band on the gel.

forms via the spectrophotometric assay, one would have to fractionate the cells, which leads to the possibility of contamination of the mitochondrial and cytosolic fractions, reducing interpretability. The spectroscopic assay can be particularly useful for measuring the enzymatic activity of isolated protein.

2.1 Spectrophotometric Aconitase Assay

Using a UV–vis spectrophotometer, aconitase activity can be quantified according to the formation of NADPH (Kennedy et al., 1983; Rose & O'Connell, 1967). The reaction is best monitored in an instrument that can perform kinetic measurements, as the reaction is typically complete within 10 min. The reaction should be carried out in a 50 m*M* Tris–HCl buffer with 5 m*M* $MgCl_2$. For the reaction to proceed, isocitrate dehydrogenase (0.2 units/mL final concentration) and $NADP^+$ (0.2 m*M* final concentration) should be added. The sample (containing aconitase) should be added to the cuvette and this should be used as the blank. To start the reaction the substrate intermediate, *cis*-aconitate (0.5 m*M* final concentration), should be added and the reaction should be monitored at 340 nm. NADPH has an extinction coefficient (ε) of 6300 $M^{-1}\,cm^{-1}$ (Bergmeyer, 1975); with a known value for ε, one can determine the concentration of

NADPH that has been made, and this should correspond to the units of aconitase activity in the sample (after correction for the dilution of the sample).

2.2 In-Gel Aconitase Activity Assay

Although the in-gel assay is like the spectrophotometric determination of aconitase activity, there are several differences. First, it is necessary to separate the proteins in the sample. This separation is performed on a non-denaturing gel typically made from an 8% acrylamide (132 m*M* Tris base, 132 m*M* borate, 3.6 m*M* citrate) separating, and a 4% acrylamide (67 m*M* Tris base, 67 m*M* borate, 3.6 m*M* citrate) stacking gel. The separation is performed in a running buffer with 25 m*M* Tris (pH 8.3), 192 m*M* glycine, and 3.6 m*M* citrate for 2–4 h at 170 V. Earlier methodologies employed use of a starch gel for separation (Morden, Doebley, & Schertz, 1988), but these gels tended to give results that were not reproducible.

Following this separation, the gel is then incubated in a solution of 100 m*M* Tris (pH 8.0), 1 m*M* NADP, 2.5 m*M* *cis*-aconitate, 5 m*M* $MgCl_2$ 1.2 m*M* MTT, 0.3 m*M* phenazine methosulfate, and 5 μ/mL isocitrate dehydrogenase in the dark at 37°C for up to an hour. Ultimately, the area of the gel corresponding to the mitochondrial and cytosolic aconitase will display a dark blue/violet color. This color corresponds to the reduction of MTT to formazan, an insoluble purple product with an absorbance maximum at 560 nm.

3. ELECTROPHORETIC MOBILITY SHIFT ASSAY

Cell or tissue lysates are prepared in oxygen-depleted lysis buffer (by freeze–thaw method in presence of nitrogen) containing 10 m*M* HEPES (pH 7.2), 3 m*M* $MgCl_2$, 40 m*M* KCl, 5% glycerol, 0.2% Nonidet P-40, 5 m*M* DTT, 1 m*M* protease inhibitor AEBSF, 10 μg/mL Leupeptin, Complete EDTA-free protease inhibitor mixture (Roche Applied Science), and 50 μ*M* DfO (desferroxamine). Lysate (x μL) containing 10 μg of total protein is added to (12.5-x) μL of anaerobic band shift buffer containing 25 m*M* Tris–HCl (pH 7.5) and 40 m*M* KCl. The lysate samples are incubated for 5 min at room temperature with 12.5 μL of a reaction mixture containing 0.2 units/μL Super RNAsine (Ambion), 0.6 μg/μL yeast t-RNA, 5 m*M* DTT, 20% glycerol, and 20 n*M* ^{32}P-labeled ferritin IRE in 25 m*M* Tris–HCl (pH 7.5) and 40 m*M* KCl.

A 20 μL aliquot of this reaction mixture is loaded into an 8% acrylamide/TBE gel, which is run at 200 V for 2 h 15 min for mouse cell/tissue lysates or 4.5 h for human cell lysates, to allow enough time for the IRP1–IRE and IRP2–IRE complexes to resolve, and then the gel is fixed in a mixture containing 10% acetic acid and 40% methanol, dried in a gel drier, and exposed for autoradiography. This band shift assay method has been derived from the published works (Allerson et al., 1999; Ghosh et al., 2008; Haile, Hentze, Rouault, Harford, & Klausner, 1989; Samaniego et al., 1994).

A representative band shift gel is shown in Fig. 2. In iron-depleted conditions, i.e., when the cells are treated with iron chelator DfO, both Irp1 and Irp2 show increased binding to IRE probes. As a result, the expression of the iron import protein, TfR1, increases, whereas that of the iron storage protein, ferritin, concomitantly decreases. On the contrary, under iron-replete conditions, i.e., when the cells have been treated with ferric

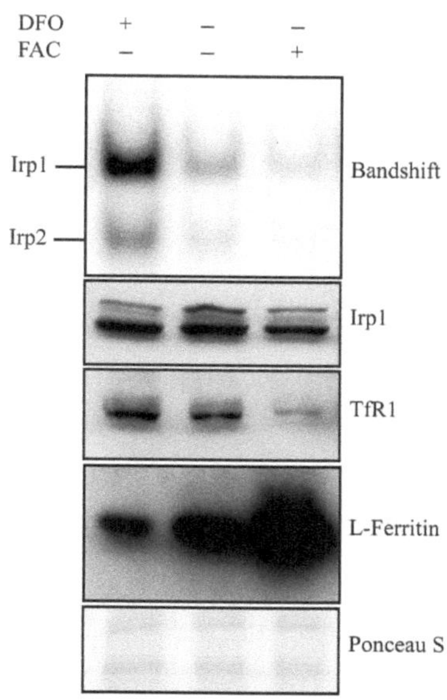

Fig. 2 MEF (mouse embryonic fibroblast) cells treated with iron chelator (DFO, desferroxamine, 0.1 m*M*) or iron supplement (FAC, ferric ammonium citrate 0.1 m*M*). *Top*: EMSA or band shift assay showing IRE-binding activities of Irp1 and Irp2, *second from top*: Irp1 protein levels, *third from top*: Transferrin receptor 1 protein levels, *second from bottom*: L-Ferritin protein levels, *Bottom*: Ponceau S stain was used as a loading control. The data in this figure are unpublished but representative experiments from G. Holmes-Hampton and M. Ghosh.

ammonium citrate (FAC) or other iron sources, both Irps respond by decreasing their binding to IRE, which causes downregulation of TfR1 expression and upregulation of ferritin protein expression.

IREs have also been probed using fluorescein-labeled RNA (Ma et al., 2012). While this technique does not require the use of a radioactive probe, it has not been performed in conjunction with an electrophoretic separation, so individual quantification of IRP1 vs IRP2 activities is not possible. In addition, biotinylated IRE probes have been used in conjunction with electrophoretic mobility shift assays (Cloonan et al., 2016). Again, the inherent advantage of this technique is the lack of requirement for use of radioactive materials. While both a fluorescent and biotinylated probe may work with proper separation, additional care must be taken in creating these probes, because if the marker is placed in the wrong position on the probe, it could diminish target binding. The ^{32}P-labeled probe is free from this potential source of interference as you are making a probe that mimics the native structure. Finally, the radioactive probe provides a good range of sensitivity; with a short exposure you can see high levels of IRP1 or IRP2 levels or you can alternatively expose the film for prolonged periods of time to see weaker signals, a range which might be harder to catch using electrochemiluminescent probes.

4. TARGETS OF IRP1

Table 1 lists 12 proteins each of which is encoded by an mRNA that has an IRE-binding site; of these, one is a gene specific to *Drosophila melanogaster* (dSDH), and another is specific to mice (Hao1). The remaining 10 are found in human genes and can be used to probe the iron content of the cell. Those with a 5′UTR IRE will have increased protein expression when iron content is high and decreased expression when iron content is low. Alternatively, those with a 3′UTR will have increased protein expression when iron content is low and decreased expression when iron content is high. By probing these proteins by Western blot, we can learn about the iron content of the cell based on the expression level of these proteins that are under the control of the IRE-binding sites in their mRNA.

The middle three panels of Fig. 2 demonstrate the response of Irp1, TfR1, and L-ferritin proteins to iron status in mouse embryonic fibroblast (MEF) cells. The second panel shows that Irp1 protein levels remain unchanged regardless of treatment with the iron chelator deferoxamine

Table 1 Known mRNA Sequences With IRE-Binding Sites

Gene	Description	IRE Location	References
ACO2	Mitochondrial aconitase	5′UTR	Kim et al. (1996) and Campillos, Cases, Hentze, and Sanchez (2010) and Mirel et al. (1998)
ALAS2	First enzyme in heme biosynthetic pathway	5′UTR	Dandekar et al. (1991)
EPAS1	Hypoxia inducible factor-2α	5′UTR	Sanchez et al. (2007)
H-Ferritin	Heavy chain ferritin (iron storage protein)	5′UTR	Hentze et al. (1987) and Leibold and Munro (1988)
L-Ferritin	Light chain ferritin (iron storage protein)	5′UTR	Hentze et al. (1987) and Leibold and Munro (1988)
SLC40A1	Solute carrier family 40 member 1 (ferroportin-1)	5′UTR	Abboud and Haile (2000) and Donovan et al. (2000)
dSDH	*Drosophila melanogaster* succinate dehydrogenase	5′UTR	Kohler, Henderson, and Kuhn (1995) and Melefors (1996)
CDC14A	Cell division cycle	3′UTR	Sanchez et al. (2006)
CDC42BPA	CDC42-binding protein alpha	3′UTR	Cmejla, Petrak, and Cmejlova (2006)
SLC11A2	Solute carrier family 11 member 2 (DMT1)	3′UTR	Gunshin et al. (1997)
TFR1	Transferrin receptor 1	3′UTR	Casey et al. (1989)
Hao1	Mouse glycolate oxidase	3′UTR	Kohler, Menotti, and Kuhn (1999)

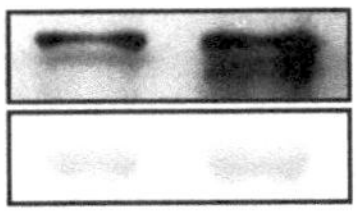

Fig. 3 Western blot of Hif2α in wild-type and *Irp1*$^{-/-}$ MEF cells. Ponceau S was included as a loading control. The data in this figure are unpublished but representative experiments from G. Holmes-Hampton and M. Ghosh.

(DFO) or the iron salt FAC. The third panel shows increased TfR1 protein expression in DFO-treated cells and decreased expression in FAC-treated cells, consistent with the behavior of a 3′UTR containing IRE gene. The fourth panel shows decreased L-ferritin protein expression in DFO-treated cells and increased expression in FAC-treated cells, consistent with the behavior of a 5′UTR containing IRE gene. These experiments were performed with standard Western blotting procedures.

Immunoblot assays of most of the Irp targets are quite straightforward. However, successful immunoblot assays of Hif2α remain technically challenging. We have been more successful with the following method (Ghosh et al., 2013); cells or tissues are harvested in absence of air, and the fresh tissues are used. All the operations before the denaturation of the proteins are done in an anaerobic chamber. Nuclear fractions are separated from cells and tissues following the manufacturer's protocol (Active Motif). The protein blot is incubated overnight at 4°C with Hif2α antibody (affinity-purified goat antihuman from R&D Systems) at a final concentration of 1.0 μg/mL. Western blots are treated with secondary peroxidase-conjugated bovine antigoat IgG antibody from Santa Cruz biotechnology, Inc. at 1:1000 dilution for 2 h at room temperature. Hif2α Western blots are developed using 1:3 diluted (with water) supersignal west femto maximum sensitivity substrate (Thermoscienific). A representative Hif2α western with WT and Irp1-KO MEF (mouse embryonic fibroblast) cells is shown in Fig. 3. Since the IRE in the Hif2α transcript is located in the 5′UTR of the mRNA, deletion of Irp1 results in derepression of Hif2α.

5. RECONSTITUTION, BIOANALYTICAL, AND BIOPHYSICAL ANALYSIS OF IRP1

Apo-IRP1 can be reconstituted with iron and sulfide to convert from the IRE-binding form to the iron–sulfur cluster containing cytosolic aconitase form via a modified protocol (Kennedy & Beinert, 1988) briefly summarized later. For reconstitution, the following steps should be carried out in an anaerobic environment (<1 ppm O_2), starting with purified IRP1 protein (Allerson, Martinez, Yikilmaz, & Rouault, 2003). First, a solution of ferric chloride should be mixed with equimolar concentration of dithiothreitol. This solution should be added to the protein, and the solution mixture should be incubated for 30–60 min, followed by incubation with sodium sulfide for 30 min. In the final step, iron that is not bound to the protein can be removed by running the sample on a PD-10 column (GE Healthcare). Replacing the sodium sulfide in this reaction with NifS and cysteine can lead to a more robust reconstitution (Deck et al., 2009) as it limits the possibility of forming a non-proteinacious iron-sulfide moiety (Pohl, 1962). Specific steps can also be taken to purify the protein in the holo form. For example, IRP1 can be overexpressed in a bacterial strain engineered to over-express the ISC (bacterial iron–sulfur cluster) operon (Johnson, Dean, Smith, & Johnson, 2005). Stability of the cluster can be maximized by subsequent isolation and purification of the protein under anaerobic conditions (Rouault, 2015).

To confirm the successful reconstitution of the FeS cluster, several of the aforementioned techniques can be employed. First, the aconitase activity can be compared in the reconstituted (Holo-IRP1) and the Apo-IRP1 (Fig. 4, top). The holo and apo forms can also be assayed for their IRE-binding activity (Fig. 4, bottom left). Taken together, these techniques indicate a gain of aconitase activity and a loss of IRE-binding activity. However, with bioanalytical techniques one can confirm that these changes are a result of the addition of iron. By assaying the samples with ICP-MS or a similar analytical technique (Holmes-Hampton, Tong, & Rouault, 2014) after measuring the protein concentration of the sample, one can find the ratio of iron to protein (Fig. 4, bottom right). In general, reconstitution levels similar to the hypothetical level (four irons per protein) are indicative of the expected species. However to further explore whether a cubane FeS cluster has been reconstituted, biophysical methods can be used to confirm the iron-containing moiety associated with the protein. For instance, EXAFS can give information on the ligands surrounding the iron, EPR and Mössbauer spectroscopies can confirm that the structure is an iron–sulfur cluster (Emptage, Kent, Kennedy, Beinert, & Munck, 1983).

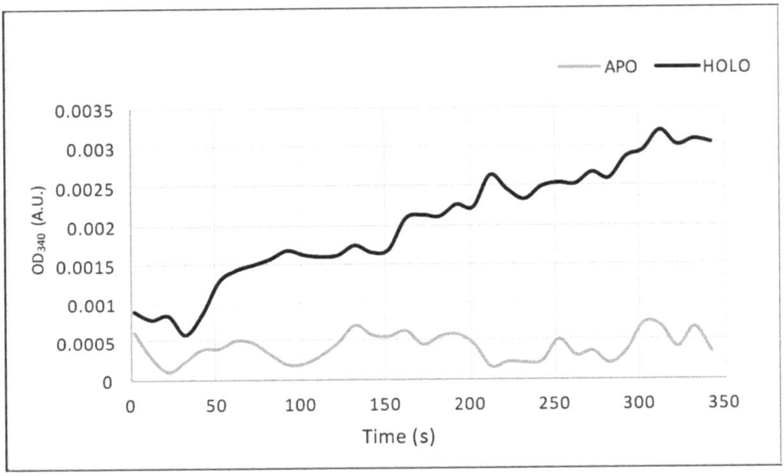

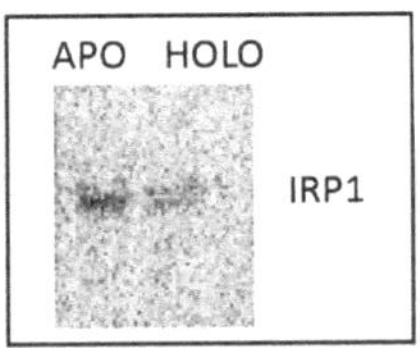

Sample	Iron Content	Protein Content	Fe:Protein Ratio
Apo-IRP1	0.0285 (μ*M*)	0.50 mg/mL (~5.10 μ*M*)	0.006
Holo-IRP1	0.6808 (μ*M*)	0.02 mg/mL (~0.204 μ*M*)	3.337

Fig. 4 Aconitase activity, electrophoretic mobility shift assay, and iron content of apo- and holo-IRP1 protein. *Top*: kinetic traces of apo- and holo-IRP1 monitored at 340 nm. *Bottom Left*: an electrophoretic mobility shift assay (EMSA) demonstrating the apo and holo forms. *Bottom right*: ICP-MS data and protein concentrations for the apo and holo forms of the protein. The data in this figure are unpublished but representative experiments from G. Holmes-Hampton and M. Ghosh.

6. CONCLUSIONS

In this discussion, we have highlighted the analysis of activities of the bifunctional protein IRP1, which serves a major regulatory role in the cell. This protein is a translation factor in its apo form and a cytosolic aconitase in its holo form. The protein shuffles between these forms depending on the iron content of the cell (Fig. 5). Specifically, we have discussed two assays for aconitase activity, a band shift assay for determining IRE binding, and the use of target proteins that are regulated by IRP1 to determine its status. In addition, we have discussed a simple method that permits reconstitution of the FeS cluster on apo-IRP1, and methods that can be used to verify that the conversion of apo-IRP1 to the holo-IRP1 form.

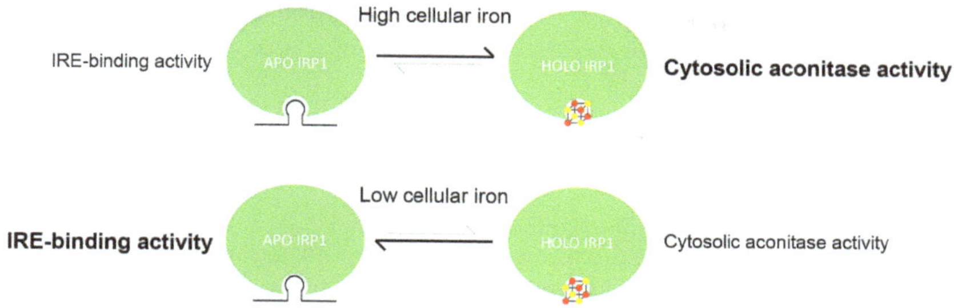

Fig. 5 Schematic of IRP1 status under high and low cellular iron concentrations.

ACKNOWLEDGMENT

This work was supported by the *Eunice Kennedy Shriver* NICHD Intramural Research Program of the National Institutes of Health, Bethesda, MD, USA.

REFERENCES

Abboud, S., & Haile, D. J. (2000). A novel mammalian iron-regulated protein involved in intracellular iron metabolism. *The Journal of Biological Chemistry*, *275*(26), 19906–19912. https://doi.org/10.1074/jbc.M000713200.

Allerson, C. R., Cazzola, M., & Rouault, T. A. (1999). Clinical severity and thermodynamic effects of iron-responsive element mutations in hereditary hyperferritinemia-cataract syndrome. *The Journal of Biological Chemistry*, *274*(37), 26439–26447.

Allerson, C. R., Martinez, A., Yikilmaz, E., & Rouault, T. A. (2003). A high-capacity RNA affinity column for the purification of human IRP1 and IRP2 overexpressed in Pichia pastoris. *RNA*, *9*(3), 364–374.

Anderson, S. A., Nizzi, C. P., Chang, Y. I., Deck, K. M., Schmidt, P. J., Galy, B., et al. (2013). The IRP1-HIF-2α axis coordinates iron and oxygen sensing with erythropoiesis and iron absorption. *Cell Metabolism*, *17*(2), 282–290. https://doi.org/10.1016/j.cmet.2013.01.007.

Aziz, N., & Munro, H. N. (1987). Iron regulates ferritin mRNA translation through a segment of its 5′ untranslated region. *Proceedings of the National Academy of Sciences of the United States of America*, *84*(23), 8478–8482.

Beinert, H., Kennedy, M. C., & Stout, C. D. (1996). Aconitase as iron-sulfur protein, enzyme, and iron-regulatory protein. *Chemical Reviews*, *96*(7), 2335–2374.

Bergmeyer, H. U. (1975). New values for the molar extinction coefficients of NADH and NADPH for the use in routine laboratories (author's transl). *Zeitschrift für Klinische Chemie und Klinische Biochemie*, *13*(11), 507–508.

Brown, N. M., Kennedy, M. C., Antholine, W. E., Eisenstein, R. S., & Walden, W. E. (2002). Detection of a [3Fe-4S] cluster intermediate of cytosolic aconitase in yeast expressing iron regulatory protein 1. Insights into the mechanism of Fe-S cluster cycling. *Journal of Biological Chemistry*, *277*(9), 7246–7254. https://doi.org/10.1074/jbc.M110282200.

Butt, J., Kim, H. Y., Basilion, J. P., Cohen, S., Iwai, K., Philpott, C. C., et al. (1996). Differences in the RNA binding sites of iron regulatory proteins and potential target diversity. *Proceedings of the National Academy of Sciences of the United States of America*, *93*(9), 4345–4349.

Campillos, M., Cases, I., Hentze, M. W., & Sanchez, M. (2010). SIREs: Searching for iron-responsive elements. *Nucleic Acids Research*, *38*(web server issue), W360–W367. https://doi.org/10.1093/nar/gkq371.

Casey, J. L., Koeller, D. M., Ramin, V. C., Klausner, R. D., & Harford, J. B. (1989). Iron regulation of transferrin receptor mRNA levels requires iron-responsive elements and a rapid turnover determinant in the 3′ untranslated region of the mRNA. *The EMBO Journal*, *8*(12), 3693–3699.

Caughman, S. W., Hentze, M. W., Rouault, T. A., Harford, J. B., & Klausner, R. D. (1988). The iron-responsive element is the single element responsible for iron-dependent translational regulation of ferritin biosynthesis. Evidence for function as the binding site for a translational repressor. *The Journal of Biological Chemistry*, *263*(35), 19048–19052.

Cherbavaz, D. B., Lee, M. E., Stroud, R. M., & Koshland, D. E., Jr. (2000). Active site water molecules revealed in the 2.1 A resolution structure of a site-directed mutant of isocitrate dehydrogenase. *Journal of Molecular Biology*, *295*(3), 377–385. https://doi.org/10.1006/jmbi.1999.3195.

Cloonan, S. M., Glass, K., Laucho-Contreras, M. E., Bhashyam, A. R., Cervo, M., Pabon, M. A., et al. (2016). Mitochondrial iron chelation ameliorates cigarette smoke-induced bronchitis and emphysema in mice. *Nature Medicine*, *22*(2), 163–174. https://doi.org/10.1038/nm.4021.

Cmejla, R., Petrak, J., & Cmejlova, J. (2006). A novel iron responsive element in the 3′UTR of human MRCKalpha. *Biochemical and Biophysical Research Communications*, *341*(1), 158–166. https://doi.org/10.1016/j.bbrc.2005.12.155.

Dandekar, T., Stripecke, R., Gray, N. K., Goossen, B., Constable, A., Johansson, H. E., et al. (1991). Identification of a novel iron-responsive element in murine and human erythroid delta-aminolevulinic acid synthase mRNA. *The EMBO Journal*, *10*(7), 1903–1909.

Deck, K. M., Vasanthakumar, A., Anderson, S. A., Goforth, J. B., Kennedy, M. C., Antholine, W. E., et al. (2009). Evidence that phosphorylation of iron regulatory protein 1 at serine 138 destabilizes the [4Fe-4S] cluster in cytosolic aconitase by enhancing 4Fe-3Fe cycling. *The Journal of Biological Chemistry*, *284*(19), 12701–12709. https://doi.org/10.1074/jbc.M807717200.

DeRusso, P. A., Philpott, C. C., Iwai, K., Mostowski, H. S., Klausner, R. D., & Rouault, T. A. (1995). Expression of a constitutive mutant of iron regulatory protein 1 abolishes iron homeostasis in mammalian cells. *The Journal of Biological Chemistry*, *270*(26), 15451–15454.

Donovan, A., Brownlie, A., Zhou, Y., Shepard, J., Pratt, S. J., Moynihan, J., et al. (2000). Positional cloning of zebrafish ferroportin1 identifies a conserved vertebrate iron exporter. *Nature*, *403*(6771), 776–781. https://doi.org/10.1038/35001596.

Duncan, R., Faggart, M. A., Roger, A. J., & Cornell, N. W. (1999). Phylogenetic analysis of the 5-aminolevulinate synthase gene. *Molecular Biology and Evolution*, *16*(3), 383–396.

Emptage, M. H., Kent, T. A., Kennedy, M. C., Beinert, H., & Munck, E. (1983). Mossbauer and EPR studies of activated aconitase: Development of a localized valence state at a subsite of the [4Fe-4S] cluster on binding of citrate. *Proceedings of the National Academy of Sciences of the United States of America*, *80*(15), 4674–4678.

Ghosh, M. C., Tong, W. H., Zhang, D., Ollivierre-Wilson, H., Singh, A., Krishna, M. C., et al. (2008). Tempol-mediated activation of latent iron regulatory protein activity prevents symptoms of neurodegenerative disease in IRP2 knockout mice. *Proceedings of the National Academy of Sciences of the United States of America*, *105*(33), 12028–12033. https://doi.org/10.1073/pnas.0805361105.

Ghosh, M. C., Zhang, D. L., Jeong, S. Y., Kovtunovych, G., Ollivierre-Wilson, H., Noguchi, A., et al. (2013). Deletion of iron regulatory protein 1 causes polycythemia and pulmonary hypertension in mice through translational derepression of HIF2alpha. *Cell Metabolism*, *17*(2), 271–281. https://doi.org/10.1016/j.cmet.2012.12.016.

Ghosh, M. C., Zhang, D. L., & Rouault, T. A. (2015). Iron misregulation and neurodegenerative disease in mouse models that lack iron regulatory proteins. *Neurobiology of Disease*, *81*, 66–75. https://doi.org/10.1016/j.nbd.2015.02.026.

Gunshin, H., Mackenzie, B., Berger, U. V., Gunshin, Y., Romero, M. F., Boron, W. F., et al. (1997). Cloning and characterization of a mammalian proton-coupled metal-ion transporter. *Nature*, *388*(6641), 482–488.

Guo, B., Yu, Y., & Leibold, E. A. (1994). Iron regulates cytoplasmic levels of a novel iron-responsive element-binding protein without aconitase activity. *Journal of Biological Chemistry*, *269*(39), 24252–24260.

Haile, D. J., Hentze, M. W., Rouault, T. A., Harford, J. B., & Klausner, R. D. (1989). Regulation of interaction of the iron-responsive element binding protein with iron-responsive RNA elements. *Molecular and Cellular Biology*, *9*(11), 5055–5061.

Haile, D. J., Rouault, T. A., Tang, C. K., Chin, J., Harford, J. B., & Klausner, R. D. (1992). Reciprocal control of RNA-binding and aconitase activity in the regulation of the iron-responsive element binding protein: Role of the iron-sulfur cluster. *Proceedings of the National Academy of Sciences of the United States of America*, *89*(16), 7536–7540.

Henderson, B. R., Menotti, E., & Kuhn, L. C. (1996). Iron regulatory proteins 1 and 2 bind distinct sets of RNA target sequences. *The Journal of Biological Chemistry*, *271*(9), 4900–4908.

Hentze, M. W., Caughman, S. W., Rouault, T. A., Barriocanal, J. G., Dancis, A., Harford, J. B., et al. (1987). Identification of the iron-responsive element for the translational regulation of human ferritin mRNA. *Science*, *238*(4833), 1570–1573.

Holmes-Hampton, G. P., Tong, W. H., & Rouault, T. A. (2014). Biochemical and biophysical methods for studying mitochondrial iron metabolism. *Methods in Enzymology*, *547*, 275–307. https://doi.org/10.1016/B978-0-12-801415-8.00015-1.

Johnson, D. C., Dean, D. R., Smith, A. D., & Johnson, M. K. (2005). Structure, function, and formation of biological iron-sulfur clusters. *Annual Review of Biochemistry*, *74*, 247–281. https://doi.org/10.1146/annurev.biochem.74.082803.133518.

Kennedy, M. C., & Beinert, H. (1988). The state of cluster SH and S^{2-} of aconitase during cluster interconversions and removal. A convenient preparation of apoenzyme. *The Journal of Biological Chemistry*, *263*(17), 8194–8198.

Kennedy, M. C., Emptage, M. H., Dreyer, J. L., & Beinert, H. (1983). The role of iron in the activation-inactivation of aconitase. *The Journal of Biological Chemistry*, *258*(18), 11098–11105.

Kennedy, M. C., Mende-Mueller, L., Blondin, G. A., & Beinert, H. (1992). Purification and characterization of cytosolic aconitase from beef liver and its relationship to the iron-responsive element binding protein. *Proceedings of the National Academy of Sciences of the United States of America*, *89*(24), 11730–11734.

Kim, H. Y., LaVaute, T., Iwai, K., Klausner, R. D., & Rouault, T. A. (1996). Identification of a conserved and functional iron-responsive element in the 5′-untranslated region of mammalian mitochondrial aconitase. *The Journal of Biological Chemistry*, *271*(39), 24226–24230.

Klausner, R. D., & Rouault, T. A. (1993). A double life: Cytosolic aconitase as a regulatory RNA binding protein. *Molecular Biology of the Cell*, *4*(1), 1–5.

Kohler, S. A., Henderson, B. R., & Kuhn, L. C. (1995). Succinate dehydrogenase b mRNA of Drosophila melanogaster has a functional iron-responsive element in its 5′-untranslated region. *The Journal of Biological Chemistry*, *270*(51), 30781–30786.

Kohler, S. A., Menotti, E., & Kuhn, L. C. (1999). Molecular cloning of mouse glycolate oxidase. High evolutionary conservation and presence of an iron-responsive element-like sequence in the mRNA. *Journal of Biological Chemistry*, *274*(4), 2401–2407.

Leibold, E. A., & Munro, H. N. (1988). Cytoplasmic protein binds in vitro to a highly conserved sequence in the 5′ untranslated region of ferritin heavy- and light-subunit mRNAs. *Proceedings of the National Academy of Sciences of the United States of America*, *85*(7), 2171–2175.

Linder, M., Munro, H. N., & Morris, H. P. (1970). Rat ferritin isoproteins and their response to iron administration in a series of hepatic tumors and in normal and regenerating liver. *Cancer Research, 30*(8), 2231.

Ma, J., Haldar, S., Khan, M. A., Sharma, S. D., Merrick, W. C., Theil, E. C., et al. (2012). Fe2+ binds iron responsive element-RNA, selectively changing protein-binding affinities and regulating mRNA repression and activation. *Proceedings of the National Academy of Sciences of the United States of America, 109*(22), 8417–8422. https://doi.org/10.1073/pnas.1120045109.

Melefors, O. (1996). Translational regulation in vivo of the Drosophila melanogaster mRNA encoding succinate dehydrogenase iron protein via iron responsive elements. *Biochemical and Biophysical Research Communications, 221*(2), 437–441. https://doi.org/10.1006/bbrc.1996.0613.

Menotti, E., Henderson, B. R., & Kuhn, L. C. (1998). Translational regulation of mRNAs with distinct IRE sequences by iron regulatory proteins 1 and 2. *The Journal of Biological Chemistry, 273*(3), 1821–1824.

Mirel, D. B., Marder, K., Graziano, J., Freyer, G., Zhao, Q., Mayeux, R., et al. (1998). Characterization of the human mitochondrial aconitase gene (ACO2). *Gene, 213*(1–2), 205–218.

Morden, C. W., Doebley, J. F., & Schertz, K. F. (1988). Genetic-control and subcellular-localization of aconitase isozymes in Sorghum. *Journal of Heredity, 79*(4), 294–299.

Pantopoulos, K. (2004). Iron metabolism and the IRE/IRP regulatory system: An update. *Annals of the New York Academy of Sciences, 1012*, 1–13.

Pohl, H. A. (1962). Solubility of iron sulfides. *Journal of Chemical and Engineering Data*, 7(2), 295–306. https://doi.org/10.1021/je60013a042.

Rose, I. A., & O'Connell, E. L. (1967). Mechanism of aconitase action: I. The hydrogen transfer reaction. *Journal of Biological Chemistry, 242*(8), 1870–1879.

Rouault, T. A. (2006). The role of iron regulatory proteins in mammalian iron homeostasis and disease. *Nature Chemical Biology, 2*(8), 406–414. https://doi.org/10.1038/nchembio807.

Rouault, T. A. (2015). Iron-sulfur proteins hiding in plain sight. *Nature Chemical Biology, 11*(7), 442–445. https://doi.org/10.1038/nchembio.1843.

Rouault, T. A., Hentze, M. W., Caughman, S. W., Harford, J. B., & Klausner, R. D. (1988). Binding of a cytosolic protein to the iron-responsive element of human ferritin messenger RNA. *Science, 241*(4870), 1207–1210.

Rouault, T. A., Stout, C. D., Kaptain, S., Harford, J. B., & Klausner, R. D. (1991). Structural relationship between an iron-regulated RNA-binding protein (IRE-BP) and aconitase: Functional implications. *Cell, 64*(5), 881–883.

Samaniego, F., Chin, J., Iwai, K., Rouault, T. A., & Klausner, R. D. (1994). Molecular characterization of a second iron-responsive element binding protein, iron regulatory protein 2. Structure, function, and post-translational regulation. *The Journal of Biological Chemistry, 269*(49), 30904–30910.

Sanchez, M., Galy, B., Dandekar, T., Bengert, P., Vainshtein, Y., Stolte, J., et al. (2006). Iron regulation and the cell cycle: Identification of an iron-responsive element in the 3′-untranslated region of human cell division cycle 14A mRNA by a refined microarray-based screening strategy. *The Journal of Biological Chemistry, 281*(32), 22865–22874. https://doi.org/10.1074/jbc.M603876200.

Sanchez, M., Galy, B., Muckenthaler, M. U., & Hentze, M. W. (2007). Iron-regulatory proteins limit hypoxia-inducible factor-2alpha expression in iron deficiency. *Nature Structural & Molecular Biology, 14*(5), 420–426.

Sanchez, M., Galy, B., Schwanhaeusser, B., Blake, J., Bahr-Ivacevic, T., Benes, V., et al. (2011). Iron regulatory protein-1 and -2: Transcriptome-wide definition of binding mRNAs and shaping of the cellular proteome by iron regulatory proteins. *Blood, 118*(22), e168–e179. https://doi.org/10.1182/blood-2011-04-343541.

Schalinske, K. L., Chen, O. S., & Eisenstein, R. S. (1998). Iron differentially stimulates translation of mitochondrial aconitase and ferritin mRNAs in mammalian cells. Implications for iron regulatory proteins as regulators of mitochondrial citrate utilization. *Journal of Biological Chemistry*, *273*(6), 3740–3746.

Smith, S. R., Ghosh, M. C., Ollivierre-Wilson, H., Hang Tong, W., & Rouault, T. A. (2006). Complete loss of iron regulatory proteins 1 and 2 prevents viability of murine zygotes beyond the blastocyst stage of embryonic development. *Blood Cells, Molecules & Diseases*, *36*(2), 283–287. https://doi.org/10.1016/j.bcmd.2005.12.006.

Villafranca, J. J. (1974). The mechanism of aconitase action. Evidence for an enzyme isomerization by studies of inhibition by tricarboxylic acids. *The Journal of Biological Chemistry*, *249*(19), 6149–6155.

Wilkinson, N., & Pantopoulos, K. (2013). IRP1 regulates erythropoiesis and systemic iron homeostasis by controlling HIF2alpha mRNA translation. *Blood*, *122*(9), 1658–1668. https://doi.org/10.1182/blood-2013-03-492454.

Zhang, D. L., Ghosh, M. C., & Rouault, T. A. (2014). The physiological functions of iron regulatory proteins in iron homeostasis—An update. *Frontiers in Pharmacology*, *5*, 124. https://doi.org/10.3389/fphar.2014.00124.

CHAPTER SIX

Robust Production, Crystallization, Structure Determination, and Analysis of [Fe–S] Proteins: Uncovering Control of Electron Shuttling and Gating in the Respiratory Metabolism of Molybdopterin Guanine Dinucleotide Enzymes

Chi-Lin Tsai*, John A. Tainer*,†,1

*The University of Texas M. D. Anderson Cancer Center, Houston, TX, United States
†Molecular Biophysics and Integrated Bioimaging, Lawrence Berkeley National Laboratory, Berkeley, CA, United States
[1]Corresponding author: e-mail address: jtainer@mdanderson.org

Contents

Methods in Enzymology, Volume 599
ISSN 0076-6879
https://doi.org/10.1016/bs.mie.2017.11.006

Abstract

[Fe–S] clusters are essential cofactors in all domains of life. They play many biological roles due to their unique abilities for electron transfer and conformational control. Yet, producing and analyzing Fe-S proteins can be difficult and even misleading if not done anaerobically. Due to unique redox properties of [Fe–S] clusters and their oxygen sensitivity, they pose multiple challenges and can lose enzymatic activity or cause their component proteins to be structurally disordered due to [Fe–S] cluster oxidation and loss in air. Here we highlight tested protocols and strategies enabling efficient and stable [Fe–S] protein production, purification, crystallization, X-ray diffraction data collection, and structure determination. From multiple high-resolution anaerobic crystal structures, we furthermore analyze exemplary data defining [Fe–S] clusters, substrate entry, and product exit for the functional oxidation states of type II molybdo-bis(-molybdopterin guanine dinucleotide) (Mo-bisMGD) enzymes. Notably, these enzymes perform electron shuttling between quinone pools and specific substrates to catalyze respiratory metabolism. The identified structure–activity relationships for this enzyme class have broad implications germane to perchlorate environments on Earth and Mars extending to an alternative mechanism underlying metabolic origins for the evolution of the oxygen atmosphere. Integrated structural analyses of type II Mo-bisMGD enzymes unveil novel distinctive shared molecular mechanisms for dynamic control of substrate entry and product release gated by hydrophobic residues. Collective findings support a prototypic model for type II Mo-bisMGD enzymes including insights for a fundamental molecular mechanistic understanding of selectivity and regulation by a conformationally gated channel with general implications for [Fe–S] cluster respiratory enzymes.

1. INTRODUCTION

Life evolved in an anaerobic world, which can frequently be overlooked from the focus of most funded research on aerobic eukaryotes. Due to this anaerobic origin, our experimental approaches must consider fundamental enzymatic mechanisms and biochemical pathways that were developed and integrated into metabolism in the absence of selective pressure to avoid oxygen reactivity (Imlay, 2008). The [Fe–S] cluster is prone to oxygen sensitivity due to its redox properties enabling conversion from Fe^{2+} to Fe^{3+}, and the fact that its redox potential can range from −700 to +450 mV (Bak & Elliott, 2014; Imlay, 2006). As the redox potential of molecular oxygen ranges from +300 to +800 mV (Segel, 1976), the aerobic purification of [Fe–S] proteins can make [Fe–S] clusters prone to oxidation resulting in the loss of [Fe–S] clusters and consequently protein dysfunction and even domain misfolding, as seen for XPD (Fan et al., 2008). Such oxidative damage has biological and practical relevance as oxidation causing mis-metalation can occur in cells, especially under conditions of oxidative stress (Cotruvo & Stubbe, 2012; Imlay, 2014). In aerobic cells, reactive oxygen species are scavenged by oxygen into superoxide that can accelerate DNA and other damage by destroying [Fe–S] clusters with a rate constant of 10^6–$10^7 M^{-1} s^{-1}$ and elevating free-iron levels (Imlay, 2008; Keyer & Imlay, 1996). As a result, superoxide dismutase metalloenzymes, which convert superoxide radicals to molecular oxygen and hydrogen peroxide that is further reduced to oxygen by peroxidases and catalases, are critical to the stability of [Fe–S] cluster enzymes in many cells including many anaerobes that may generate oxygen during metabolism (Imlay, 2008; Perry, Shin, Getzoff, & Tainer, 2010). The fundamental sensitivity of [Fe–S] clusters to superoxide and nitric oxide, which is tightly controlled by its synthesis from arginine in eukaryotes (Crane et al., 1998) and some microbes (Adak et al., 2002), drove the development of multiple protective superoxide dismutases with several different folds and metal ion active sites (Perry et al., 2010). In human cells, DNA sequence regions without stable double-helical conformations are especially vulnerable to iron-mediated damage, and translocation breakpoints in patient tumor samples are associated with these regions (Bacolla, Tainer, Vasquez, & Cooper, 2016). Strikingly, in higher eukaryotes both growth defects and DNA damage caused by superoxide develop primarily from its toxic damage to iron–sulfur clusters (ISCs) (Keyer & Imlay, 1996).

Whenever doing experiments with [Fe–S] cluster proteins, it is therefore crucial to appreciate dioxygen toxicity to [Fe–S] clusters, despite the fact that they are truly ancient cofactors essential for all domains of life (Johnson, Dean, Smith, & Johnson, 2005; Lill, 2009). Cubane [4Fe–4S] and rhombic [2Fe–2S] clusters are those found most frequently in proteins (Beinert, 1997; Johnson et al., 2005) although more complex [Fe–S] clusters such as H-cluster, P-cluster, and FeMo cofactor also exist (Broderick et al., 2014; Burgess & Lowe, 1996; Hoffman, Lukoyanov, Yang, Dean, & Seefeldt, 2014). They play essential roles for protein function and are even required for DNA replication, transcription, and repair critical for genome integrity and stability (Fuss, Tsai, Ishida, & Tainer, 2015) where their function may include detection of non-B-DNA areas, which are generally associated with increased breaks and translocations linked with cancer (Bacolla et al., 2016). In prokaryotes, the formation, regulation, and diversity of [Fe–S] clusters have been most studied for *Escherichia coli* and *Azotobacter vinelandii* (Frazzon & Dean, 2003; Roche et al., 2013). In eukaryotes, maturation of both cytosolic and nuclear Fe–S proteins requires the cooperation of the mitochondrial systems derived from microbial ancestors (Lill & Mühlenhoff, 2006; Paul & Lill, 2015). In fact, [Fe–S] cluster proteins are key to mitochondrial respiration including the assembly, stability, and function of respiratory complexes I, II, and III, which underscores the fundamental roles of Fe–S clusters in biological respiration (Melber et al., 2016). The maturation and insertion of [Fe–S] clusters into apo-target proteins require mitochondrial ISC assembly machinery that contains scaffold protein IscU, cysteine desulfurase (Nfs1), accessory protein Isd11, activator/iron donor Frataxin, and Ferredoxin (Agar et al., 2000; Colin et al., 2013; Cory et al., 2017; Johnson et al., 2005; Lill, 2009; Pandey et al., 2013; Raulfs, O'Carroll, Santos Dos, Unciuleac, & Dean, 2008; Schmucker et al., 2011; Tsai & Barondeau, 2010; Webert et al., 2014; Yoon & Cowan, 2003; Zheng, Cash, Flint, & Dean, 1998). [Fe–S] clusters are assembled in mitochondrial and cytosolic machineries (Braymer & Lill, 2017; Bridwell-Rabb, Fox, Tsai, Winn, & Barondeau, 2014; Lill et al., 2006), and clusters are transferred to apo-target proteins aided by chaperones HscA/HscB upon maturation (Bonomi, Iametti, Morleo, Ta, & Vickery, 2011; Chandramouli & Johnson, 2006; Fox, Chakrabarti, McCormick, Lindahl, & Barondeau, 2015; Fox, Das, Chakrabarti, Lindahl, & Barondeau, 2015; Uzarska, Dutkiewicz, Freibert, Lill, & Mühlenhoff, 2013; Vranish et al., 2015). Defects in Fe–S cluster assembly and maturation lead to multiple human diseases and protein dysfunctions (Rouault, 2015; Stehling, Wilbrecht, & Lill, 2014) that encompass many processes including metabolism, DNA maintenance, Fe–S cluster assembly machineries, respiratory chain metabolism, ribosome function, and tRNA modification (Andreini, Banci, & Rosato, 2016). The roles of [Fe–S] clusters include key functions in gene expression activation, substrate binding, electron transfer, sulfur donation, and structural stabilization when ligated to protein cysteine residues (Johnson et al., 2005). Of these many functions, a remarkable and exemplary function in [Fe–S] proteins is electron shuttling, which was proposed as the rate-limiting step, between multi-[Fe–S] clusters functioning in respiratory metabolism (de Vries, Dörner, Strampraad, & Friedrich, 2015; Martin & Matyushov, 2017; Rikken, Kroon, & van Ginkel, 1996).

For such respiratory electron transfer, an underappreciated and critical [Fe–S] function involves the respiration of perchlorate (ClO_4^-) and chlorate (ClO_3^-) [collectively, (per)chlorate] that occurs in dissimilatory (per)chlorate-reducing bacteria (Youngblut, Wang, Barnum,& Coates, 2016). Perchlorate is a water-soluble chemical that can be reduced to chlorate and subsequently reduced to chlorite (ClO_2^-) by (per)chlorate reductase (PcrAB). The product chlorite, a harmful molecule, is then dismutated to chloride and oxygen by chlorite dismutase (Cld) (Lee, Streit, Zdilla, Abu-Omar, & DuBois, 2008) in a manner somewhat reminiscent of the superoxide dismutase reaction (Perry et al., 2010). The oxygen is further used by cytochrome *cbb*$_3$ oxidase for quinone oxidation (Bak & Elliott, 2014; Buschmann et al., 2010; Coates & Achenbach, 2004; Hosler, Ferguson-Miller, & Mills, 2006; Imlay, 2006). A remarkable feature of (per)chlorate respiration is electron shuttling ($2e^-$ at a time) from the quinone pool to heme to five [Fe–S] clusters, and finally to the Mo-bisMGD cofactor for (per)chlorate reduction (Youngblut, Tsai, et al., 2016; Youngblut, Wang, et al., 2016). The structure–activity relationships for these enzymes are relevant to removal of perchlorate (ClO_4^-) and nitrate (NO_3^-) from drinking water (Matos, Velizarov, Crespo, & Reis, 2006), to environmental controls of perchlorate and nitrate co-occurrence in arid and semiarid environments (Jackson, Böhlke, et al., 2015), to the control of H_2S production from industrial oil recovery and mining (Carlson et al., 2015), and even to the generation of water and oxygen from perchlorate on Mars (Carlson et al., 2015; Ojha et al., 2015). Given its emblematic and understudied features, we focus on PcrAB to examine relationships of protein structure to the regulation of [Fe–S] cluster function.

Importantly, examination of PcrAB enzyme structures discovered distinct conformational states in reduced and oxidized [Fe–S] cluster forms that inform a detailed mechanism for (per)chlorate reduction. These structures showcase the fundamental importance of [Fe–S] cluster enzymes for cellular energy and their dramatic effects on altering protein functional conformations. The identified aromatic gate residues guard substrate entry and regulate product release to control the electron flow for (per)chlorate reduction (Youngblut, Tsai, et al., 2016). Yet, we find that in light of comparative structural analyses, the seemingly unique aromatic gate opening and closing for PcrAB may in fact serve as a prototypic model for Mo-bisMGD- and [Fe–S] cluster-containing respiratory proteins, which can be usefully grouped into the DMSO reductase family (Schindelin, Kisker, Hilton, Rajagopalan, & Rees, 1996; Youngblut, Tsai, et al., 2016). To furthermore robustly enable research on [Fe–S] cluster-containing proteins, we herein share our tactics, protocols, and strategies for [Fe–S] protein production, purification, crystallization, data collection, and X-ray structure determination and use PcrAB as a prototypic system for structure analysis.

2. RECOMBINANT [Fe–S] PROTEIN PRODUCTION AND PURIFICATION IN GENERAL: WHAT IS NEEDED AND WHAT NEEDS TO BE CONSIDERED?

Aerobic oxidative damage for eukaryotic DNA polymerases α, δ, and ε provides a sobering warning regarding metalloenzyme studies. These were only found to contain [4Fe–4S] clusters in 2011 (Netz et al., 2011) about 30 years after DNA polymerase α was isolated (Johnson, Snyder, Chang, & Davis, 1985). This unrecognized existence of [Fe–S] clusters in eukaryotic DNA polymerases, which delayed progress for decades, resulted from their aerobic protein production and purification, which caused oxidation damage and loss of the clusters resulting in dysfunctional polymerases. This is a general problem for native and recombinant proteins, as the metalloprotein remains largely undefined (Yannone, Hartung, Menon, Adams, & Tainer, 2012) even for microbes (Cvetkovic et al., 2010). Here we share general protocols and strategies that have proven successful for working with and analyzing proteins containing [Fe–S] clusters in ways that can protect them from oxidation damage and thus allow detailed structure determinations and analyses, as seen for multiple DNA repair proteins (Banda, Nuñez, Burnside, Bradshaw, & David, 2017), for sulfite reductase (Crane, Siegel, & Getzoff, 1995), and for the XPD helicase that acts in transcription and repair (Fan et al., 2008).

2.1 [Fe–S] Protein Production

Recombinant [Fe–S] protein overexpression in *E. coli* is often a first choice for large-scale protein production at low cost, although baculovirus methods continue to improve especially for multiprotein complexes such as TFIIH (Gradia, Ishida, Tsai, Jeans, & Tainer, 2017), and an exciting novel system has been developed in *Pyrococcus furiosus* that should be considered as well (McTernan et al., 2014). Importantly, *E. coli* contains the ISC biosynthetic system (*iscS*, *iscU*, *iscA*, *hscB*, *hscA*, and *fdx*) for general housekeeping Fe–S cluster assembly (Johnson et al., 2005; Takahashi & Nakamura, 1999; Zheng et al., 1998) and the sulfur-utilization factor (SUF) system (*sufA*, *sufB*, *sufC*, *sufD*, *sufS*, and *sufE*) for Fe–S cluster assembly when iron or sulfur sources are scarce (Barras, Loiseau, & Py, 2005; Outten, Djaman, & Storz, 2004). Thus, functional [Fe–S] clusters can be incorporated into *E. coli* that recombinantly expressed [Fe–S] proteins. However, in some cases, overexpressed [Fe–S] proteins lack [Fe–S] clusters or have reduced levels of clusters. In such cases, the [Fe–S] cluster maturation can be facilitated by coexpressing the *isc* operon in *E. coli* strains (Nakamura, Saeki, & Takahashi, 1999) or by adding cysteine and iron supplement (Jaganaman, Pinto, Tarasev, & Ballou, 2007; Kuchenreuther et al., 2010) before induction or by expressing the bacteria anaerobically (Kuchenreuther et al., 2010). For example, in the case of perchlorate reductase (PcrAB) from *Azospira suillum* PS, overexpression of PcrAB in *E. coli* resulted the loss of enzymatic activity due to the lack of Mo-bisMGD cofactor incorporation, which required expression of the PcrAB in the *A. suillum* PS native host (Youngblut, Tsai, et al., 2016; Youngblut, Wang, et al., 2016).

Five tips for [Fe–S] protein production

1. Subcloning the gene of interest into the vector of choice. pET vector is one of the most common vectors used for protein expression in bacteria. Many [Fe–S] proteins were expressed using His-tag as their affinity tag of choice. However, the potential risk in employing His-tag constructs for [Fe–S] proteins is that the metal chelate affinity purification process may chelate iron and thereby lower the occupancy of [Fe–S] cluster bound to the target protein (Daniels et al., 2000). Furthermore, the nonspecific binding of untagged protein may increase the complexity and time of purification process (Bornhorst & Falke, 2000). Alternatively, GST (glutathione *S*-transferase) tag and MBP (maltose-binding protein) tags may have more specific binding that can increase the purity of [Fe–S] protein, decrease the time of purification process, and increase the solubility of the [Fe–S] protein (Lebendiker & Danieli, 2017; Schäfer, Seip, Maertens, Block, & Kubicek, 2015). Moreover, addition of 1–10 m*M* reducing agents like dithiothreitol (DTT) or β-mercaptoethanol (BME) or tris(2-carboxyethyl)phosphine (TCEP) may help to stabilize [Fe–S] protein and increase binding affinity of GST fusion (Schäfer et al., 2015). Glutathione used in GST elution buffer can provide additional oxidation resistance protection for [Fe–S] protein. If the binding affinity of GST-tagged protein is low, incubation of GST-fusion protein with the GST resin for 1 h or slow application rate (0.5 mL/min) or addition of stabilization agents (500 m*M* NaCl and 10% glycerol) may help to increase the binding affinity.
2. Choosing host strains. BL21(DE3) strains are often a good candidate for bacterial host strain choice. For human [Fe–S] protein expression, Rosetta 2(DE3) host strain (Novagen) is a BL21 derivative designed to increase eukaryotic protein expression. Alternatively, yeast strains or other native strains can help the expression and solubility of [Fe–S]

proteins (Pierik, Netz, & Lill, 2009; Youngblut, Tsai, et al., 2016). For difficult eukaryotic [Fe–S] protein or protein complex, insect cells Sf9 or Hi5 are the commonly used for protein expression and the Hi-5 insect cell is designed to enhance the protein expression level (Kidd & Emery, 1993). QB3 Macrolab from University of California, Berkeley developed baculovirus expression vectors with biobrick polypromotor, called MacroBac (Gradia et al., 2017), that is suitable for difficult eukaryotic [Fe–S] protein or multiprotein complex expression. The concept of MacroBac system was adapted from MultiBac system (Berger, Fitzgerald, & Richmond, 2004; Fitzgerald et al., 2006) along with the ligation-independent cloning technique (Aslanidis & de Jong, 1990) to provide efficient and robust cloning system for multiprotein complex expression. Other baculovirus expression system such as biGBac system (Weissmann et al., 2016) adapted from the Gibson assembly cloning concept (Gibson et al., 2008, 2009) was also developed.

3. Choosing the growth medium. Terrific broth (TB) was tested to produce at least 25% more [Fe–S] protein than minimal media (M9) and Luria-Broth (LB) for bacteria growth due to its higher buffering capacity when induced at high OD ($OD_{600} > 1$) (Jaganaman et al., 2007).
4. Adding the supplement. When [Fe–S] protein expression level or [Fe–S] cluster occupancy is low, addition of 1 mM cysteine and 1 mM ferric ammonium citrate or ferric citrate during induction may help to boost the protein expression (Jaganaman et al., 2007).
5. Selecting the temperature for protein expression. We routinely screen temperature for protein expression in bacteria from 16°C to 37°C. We normally pick three to four temperatures to test, e.g., 16°C, 22°C, 28°C, and 37°C. For eukaryotic protein expression, we generally see better solubility and folding (i.e., less protein aggregation) at lower temperature (16°C or 22°C).

2.2 [Fe–S] Protein Purification: Use Anaerobic Hood and Check for Full Fe–S Occupancy

A dark brown color for concentrated sample is usually characteristic of [Fe–S] clusters ligated to cysteine residues. If an [Fe–S] protein solution turns a light yellow color, it is typically a warning sign of [Fe–S] cluster oxidation or loss. Although some [Fe–S] proteins were purified aerobically from many laboratories with some success due to the lack of an anaerobic chamber, the purification of [Fe–S] proteins anaerobically could be critical to successfully maintain the [Fe–S] cluster and enzyme activity, especially for more oxygen-labile [Fe–S] proteins like FNR (Yan & Kiley, 2009) and XPD (Fan et al., 2008). Here we share our tips to maintain full [Fe–S] cluster occupancy during purification.

1. The ideal way to purify an [Fe–S] protein is to purify it anaerobically in a well-designed anaerobic chamber with low oxygen level (<0.1 ppm) based upon a reliable oxygen monitor. We have found that the vinyl anaerobic chamber from Coy Labs can provide a suitable anaerobic environment for reliable oxygen-sensitive protein preparation (Fig. 1A). The anaerobically chamber is typically equipped with an infuser (Coy Labs), which can constantly maintain H_2 level (usually set at 3%) to ensure that the active palladium catalyst converts oxygen to water. Cells can be

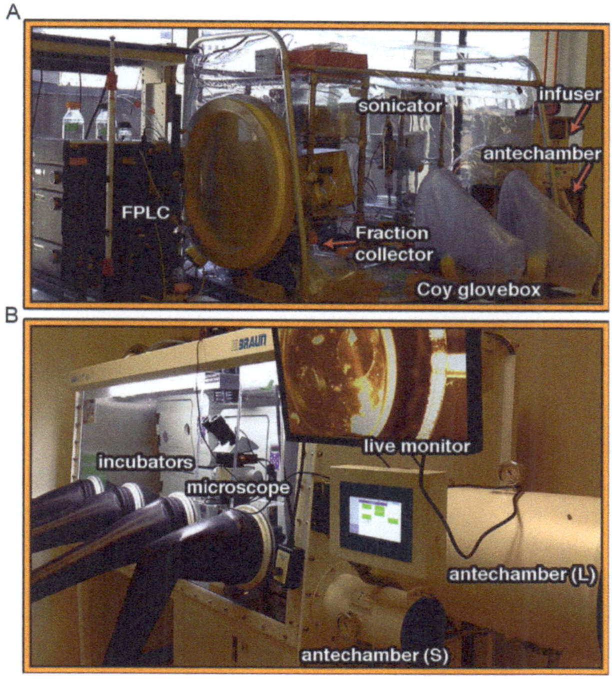

Fig. 1 Glovebox setup for anaerobic purification and crystallization. (A) Cells can be homogenized inside the glovebox by using a sonicator. The supernatant is loaded onto FPLC columns through a superloop. Fractions are all collected inside the glovebox. An infuser from the CoyLab is used to automatically maintain the $3\%H_2/97\%N_2$ level. (B) Exemplary setup for crystallizations and screening in the glovebox (<0.1 ppm O_2). Three incubators are included to maintain three different temperatures. Crystals, which are visualized and harvested under the microscope, can be immediately transferred out of the glovebox through the small antechamber for flash cooling.

homogenized inside the glovebox by using a sonicator, and protein can be purified by using an AKTA FPLC system that is connected to the anaerobic chamber, as shown for our laboratory setup (Fig. 1A).

2. All buffer should be purged with Argon or N_2 ($\geq$99.999% purity with $\leq$1 ppm O_2) inert gas to minimize the oxygen level by using a Schlenk line manifold system with or even without an anaerobic chamber. Moreover, we suggest the use of the reducing agent TCEP, which has more resistance to oxidation in air (Burns, Butler, Moran, & Whitesides, 1991) compared to DTT ($E = -0.33$ V at pH 7) (Cleland, 1964), BME ($E = -0.28$ V at pH 7), and glutathione ($E = -0.28$ V at pH 7) (Mickey & Howard, 1995). It is a good idea to include reducing agents in the buffer to scavenge oxygen and keep cysteine ligands reduced unless the [Fe–S] protein is sensitive to reducing agents, such as the disulfide bond-dependent protein pyruvate–ferredoxin oxidoreductase (Vita, Hatchikian, Nouailler, Dolla, & Pieulle, 2008).
3. The iron content of a purified [Fe–S] protein can be usefully measured by colorimetric assay (Cowart, Singleton, & Hind, 1993; Fish, 1988) or inductively coupled plasma mass spectrometry (ICP-MS) (Fox, Chakrabarti, et al., 2015; Yan & Kiley, 2009) to calculate the occupancy of [Fe–S] clusters.
4. In the case of low [Fe–S] occupancy, the [Fe–S] cluster may be reconstituted using chemical or enzymatic methods in vitro as noted here.

   For the chemical method, ferrous ammonium sulfate $Fe(NH_4)_2(SO_4)_2$ (>100 *M* excess) and DTT (>100 *M* excess) are added to the apoprotein, followed by adding sodium sulfide Na_2S (>100 *M* excess) dropwise into the mixture with stirring to prevent the immediate generation of insoluble Fe–S colloids. After the mixture is stirred for an additional hour, the excess iron and sulfide materials can be removed on a desalting column (Coghlan & Vickery, 1991; Malkin & Rabinowitz, 1966; Venkateswara Rao & Holm, 2004).

   Alternatively, the enzymatic method is cleaner and more efficient. Basically, excess Fe^{2+} or Fe^{3+} (>10 *M* excess), L-cysteine (>10 *M* excess), and DTT (>100 *M* excess) are added to the apoprotein. The assembly reaction is initiated by adding catalytic amount of cysteine desulfurase (IscS). IscS generates persulfide from L-cysteine, followed by transferring to apoprotein for [Fe–S] cluster assembly process (Agar et al., 2000; Fox, Chakrabarti, et al., 2015; Fox, Das, et al., 2015; Kiley & Beinert, 2003). Excess materials can then be removed on a desalting column.

3. [Fe–S] PROTEIN CRYSTALLIZATION, DATA COLLECTION, AND STRUCTURE DETERMINATION

3.1 Considerations and Strategies for Anaerobic Crystallization

High-resolution X-ray structures provide unmatched detail that often uniquely enables insights into the biological activity and function of a given [Fe–S] cluster protein. However, the loss of [Fe–S] cluster due to oxidation may result in structure disorder that impacts the protein function. For example, [4Fe–4S] cluster acts in the structure ordering role in [Fe–S] cluster domain of XPD helicase. The loss of [4Fe–4S] cluster reveals a disordered structure that significantly decreases XPD catalytic activity and DNA-binding affinity (Fan et al., 2008). Therefore, anaerobic crystallization of an [Fe–S] protein is a key step to obtain the intact protein structure, especially for oxygen-labile [Fe–S] proteins such as XPD, IRP1, and PcrAB (Dupuy et al., 2005; Fan et al., 2008; Youngblut, Tsai, et al., 2016). The order or full occupancy of [Fe–S] cluster in the structure depends on how sensitive of [Fe–S] cluster is to oxygen. And the level of oxygen sensitivity can depend upon shifts in the [Fe–S] redox potential, such as occurring from changes in the electrostatic environment of the cluster upon DNA binding, as seen for XPD, MutY, and endonuclease III (Boon, Livingston, Chmiel, David, & Barton, 2003; Gorodetsky, Boal, & Barton, 2006; Sontz, Mui, Fuss, Tainer, & Barton, 2012). When the purification and crystallization are performed in air, there is always a question whether [Fe–S] protein remains stable and functional. E.g., perchlorate reductase PcrAB lost its brown color and activity (i.e., loss of [Fe–S] clusters) after overnight incubation at 4°C in air. For oxygen-sensitive [Fe–S] proteins, such as PcrAB, the crystallization buffer must be purged with N_2 or argon, and crystallization trials must be set up in the anaerobic chamber (Fig. 1B).

Tips for setting up anaerobic crystallization are outlined below based upon our experience.

1. Without an anaerobic chamber, the crystallization buffer should minimally be purged with high purity N_2 or argon to prolong the half-life of [Fe–S] clusters.
2. It is usually a good idea to include 0.5–2.0 m*M* TCEP and/or 1–10 m*M* DTT and/or 5–20 m*M* BME in your protein buffer to help keep cysteines reduced and slow down the oxidation of any [Fe–S] clusters.

3. With an anaerobic chamber, the crystallization screening solution in a 96-deep-well block and any necessary crystallization plates, pipette tips, and reagents should be brought into the anaerobic chamber at least 3 days in advance. This is important to allow the chamber catalyst to remove most of the dioxygen from the plastic or solution before setting up crystallization trials. If the screening solution cannot be preequilibrated in the glovebox a few days in advance, you can equilibrate the screening solution (50–100 μL) by exposing it to high purity N_2 in the glovebox for 30 min to exchange the oxygen after being pipetted into the crystallization plate.
4. For setting up sparse-matrix crystallization screening trials, a multichannel pipette is usually efficient, which usually takes about 10 min per plate, and 0.5–1 μL protein per drop for good accuracy on pipetting. For some less oxygen-sensitive [Fe–S] proteins, crystallization screening trials can be setup with pipetting robot located outside the anaerobic chamber (*note*: it is preferable if the robot can locate inside the chamber), followed by immediately bringing the crystallization plate into the anaerobic chamber for incubation. The robot offers the advantage of less time (about 1–2 min per plate) and lower amounts of protein (0.1–0.2 μL per drop, 10–20 μL protein per plate) if the protein sample is limited. For this method, crystallization screens should be preequilibrated in the glovebox using tip #3 above. After the plates have been set up by the robot, the plates should be sealed tightly with the optical transparent sealing film and clamped with large paper clips to prevent the evaporation during evacuation cycle using the vacuum pump (Fig. 2A).
5. For growing crystals in the reduced state, a glovebox is optimal. E.g., PcrAB crystals were grown in an oxidized state where Mo atom is in +6 oxidation state (Fig. 2B). To reduce the oxidation state, a reducing agent like sodium dithionite ($Na_2S_2O_4$, $E = -0.66$ V at pH 7) or zinc powder ($E = -0.76$ V at pH 7) combined with the electron carrier methyl viologen ($E = -0.446$ V) can be used to reduce [Fe–S] clusters (Mayhew, 1978; Najmudin et al., 2008; Youngblut, Tsai, et al., 2016). To prepare reduced methyl viologen solution (10 m*M*), sodium dithionite (100 m*M*) or excess Zn powder (100 mg) is added into 1 mL of methyl viologen solution in tube. The solution should turn deep-blue color (absorption 600 nm, $\varepsilon_{600} = 8.25\ \mathrm{m}M^{-1}\ \mathrm{cm}^{-1}$) when the methyl viologen is reduced (Yu & Wolin, 1969); it turns colorless when it is oxidized. The excess Zn powder can be removed by centrifugation, and the supernatant containing reduced methyl viologen is then added into the

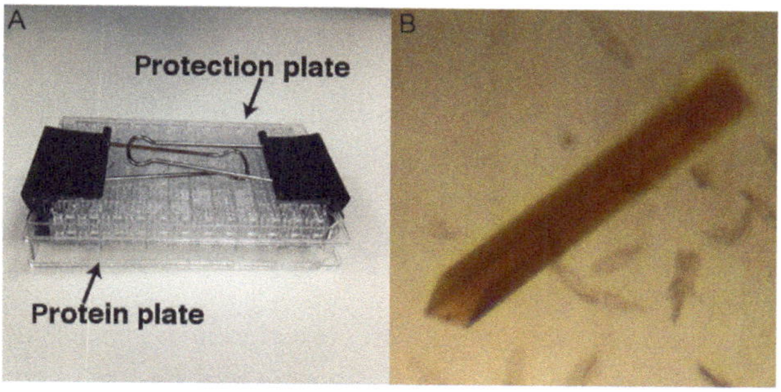

Fig. 2 Anaerobic crystallization setup and transfer. (A) Simple but effective setup to bring a setup crystallization plate into the glovebox for less oxygen-sensitive [Fe–S] proteins using pipetting robot located outside the glovebox. The protection plate was placed on top of a protein crystallization plate and clamped down with paper clips to prevent evaporation from the drops and reservoir during the vacuum pump evacuation cycle. (B) Oxygen-sensitive PcrAB crystal grown in the glovebox at ~20°C. The *brown crystal color* shows the presence of intact [Fe–S] clusters.

protein solution or crystals for reduction (Najmudin et al., 2008; Youngblut, Tsai, et al., 2016). Reduced methyl viologen can be added to the purification or assay buffer to scavenge oxygen and maintain the anaerobic condition by monitoring the blue color change.

6. A microscope can be brought into the anaerobic chamber or mounted on the panel of the anaerobic chamber for efficient screening of crystallization trays (Fig. 1B). Automatic crystallization plate tracking and imaging system provides periodically recorded images that can be stored and analyzed from each crystallization drop to monitor the crystal growth over time (Mayo et al., 2005).

3.2 Protocols for Anaerobic Crystal Mounting

Mounting crystals with those thick gloves attached to the anaerobic chamber are quite challenging. Often crystallographers replace the standard gloves with more flexible ones that can be clamped off when not in use. They also may bring in liquid nitrogen into the glovebox to freeze the protein crystals. Normally, the liquid nitrogen in the dewar is placed in the antechamber. The antechamber is then evacuated to about halfway of the vacuum pressure gauge reads, followed by refilling. By repeating this process for ~10–15 times, we found it challenging to bring liquid nitrogen into the glovebox without contaminating the glovebox with some dioxygen. An alternative is to employ the room temperature mounting kit from MiTeGen (MicroRT capillary system), which is usually used to test diffraction without cryoprotectant or to do controllable crystal dehydration experiments. We found the MiTeGen kit to be a suitable tool to mount crystals anaerobically. Drawbacks are that only one crystal is mounted and taken out at a time from the glovebox, and it may not work well if you are doing time-resolved crystallography due to the longer transfer time.

Protocol for mounting crystals anaerobically with a room temperature mounting kit

1. To harvest the crystals from 96-well crystallization plates is usually harder anaerobically. You need to cut off the tape and mount your crystal with thick gloves or with replaced more sensitive gloves that may allow oxygen into the chamber. Alternatively, if you can optimize the crystal conditions for a screw-cap crystallization plate (Qiagen EasyXtal 15-Well Tool). This will then be much easier to work with when mounting the crystals.
2. Before mounting the crystals, pipette 40 μl of artificial cryoprotectant solution into the MicroRT capillary with a gel-loading pipette tip to maintain the crystal hydration.
3. Pipette 10 μL of cryoprotectant solution on the siliconized glass cover slide (Hampton Research). Transfer single crystals from the drop into cryoprotectant solution (only if your crystals need added cryoprotectant).
4. Mount the crystal with a loop that has been inserted into an 18-mm MiTeGen goniometer base (MiTeGen B3S) with a magnetic wand. Slide the capillary tubing past your crystal and onto the goniometer base with the MicroRT aligner. The cryoprotectant solution inside the MicroRT capillary should prevent the dehydration of the mounted crystal for a few hours, which allows the time to transfer the mounted crystal out of the glovebox.
5. Fasten the magnetic wand on the aligner with a laboratory labeling tape to secure the position and transfer the crystal out of the anaerobic chamber.
6. Carefully remove the capillary tubing without touching the crystal, and flash cool it in liquid nitrogen. This procedure was used for anaerobic [Fe–S] protein crystals that have yielded a 1.86 Å resolution structure with low mosaic spread (0.2 degrees) (Youngblut, Tsai, et al., 2016).

3.3 Data Collection Strategy for [Fe–S] Proteins Employing Fe Phasing With Minimum Exposure and Maximum Redundancy

A key benefit of [Fe–S] proteins for structure determination is that you may be able to use the Fe anomalous signal to solve the phase problem (Bowman, Bridwell-Rabb, & Drennan, 2016), provided you are careful to avoid loss of the damage-sensitive anomalous signal. The Fe anomalous absorption peak is at ~7112 eV (1.7433 Å), which is accessible at most of synchrotron beamlines, such as SIBYLS at the Advanced Light Source (Classen et al., 2013).

Protein crystals with heavy atoms are often more sensitive to radiation damage. In the case of PcrAB, each PcrAB heterodimer contains 20 Fe atoms and 1 Mo atom. When the protein crystal suffers radiation damage, diffraction spots at a high-resolution shell start to fade away quickly, but even more importantly, the anomalous signal is lost before the diffraction peak disappears. You can compare images to see if the diffraction spots in the higher resolution shells are fading. Image quality including radiation damage should be carefully assessed by checking the R_{sym} to see if the number increases dramatically. Additionally, X-ray absorption spectroscopy (XAS) can be measured to assess the dose-dependent photoreduction of crystals (Corbett et al., 2007). We recommend that diffraction data are collected using minimal exposure and with high redundancy ~20-fold. Notably, collecting the data repeatedly at the same spot in the crystal for over 360 degrees will increase redundancy, but may not improve data quality. Ideally, multiple datasets from different spots in the same single crystal can be merged to increase redundancy and improve data quality. For an optimal anomalous data collection strategy, collecting 30 degrees wedges for peak wavelength with a reverse beam for total 360 degrees is followed by remote and inflection wavelength data collection. This strategy ensures at least a complete SAD dataset at peak wavelength in case radiation damage happens more quickly than completion of multiple wavelength datasets. Alternatively, you can use helical (Flot et al., 2010) or vector (Hilgart et al., 2011) data collection strategies. These minimize radiation damage by collecting images helically around your crystals or by collecting from two endpoints of your crystals. These strategies work if your crystal is long enough to obtain a complete dataset by moving along the crystal length, and the beamline is capable of these data collection strategies. Overall, the key to collect good diffraction data from [Fe–S] protein crystals is to minimize the radiation damage by collecting data quickly (with the shortest practical exposures) and with high redundancy.

3.4 Structure Determination: Proper Refinement to Maintain Cluster Geometry and Occupancy

Improper structure determination and refinement may distort the geometry of [Fe–S] clusters. Proper refinement is important as active sites may show actual functional distortions in the [Fe–S] clusters: you will wish to distinguish possible small functional distortions from poor refinement practices. Experimental phases can be obtained from an anomalous dataset by using heavy-atom phasing and refinement with Phenix.autosol (Terwilliger et al., 2009), SHARP (Bricogne, Vonrhein, Flensburg, Schiltz, & Paciorek, 2003), or SHELX (Sheldrick, 2008). Alternatively, molecular replacement (MR) using homology models (generally >30% sequence identity) by PHASER (McCoy et al., 2007) or other MR programs in CCP4 suite (Winn et al., 2011) may be a good method to develop structure solutions. For difficult structure solution, especially for a low-resolution dataset or poor homology models (<30% sequence identity), MR combined with structure-modeling software Rosetta (Das & Baker, 2008) may help to generate a better solution (DiMaio et al., 2011). Once an electron density map is generated, proper refinement with correct cluster geometry needs to be carefully done.

Restraints of geometry for the [Fe–S] cluster should be included in programs like refmac (Murshudov, Vagin, & Dodson, 1997) or Phenix.refine (Afonine et al., 2012), where typically the ideal Fe–S bond distance is set as 2.30 Å with sigma 0.05 for bond distance and no restraint is set for bond angle. Besides the geometry of [Fe–S] clusters, occupancy and *B*-factor of [Fe–S] clusters should be refined as well. Changing the occupancy of each [Fe–S] cluster to below 1 (e.g., 0.95) will allow the Phenix.refine to refine the occupancy automatically based on your experimental data or you can specifically refine the occupancy of a particular cluster by selecting its atoms in the Phenix.refine parameter file. Observation of electron density maps to check positive and negative difference electron density peaks around the [Fe–S] clusters allows checks of Fourier truncation effects and the Fe atom position, occupancy, and temperature factor. More refinement cycles with refinement of occupancy and anisotropic atomic displacement parameters of the metal ions may help to resolve these parameters, but do not resolve Fourier truncation ripples. Furthermore, the electron density map of [Fe–S] cluster should show good to strong peaks that indicate little to no radiation damage was produced (Fig. 3). The proper refinement of [Fe–S] cluster structures may reveal the extend to which their distortion impacts enzymatic activity and biological function (Bowman et al., 2016). Such distortion can be further measured by Mössbauer spectroscopy (Pandelia, Lanz, Booker, & Krebs, 2015) and XAS (Einsle, Andrade, Dobbek, Meyer, & Rees, 2007; Ha et al., 2017) to evaluate the coordination geometry, solvation, and redox states of [Fe–S] clusters. These data can then be tested by computational

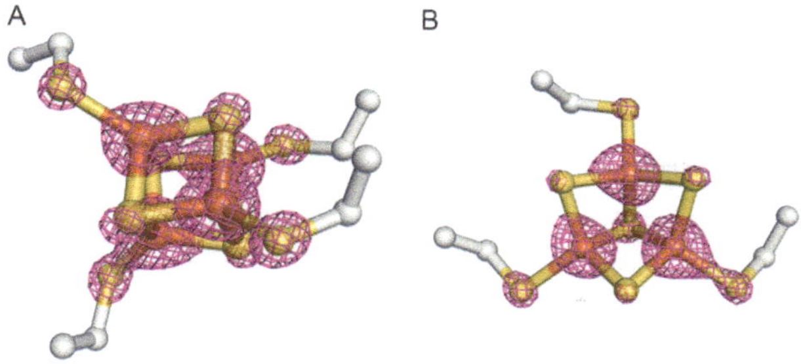

Fig. 3 Electron density maps of (A) [4Fe–4S] (FS1) and (B) [3Fe–4S] (FS4) cluster of PcrAB (PDB: 4ydd). The 2Fo–Fc CCP4 map was generated using Phenix.fft with a grid resolution factor of 0.167. The graphic figure was contoured at 7.0 sigma and prepared on PyMOL. Comparatively, the electron density of the [4Fe–4S] cluster shown here suggests that it may suffer less radiation damage than the [3Fe–4S] cluster, reflecting the direct solvent exposure of the [3Fe–4S] cluster.

modeling to investigate the electronic structures of the [Fe–S] clusters (Fee et al., 2003; Sandala, Hopmann, Ghosh, & Noodleman, 2011; Shomura, Yoon, Nishihara, & Higuchi, 2011).

4. STRUCTURAL ANALYSES: WHAT TO LOOK FOR IN Fe–S CLUSTER STRUCTURES: CASE STUDIES IN TYPE II MO-BISMGD ENZYMES

Structural analysis can be critical to provide insight into protein function and mechanism. For example, DNA binding to EndoIII and MutY glycosylases shifts the redox potential of their [4Fe–4S] clusters into physiological range to form a redox-active state (Boal et al., 2005; Boon et al., 2003). This shift results in part by the position of the DNA polyanion near the [Fe–S] cluster with accompanying structural changes. These changes include significantly altering Fe–S cluster loop (FCL) between the first two ligating cysteines that reside along the DNA-bound interface, as were revealed by the crystal structures of EndoIII and MutY (Chepanoske, Golinelli, Williams, & David, 2000; Fromme & Verdine, 2003; Guan et al., 1998; Kuo et al., 1992; Thayer, Ahern, Xing, Cunningham, & Tainer, 1995). These electrostatic and protein structural changes create a large increase in the covalency of the Fe–S bond upon DNA binding, which can be measured by sulfur K-edge XAS, despite the similarity of their [Fe–S] cluster structures with and without DNA (Ha et al., 2017). At the same time, some [Fe–S] protein structure domain folds depend upon their [Fe–S] clusters, such as *Sulfolobus acidocaldarius* XPD (Fan et al., 2008), human primase (Baranovskiy et al., 2015; Vaithiyalingam, Warren, Eichman, & Chazin, 2010), *Bacillus subtilis* AddAB (Saikrishnan et al., 2012), human exonuclease 5 (Sparks et al., 2012), and *Mus musculus* DNA2 (Zhou, Pourmal, & Pavletich, 2015). Loss of such [Fe–S] clusters may cause domain disorders that decrease DNA-binding affinity and lead to functional defects. Indeed, metal site loss or defects may in general cause decreased fold stability and even human diseases, as seen, for example, for human superoxide dismutase (Pratt et al., 2014).

For [Fe–S] cluster-containing respiratory protein complexes, we herein compare three enzymes (perchlorate reductase PcrAB, nitrate reductase NarGHI, and ethylbenzene dehydrogenase, EbdABC). These enzymes are key members of the type II DMSO reductase family, which commonly contains the signature bis (molybdopterin guanine dinucleotide)Mo cofactor (Mo-bisMGD) and an Asp-Mo ligand (Iobbi-Nivol & Leimkühler, 2013; Jormakka, Richardson, Byrne, & Iwata, 2004; McDevitt, Hugenholtz, Hanson, & McEwan, 2002; Moura, Brondino, Trincão, & Romão, 2004; Schindelin et al., 1996; Schwarz, Mendel, & Ribbe, 2009).

Interestingly, these three protein complexes furthermore have as a hallmark five [Fe–S] clusters that are responsible for electron shuttling to Mo atom or to heme cofactor (Figs. 4, 6, and 7). PcrAB is an anaerobic respiratory protein found in dissimilatory (per)chlorate-reducing bacteria that use (per)chlorate as terminal electron acceptors by converting perchlorate to chlorate and then to chlorite (Bender et al., 2005; Coates & Achenbach, 2004; Youngblut, Wang, et al., 2016). PcrAB from *A. suillum* PS is a periplasmic heterodimeric protein complex that contains a Mo-bisMGD and [4Fe–4S] cluster (FS0) in PcrA, plus three [4Fe–4S] clusters (FS1–3) and one [3Fe–4S] cluster (FS4) in PcrB, respectively (Fig. 4A). The reduction of perchlorate to chlorite requires two separate $2e^-$ transfer reactions that may be transferred from heme-containing proteins PcrQ, PcrO, and PcrC to PcrB and PcrA (Fig. 4A) (Youngblut, Tsai, et al., 2016; Youngblut, Wang, et al., 2016). It remains elusive if the PcrAB forms a complex with heme proteins PcrC, PcrO, and PcrQ to thereby resemble the heterotrimeric cytosolic nitrate reductase cNarGHI (Bertero et al., 2003). The distances between the five [Fe–S] clusters and the Mo atom can range from 9.41 to 12.8 Å (Fig. 4B), which are similar to the electron transfer distance in respiratory complex I (Hayashi & Stuchebrukhov, 2010) and fall within the proposed limit of 14 Å for an electron transfer distance (Page, Moser,

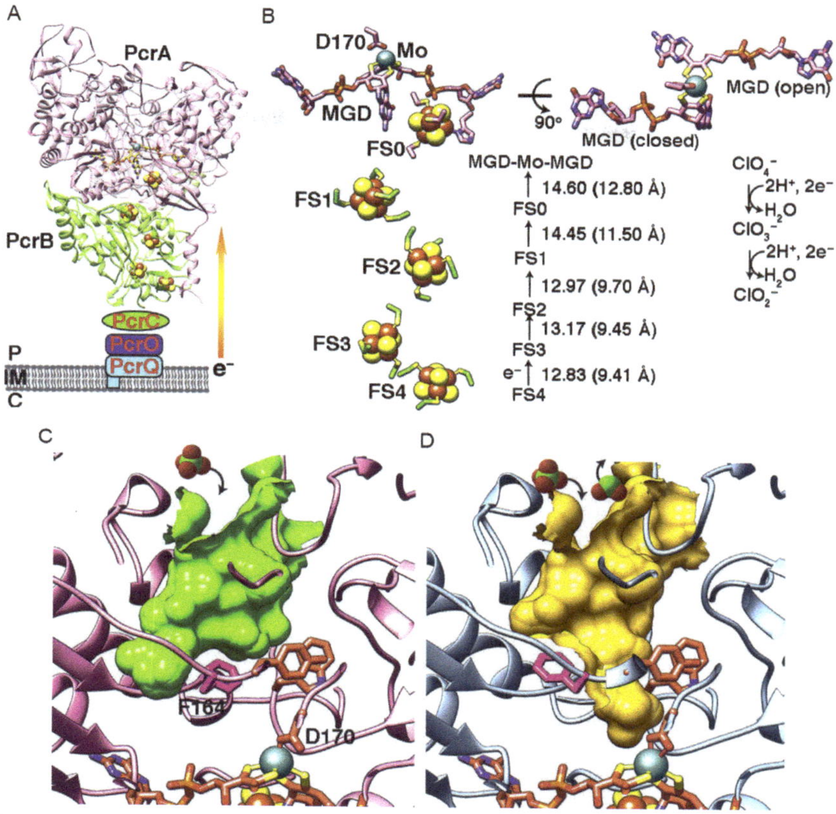

Fig. 4 PcrAB structure from *Azospira suillum* PS containing a Mo-bisMGD cofactor plus five [Fe–S] clusters (PDB: 4ydd and 5ch7). (A) Periplasmic PcrAB is proposed to receive electrons from membrane-bound heme protein PcrQ and two heme-binding proteins PcrO and PcrC via quinone oxidation. The electrons are then transferred to downstream [Fe–S] clusters and Mo-bisMGD for (per)chlorate catalysis. *Labels* show P (periplasm), IM (inner membrane), and C (cytoplasm). (B) The layout of the active site Mo-bisMGD, conserved Asp170 ligand, and [Fe–S] clusters. MGD exhibits tricyclic pterin (closed) and bicyclic dihydropterin (open) forms in both oxidized and reduced Mo states. The distances between each cofactor are shown. The edge-to-edge cluster distance is calculated from the closest two atoms of two neighboring cofactors; center-to-center distance is calculated from the average of the longest + shortest distance between two cofactors. (C) The substrate tunnel (*green surface*) is unveiled in oxidized PcrA (PDB: 4ydd), while F164 is in the closed conformation. (D) PcrA is in the reduced state (PDB: 5ch7); D170 was shifted from monodentate to the bidentate coordination state. F164 rotates to a different conformation and opens the substrate tunnel (*yellow surface*) into the active site for (per)chlorate catalysis. Substrate perchlorate and product chlorite are shown in space-filling models.

Chen, & Dutton, 1999). In the oxidized form of PcrA, the substrate tunnel to the Mo-bisMGD active site is closed by the gate residue F164 (Fig. 4C). Furthermore, here we identified the surface topography of these proteins through the CASTp server (http://sts.bioe.uic.edu/castp/index.php), which also shows volume and area of the protein pockets and cavities (Liang, Edelsbrunner, Fu, Sudhakar, & Subramaniam, 1998a, 1998b; Liang, Edelsbrunner, & Woodward, 1998). We also identified gate residues by carefully examining the residues lining the tunnel to the active site. During catalysis, ligand D170 shifts from a mondentate to the bidentate coordination state when the Mo atom is reduced, and the substrate tunnel is opened via the shifting of F164 to different rotamers (Figs. 4C–D and 5). The F164 gate residue controls substrate entry and product release to enhance the catalytic efficiency when the (per)chlorate concentration is limited in the bacterial environment (Youngblut, Tsai, et al., 2016; Youngblut, Wang, et al., 2016). Indeed, this explains why PcrAB has a relatively high perchlorate affinity ($K_m = 6\,\mu M$) and substrate inhibition over $200\,\mu M$ compared to nitrate reductase NarGHI, which has a lower perchlorate affinity ($K_m = 1.1\,mM$). In fact, mutation of F164A in *A. suillum* PcrA resulted in a loss of its ability to grow by (per)chlorate respiration (Youngblut, Tsai, et al., 2016).

Comparative examination of [Fe–S] protein structures aids the identification of functional vs circumstantial elements. We therefore examined the [Fe–S] clusters, Mo-bisMGD cofactor, and substrate tunnels in cytosolic nitrate reductase cNarGHI (Bertero et al., 2003) and periplasmic ethylbenzene dehydrogenase EbdABC (Kloer, Hagel, Heider, & Schulz, 2006) (Figs. 6 and 7). For cNarGHI, the layout of the [Fe–S] clusters and Mo-bisMGD active site in cNarGH resembles PcrAB (Fig. 6A and B). In fact, nitrate reductase can also reduce perchlorate and chlorate at a slower rate than nitrate (Marangon et al., 2012; Youngblut, Tsai, et al., 2016). The reduction of nitrate requires $2e^-$ from quinone pool oxidation that generates a $2H^+$ proton-motive force from cytoplasm to periplasm and delivers $2e^-$ to cNarI heme protein, to cNarH, and then to cNarG (Bertero et al., 2003). It was proposed that formate dehydrogenase transfers $2e^-$ to the quinone pool and forms a redox loop to cytosolic nitrate reductase (Bertero et al., 2003; Jormakka, Törnroth, Byrne, & Iwata, 2002). Using comparative analyses and the above methodology, we identify here the gate residue F216 in *E. coli* cNarG. Shift of the F216 rotamer conformation is suitable to enable access of nitrate substrate to the Mo-bisMGD active site for catalysis (Fig. 6C and D).

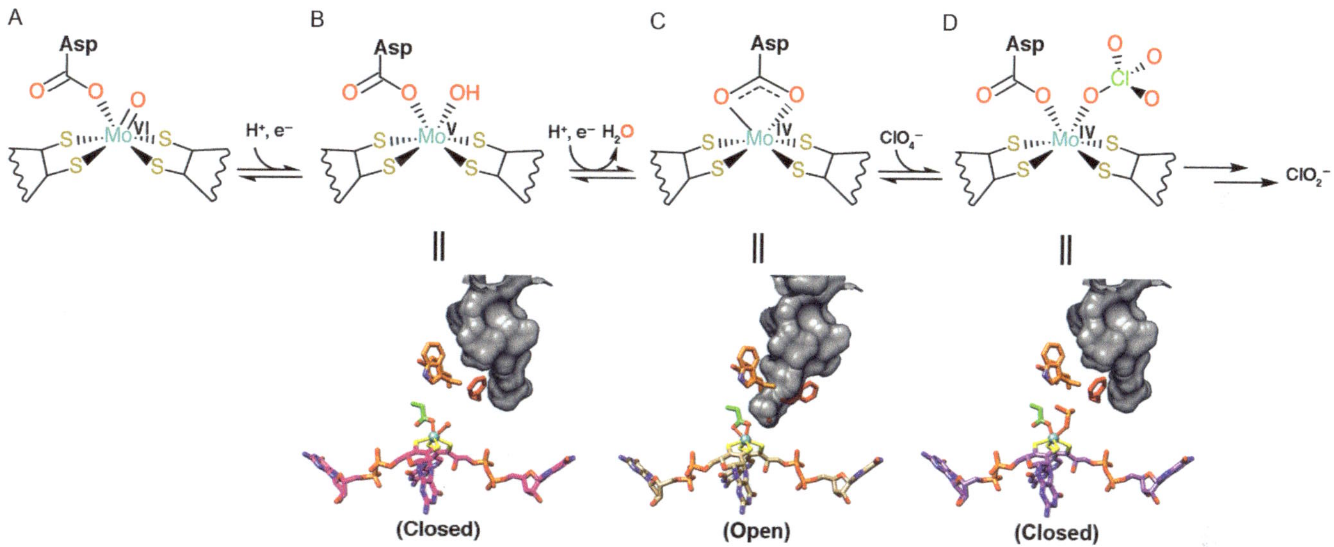

Fig. 5 Dissecting the mechanism of (per)chlorate catalysis based on PcrAB structures. PcrAB structures reveal different redox states corresponding to each catalytic step (A–D labels). The gate residue F164 (*red sticks*) controls the substrate/product access (shown in *gray surface*) to the Mo-bisMGD active site. In step (D) the substrate analog SeO_3^{2-} bound to Mo atom was visible in the structure. Notably, the Asp side chain (*green sticks* with *red* oxygen atoms) shifts coordination states from monodentate to bidentate and back to monodentate in order to maintain a six-coordination geometry with the Mo atom.

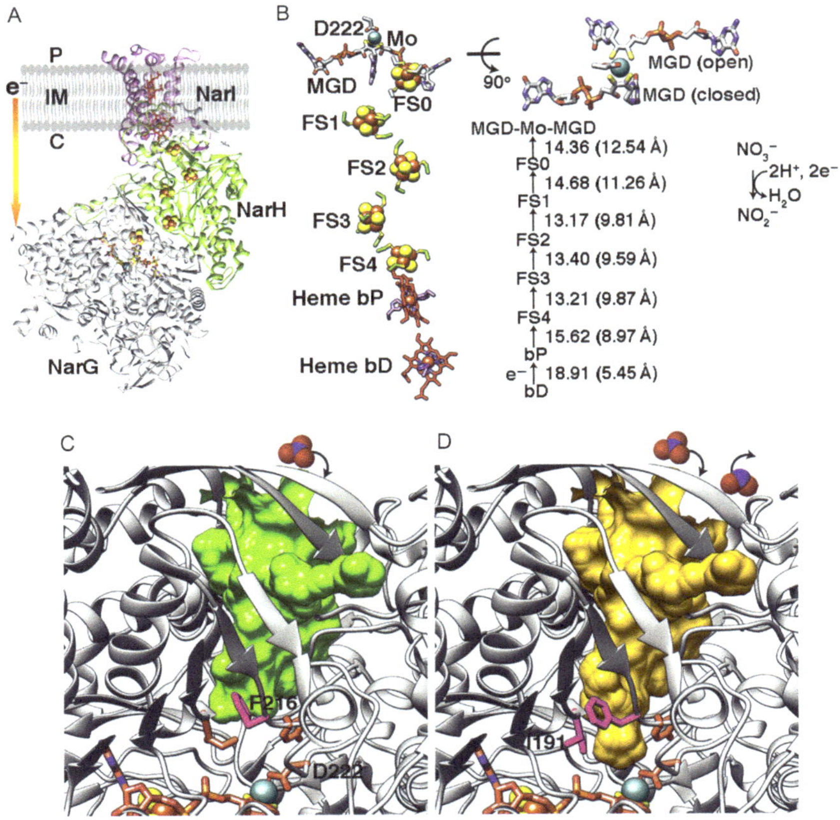

Fig. 6 Cytosolic NarGHI structure from *E. coli* (PDB: 1q16) and comparative implications from PcrAB structures. (A) Overall structure of cNarGHI. The membrane-bound b-type heme protein NarI transfers electrons from the quinone pool to the cytosolic-facing nitrate reductase NarGH via the [Fe–S] clusters. (B) The layout and cofactor distances in cNarGHI are shown. MGD cofactor shows open and closed forms resembling those found in PcrA. (C) cNarG is in the closed state. Based upon our data and analyses, residue F216 is predicted to control the substrate tunnel access. (D) I191 may also be required to change conformation along with F216 to open full access to the active site. Substrate nitrate and product nitrite are shown in space-filling models (*top*).

The crystal structure of ethylbenzene dehydrogenase EbdABC from *Aromatoleum aromaticum* forms a heterotrimeric complex that is located in periplasm (Johnson, Pelletier, & Spormann, 2001; Kloer et al., 2006) (Fig. 7). EbdABC catalyzes oxygen-independent oxidation on hydrocarbon by hydroxylating ethylbenzene to (*S*)-1-phenylethanol. This generates $2H^+$

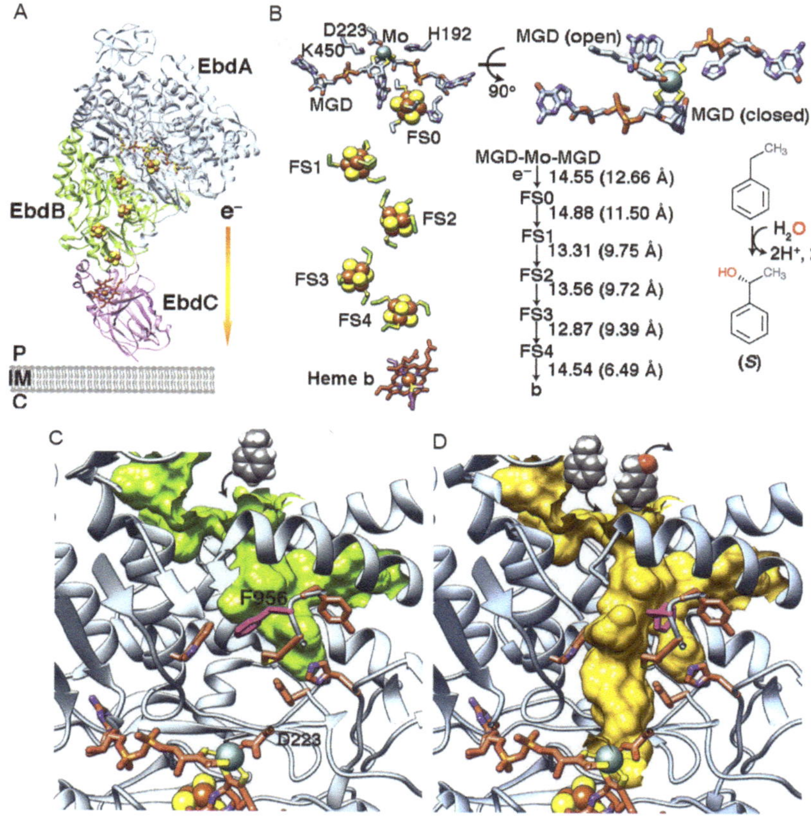

Fig. 7 EdbABC structure from *Aromatoleum aromaticum* (PDB: 2ivf). (A) The overall structure of EdbABC. EbdABC is a periplasmic heterotrimer. The terminal electron acceptor is still unknown. (B) The layout and distances of cofactors and active site. Residues D223, K450, and H192 that act in catalysis are shown. MGD cofactors show the bicyclic dihydropterin open state and tricyclic pterin closed forms in the crystal structure. The electrons are transferred from Mo-bisMGD to the [Fe–S] clusters to heme b cofactor by converting ethylbenzene to (*S*)-1-phenylethanol. (C) Residue F956 is predicted to gate the substrate tunnel to active site. (D) The switch of F956 to a different conformation allows substrate access to the active site for catalysis. Substrate ethylbenzene and product (*S*)-1-phenylethanol are shown in space-filling models.

and $2e^-$ that are transferred to unknown terminal electron acceptors (Szaleniec et al., 2007). It was proposed that EbdABC generates a redox loop by transferring electrons to the quinone pool and then to cytosolic nitrate reductase for nitrate respiration (Kloer et al., 2006). Using CASTp surface topography search, we predict here that the substrate tunnel is gated by residue F956 (Fig. 7C). A different rotamer conformation of F956 is suitable to open access of the substrate tunnel from outside the active site (Fig. 7D). This observation supports the feasibility of identifying the gate residue for type II Mo-bisMGD enzymes based on comparative structural analyses that include the [Fe–S] sites and protein surface topography.

It remains difficult to predict the existence and function of [Fe–S] clusters based on the protein sequence, as there is no conserved protein motif or domain for all [Fe–S] clusters (Fuss et al., 2015; Lill, 2009). In fact, the prediction of all metal-binding sites remains challenging (Cvetkovic et al., 2010; Yannone et al., 2012). Therefore, structures become a critical tool to identify, interpret, and investigate the existence and function of [Fe–S] clusters. Here, we demonstrate the power of comparative metalloenzyme structural analyses. We superimpose the structures of type II Mo-bisMGD enzymes PcrAB, cNarGHI, and EbdABC to assess the function from ligands and [Fe–S] cluster environments (Fig. 8). A common feature of [Fe–S] clusters consists of conserved cysteine ligands. In type II Mo-bisMGD enzymes, FS0 contains specific H/CxxxCxxxC(x)$_n$C sequence motif except EbdA (Coelho & Romão, 2015; Jormakka et al., 2004) (Figs. 4 and 6–8). The non-cysteine ligands in the [Fe–S] protein are seen in many proteins, and their unique features tune the redox potential of the clusters for diverse protein functions (Bak & Elliott, 2014). For example, the His to Cys mutation of FS0 in cNarG shows that enzyme activity is compromised due to a decrease in the redox potential by at least 500 mV (Magalon et al., 1998; Rothery et al., 2010). The neighboring Arg/Lys residue between FS0 and Mo-bisMGD also plays a critical role for mediating the electron transfer and enzyme activity (Rothery et al., 2010) (Fig. 8A). Interestingly, the loop containing H192 in the EbdA structure shifts closer to the Mo-bisMGD cofactor for catalysis, while the loops in PcrA and cNarG stay away (Fig. 8A).

The residues around [Fe–S] clusters in the respiratory protein shown here are mostly hydrophobic with positively charged His or Lys or Arg residues between the clusters (Fig. 8). Interestingly, many His/Arg/Lys residues are positioned between [3Fe–4S] cluster and heme b cofactor that may facilitate the electron transfer (Kloer et al., 2006; Rothery, Blasco, & Weiner, 2001), although the roles of those residues remain to be tested (Fig. 8D). Therefore, we show here how attentive examination of the [Fe–S] cluster environment including geometry, ligands, fold, surrounding residues, and surface topography may prove critical to understanding the function of [Fe–S] clusters in proteins.

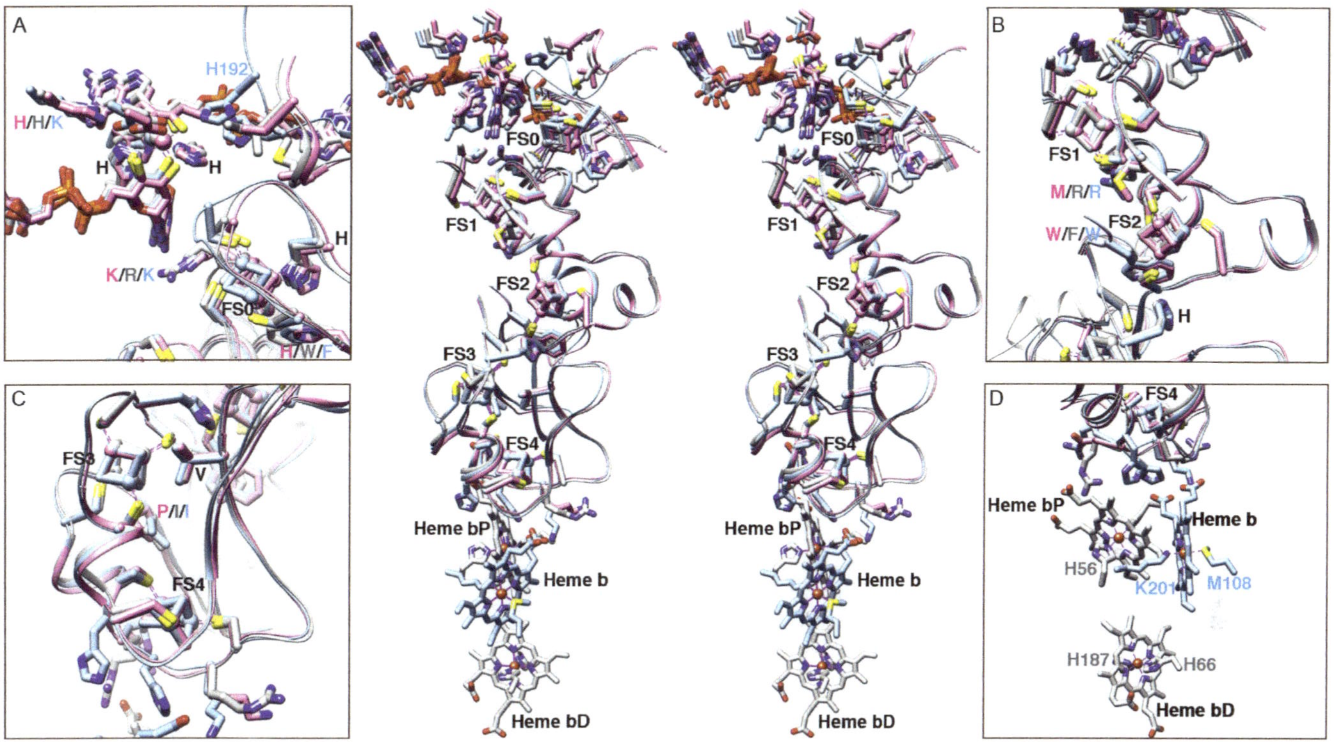

Fig. 8 Unified insights for a fundamental mechanistic understanding of selectivity and regulation come from comparative superimposition of structures of PcrAB (PDB: 4ydd, *purple*), cNarGHI (PDB: 1q16, *light gray*), and EdbABC (PDB: 2ivf, *sky blue*). Stereo view (*center panel*) of the superimposed structures (*center*) reveals striking structural conservation. Close-up views of the Mo-bisMGD and [Fe–S] clusters are shown in boxes A–D. Superimposition of the structures was based on the PcrB as the reference structure and used Chimera software. RMSD between 228 pruned atom pairs out of 326 pairs is 0.779 Å for cNarH; RMSD between 281 pruned atom pairs out of 324 pairs is 0.786 Å for EbdB.

5. SUMMARY AND PROSPECTS FOR ADVANCES

The great importance of [Fe–S] cluster-containing enzymes for cellular energy and for controlling protein conformations (Beinert, 1997) makes the ability to prepare and study these proteins with intact [Fe–S] clusters central to biochemistry, bioinorganic chemistry, and cell biology. As [Fe–S] clusters predate the origin of an oxygen atmosphere, they became critical in cell biology before dioxygen sensitivity became a possible evolutionary selection. Yet, humans and other eukaryotes have surprisingly not only retained [Fe–S] clusters but furthermore made them a requirement for DNA replication, transcription, and repair as well as for genome integrity (Fuss et al., 2015).

These eukaryotic [Fe–S] clusters are formed in the mitochondria and are critical for mitochondrial metabolism (Lill et al., 2012; Rouault, 2015), showing that eukaryotes captured and kept the ability to make and use these clusters from microbes. Indeed, disruption of [Fe–S] cluster assembly biosynthesis results in human diseases, as typified by Friedreich ataxia (Martelli & Puccio, 2014; Rötig et al., 1997; Stemmler, Lesuisse, Pain, & Dancis, 2010; Tsai, Bridwell-Rabb, & Barondeau, 2011). Furthermore, the proper formation of [Fe–S] cluster assembly complexes in mitochondria is connected via an allosteric regulation between respiratory chain complexes and apoptosis-inducing factor that links mitochondrial redox state to cell growth vs apoptosis (Brosey et al., 2016). These observations underscore the unique and powerful features of [Fe–S] clusters and their electron transfer abilities that have evidently provided strong evolutionary pressure to not only keep and furthermore to expand [Fe–S] cluster uses for cell biology, despite associated oxygen sensitivity.

Ironically, PcrAB [Fe–S] clusters function to create oxygen, which can damage the PcrAB [Fe–S] cluster, by reducing perchlorate to chlorite such that chlorite dismutase transforms the chlorite into chloride and oxygen. Notably, oxygen biogenesis initiated by PcrAB thus provides an intriguing early direct metabolic link between aerobic and anaerobic metabolisms (Coates & Achenbach, 2004; Coates et al., 1999; Rikken et al., 1996). As perchlorate respiration likely developed before oxygen production by photosynthesis (and oxygen respiration), PcrAB reveals a chemistry suitable for the initial origin of an oxygen atmosphere and for providing provocative insights into when and how life may have adapted to oxidizing conditions (Nerenberg, 2013).

Besides possibly contributing to the evolution of the oxygen atmosphere, the widespread occurrence of perchlorate as a pollutant, in arid regions on Earth, and on Mars (Jackson, Davila, et al., 2015; Ming et al., 2014) highlights the need to understand the basic geochemical cycling of perchlorate as well as the related chemistries of other Mo-bisMGD enzymes. Furthermore, the use of microbes and their PcrAB enzymes provides an effective means for environmental removal of perchlorate (Xu, Song, Min, Steinberg, & Logan, 2003). On Mars, the activity of PcrAB makes the ClO_4^- ion with its central chlorine atom surrounded by four tetrahedral oxygen atoms, a possible rich oxygen source. Bio-beads or other stabilized constructs with purified PcrAB plus chlorite dismutase could transform perchlorate on Mars to form O_2 for astronauts and fuel as the measured amounts of perchlorate in Martian regolith appear sufficient to make this process feasible (Davila, Willson, Coates, & McKay, 2013).

Here exemplary structural analyses of multiple structures of the five [Fe–S] clusters and molybdenum cofactor enzyme perchlorate reductase uncovered new biology and aided the accurate definition of the mechanism for multi-[Fe–S] cluster electron transfer in this and related systems. In considering these and other metalloenzyme results, it is important to recall the need for rigorous refinement of [Fe–S] clusters and for considering the coupled challenges from accurately modeling bound solvent and flexible regions. All existing crystal structures contain errors in refined models for flexible regions and bound solvent that greatly exceed the errors in the crystallographic data, so modeling and refining these aspects merit careful attention (Holton, Classen, Frankel, & Tainer, 2014). Going forward, the combination of X-ray crystallography with other methods such as X-ray scattering in solution may be key to defining the protein structure with flexibility and bound solvent (Rambo & Tainer, 2013a, 2013b). X-ray scattering from [Fe–S] cluster sites, as well as other metal sites such as gold, may allow direct examination of metal site distances (Grishaev, Anthis, & Clore, 2012; Hura et al., 2013). In part, bound waters and ligands, two of the parameters analyzed here for [Fe–S] cluster proteins, depend upon protein surface topology (Kuhn et al., 1992) that is influenced by side-chain flexibility (Schnecke, Swanson, Getzoff, Tainer, & Kuhn, 1998). Importantly, the high-resolution structures and combined mutational analyses, as obtained with the anaerobic approaches described here, are sufficient to discover an informed mechanistic model. Such detailed structures are furthermore key to the design of metal sites in proteins including antibodies (Roberts et al., 1990; Tainer, Roberts, & Getzoff, 1991), which has recently become a rapidly advancing area of science and engineering, as seen for protein cages (Lai et al., 2016) and repeat protein folds (Brunette et al., 2015). Indeed as part of this report on methods including data interpretation and analysis, we were able to delineate novel comparative structural analyses of key Mo-bisMGD- and [Fe–S] cluster-containing proteins that disclose specific gate residues regulating substrate entry and product release mechanisms to inform biology and possible design.

With the above points in mind, we maintain that PcrAB structures and mutations serve as an archetype for Mo-bisMGD- and [Fe–S] cluster-containing respiratory proteins. In particular, our described methods and findings provide a prototypic model for substrate entry and product release that can be tested and applied to other Mo-bisMGD multi-[Fe–S] cluster enzymes. By presenting analyses of key examples including cNarGHI and EbdABC to identify proposed gating residues (Figs. 6 and 7), these ideas can now be functionally tested by other researchers. Importantly, we show how to implement PcrAB analyses in order to bring out a unified testable model for other Mo-bisMGD enzymes. Of course, all such analyses may require the successful anaerobic [Fe–S] protein production, purification, crystallization, data collection, and structure determination methods noted here. By noting tested strategies and protocols for the anaerobic handling of [Fe–S] proteins, we hope that these procedures and ideas will benefit the rigorous study of anaerobic [Fe–S] protein and Mo-bisMGD-containing enzymes and help other researchers advance our knowledge of these biologically and environmentally critical systems.

ACKNOWLEDGMENTS

We acknowledge that the work and ideas reported here build upon the many pioneering researchers in the [Fe–S] field, especially Roland Lill, Michael Johnson, Helmut Beinert, Richard Holm, Eckard Münck, Patricia Kiley, Tracey Rouault, James Imlay, F. Wayne Outten, David Stout, Lou Noodleman, Barbara Burgess, Joan Broderick, John Peters, Brian Hoffman, Fraser Armstrong, Dennis Winge, Brian Crane, Harry Gray, Elizabeth Getzoff, Michael Adams, David Barondeau, Timothy Stemmler, Cathy Drennan, JoAnne Stubbe, and Doug Rees. These investigators taught us many interesting things about metalloproteins and how to study them properly. Unfortunately this expertise is disappearing, so we hope to have recorded some useful points we developed from others as well as from our own efforts. This work was supported in part by the Robert A. Welch Distinguished Chair in Chemistry, startup funds from the Cancer Prevention and Research Institute of Texas, and the University of Texas STARs program. Integrated X-ray structural efforts at the Advanced Light Source SIBYLS beamline are supported in part by the United States Department of Energy program Integrated Diffraction Analysis Technologies (IDAT).

REFERENCES

Adak, S., Bilwes, A. M., Panda, K., Hosfield, D., Aulak, K. S., McDonald, J. F., et al. (2002). Cloning, expression, and characterization of a nitric oxide synthase protein from Deinococcus radiodurans. *Proceedings of the National Academy of Sciences of the United States of America*, *99*(1), 107–112. https://doi.org/10.1073/pnas.012470099.

Afonine, P. V., Grosse-Kunstleve, R. W., Echols, N., Headd, J. J., Moriarty, N. W., Mustyakimov, M., et al. (2012). Towards automated crystallographic structure refinement with phenix.refine. *Acta Crystallographica. Section D, Biological Crystallography*, *68*(Pt. 4), 352–367. https://doi.org/10.1107/S0907444912001308.

Agar, J. N., Krebs, C., Frazzon, J., Huynh, B. H., Dean, D. R., & Johnson, M. K. (2000). IscU as a scaffold for iron-sulfur cluster biosynthesis: Sequential assembly of [2Fe-2S] and [4Fe-4S] clusters in IscU. *Biochemistry*, *39*(27), 7856–7862.

Andreini, C., Banci, L., & Rosato, A. (2016). Exploiting bacterial operons to illuminate human iron-sulfur proteins. *Journal of Proteome Research*, *15*(4), 1308–1322. https://doi.org/10.1021/acs.jproteome.6b00045.

Aslanidis, C., & de Jong, P. J. (1990). Ligation-independent cloning of PCR products (LIC-PCR). *Nucleic Acids Research*, *18*(20), 6069–6074.

Bacolla, A., Tainer, J. A., Vasquez, K. M., & Cooper, D. N. (2016). Translocation and deletion breakpoints in cancer genomes are associated with potential non-B DNA-forming sequences. *Nucleic Acids Research*, *44*(12), 5673–5688. https://doi.org/10.1093/nar/gkw261.

Bak, D. W., & Elliott, S. J. (2014). Alternative FeS cluster ligands: Tuning redox potentials and chemistry. *Current Opinion in Chemical Biology*, *19*, 50–58. https://doi.org/10.1016/j.cbpa.2013.12.015.

Banda, D. M., Nuñez, N. N., Burnside, M. A., Bradshaw, K. M., & David, S. S. (2017). Repair of 8-oxoG:A mismatches by the MUTYH glycosylase: Mechanism, metals and medicine. *Free Radical Biology & Medicine*, *107*, 202–215. https://doi.org/10.1016/j.freeradbiomed.2017.01.008.

Baranovskiy, A. G., Zhang, Y., Suwa, Y., Babayeva, N. D., Gu, J., Pavlov, Y. I., et al. (2015). Crystal structure of the human primase. *The Journal of Biological Chemistry*, *290*(9), 5635–5646. https://doi.org/10.1074/jbc.M114.624742.

Barras, F., Loiseau, L., & Py, B. (2005). How Escherichia coli and Saccharomyces cerevisiae build Fe/S proteins. *Advances in Microbial Physiology*, *50*, 41–101.

Beinert, H. (1997). Iron-sulfur clusters: Nature's modular, multipurpose structures. *Science*, *277*(5326), 653–659. https://doi.org/10.1126/science.277.5326.653.

Bender, K. S., Shang, C., Chakraborty, R., Belchik, S. M., Coates, J. D., & Achenbach, L. A. (2005). Identification, characterization, and classification of genes encoding perchlorate reductase. *Journal of Bacteriology*, *187*(15), 5090–5096. https://doi.org/10.1128/JB.187.15.5090-5096.2005.

Berger, I., Fitzgerald, D. J., & Richmond, T. J. (2004). Baculovirus expression system for heterologous multiprotein complexes. *Nature Biotechnology*, *22*(12), 1583–1587. https://doi.org/10.1038/nbt1036.

Bertero, M. G., Rothery, R. A., Palak, M., Hou, C., Lim, D., Blasco, F., et al. (2003). Insights into the respiratory electron transfer pathway from the structure of nitrate reductase A. *Nature Structural Biology*, *10*(9), 681–687. https://doi.org/10.1038/nsb969.

Boal, A. K., Yavin, E., Lukianova, O. A., O'Shea, V. L., David, S. S., & Barton, J. K. (2005). DNA-bound redox activity of DNA repair glycosylases containing [4Fe-4S] clusters. *Biochemistry*, *44*(23), 8397–8407. https://doi.org/10.1021/bi047494n.

Bonomi, F., Iametti, S., Morleo, A., Ta, D., & Vickery, L. E. (2011). Facilitated transfer of IscU-[2Fe2S] clusters by chaperone-mediated ligand exchange. *Biochemistry*, *50*(44), 9641–9650. https://doi.org/10.1021/bi201123z.

Boon, E. M., Livingston, A. L., Chmiel, N. H., David, S. S., & Barton, J. K. (2003). DNA-mediated charge transport for DNA repair. *Proceedings of the National Academy of Sciences of the United States of America*, *100*(22), 12543–12547. https://doi.org/10.1073/pnas.2035257100.

Bornhorst, J. A., & Falke, J. J. (2000). Purification of proteins using polyhistidine affinity tags. *Methods in Enzymology*, *326*, 245–254.

Bowman, S. E. J., Bridwell-Rabb, J., & Drennan, C. L. (2016). Metalloprotein crystallography: More than a structure. *Accounts of Chemical Research*, *49*(4), 695–702. https://doi.org/10.1021/acs.accounts.5b00538.

Braymer, J. J., & Lill, R. (2017). Iron-sulfur cluster biogenesis and trafficking in mitochondria. *The Journal of Biological Chemistry*, *292*(31), 12754–12763. https://doi.org/10.1074/jbc.R117.787101.

Bricogne, G., Vonrhein, C., Flensburg, C., Schiltz, M., & Paciorek, W. (2003). Generation, representation and flow of phase information in structure determination: Recent developments in and around SHARP 2.0. *Acta Crystallographica. Section D, Biological Crystallography*, *59*(Pt. 11), 2023–2030.

Bridwell-Rabb, J., Fox, N. G., Tsai, C.-L., Winn, A. M., & Barondeau, D. P. (2014). Human frataxin activates Fe–S cluster biosynthesis by facilitating sulfur transfer chemistry. *Biochemistry*, *53*(30), 4904–4913. https://doi.org/10.1021/bi500532e.

Broderick, J. B., Byer, A. S., Duschene, K. S., Duffus, B. R., Betz, J. N., Shepard, E. M., et al. (2014). H-cluster assembly during maturation of the [FeFe]-hydrogenase. *Journal of Biological Inorganic Chemistry*, *19*(6), 747–757. https://doi.org/10.1007/s00775-014-1168-8.

Brosey, C. A., Ho, C., Long, W. Z., Singh, S., Burnett, K., Hura, G. L., et al. (2016). Defining NADH-driven allostery regulating apoptosis-inducing factor. *Structure (London, England: 1993)*, *24*(12), 2067–2079. https://doi.org/10.1016/j.str.2016.09.012.

Brunette, T. J., Parmeggiani, F., Huang, P.-S., Bhabha, G., Ekiert, D. C., Tsutakawa, S. E., et al. (2015). Exploring the repeat protein universe through computational protein design. *Nature*, *528*(7583), 580–584. https://doi.org/10.1038/nature16162.

Burgess, B. K., & Lowe, D. J. (1996). Mechanism of molybdenum nitrogenase. *Chemical Reviews*, *96*(7), 2983–3012.

Burns, J. A., Butler, J. C., Moran, J., & Whitesides, G. M. (1991). Selective reduction of disulfides by tris(2-carboxyethyl)phosphine. *Journal of Organic Chemistry*, *56*(8), 2648–2650.

Buschmann, S., Warkentin, E., Xie, H., Langer, J. D., Ermler, U., & Michel, H. (2010). The structure of cbb3 cytochrome oxidase provides insights into proton pumping. *Science*, *329*(5989), 327–330. https://doi.org/10.1126/science.1187303.

Carlson, H. K., Kuehl, J. V., Hazra, A. B., Justice, N. B., Stoeva, M. K., Sczesnak, A., et al. (2015). Mechanisms of direct inhibition of the respiratory sulfate-reduction pathway by (per)chlorate and nitrate. *The ISME Journal*, *9*(6), 1295–1305. https://doi.org/10.1038/ismej.2014.216.

Chandramouli, K., & Johnson, M. K. (2006). HscA and HscB stimulate [2Fe-2S] cluster transfer from IscU to apoferredoxin in an ATP-dependent reaction. *Biochemistry*, *45*(37), 11087–11095. https://doi.org/10.1021/bi061237w.

Chepanoske, C. L., Golinelli, M. P., Williams, S. D., & David, S. S. (2000). Positively charged residues within the iron-sulfur cluster loop of E. coli MutY participate in damage recognition and removal. *Archives of Biochemistry and Biophysics*, *380*(1), 11–19. https://doi.org/10.1006/abbi.2000.1890.

Classen, S., Hura, G. L., Holton, J. M., Rambo, R. P., Rodic, I., McGuire, P. J., et al. (2013). Implementation and performance of SIBYLS: A dual endstation small-angle X-ray scattering and macromolecular crystallography beamline at the advanced light source. *Journal of Applied Crystallography*, *46*(Pt. 1), 1–13. https://doi.org/10.1107/S0021889812048698.

Cleland, W. W. (1964). Dithiothreitol, a new protective reagent for SH groups. *Biochemistry*, *3*, 480–482.

Coates, J. D., & Achenbach, L. A. (2004). Microbial perchlorate reduction: Rocket-fuelled metabolism. *Nature Reviews. Microbiology*, *2*(7), 569–580. https://doi.org/10.1046/j.1462-2920.2000.00078.x.

Coates, J. D., Michaelidou, U., Bruce, R. A., O'Connor, S. M., Crespi, J. N., & Achenbach, L. A. (1999). Ubiquity and diversity of dissimilatory (per)chlorate-reducing bacteria. *Applied and Environmental Microbiology*, *65*(12), 5234–5241.

Coelho, C., & Romão, M. J. (2015). Structural and mechanistic insights on nitrate reductases. *Protein Science: A Publication of the Protein Society*, *24*(12), 1901–1911. https://doi.org/10.1002/pro.2801.

Coghlan, V. M., & Vickery, L. E. (1991). Site-specific mutations in human ferredoxin that affect binding to ferredoxin reductase and cytochrome P450scc. *The Journal of Biological Chemistry*, *266*(28), 18606–18612.

Colin, F., Martelli, A., Clémancey, M., Latour, J.-M., Gambarelli, S., Zeppieri, L., et al. (2013). Mammalian frataxin controls sulfur production and iron entry during de novo Fe4S4 cluster assembly. *Journal of the American Chemical Society*, *135*(2), 733–740. https://doi.org/10.1021/ja308736e.

Corbett, M. C., Latimer, M. J., Poulos, T. L., Sevrioukova, I. F., Hodgson, K. O., & Hedman, B. (2007). Photoreduction of the active site of the metalloprotein putidaredoxin by synchrotron radiation. *Acta Crystallographica. Section D, Biological Crystallography*, *63*(Pt. 9), 951–960. https://doi.org/10.1107/S0907444907035160.

Cory, S. A., Van Vranken, J. G., Brignole, E. J., Patra, S., Winge, D. R., Drennan, C. L., et al. (2017). Structure of human Fe-S assembly subcomplex reveals unexpected cysteine desulfurase architecture and acyl-ACP-ISD11 interactions. *Proceedings of the National Academy of Sciences of the United States of America*, *114*(27), E5325–E5334. https://doi.org/10.1073/pnas.1702849114.

Cotruvo, J. A., & Stubbe, J. (2012). Metallation and mismetallation of iron and manganese proteins in vitro and in vivo: The class I ribonucleotide reductases as a case study. *Metallomics: Integrated Biometal Science*, *4*(10), 1020–1036. https://doi.org/10.1039/c2mt20142a.

Cowart, R. E., Singleton, F. L., & Hind, J. S. (1993). A comparison of bathophenanthrolinedisulfonic acid and ferrozine as chelators of iron(II) in reduction reactions. *Analytical Biochemistry*, *211*(1), 151–155. https://doi.org/10.1006/abio.1993.1246.

Crane, B. R., Arvai, A. S., Ghosh, D. K., Wu, C., Getzoff, E. D., Stuehr, D. J., et al. (1998). Structure of nitric oxide synthase oxygenase dimer with pterin and substrate. *Science*, *279*(5359), 2121–2126. https://doi.org/10.1126/science.279.5359.2121.

Crane, B. R., Siegel, L. M., & Getzoff, E. D. (1995). Sulfite reductase structure at 1.6 A: Evolution and catalysis for reduction of inorganic anions. *Science*, *270*(5233), 59–67.

Cvetkovic, A., Menon, A. L., Thorgersen, M. P., Scott, J. W., Poole, F. L., Jenney, F. E., et al. (2010). Microbial metalloproteomes are largely uncharacterized. *Nature*, *466*(7307), 779–782. https://doi.org/10.1038/nature09265.

Daniels, D. S., Mol, C. D., Arvai, A. S., Kanugula, S., Pegg, A. E., & Tainer, J. A. (2000). Active and alkylated human AGT structures: A novel zinc site, inhibitor and extrahelical base binding. *The EMBO Journal*, *19*(7), 1719–1730. https://doi.org/10.1093/emboj/19.7.1719.

Das, R., & Baker, D. (2008). Macromolecular modeling with rosetta. *Annual Review of Biochemistry*, 77, 363–382. https://doi.org/10.1146/annurev.biochem.77.062906.171838.

Davila, A. F., Willson, D., Coates, J. D., & McKay, C. P. (2013). Perchlorate on Mars: A chemical hazard and a resource for humans. *International Journal of Astrobiology*, *12*(4), 321–325. https://doi.org/10.1017/S1473550413000189.

de Vries, S., Dörner, K., Strampraad, M. J. F., & Friedrich, T. (2015). Electron tunneling rates in respiratory complex I are tuned for efficient energy conversion. *Angewandte Chemie (International Ed in English)*, *54*(9), 2844–2848. https://doi.org/10.1002/anie.201410967.

DiMaio, F., Terwilliger, T. C., Read, R. J., Wlodawer, A., Oberdorfer, G., Wagner, U., et al. (2011). Improved molecular replacement by density- and energy-guided protein structure optimization. *Nature*, *473*(7348), 540–543. https://doi.org/10.1038/nature09964.

Dupuy, J., Darnault, C., Brazzolotto, X., Kühn, L. C., Moulis, J. M., Volbeda, A., et al. (2005). Crystallization and preliminary X-ray diffraction data for the aconitase form of human iron-regulatory protein 1. *Acta Crystallographica. Section F, Structural Biology and Crystallization Communications*, *61*(Pt. 5), 482–485. https://doi.org/10.1107/S1744309105010444.

Einsle, O., Andrade, S. L. A., Dobbek, H., Meyer, J., & Rees, D. C. (2007). Assignment of individual metal redox states in a metalloprotein by crystallographic refinement at multiple X-ray wavelengths. *Journal of the American Chemical Society*, *129*(8), 2210–2211. https://doi.org/10.1021/ja067562o.

Fan, L., Fuss, J. O., Cheng, Q. J., Arvai, A. S., Hammel, M., Roberts, V. A., et al. (2008). XPD helicase structures and activities: Insights into the cancer and aging phenotypes from XPD mutations. *Cell*, *133*(5), 789–800. https://doi.org/10.1016/j.cell.2008.04.030.

Fee, J. A., Castagnetto, J. M., Case, D. A., Noodleman, L., Stout, C. D., & Torres, R. A. (2003). The circumsphere as a tool to assess distortion in [4Fe-4S] atom clusters. *Journal of Biological Inorganic Chemistry*, *8*(5), 519–526. https://doi.org/10.1007/s00775-003-0445-8.

Fish, W. W. (1988). Rapid colorimetric micromethod for the quantitation of complexed iron in biological samples. *Methods in Enzymology*, *158*, 357–364.

Fitzgerald, D. J., Berger, P., Schaffitzel, C., Yamada, K., Richmond, T. J., & Berger, I. (2006). Protein complex expression by using multigene baculoviral vectors. *Nature Methods*, *3*(12), 1021–1032. https://doi.org/10.1038/nmeth983.

Flot, D., Mairs, T., Giraud, T., Guijarro, M., Lesourd, M., Rey, V., et al. (2010). The ID23-2 structural biology microfocus beamline at the ESRF. *Journal of Synchrotron Radiation*, *17*(1), 107–118. https://doi.org/10.1107/S0909049509041168.

Fox, N. G., Chakrabarti, M., McCormick, S. P., Lindahl, P. A., & Barondeau, D. P. (2015). The human iron-sulfur assembly complex catalyzes the synthesis of [2Fe-2S] clusters on ISCU2 that can be transferred to acceptor molecules. *Biochemistry*, *54*(25), 3871–3879. https://doi.org/10.1021/bi5014485.

Fox, N. G., Das, D., Chakrabarti, M., Lindahl, P. A., & Barondeau, D. P. (2015). Frataxin accelerates [2Fe-2S] cluster formation on the human Fe-S assembly complex. *Biochemistry*, *54*(25), 3880–3889. https://doi.org/10.1021/bi5014497.

Frazzon, J., & Dean, D. R. (2003). Formation of iron-sulfur clusters in bacteria: An emerging field in bioinorganic chemistry. *Current Opinion in Chemical Biology*, 7(2), 166–173.

Fromme, J. C., & Verdine, G. L. (2003). Structure of a trapped endonuclease III-DNA covalent intermediate. *The EMBO Journal*, *22*(13), 3461–3471. https://doi.org/10.1093/emboj/cdg311.

Fuss, J. O., Tsai, C.-L., Ishida, J. P., & Tainer, J. A. (2015). Emerging critical roles of Fe-S clusters in DNA replication and repair. *Biochimica et Biophysica Acta*, *1853*(6), 1253–1271. https://doi.org/10.1016/j.bbamcr.2015.01.018.

Gibson, D. G., Benders, G. A., Andrews-Pfannkoch, C., Denisova, E. A., Baden-Tillson, H., Zaveri, J., et al. (2008). Complete chemical synthesis, assembly, and cloning of a Mycoplasma genitalium genome. *Science*, *319*(5867), 1215–1220. https://doi.org/10.1126/science.1151721.

Gibson, D. G., Young, L., Chuang, R.-Y., Venter, J. C., Hutchison, C. A., & Smith, H. O. (2009). Enzymatic assembly of DNA molecules up to several hundred kilobases. *Nature Methods*, *6*(5), 343–345. https://doi.org/10.1038/nmeth.1318.

Gorodetsky, A. A., Boal, A. K., & Barton, J. K. (2006). Direct electrochemistry of endonuclease III in the presence and absence of DNA. *Journal of the American Chemical Society, 128*(37), 12082–12083. https://doi.org/10.1021/ja064784d.

Gradia, S. D., Ishida, J. P., Tsai, M. S., Jeans, C., & Tainer, J. A. (2017). Chapter one—MacroBac: New technologies for robust and efficient large-scale production of recombinant multiprotein complexes. *Methods in Enzymology, 592*, 1–26.

Grishaev, A., Anthis, N. J., & Clore, G. M. (2012). Contrast-matched small-angle X-ray scattering from a heavy-atom-labeled protein in structure determination: Application to a lead-substituted calmodulin-peptide complex. *Journal of the American Chemical Society, 134*(36), 14686–14689.

Guan, Y., Manuel, R. C., Arvai, A. S., Parikh, S. S., Mol, C. D., Miller, S. L., et al. (1998). MutY catalytic core, mutant and bound adenine structures define specificity for DNA repair enzyme superfamily. *Nature Structural Biology, 5*(12), 1058–1064. https://doi.org/10.1038/4168.

Ha, Y., Arnold, A. R., Nuñez, N. N., Bartels, P. L., Zhou, A., David, S. S., et al. (2017). Sulfur K-edge XAS studies of the effect of DNA binding on the [Fe4S4] site in EndoIII and MutY. *Journal of the American Chemical Society, 139*(33), 11434–11442. https://doi.org/10.1021/jacs.7b03966.

Hayashi, T., & Stuchebrukhov, A. A. (2010). Electron tunneling in respiratory complex I. *Proceedings of the National Academy of Sciences of the United States of America, 107*(45), 19157–19162. https://doi.org/10.1073/pnas.1009181107.

Hilgart, M. C., Sanishvili, R., Ogata, C. M., Becker, M., Venugopalan, N., Stepanov, S., et al. (2011). Automated sample-scanning methods for radiation damage mitigation and diffraction-based centering of macromolecular crystals. *Journal of Synchrotron Radiation, 18*(Pt. 5), 717–722. https://doi.org/10.1107/S0909049511029918.

Hoffman, B. M., Lukoyanov, D., Yang, Z.-Y., Dean, D. R., & Seefeldt, L. C. (2014). Mechanism of nitrogen fixation by nitrogenase: The next stage. *Chemical Reviews, 114*(8), 4041–4062. https://doi.org/10.1021/cr400641x.

Holton, J. M., Classen, S., Frankel, K. A., & Tainer, J. A. (2014). The R-factor gap in macromolecular crystallography: An untapped potential for insights on accurate structures. *The FEBS Journal, 281*(18), 4046–4060. https://doi.org/10.1111/febs.12922.

Hosler, J. P., Ferguson-Miller, S., & Mills, D. A. (2006). Energy transduction: Proton transfer through the respiratory complexes. *Annual Review of Biochemistry, 75*, 165–187. https://doi.org/10.1146/annurev.biochem.75.062003.101730.

Hura, G. L., Tsai, C.-L., Claridge, S. A., Mendillo, M. L., Smith, J. M., Williams, G. J., et al. (2013). DNA conformations in mismatch repair probed in solution by X-ray scattering from gold nanocrystals. *Proceedings of the National Academy of Sciences of the United States of America, 110*(43), 17308–17313.

Imlay, J. A. (2006). Iron-sulphur clusters and the problem with oxygen. *Molecular Microbiology, 59*(4), 1073–1082. https://doi.org/10.1111/j.1365-2958.2006.05028.x.

Imlay, J. A. (2008). Cellular defenses against superoxide and hydrogen peroxide. *Annual Review of Biochemistry, 77*, 755–776. https://doi.org/10.1146/annurev.biochem.77.061606.161055.

Imlay, J. A. (2014). The mismetallation of enzymes during oxidative stress. *The Journal of Biological Chemistry, 289*(41), 28121–28128. https://doi.org/10.1074/jbc.R114.588814.

Iobbi-Nivol, C., & Leimkühler, S. (2013). Molybdenum enzymes, their maturation and molybdenum cofactor biosynthesis in Escherichia coli. *Biochimica et Biophysica Acta, 1827*(8–9), 1086–1101. https://doi.org/10.1016/j.bbabio.2012.11.007.

Jackson, W. A., Böhlke, J. K., Andraski, B. J., Fahlquist, L., Bexfield, L., Eckardt, F. D., et al. (2015). Global patterns and environmental controls of perchlorate and nitrate co-occurrence in arid and semi-arid environments. *Geochimica et Cosmochimica Acta, 164*, 502–522. https://doi.org/10.1016/j.gca.2015.05.016.

Jackson, W. A., Davila, A. F., Sears, D. W. G., Coates, J. D., McKay, C. P., Brundrett, M., et al. (2015). Widespread occurrence of (per)chlorate in the solar system. *Earth and Planetary Science Letters*, *430*, 470–476. https://doi.org/10.1016/j.epsl.2015.09.003.
Jaganaman, S., Pinto, A., Tarasev, M., & Ballou, D. P. (2007). High levels of expression of the iron-sulfur proteins phthalate dioxygenase and phthalate dioxygenase reductase in Escherichia coli. *Protein Expression and Purification*, *52*(2), 273–279. https://doi.org/10.1016/j.pep.2006.09.004.
Johnson, D. C., Dean, D. R., Smith, A. D., & Johnson, M. K. (2005). Structure, function, and formation of biological iron-sulfur clusters. *Annual Review of Biochemistry*, *74*, 247–281. https://doi.org/10.1146/annurev.biochem.74.082803.133518.
Johnson, H. A., Pelletier, D. A., & Spormann, A. M. (2001). Isolation and characterization of anaerobic ethylbenzene dehydrogenase, a novel Mo-Fe-S enzyme. *Journal of Bacteriology*, *183*(15), 4536–4542. https://doi.org/10.1128/JB.183.15.4536-4542.2001.
Johnson, L. M., Snyder, M., Chang, L., & Davis, R. W. (1985). Isolation of the gene encoding yeast DNA polymerase I. *Cell*, *43*, 369–377.
Jormakka, M., Richardson, D., Byrne, B., & Iwata, S. (2004). Architecture of NarGH reveals a structural classification of Mo-bisMGD enzymes. *Structure (London, England: 1993)*, *12*(1), 95–104.
Jormakka, M., Törnroth, S., Byrne, B., & Iwata, S. (2002). Molecular basis of proton motive force generation: Structure of formate dehydrogenase-N. *Science*, *295*(5561), 1863–1868. https://doi.org/10.1126/science.1068186.
Keyer, K., & Imlay, J. A. (1996). Superoxide accelerates DNA damage by elevating free-iron levels. *Proceedings of the National Academy of Sciences of the United States of America*, *93*(24), 13635–13640.
Kidd, I. M., & Emery, V. C. (1993). The use of baculoviruses as expression vectors. *Applied Biochemistry and Biotechnology*, *42*(2–3), 137–159. https://doi.org/10.1007/BF02788049.
Kiley, P. J., & Beinert, H. (2003). The role of Fe–S proteins in sensing and regulation in bacteria. *Current Opinion in Microbiology*, *6*, 181–185.
Kloer, D. P., Hagel, C., Heider, J., & Schulz, G. E. (2006). Crystal structure of ethylbenzene dehydrogenase from Aromatoleum aromaticum. *Structure (London, England: 1993)*, *14*(9), 1377–1388. https://doi.org/10.1016/j.str.2006.07.001.
Kuchenreuther, J. M., Grady-Smith, C. S., Bingham, A. S., George, S. J., Cramer, S. P., & Swartz, J. R. (2010). High-yield expression of heterologous [FeFe] hydrogenases in Escherichia coli. *PLoS One*, *5*(11), e15491. https://doi.org/10.1371/journal.pone.0015491.
Kuhn, L. A., Siani, M. A., Pique, M. E., Fisher, C. L., Getzoff, E. D., & Tainer, J. A. (1992). The interdependence of protein surface topography and bound water molecules revealed by surface accessibility and fractal density measures. *Journal of Molecular Biology*, *228*(1), 13–22.
Kuo, C. F., McRee, D. E., Fisher, C. L., O'Handley, S. F., Cunningham, R. P., & Tainer, J. A. (1992). Atomic structure of the DNA repair [4Fe-4S] enzyme endonuclease III. *Science*, *258*(5081), 434–440.
Lai, Y.-T., Hura, G. L., Dyer, K. N., Tang, H. Y. H., Tainer, J. A., & Yeates, T. O. (2016). Designing and defining dynamic protein cage nanoassemblies in solution. *Science Advances*, *2*(12), e1501855. https://doi.org/10.1126/sciadv.1501855.
Lebendiker, M., & Danieli, T. (2017). Purification of proteins fused to maltose-binding protein. *Methods in Molecular Biology (Clifton, N.J.)*, *1485*, 257–273. https://doi.org/10.1007/978-1-4939-6412-3_13.
Lee, A. Q., Streit, B. R., Zdilla, M. J., Abu-Omar, M. M., & DuBois, J. L. (2008). Mechanism of and exquisite selectivity for O-O bond formation by the heme-dependent chlorite dismutase. *Proceedings of the National Academy of Sciences of the United States of America*, *105*(41), 15654–15659. https://doi.org/10.1073/pnas.0804279105.

Liang, J., Edelsbrunner, H., Fu, P., Sudhakar, P. V., & Subramaniam, S. (1998a). Analytical shape computation of macromolecules: I. Molecular area and volume through alpha shape. *Proteins*, *33*(1), 1–17.

Liang, J., Edelsbrunner, H., Fu, P., Sudhakar, P. V., & Subramaniam, S. (1998b). Analytical shape computation of macromolecules: II. Inaccessible cavities in proteins. *Proteins*, *33*(1), 18–29.

Liang, J., Edelsbrunner, H., & Woodward, C. (1998). Anatomy of protein pockets and cavities: Measurement of binding site geometry and implications for ligand design. *Protein Science*, 7(9), 1884–1897. https://doi.org/10.1002/pro.5560070905.

Lill, R. (2009). Function and biogenesis of iron-sulphur proteins. *Nature*, *460*(7257), 831–838. https://doi.org/10.1038/nature08301.

Lill, R., Dutkiewicz, R., Elsässer, H.-P., Hausmann, A., Netz, D. J. A., Pierik, A. J., et al. (2006). Mechanisms of iron-sulfur protein maturation in mitochondria, cytosol and nucleus of eukaryotes. *Biochimica et Biophysica Acta*, *1763*(7), 652–667. https://doi.org/10.1016/j.bbamcr.2006.05.011.

Lill, R., Hoffmann, B., Molik, S., Pierik, A. J., Rietzschel, N., Stehling, O., et al. (2012). The role of mitochondria in cellular iron-sulfur protein biogenesis and iron metabolism. *Biochimica et Biophysica Acta*, *1823*(9), 1491–1508. https://doi.org/10.1016/j.bbamcr.2012.05.009.

Lill, R., & Mühlenhoff, U. (2006). Iron-sulfur protein biogenesis in eukaryotes: Components and mechanisms. *Annual Review of Cell and Developmental Biology*, *22*, 457–486. https://doi.org/10.1146/annurev.cellbio.22.010305.104538.

Magalon, A., Asso, M., Guigliarelli, B., Rothery, R. A., Bertrand, P., Giordano, G., et al. (1998). Molybdenum cofactor properties and [Fe-S] cluster coordination in Escherichia coli nitrate reductase A: Investigation by site-directed mutagenesis of the conserved his-50 residue in the NarG subunit. *Biochemistry*, *37*(20), 7363–7370. https://doi.org/10.1021/bi972858f.

Malkin, R., & Rabinowitz, J. C. (1966). The reconstitution of clostridial ferredoxin. *Biochemical and Biophysical Research Communications*, *23*(6), 822–827.

Marangon, J., Paes de Sousa, P. M., Moura, I., Brondino, C. D., Moura, J. J. G., & González, P. J. (2012). Substrate-dependent modulation of the enzymatic catalytic activity: Reduction of nitrate, chlorate and perchlorate by respiratory nitrate reductase from Marinobacter hydrocarbonoclasticus 617. *Biochimica et Biophysica Acta*, *1817*(7), 1072–1082. https://doi.org/10.1016/j.bbabio.2012.04.011.

Martelli, A., & Puccio, H. (2014). Dysregulation of cellular iron metabolism in Friedreich ataxia: From primary iron-sulfur cluster deficit to mitochondrial iron accumulation. *Frontiers in Pharmacology*, *5*, 130. https://doi.org/10.3389/fphar.2014.00130.

Martin, D. R., & Matyushov, D. V. (2017). Electron-transfer chain in respiratory complex I. *Scientific Reports*, 7(1), 5495. https://doi.org/10.1038/s41598-017-05779-y.

Matos, C. T., Velizarov, S., Crespo, J. G., & Reis, M. A. M. (2006). Simultaneous removal of perchlorate and nitrate from drinking water using the ion exchange membrane bioreactor concept. *Water Research*, *40*(2), 231–240. https://doi.org/10.1016/j.watres.2005.10.022.

Mayhew, S. G. (1978). The redox potential of dithionite and SO-2 from equilibrium reactions with flavodoxins, methyl viologen and hydrogen plus hydrogenase. *European Journal of Biochemistry*, *85*(2), 535–547.

Mayo, C. J., Diprose, J. M., Walter, T. S., Berry, I. M., Wilson, J., Owens, R. J., et al. (2005). Benefits of automated crystallization plate tracking, imaging, and analysis. *Structure (London, England: 1993)*, *13*(2), 175–182. https://doi.org/10.1016/j.str.2004.12.010.

McCoy, A. J., Grosse-Kunstleve, R. W., Adams, P. D., Winn, M. D., Storoni, L. C., & Read, R. J. (2007). Phaser crystallographic software. *Journal of Applied Crystallography*, *40*(Pt. 4), 658–674. https://doi.org/10.1107/S0021889807021206.

McDevitt, C. A., Hugenholtz, P., Hanson, G. R., & McEwan, A. G. (2002). Molecular analysis of dimethyl sulphide dehydrogenase from Rhodovulum sulfidophilum: Its place in the dimethyl sulphoxide reductase family of microbial molybdopterin-containing enzymes. *Molecular Microbiology*, *44*(6), 1575–1587.

McTernan, P. M., Chandrayan, S. K., Wu, C.-H., Vaccaro, B. J., Lancaster, W. A., Yang, Q., et al. (2014). Intact functional fourteen-subunit respiratory membrane-bound [NiFe]-hydrogenase complex of the hyperthermophilic archaeon Pyrococcus furiosus. *The Journal of Biological Chemistry*, *289*(28), 19364–19372. https://doi.org/10.1074/jbc.M114.567255.

Melber, A., Na, U., Vashisht, A., Weiler, B. D., Lill, R., Wohlschlegel, J. A., et al. (2016). Role of Nfu1 and Bol3 in iron-sulfur cluster transfer to mitochondrial clients. *eLife*, *5*, e15991. https://doi.org/10.7554/eLife.15991.

Mickey, B., & Howard, J. (1995). Rigidity of microtubules is increased by stabilizing agents. *The Journal of Cell Biology*, *130*(4), 909–917.

Ming, D. W., Archer, P. D., Glavin, D. P., Eigenbrode, J. L., Franz, H. B., Sutter, B., et al. (2014). Volatile and organic compositions of sedimentary rocks in Yellowknife Bay, Gale crater, Mars. *Science*, *343*(6169), 1245267. https://doi.org/10.1126/science.1245267.

Moura, J. J. G., Brondino, C. D., Trincão, J., & Romão, M. J. (2004). Mo and W bis-MGD enzymes: Nitrate reductases and formate dehydrogenases. *Journal of Biological Inorganic Chemistry*, *9*(7), 791–799. https://doi.org/10.1007/s00775-004-0573-9.

Murshudov, G. N., Vagin, A. A., & Dodson, E. J. (1997). Refinement of macromolecular structures by the maximum-likelihood method. *Acta Crystallographica. Section D, Biological Crystallography*, *53*(Pt. 3), 240–255. https://doi.org/10.1107/S0907444996012255.

Najmudin, S., González, P. J., Trincão, J., Coelho, C., Mukhopadhyay, A., Cerqueira, N. M. F. S. A., et al. (2008). Periplasmic nitrate reductase revisited: A sulfur atom completes the sixth coordination of the catalytic molybdenum. *Journal of Biological Inorganic Chemistry*, *13*(5), 737–753. https://doi.org/10.1007/s00775-008-0359-6.

Nakamura, M., Saeki, K., & Takahashi, Y. (1999). Hyperproduction of recombinant ferredoxins in Escherichia coli by coexpression of the ORF1-ORF2-iscS-iscU-iscA-hscB-hscA-fdx-ORF3 gene cluster. *Journal of Biochemistry*, *126*(1), 10–18.

Nerenberg, R. (2013). Breathing perchlorate. *Science*, *340*(6128), 38–39. https://doi.org/10.1126/science.1236336.

Netz, D. J. A., Stith, C. M., Stümpfig, M., Köpf, G., Vogel, D., Genau, H. M., et al. (2011). Eukaryotic DNA polymerases require an iron-sulfur cluster for the formation of active complexes. *Nature Chemical Biology*, *8*(1), 125–132. https://doi.org/10.1038/nchembio.721.

Ojha, L., Wilhelm, M. B., Murchie, S. L., McEwen, A. S., Wray, J. J., Hanley, J., et al. (2015). Spectral evidence for hydrated salts in recurring slope lineae on Mars. *Nature Geoscience*, *8*(1), 829–832. https://doi.org/10.1038/ngeo2546.

Outten, F. W., Djaman, O., & Storz, G. (2004). A suf operon requirement for Fe-S cluster assembly during iron starvation in Escherichia coli. *Molecular Microbiology*, *52*(3), 861–872. https://doi.org/10.1111/j.1365-2958.2004.04025.x.

Page, C. C., Moser, C. C., Chen, X., & Dutton, P. L. (1999). Natural engineering principles of electron tunnelling in biological oxidation-reduction. *Nature*, *402*(6757), 47–52. https://doi.org/10.1038/46972.

Pandelia, M. E., Lanz, N. D., Booker, S. J., & Krebs, C. (2015). Mössbauer spectroscopy of Fe/S proteins. *Biochimica et Biophysica Acta*, *1853*(6), 1395–1405.

Pandey, A., Gordon, D. M., Pain, J., Stemmler, T. L., Dancis, A., & Pain, D. (2013). Frataxin directly stimulates mitochondrial cysteine desulfurase by exposing substrate-binding sites, and a mutant Fe-S cluster scaffold protein with frataxin-bypassing ability acts similarly. *The Journal of Biological Chemistry*, *288*(52), 36773–36786. https://doi.org/10.1074/jbc.M113.525857.

Paul, V. D., & Lill, R. (2015). Biogenesis of cytosolic and nuclear iron-sulfur proteins and their role in genome stability. *Biochimica et Biophysica Acta, 1853*(6), 1528–1539. https://doi.org/10.1016/j.bbamcr.2014.12.018.

Perry, J. J. P., Shin, D. S., Getzoff, E. D., & Tainer, J. A. (2010). The structural biochemistry of the superoxide dismutases. *Biochimica et Biophysica Acta, 1804*(2), 245–262. https://doi.org/10.1016/j.bbapap.2009.11.004.

Pierik, A. J., Netz, D. J. A., & Lill, R. (2009). Analysis of iron-sulfur protein maturation in eukaryotes. *Nature Protocols, 4*(5), 753–766. https://doi.org/10.1038/nprot.2009.39.

Pratt, A. J., Shin, D. S., Merz, G. E., Rambo, R. P., Lancaster, W. A., Dyer, K. N., et al. (2014). Aggregation propensities of superoxide dismutase G93 hotspot mutants mirror ALS clinical phenotypes. *Proceedings of the National Academy of Sciences of the United States of America, 111*(43), E4568–4576.

Rambo, R. P., & Tainer, J. A. (2013a). Accurate assessment of mass, models and resolution by small-angle scattering. *Nature, 496*(7446), 477–481. https://doi.org/10.1038/nature12070.

Rambo, R. P., & Tainer, J. A. (2013b). Super-resolution in solution X-ray scattering and its applications to structural systems biology. *Annual Review of Biophysics, 42*, 415–441. https://doi.org/10.1146/annurev-biophys-083012-130301.

Raulfs, E. C., O'Carroll, I. P., Santos Dos, P. C., Unciuleac, M.-C., & Dean, D. R. (2008). In vivo iron-sulfur cluster formation. *Proceedings of the National Academy of Sciences of the United States of America, 105*(25), 8591–8596. https://doi.org/10.1073/pnas.0803173105.

Rikken, G. B., Kroon, A. G. M., & van Ginkel, C. G. (1996). Transformation of (per)chlorate into chloride by a newly isolated bacterium: Reduction and dismutation. *Applied Microbiology and Biotechnology, 45*(3), 420–426. https://doi.org/10.1007/s002530050707.

Roberts, V. A., Iverson, B. L., Iverson, S. A., Benkovic, S. J., Lerner, R. A., Getzoff, E. D., et al. (1990). Antibody remodeling: A general solution to the design of a metal-coordination site in an antibody binding pocket. *Proceedings of the National Academy of Sciences of the United States of America, 87*(17), 6654–6658.

Roche, B., Aussel, L., Ezraty, B., Mandin, P., Py, B., & Barras, F. (2013). Iron/sulfur proteins biogenesis in prokaryotes: Formation, regulation and diversity. *Biochimica et Biophysica Acta, 1827*(3), 455–469. https://doi.org/10.1016/j.bbabio.2012.12.010.

Rothery, R. A., Bertero, M. G., Spreter, T., Bouromand, N., Strynadka, N. C. J., & Weiner, J. H. (2010). Protein crystallography reveals a role for the FS0 cluster of Escherichia coli nitrate reductase A (NarGHI) in enzyme maturation. *The Journal of Biological Chemistry, 285*(12), 8801–8807. https://doi.org/10.1074/jbc.M109.066027.

Rothery, R. A., Blasco, F., & Weiner, J. H. (2001). Electron transfer from heme bL to the [3Fe-4S] cluster of Escherichia coli nitrate reductase A (NarGHI). *Biochemistry, 40*(17), 5260–5268.

Rötig, A., de Lonlay, P., Chretien, D., Foury, F., Koenig, M., Sidi, D., et al. (1997). Aconitase and mitochondrial iron-sulphur protein deficiency in Friedreich ataxia. *Nature Genetics, 17*(2), 215–217. https://doi.org/10.1038/ng1097-215.

Rouault, T. A. (2015). Mammalian iron-sulphur proteins: Novel insights into biogenesis and function. *Nature Reviews. Molecular Cell Biology, 16*(1), 45–55. https://doi.org/10.1038/nrm3909.

Saikrishnan, K., Yeeles, J. T., Neville, S. G., Krajewski, W. W., Dillingham, M. S., & Wigley, D. B. (2012). Insights into Chi recognition from the structure of an AddAB-type helicase-nuclease complex. *The EMBO Journal, 31*(6), 1568–1578.

Sandala, G. M., Hopmann, K. H., Ghosh, A., & Noodleman, L. (2011). Calibration of DFT functionals for the prediction of Fe Mössbauer spectral parameters in iron-nitrosyl and iron-sulfur complexes: Accurate geometries prove essential. *Journal of Chemical Theory and Computation*, 7(10), 3232–3247. https://doi.org/10.1021/ct200187d.

Schäfer, F., Seip, N., Maertens, B., Block, H., & Kubicek, J. (2015). Purification of GST-tagged proteins. *Methods in Enzymology*, *559*, 127–139. https://doi.org/10.1016/bs.mie.2014.11.005.

Schindelin, H., Kisker, C., Hilton, J., Rajagopalan, K. V., & Rees, D. C. (1996). Crystal structure of DMSO reductase: Redox-linked changes in molybdopterin coordination. *Science*, *272*(5268), 1615–1621. https://doi.org/10.1126/science.272.5268.1615.

Schmucker, S., Martelli, A., Colin, F., Page, A., Wattenhofer-Donzé, M., Reutenauer, L., et al. (2011). Mammalian frataxin: An essential function for cellular viability through an interaction with a preformed ISCU/NFS1/ISD11 iron-sulfur assembly complex. *PLoS One*, *6*(1), e16199. https://doi.org/10.1371/journal.pone.0016199.

Schnecke, V., Swanson, C. A., Getzoff, E. D., Tainer, J. A., & Kuhn, L. A. (1998). Screening a peptidyl database for potential ligands to proteins with side-chain flexibility. *Proteins*, *33*(1), 74–87.

Schwarz, G., Mendel, R. R., & Ribbe, M. W. (2009). Molybdenum cofactors, enzymes and pathways. *Nature*, *460*(7257), 839–847. https://doi.org/10.1038/nature08302.

Segel, I. H. (1976). *Biochemical calculations: How to solve mathematical problems in general biochemistry* (2nd ed.). . New York: John Wiley & Sons. Retrieved from http://www.wiley.com/WileyCDA/WileyTitle/productCd-0471774219.html.

Sheldrick, G. M. (2008). A short history of SHELX. *Acta Crystallographica. Section A: Foundations of Crystallography*, *64*(Pt. 1), 112–122. https://doi.org/10.1107/S0108767307043930.

Shomura, Y., Yoon, K.-S., Nishihara, H., & Higuchi, Y. (2011). Structural basis for a [4Fe-3S] cluster in the oxygen-tolerant membrane-bound [NiFe]-hydrogenase. *Nature*, *479*(7372), 253–256. https://doi.org/10.1038/nature10504.

Sontz, P. A., Mui, T. P., Fuss, J. O., Tainer, J. A., & Barton, J. K. (2012). DNA charge transport as a first step in coordinating the detection of lesions by repair proteins. *Proceedings of the National Academy of Sciences of the United States of America*, *109*(6), 1856–1861. https://doi.org/10.1073/pnas.1120063109.

Sparks, J. L., Kumar, R., Singh, M., Wold, M. S., Pandita, T. K., & Burgers, P. M. (2012). Human exonuclease 5 is a novel sliding exonuclease required for genome stability. *The Journal of Biological Chemistry*, *287*(51), 42773–42783.

Stehling, O., Wilbrecht, C., & Lill, R. (2014). Mitochondrial iron-sulfur protein biogenesis and human disease. *Biochimie*, *100*, 61–77. https://doi.org/10.1016/j.biochi.2014.01.010.

Stemmler, T. L., Lesuisse, E., Pain, D., & Dancis, A. (2010). Frataxin and mitochondrial FeS cluster biogenesis. *The Journal of Biological Chemistry*, *285*(35), 26737–26743. https://doi.org/10.1074/jbc.R110.118679.

Szaleniec, M., Hagel, C., Menke, M., Nowak, P., Witko, M., & Heider, J. (2007). Kinetics and mechanism of oxygen-independent hydrocarbon hydroxylation by ethylbenzene dehydrogenase. *Biochemistry*, *46*(25), 7637–7646. https://doi.org/10.1021/bi700633c.

Tainer, J. A., Roberts, V. A., & Getzoff, E. D. (1991). Metal-binding sites in proteins. *Current Opinion in Biotechnology*, *2*(4), 582–591.

Takahashi, Y., & Nakamura, M. (1999). Functional assignment of the ORF2-iscS-iscU-iscA-hscB-hscA-fdx-ORF3 gene cluster involved in the assembly of Fe-S clusters in Escherichia coli. *Journal of Biochemistry*, *126*(5), 917–926.

Terwilliger, T. C., Adams, P. D., Read, R. J., McCoy, A. J., Moriarty, N. W., Grosse-Kunstleve, R. W., et al. (2009). Decision-making in structure solution using Bayesian estimates of map quality: The PHENIX AutoSol wizard. *Acta Crystallographica. Section D, Biological Crystallography*, *65*(Pt. 6), 582–601. https://doi.org/10.1107/S0907444909012098.

Thayer, M. M., Ahern, H., Xing, D., Cunningham, R. P., & Tainer, J. A. (1995). Novel DNA binding motifs in the DNA repair enzyme endonuclease III crystal structure. *The EMBO Journal*, *14*(16), 4108–4120.

Tsai, C.-L., & Barondeau, D. P. (2010). Human frataxin is an allosteric switch that activates the Fe-S cluster biosynthetic complex. *Biochemistry*, *49*(43), 9132–9139. https://doi.org/10.1021/bi1013062.

Tsai, C.-L., Bridwell-Rabb, J., & Barondeau, D. P. (2011). Friedreich's ataxia variants I154F and W155R diminish frataxin-based activation of the iron-sulfur cluster assembly complex. *Biochemistry*, *50*(29), 6478–6487. https://doi.org/10.1021/bi200666h.

Uzarska, M. A., Dutkiewicz, R., Freibert, S.-A., Lill, R., & Mühlenhoff, U. (2013). The mitochondrial Hsp70 chaperone Ssq1 facilitates Fe/S cluster transfer from Isu1 to Grx5 by complex formation. *Molecular Biology of the Cell*, *24*(12), 1830–1841. https://doi.org/10.1091/mbc.E12-09-0644.

Vaithiyalingam, S., Warren, E. M., Eichman, B. F., & Chazin, W. J. (2010). Insights into eukaryotic DNA priming from the structure and functional interactions of the 4Fe-4S cluster domain of human DNA primase. *Proceedings of the National Academy of Sciences of the United States of America*, *107*(31), 13684–13689. https://doi.org/10.1073/pnas.1002009107.

Venkateswara Rao, P., & Holm, R. H. (2004). Synthetic analogues of the active sites of iron-sulfur proteins. *Chemical Reviews*, *104*(2), 527–559. https://doi.org/10.1021/cr020615.

Vita, N., Hatchikian, E. C., Nouailler, M., Dolla, A., & Pieulle, L. (2008). Disulfide bond-dependent mechanism of protection against oxidative stress in pyruvate-ferredoxin oxidoreductase of anaerobic Desulfovibrio bacteria. *Biochemistry*, *47*(3), 957–964. https://doi.org/10.1021/bi7014713.

Vranish, J. N., Russell, W. K., Yu, L. E., Cox, R. M., Russell, D. H., & Barondeau, D. P. (2015). Fluorescent probes for tracking the transfer of iron-sulfur cluster and other metal cofactors in biosynthetic reaction pathways. *Journal of the American Chemical Society*, *137*(1), 390–398. https://doi.org/10.1021/ja510998s.

Webert, H., Freibert, S.-A., Gallo, A., Heidenreich, T., Linne, U., Amlacher, S., et al. (2014). Functional reconstitution of mitochondrial Fe/S cluster synthesis on Isu1 reveals the involvement of ferredoxin. *Nature Communications*, *5*, 5013. https://doi.org/10.1038/ncomms6013.

Weissmann, F., Petzold, G., VanderLinden, R., Huis In 't Veld, P. J., Brown, N. G., Lampert, F., et al. (2016). biGBac enables rapid gene assembly for the expression of large multisubunit protein complexes. *Proceedings of the National Academy of Sciences of the United States of America*, *113*(19), E2564–9. https://doi.org/10.1073/pnas.1604935113.

Winn, M. D., Ballard, C. C., Cowtan, K. D., Dodson, E. J., Emsley, P., Evans, P. R., et al. (2011). Overview of the CCP4 suite and current developments. *Acta Crystallographica. Section D, Biological Crystallography*, *67*(Pt. 4), 235–242. https://doi.org/10.1107/S0907444910045749.

Xu, J., Song, Y., Min, B., Steinberg, L., & Logan, B. E. (2003). Microbial degradation of perchlorate: Principles and applications. *Environmental Engineering Science*, *20*(5), 405–422. https://doi.org/10.1089/109287503768335904.

Yan, A., & Kiley, P. J. (2009). Techniques to isolate O2-sensitive proteins: [4Fe-4S]-FNR as an example. *Methods in Enzymology*, *463*, 787–805. https://doi.org/10.1016/S0076-6879(09)63042-1.

Yannone, S. M., Hartung, S., Menon, A. L., Adams, M. W. W., & Tainer, J. A. (2012). Metals in biology: Defining metalloproteomes. *Current Opinion in Biotechnology*, *23*(1), 89–95. https://doi.org/10.1016/j.copbio.2011.11.005.

Yoon, T., & Cowan, J. A. (2003). Iron-sulfur cluster biosynthesis. Characterization of frataxin as an iron donor for assembly of [2Fe-2S] clusters in ISU-type proteins. *Journal of the American Chemical Society*, *125*(20), 6078–6084. https://doi.org/10.1021/ja027967i.

Youngblut, M. D., Tsai, C.-L., Clark, I. C., Carlson, H. K., Maglaqui, A. P., Gau-Pan, P. S., et al. (2016). Perchlorate reductase is distinguished by active site aromatic gate residues. *The Journal of Biological Chemistry*, *291*(17), 9190–9202. https://doi.org/10.1074/jbc.M116.714618.

Youngblut, M. D., Wang, O., Barnum, T. P., & Coates, J. D. (2016). (Per)chlorate in biology on earth and beyond. *Annual Review of Microbiology*, *70*, 435–457. https://doi.org/10.1146/annurev-micro-102215-095406.

Yu, L., & Wolin, M. J. (1969). Hydrogenase measurement with photochemically reduced methyl viologen. *Journal of Bacteriology*, *98*(1), 51–55.

Zheng, L., Cash, V. L., Flint, D. H., & Dean, D. R. (1998). Assembly of iron-sulfur clusters: Identification of an iscSUA-hscBA-fdx gene cluster from Azotobacter vinelandii. *The Journal of Biological Chemistry*, *273*(21), 13264–13272. https://doi.org/10.1074/jbc.273.21.13264.

Zhou, C., Pourmal, S., & Pavletich, N. P. (2015). Dna2 nuclease-helicase structure, mechanism and regulation by rpa. *eLife*, *4*, e09832.

CHAPTER SEVEN

Biochemical Reconstitution and Spectroscopic Analysis of Iron–Sulfur Proteins

Sven-Andreas Freibert*,[2], Benjamin D. Weiler*,[2], Eckhard Bill[†], Antonio J. Pierik[‡], Ulrich Mühlenhoff*, Roland Lill*,[§,1]

*Institut für Zytobiologie, Philipps-Universität, Marburg, Germany
[†]Max-Planck-Institut für Chemische Energiekonversion, Mülheim an der Ruhr, Germany
[‡]Chemistry and Biochemistry, Technical University of Kaiserlautern, Kaiserlautern, Germany
[§]LOEWE Zentrum für Synthetische Mikrobiologie SynMikro, Marburg, Germany
[1]Corresponding author: e-mail address: lill@staff.uni-marburg.de

Contents

[2] These authors contributed equally.

Methods in Enzymology, Volume 599
ISSN 0076-6879
https://doi.org/10.1016/bs.mie.2017.11.034

Abstract

Iron–sulfur (Fe/S) proteins are involved in numerous key biological functions such as respiration, metabolic processes, protein translation, DNA synthesis, and DNA repair. The simplest types of Fe/S clusters include [2Fe–2S], [3Fe–4S], and [4Fe–4S] forms that sometimes are present in multiple copies. De novo assembly of Fe/S cofactors and their insertion into apoproteins in living cells requires complex proteinaceous machineries that are frequently highly conserved. In eukaryotes such as yeast and mammals, the mitochondrial iron–sulfur cluster assembly machinery and the cytosolic iron–sulfur protein assembly system consist of more than 30 components that cooperate in the generation of some 50 cellular Fe/S proteins. Both the mechanistic dissection of the intracellular Fe/S protein assembly pathways and the identification and characterization of Fe/S proteins rely on tool boxes of in vitro and in vivo methods. These cell biological, biochemical, and biophysical techniques help to determine the extent, stability, and type of bound Fe/S cluster. They also serve to distinguish bona fide Fe/S proteins from other metal-binding proteins containing similar cofactor coordination motifs. Here, we present a collection of in vitro methods that have proven useful for basic biochemical and biophysical characterization of Fe/S proteins. First, we describe the chemical assembly of [2Fe–2S] or [4Fe–4S] clusters on purified apoproteins. Then, we summarize a reconstitution system reproducing the de novo synthesis of a [2Fe–2S] cluster in mitochondria. Finally, we explain the use of UV–vis, CD, electron paramagnetic resonance, and Mössbauer spectroscopy for the routine characterization of Fe/S proteins.

1. INTRODUCTION

1.1 Fe/S Clusters and Their Coordination

Iron–sulfur (Fe/S) clusters are ancient, ubiquitous, and versatile protein cofactors present in all kingdoms of life (Beinert, Holm, & Münck, 1997). The simplest forms of Fe/S clusters are of the [2Fe–2S], [3Fe–4S], and [4Fe–4S] types, which together represent more than 90% of all Fe/S cofactors (Andreini, Rosato, & Banci, 2017). The most common type is the cubane [4Fe–4S] cluster present in more than 60% of the cellular Fe/S proteins, followed by the dinuclear [2Fe–2S] cluster with a 10%–30% incidence in various species. A number of Fe/S proteins carry more than one of these cofactors, with bacterial or mitochondrial respiratory complexes I being particularly impressive examples harboring nine (bacteria) or eight (mitochondria) Fe/S clusters. Fe/S clusters are usually bound to the polypeptide chains via iron ligation to Cys or His residues, but also Asp, Arg, Ser, Glu, or Tyr amino acid residues have been described as ligands of one of the iron ions. Especially in complex Fe/S proteins such as nitrogenases or hydrogenases CO or CN^- can serve as additional ligands (Hu & Ribbe, 2016; Peters et al., 2015).

Several consensus sequence motifs are known for Fe/S cluster coordination, yet many Fe/S proteins contain hitherto undescribed, unique motifs, making predictions of Fe/S cluster-binding rather difficult. However, a number of [4Fe–4S] proteins use the $CX_2CX_2CX_{20–40}C$ consensus motif. Remarkable conservation of binding motifs can be found in simple low potential and high potential [4Fe–4S] ferredoxins from bacteria, containing conserved $CX_2CX_2CX_3CP$ and $CX_2CX_{12–16}CX_{13}C$ consensus motifs, respectively. Mitochondrial and many bacterial [2Fe–2S] ferredoxins are recognized by a $CX_{\sim 4–5}CX_2CX_{\sim 35}C$ binding site. Rieske-type proteins coordinate a [2Fe–2S] cluster by two cysteine and two histidine ligands at a $CXHX_{13–21}CX_2H$ binding site. A highly conserved binding motif is also present in the large family of radical SAM (*S*-adenosyl methionine) Fe/S proteins, where three iron ions of the [4Fe–4S] cluster bind to a CX_3CX_2C consensus sequence, while the fourth iron is coordinated by the amino and carboxylate groups of SAM in a bidentate fashion (Landgraf, McCarthy, & Booker, 2016). Many other possibilities of cluster coordination are known, and, in most cases, these cluster arrangements are unique to the individual classes of Fe/S proteins. Occasionally, other metals such as Zn can become inserted into the Fe/S cluster binding site, especially upon overexpression or during heterologous synthesis of recombinant proteins in *Escherichia coli*. Examples include DNA polymerases and the scaffold protein IscU (Netz et al., 2012; Ramelot et al., 2004). Further, on the basis of primary sequences alone, it is sometimes difficult to distinguish Fe/S proteins from zinc finger proteins that harbor similar, yet distinct cysteine-rich binding motifs. Therefore, efficient analytical tools are needed to verify bona fide Fe/S cluster binding, both in vivo and in vitro. Further, reliable methods are required to analyze the coordination sites, types, and binding stabilities of the clusters.

1.2 The Mitochondrial ISC Machinery as a Paradigm Fe/S Protein Biogenesis System

After the first description of Fe/S proteins in the 1960s (Beinert et al., 1997; Beinert & Sands, 1960), it soon became clear that their cofactors can be reconstituted in vitro under anaerobic conditions using ionic iron, sulfide, and a thiol-specific reductant (Malkin & Rabinowitz, 1966; Meyer, Moulis, & Lutz, 1986). In cells, however, Fe/S proteins are assembled by complex proteinaceous machineries that de novo synthesize an Fe/S cluster on so-called scaffold proteins, then transfer the cluster via transient association with targeting factors to specific recipient apoproteins, and finally assist in the correct cluster insertion into the polypeptide chains (for recent reviews, see, e.g., Braymer & Lill, 2017; Lill, 2009; Paul & Lill, 2015; Py & Barras, 2015; Rouault, 2015). General biogenesis of housekeeping Fe/S proteins in bacteria is catalyzed by the iron–sulfur cluster assembly (ISC) machinery and/or the sulfur mobilization (SUF) system, while the NIF machinery present in azototrophic bacteria is specifically dedicated to assemble the complex Fe/S protein nitrogenase (Blanc, Gerez, & Ollagnier de Choudens, 2015; Hu & Ribbe, 2016; Outten, 2015; Py & Barras, 2015; Zheng, Cash, Flint, & Dean, 1998). Eukaryotes possess close relatives of the ISC system in mitochondria and mitochondria-related organelles (Freibert et al., 2017) and of the SUF machinery in plastids (Balk & Schaedler, 2014), both of which were inherited during endosymbiotic events. Moreover, eukaryotes contain the cytosolic iron–sulfur protein assembly (CIA) machinery that is specific for maturation of apoproteins located in both the cytosol and the nucleus (Lill, Srinivasan, & Mühlenhoff, 2014; Netz, Mascarenhas, Stehling, Pierik, & Lill, 2014; Paul & Lill, 2015; Sharma, Pallesen, Spang, & Walden, 2010). We will briefly describe the known mitochondrial ISC components and define the underlying mechanisms of Fe/S cluster synthesis, trafficking, and apoprotein insertion as a well-studied example of Fe/S protein biogenesis system (Fig. 1).

In any of the above-mentioned assembly systems, the critical initial step in the biosynthesis of Fe/S clusters is the activation of sulfur (Fig. 1A). This step is universally accomplished by cysteine desulfurase enzymes that belong to a small subfamily of pyridoxal 5′-phosphate (PLP)-dependent transaminases and usually work as dimers (Black & Dos Santos, 2015). The founding member is NifS of the bacterial NIF system and served to work out the biochemical mechanism of sulfur release from free cysteine and of persulfide (-SSH) generation on an active-site cysteine of the enzyme (Zheng, White, Cash, & Dean, 1994). Close homologues of NifS are encoded in the bacterial *isc* (IscS) and *suf* (SufS) operons (Zheng et al., 1998). The eukaryotic enzyme termed Nfs1 is predominantly located in mitochondria but reportedly may fulfill a yet unknown function in the nucleus and/or cytosol (Biederbick et al., 2006; Land & Rouault, 1998; Marelja et al., 2013; Mühlenhoff et al., 2004; Nakai, Nakai, Hayashi, & Kagamiyama, 2001). In contrast to bacterial desulfurases, most eukaryotic cysteine desulfurases in vivo depend on the small protein Isd11, a member of the LYRM protein family (Adam, Bornhövd, Prokisch, Neupert, & Hell, 2006; Shi, Ghosh, Tong, & Rouault, 2009; Wiedemann et al., 2006).

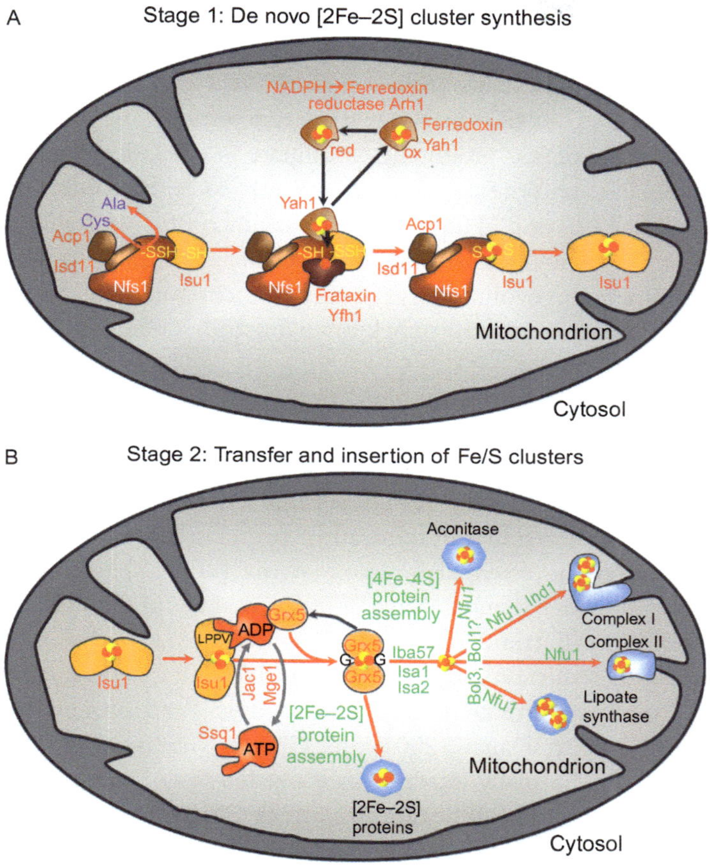

Fig. 1 A working model of Fe/S protein biogenesis in mitochondria. The pathway can be subdivided into two major parts. (A) The de novo Fe/S cluster synthesis on the scaffold protein Isu1 is catalyzed by the core ISC complex and presumably results in a [2Fe–2S] cluster in a Isu1 dimer. (B) Transfer of the [2Fe–2S] from Isu1 to Grx5 and insertion into [2Fe–2S] target proteins, as well as [4Fe–4S] cluster synthesis by the Isa1–Isa2–Iba57 complex, followed by [4Fe–4S] cluster trafficking and insertion into apoproteins by the late-acting ISC targeting factors. For details see text.

Nfs1 further binds, via Isd11, to acyl carrier protein Acp1, which is long known to be needed for mitochondrial fatty acid synthesis and lipoic acid formation (Angerer, 2013; Boniecki, Freibert, Mühlenhoff, Lill, & Cygler, 2017; Cory et al., 2017; Van Vranken et al., 2016). Different LYRM–Acp1 complexes are found associated with, e.g., mitochondrial respiratory complexes I, II, III, and V and their assembly intermediates (Angerer et al., 2014; Atkinson et al., 2011). Depletion of either Acp1 or

Isd11 results in the disappearance of Nfs1–Isd11 high-molecular-mass complexes (ca. 200 kDa) and is associated with defects in Fe/S protein biogenesis in mitochondria and cytosol (Adam et al., 2006; Van Vranken et al., 2016; Wiedemann et al., 2006). Yet, per se these two proteins are not required for Nfs1 activity as a sulfide-producing desulfurase in vitro, thus likely attributing a stabilizing and potentially regulatory role to both proteins (Boniecki et al., 2017; Li, Tong, Hughes, & Rouault, 2006; Mühlenhoff et al., 2004). Consistent with this, *Microsporidia, Giardia,* and *Entamoeba* that contain mitosomes, i.e., evolutionary reduced mitochondria which maintained Fe/S protein biogenesis as the only known function, do not encode an Acp1 homolog (Freibert et al., 2017; Goldberg et al., 2008; Tovar et al., 2003).

A central component of the mitochondrial ISC assembly system is the protein Isu1 (ISCU in humans), which serves as a scaffold for the initial de novo synthesis of a [2Fe–2S] cluster (Fig. 1A). Fe/S cluster synthesis is initiated by the transfer of the persulfide sulfur from Nfs1 to one of the three conserved cysteine residues of the scaffold protein (Parent et al., 2015) and further requires the essential function of the [2Fe–2S] ferredoxin Yah1 (FDX2 in humans) (Lange, Kaut, Kispal, & Lill, 2000; Sheftel et al., 2010) and Yfh1 (frataxin) (Pastore & Puccio, 2013). Yah1 is reduced by the ferredoxin reductase Arh1 (FdxR) and NADPH, which are indirect, yet also essential components of mitochondrial Fe/S protein biogenesis (Li, Saxena, Pain, & Dancis, 2001; Shi, Ghosh, Kovtunovych, Crooks, & Rouault, 2012). Formally, electrons are needed for reduction of the persulfide sulfur on Isu1 to sulfide during cluster synthesis; however, the assembly intermediate that is reduced by Yah1 is unknown. Yfh1 binds iron at its N-terminal helix and may facilitate the supply of the metal ion to the biosynthetic reaction center formed by Nfs1 and Isu1 (Stemmler, Lesuisse, Pain, & Dancis, 2010). Further, Yfh1 stimulates the sulfur transfer from Nfs1 to Isu1 and hence was proposed to act as an allosteric regulator of the de novo Fe/S cluster synthesis (Bridwell-Rabb, Fox, Tsai, Winn, & Barondeau, 2014; Parent et al., 2015). Recently, the synthesis of the [2Fe–2S] cluster on Isu1 was faithfully reconstituted using circular dichroism (CD) spectroscopy to follow the formation of the [2Fe–2S] cluster on Isu1 (see below) (Boniecki et al., 2017; Freibert et al., 2017; Webert et al., 2014). Six of the aforementioned ISC proteins (Nfs1-Isd11, Isu1, Yah1, Yfh1, and Arh1) were shown to be both necessary and sufficient for rapid [2Fe–2S] cluster formation in a dithiothreitol (DTT)-independent fashion. Apparently, the electron transfer chain of NADPH–FdxR–Yah1 could replace the artificial reductant DTT that is typically used in such reconstitution approaches. This system allowed initial insights into the mechanisms underlying mitochondrial [2Fe–2S] cluster formation on Isu1 (Webert et al., 2014).

After synthesis of the [2Fe–2S] cluster on Isu1, the cofactor is trafficked to the target apoproteins including [4Fe–4S] proteins (Fig. 1B). This complex pathway initially involves the function of a chaperone/cochaperone system of the Hsp70–Hsp40 (DnaK–DnaJ) type to facilitate the dissociation of the [2Fe–2S] from Isu1 and its transfer to the monothiol glutaredoxin Grx5. For details of the chaperone cycle, see excellent recent reviews by Craig and Marszalek (2017) and Kampinga and Craig (2010). Initially, the cochaperone Jac1 binds to Isu1 and then transfers it to the Hsp70 Ssq1 (or Ssc1; Pukszta et al., 2010) which attaches to the conserved hydrophobic segment LPPV of Isu1. Hsp70 binding is thought to induce Fe/S cluster release from Isu1 (Bonomi, Iametti, Morleo, Ta, & Vickery, 2008). The cluster handover from Isu1 to Grx5 may be facilitated by their simultaneous binding to the Ssq1 chaperone on different interaction sites (Uzarska, Dutkiewicz, Freibert, Lill, & Mühlenhoff, 2013). Grx5 binds the [2Fe–2S] cluster in a bridging and glutathione-dependent, labile fashion and is believed to directly insert it into recipient apoproteins without further known assistance (Rouhier, Couturier, Johnson, & Jacquot, 2010). An in vitro reconstitution system for the chaperone-dependent Fe/S cluster transfer from Isu1 to Grx5 has hitherto been described only for the corresponding bacterial system (Shakamuri, Zhang, & Johnson, 2012). In the presence of the bacterial chaperones HscA–HscB, ATP, and GSH a 700-fold faster Fe/S cluster transfer was obtained.

The synthesis of a [4Fe–4S] cluster and its insertion into apoproteins requires more than a handful of additional, "late-acting" ISC assembly components which may act in either [4Fe–4S] cluster formation or [4Fe–4S] cluster incorporation into the target polypeptide chains (Fig. 1B). The complex comprising Isa1, Isa2, and Iba57 proteins is essential for any of the analyzed mitochondrial [4Fe–4S] proteins (Gelling, Dawes, Richhardt, Lill, & Mühlenhoff, 2008; Mühlenhoff, Richter, Pines, Pierik, & Lill, 2011; Sheftel et al., 2012). In contrast, [2Fe–2S] proteins are matured without their involvement, suggesting that these three ISC components are involved in the conversion of the Grx5-bound [2Fe–2S] into a [4Fe–4S] cluster. A recent biochemical study has reported that this conversion occurs directly on Isa1–Isa2, yet this DTT-dependent in vitro system did not include Iba57, an essential in vivo component of this step (Brancaccio et al., 2014). Therefore, this conversion reaction awaits a faithful reconstitution system that fully meets all in vivo requirements.

The remaining late-acting ISC factors might be involved in facilitating a target-specific insertion of [4Fe–4S] clusters into specific subsets of apoproteins. Of these, Nfu1 transiently binds a [4Fe–4S] cluster which may be directly transferred to the main Nfu1 targets, i.e., the respiratory complexes I and II, and lipoate synthase (Cameron et al., 2011; Melber et al., 2016; Navarro-Sastre et al., 2011). The BolA family proteins Bol1 and Bol3 are auxiliary, late-acting ISC factors that have a similar target spectrum as Nfu1 (Melber et al., 2016; Uzarska et al., 2016). Both proteins interact with holo- and apo-Grx5 and form heterodimers which do or do not contain a [2Fe–2S] cluster, respectively. The physiological role of this complex formation and of the [2Fe–2S] cluster in this late step of the [4Fe–4S] cluster maturation process is still unknown. Similarly to Nfu1, the P-loop NTPase Ind1 binds a [4Fe–4S] cluster which presumably is used for specific insertion into respiratory complex I (Bych et al., 2008; Sheftel et al., 2009). The various steps of the mitochondrial Fe/S protein assembly pathway, i.e., [2Fe–2S] cluster synthesis on Isu1, transfer to Grx5, insertion into [2Fe–2S] target proteins, [4Fe–4S] cluster synthesis, transfer to targeting factors, and insertion into [4Fe–4S] target proteins, were delineated mainly by in vivo studies (Braymer & Lill, 2017). Various experimental tools for the in vivo analysis of Fe/S proteins in yeast and human cells have been published elsewhere, and we refer the readers to these assays (Molik, Lill, & Mühlenhoff, 2007; Pierik, Netz, & Lill, 2009; Stehling, Sheftel, & Lill, 2009). However, the precise biochemical understanding of the individual reactions now requires faithful in vitro reconstitution systems with suitable donor and acceptor Fe/S proteins.

In this chapter, we will first outline the experimental details of chemical and biochemical reconstitutions of Fe/S proteins in vitro. These experiments may also help in verifying the binding of Fe/S clusters to apoproteins, yet for unequivocal confirmation of physiological Fe/S cluster binding complementary in vivo approaches are needed. We will then explain the strategies for reconstituting a particular step of Fe/S protein biogenesis, i.e., the de novo [2Fe–2S] cluster formation on the mitochondrial scaffold protein Isu1 in an experimental setup that might closely mimic the physiological conditions. This CD spectroscopy-based assay may serve as a starting point for the elucidation of the biochemical mechanism by which the [2Fe–2S] cluster is formed on Isu1, because it depends on all factors defined in vivo. Finally, we will briefly describe how different spectroscopic techniques (UV–vis, CD, EPR, and Mössbauer) can be exploited for the biophysical characterization of reconstituted or purified Fe/S proteins. These types of experiments can unravel the Fe/S cluster type and its coordination geometry.

2. CHEMICAL AND BIOCHEMICAL RECONSTITUTION OF SIMPLE FE/S PROTEINS

Due to their instability against oxygen and/or metal-chelating agents, Fe/S proteins can be isolated mainly in their apoform. This is especially true when they are purified after recombinant overexpression in *E. coli*. The Fe/S cofactors of many simple Fe/S proteins can be faithfully reconstructed by chemical means in vitro under strict anaerobic conditions using simple, well-established experimental strategies.

2.1 Equipment

- Anaerobic chamber (COY Laboratory Products; SHEL LAB).

2.2 Buffers and Reagents

Chemical reconstitution protocol:

- A thiol-specific reductant (DTT; mercaptoethanol or glutathione).
- An iron source such as ferric ammonium citrate (FAC), $FeCl_3$, or $FeCl_2$.
- A sulfide salt; Li_2S (not Na_2S since it is hygroscopic and tends to oxidize).
- Reconstitution buffer (50 m*M* Tris–HCl pH 8.0–8.3, 0–150 m*M* NaCl, 0%–10% glycerol, supplemented with 3–5 m*M* DTT, or 0.1%–1% mercaptoethanol).
- Stock solutions are freshly prepared in degassed water.

Semienzymatic reconstitution protocol:

- Cysteine, DTT.
- An iron source (FAC, $FeCl_3$ or ferrous-ammonium sulfate or $FeCl_2$).
- Recombinant cysteine desulfurase, e.g., *E. coli* IscS or mitochondrial Nfs1–Isd11.
- Reconstitution buffer (see above).

2.3 Procedures

2.3.1 Chemical Reconstitution

Reconstitution protocols uniformly start with a prolonged reduction of concentrated apoprotein solutions by the thiol-specific reductants DTT (2–10 m*M* final) or mercaptoethanol (up to 1% final) under strict anaerobic conditions for at least 3 h at 4°C. For subsequent Fe/S cluster reconstitution, the reduced protein is diluted to 10–200 μ*M* final concentration (depending on the studied protein) in 50 m*M* Tris–HCl pH 8.0–8.3 at room temperature supplemented with 3–5 m*M* DTT or 0.1%–1% mercaptoethanol final concentration. The addition of 50–150 m*M* NaCl and/or 5%–10% glycerol may be beneficial, yet depends on the specific apoprotein used. Depending on the Fe/S cluster type, a two- to fourfold ([2Fe–2S]) and four- to eightfold molar excess ([4Fe–4S]) of FAC or $FeCl_3$ is added, which results in the formation of a pink solution. After 5 min incubation, lithium sulfide is added at the same molar concentration as iron, and the mixture is incubated for at least 2–3 h. During this period, the pink color fades and a red-brownish color develops. Unbound iron and sulfide are finally removed via a desalting column (PD-10) in a buffer of choice, and the reconstituted proteins are concentrated under anaerobic conditions (for classical examples, see Malkin & Rabinowitz, 1966; Meyer et al., 1986).

2.3.2 Semienzymatic Reconstitution

The reconstitution of Fe/S clusters can also be performed in the presence of a cysteine desulfurase that slowly releases sulfide in situ from cysteine and subsequently serves as substrate for chemical Fe/S cluster reconstitution. The presence of DTT is essential for this semienzymatic reconstitution approach, because DTT mediates the reductive release of sulfide from the desulfurase-bound persulfide moiety (see above). Other thiol-specific reductants are not efficient in this process (Mühlenhoff, Richhardt, Gerber, & Lill, 2002). Usually, bacterial cysteine desulfurases such as *E. coli* IscS are preferred as they are easy to handle. Eukaryotic Nfs1 tends to aggregate unless its small partner protein Isd11 is bound. As for chemical reconstitution, the apoprotein has to be reduced and subsequently diluted to 10–200 μ*M* (depending on the studied protein) in reconstitution buffer. After supplementing with DTT and cysteine at 1–10 m*M*, the desulfurase is added at 1/50 or 1/100 *M* concentration of the apoprotein. Finally, an iron source at two- to fourfold molar excess is added, and samples are incubated at room temperature for 2–3 h. This method may have some advantage over a classical chemical reconstitution, since sulfide production and therefore the rate of the reaction can be controlled by the amount of desulfurase present in the assay. Further, the freshly formed sulfide tends to be a better source of sulfide as the one from sulfide salts, as the former is not contaminated with oxidation products of sulfur (Agar, Zheng, Cash, Dean, & Johnson, 2000; Bandyopadhyay et al., 2008; Gao et al., 2013; Py et al., 2012).

2.4 Notes

Reconstitution in a buffer that stabilizes the Fe/S protein of interest is recommended. Most Fe/S proteins are stabilized by (1) slightly alkaline solutions (pH >8), (2) 50–150 m*M* NaCl or KCl, and (3) 5%–10% glycerol. For efficient removal of residual oxygen, small sample volumes with high protein concentrations are preferred for the initial reduction step. For chemical reconstitution, it is essential that sulfide is added last. Frequently, excessive amounts of iron and sulfide do not improve the efficiency of chemical Fe/S cluster reconstitution, yet increase unspecific iron binding and may promote the formation of Fe^{3+}-containing precipitates (see Fig. 3B, inset). The formation of a black precipitate during the reconstitution procedure following the addition of Li_2S is due to residual oxygen which invariably results in a failed reconstitution experiment. During chemical reconstitution, this precipitate usually forms within the first few minutes. During a semienzymatic protocol this may take longer. This Fe^{3+}-containing precipitate is frequently not fully removed by a desalting column, and its UV/vis spectrum may look similar to that of an Fe/S protein, yet typically shows significant absorption at wavelengths higher than 650 nm. Quantification of the acid-labile sulfide and Fe^{2+} content by colorimetric assays is used to assess the efficiency of Fe/S cluster reconstitution and may provide first hints concerning the type of bound Fe/S cluster (Molik et al., 2007; Rabinowitz, 1978).

3. DE NOVO SYNTHESIS OF A [2FE–2S] CLUSTERS ON THE SCAFFOLD PROTEIN ISU1 IN VITRO

The mechanistic elucidation of the various steps of Fe/S cluster synthesis and apoprotein insertion requires the development of faithful in vitro assays that closely mimic the in vivo requirements of the tested reaction. This means that reconstitution reactions should reflect the function of all biogenesis components (proteins and cofactors) known to be involved in a particular step in vivo. Further, the assay setup should not rely on artificial components such as DTT or sulfide used in chemical reconstitution assays (see above). In eukaryotes, the initial Fe/S cluster biosynthesis is carried out by the mitochondrial (or mitosomal) "core ISC complex" (Fig. 1A) (Boniecki et al., 2017; Cory et al., 2017; Freibert et al., 2017; Webert et al., 2014). This machinery is solely dedicated to the de novo generation of a [2Fe–2S] cluster on the scaffold protein Isu1 which is accomplished by six ISC proteins: the desulfurase Nfs1 and its associated partner proteins Isd11–Acp1, the ferredoxin Yah1 together with its ferredoxin reductase Arh1, and frataxin Yfh1. To study this de novo Fe/S cluster synthesis step, we developed an enzymatic reconstitution assay based on the time-dependent CD spectroscopic signal change during [2Fe–2S] cluster formation on Isu1 (Bonomi et al., 2008). This assay is independent of any artificial reductant (e.g., DTT), does not involve the generation of free sulfide, and, with the exception of the regulatory Nfs1 subunit Acp1 (see Boniecki et al., 2017), depends on all known members of the core ISC complex and the respective cofactors such as NADPH and cysteine (Webert et al., 2014). To the best of our knowledge, it closely resembles the reaction taking place in vivo. So far, this in vitro assay was used to characterize the mitochondrial core ISC complexes from the fungi *Saccharomyces cerevisiae* and *Chaetomium thermophilum*, the mitosome-containing human pathogen *Trachipleistophora hominis*, and man, as well as for reconstituting the corresponding core ISC complex from *E. coli* (Boniecki et al., 2017; Freibert et al., 2017; Webert et al., 2014; Yan, Adinolfi, & Pastore, 2015).

3.1 Equipment

- Anaerobic chamber (COY Laboratory Products; SHEL LAB).
- 5-mm square quartz cuvette, equipped with stirring bar and a soft silicon lid that can be penetrated with a Hamilton syringe.
- CD spectrometer (e.g., Jasco J-815) equipped with a stirring device.

3.2 Buffers and Reagents

- Assay buffer (35 m*M* Tris pH 8.0 with 150 m*M* NaCl, 800 μ*M* Na-ascorbate, 200 μ*M* $MgCl_2$, 300 μ*M* $FeCl_2$, 500 μ*M* NADPH).

3.3 Procedures

As for all Fe/S cluster reconstitution protocols, a successful enzymatic reconstitution of Isu1 crucially depends on the absence of oxygen. All buffers, reagents, and protein solutions, therefore, must be incubated under anaerobic conditions (e.g., in an anaerobic chamber) for at least 24 h prior to use, and all subsequent steps are performed inside the anaerobic chamber. As a buffer system we use 35 m*M* Tris pH 8.0 with 150 m*M* NaCl, a composition mimicking physiological ionic strength. Additionally, we add 200 μ*M* $MgCl_2$ as it enhances the activity of ferredoxin reductase, 300 μ*M* $FeCl_2$, 800 μ*M* sodium ascorbate, and 500 μ*M* NADPH for ferredoxin reduction via ferredoxin reductase. We use Isu1 at a concentration of 100 μ*M* because this amount gives a reliable CD signal after reconstitution. All components including the core ISC proteins (Nfs1–Isd11:Yfh1:Yah1:Arh1:Isu1 in a 1:1:1:0.1:25 ratio) are mixed to a final volume of 300 μL in buffer, transferred to a 5-mm quartz cuvette (e.g., Hellma) and sealed tightly by a silicon lid. The cuvette is placed in the tempered cuvette holder (30°C) of the CD spectrometer equipped with a stirring device. Since holo-Isu1 shows a distinct CD spectrum in the visible range with the main peak around 430 nm, we use the isosbestic point of ferredoxin at 431 nm as the wavelength of choice for monitoring holo-Isu1 formation (Fig. 2A) (Webert et al., 2014). The baseline is measured for 2 min, then an anaerobically prepared cysteine solution is injected with a Hamilton syringe to a final concentration of 500 μM. This initiates the enzymatic Fe/S cluster synthesis on Isu1 and thus will lead to an increase of the CD signal at 431 nm reaching

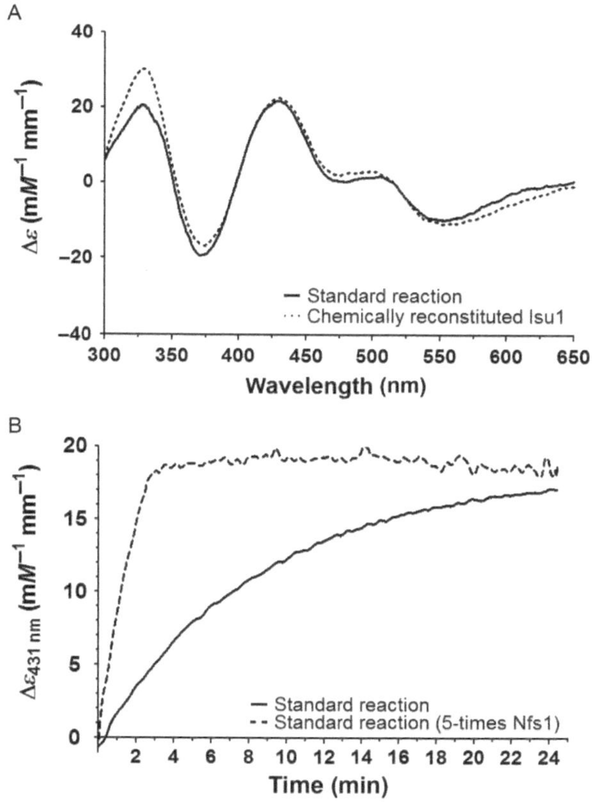

Fig. 2 In vitro reconstitution of [2Fe–2S] cluster formation on Isu1 by the core ISC complex using CD spectroscopy. (A) Comparison of the CD spectra of enzymatically (after 30 min of incubation) (—) and chemically (– –) reconstituted Isu1. (B) Time course measurement following enzymatic reconstitution of Isu1 for 25 min at 431 nm under standard conditions (—) and with a fivefold higher amount of Nfs1–Isd11 (– –).

saturation within approximately 15–20 min (Fig. 2B). After 30 min, time course measurements are stopped. To confirm successful reconstitution a CD spectrum from 300 to 650 nm is recorded. Reconstituted Isu1 will now exhibit a CD spectrum typical for a [2Fe–2S] cluster and virtually identical to the one obtained in a chemical reconstitution (Fig. 2A). When larger amounts of Nfs1–Isd11 or Yah1 compared to Isu1 are employed, a full CD spectrum has to be recorded before the time course measurements. This spectrum is later used to subtract the contributions of the PLP cofactor of Nfs1–Isd11 and of the [2Fe–2S] cluster of Yah1 from the final spectrum after reconstitution (see below).

3.4 Notes

The assay is sensitive to the amounts of ISC protein used and also offers valuable clues about rate dependencies and kinetics of the reaction. The rates of the CD signal changes linearly correlate with the amounts of Isu1 and Nfs1–Isd11. Yet, ratios higher than 1:5 of Nfs1–Isd11:Isu1 result in reaction rates that become too fast to be resolved by the experimental setup described here (Fig. 2B). In case of Yah1 and Yfh1, the concentration dependencies are coupled to the amount of Nfs1 used in the assay. While initially the rates increase linearly when raising Yah1 or Yfh1 amounts, they reach saturation at the concentration of the desulfurase complex used indicating that the core ISC complex operates as a 1:1 stoichiometric entity. This is reflected in the 3D structure of the entire complex recently resolved (Boniecki et al., 2017). Hence, the final CD signal intensity solely depends on the end product (holo-Isu1), while the rates in our setup depend on Nfs1–Isd11, which limits the persulfide production. Yah1 and Yfh1 are essential for synthesis of the [2Fe–2S] cluster on Isu1, yet they stimulate the rates of cluster formation only up to equimolar concentrations compared to Nfs1–Isd11. Consequently, since only small amounts of Nfs1–Isd11 and Yah1 are used in this assay, the contributions of both the Nfs1-bound PLP and of the [2Fe–2S] cluster of Yah1 to the final CD spectrum can be neglected under standard conditions. In summary, our in vitro enzymatic reconstitution assay is a valuable and versatile tool to study this first step of Fe/S protein biogenesis. Both signal intensities and rates can be easily adjusted to suit the experimental setup by varying protein amounts. The assay system offers convenient access to addressing various questions concerning stoichiometry and function of the core ISC complex and its individual constituents.

4. AT A GLANCE: SPECTROSCOPIC CHARACTERIZATION OF FE/S PROTEINS

For biophysical characterization of Fe/S proteins, several complementary spectroscopic techniques are available. These methods include optical absorption spectroscopy (UV–vis and CD), electron paramagnetic resonance (EPR) spectroscopy, and Mössbauer spectroscopy. The methods of choice for an in-depth characterization of Fe/S proteins are Mössbauer and EPR spectroscopy. These complementary techniques allow the assignment of Fe/S cluster types with relatively high confidence and are especially powerful when combined. For further analyses, resonance Raman spectroscopy (Spiro, 1988) and X-ray absorption spectroscopy (George & Pickering, 2013) can be used. Several excellent books on iron–sulfur proteins have been published, and many classic reviews covering much of this work may be found there (Lovenberg, 1977; Rouault, 2017; Sykes & Cammack, 1999). Here, we provide simple guidelines for the initial characterization of Fe/S proteins by UV–vis, CD, EPR, and Mössbauer spectroscopies.

4.1 Optical Absorption and CD Spectroscopy

Fe/S clusters display broad, characteristic UV/vis absorption spectra that may also provide a first hint concerning the type of Fe/S cluster (Fig. 3A). As a rule of thumb, spectra of [2Fe–2S] proteins are more complex with several characteristic absorption maxima between 410 and 430 nm, around 470 nm, and occasionally, a relatively broad peak between 550 and 600 nm. Spectra of [4Fe–4S] proteins are often less featured and display a characteristic peak around 400–420 nm. The intensity ratio of the Fe/S cluster-specific peak at ~420 nm to that of the protein at 280 nm is frequently used as a practical measure to estimate the amount of holoprotein in a preparation. The extinction coefficients of Fe/S cluster can vary a lot for different proteins. Typically, ε_{420} for [2Fe–2S] is in the range of 2500–11,000 M^{-1} cm^{-1}, and those of [3Fe–4S] and [4Fe–4S] clusters are 3000–15,000 M^{-1} cm^{-1} (Dailey, Finnegan, & Johnson, 1994; Emptage, Dreyers, Kennedy, & Beinert, 1983; Stephens et al., 1978) (Fig. 3).

Many Fe/S clusters are chiral and can be monitored by CD spectroscopy (Stephens et al., 1978). They display characteristic CD spectra in the visible region, which may be used for kinetic studies because few other biomolecules show spectra in this region (Fig. 3B). The shape and intensity of these spectra

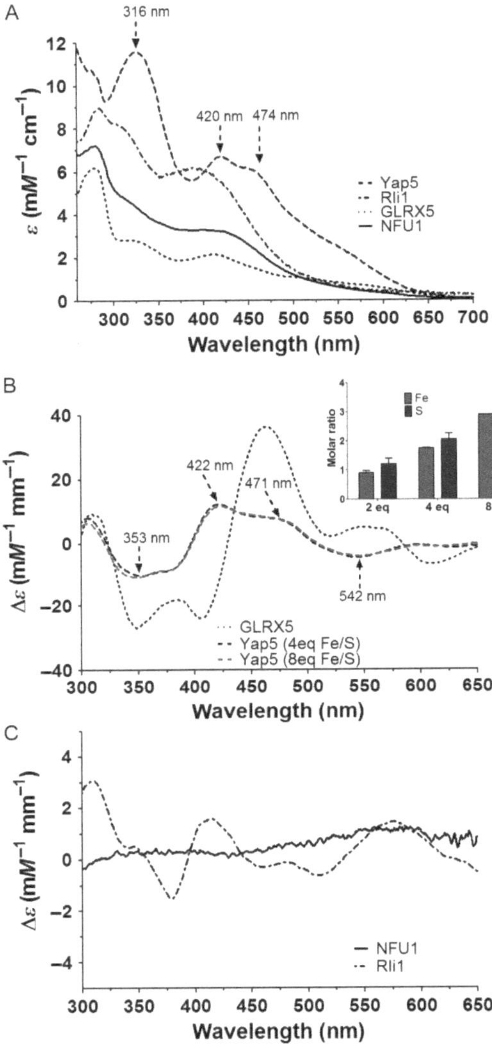

Fig. 3 UV/vis and CD spectra of representative [2Fe–2S] and [4Fe–4S] proteins. UV/vis spectra (A) and CD spectra (B and C) of two [2Fe–2S] proteins, Yap5 from *S. cerevisiae* (– –) and human GLRX5 (●●), and of two [4Fe–4S] proteins, Rli1 from *S. cerevisiae* (N-terminal domain; 1–97) (●–) and human NFU1 (—). The positions of the major peaks of the UV/vis spectrum of Yap5 are indicated in (A) and (B). *Inset* in (B): Fe^{2+} and acid-labile sulfide determination of Yap5 after reconstitution with the indicated molar excess of FAC and sulfide (Molik et al., 2007; Rabinowitz, 1978).

cannot easily be predicted. As a rule of thumb, there is a higher chance for a well-structured CD spectrum for proteins containing [2Fe–2S] clusters. In contrast, [4Fe–4S] proteins usually display CD spectra of low or even no characteristic intensity (Fig. 3C). Therefore, [4Fe–4S] proteins can be analyzed by CD spectroscopy usually only if high protein concentrations can be achieved. The lack of a CD spectrum in the visible region indicates that the Fe/S cluster is bound to the protein in a highly symmetric arrangement. Occasionally, peak maxima and minima of CD spectra correlate with those of the corresponding UV/vis spectra, but there is no clear link (Fig. 3).

4.2 EPR Spectroscopy

EPR spectroscopy is a powerful technique for the characterization of Fe/S proteins, and careful analysis of EPR spectra may reveal detailed information of the Fe/S cofactors including the cluster type, iron ligands, and Fe/S cluster environment. Here, we provide basic information that may be useful for the routine characterization of Fe/S proteins. For in-depth coverage of EPR spectroscopy, the reader is referred to the many excellent expert reviews on the topic (Cammack, 1993; Hagen, 2006, 2009). EPR spectroscopy requires the presence of unpaired electrons and usually half-integer spin. [2Fe–2S] clusters exhibit EPR resonances in the reduced +1 state ($S=1/2$), with typical average g-values around $g=1.94$ and $g=1.9$ for Rieske-type Fe/S proteins. [4Fe–4S] clusters are paramagnetic in the reduced +1 and the fully oxidized +3 state (both $S=1/2$, with $g\sim 1.94$ and $g\sim 2.01$, respectively). However, the oxidized $[2Fe–2S]^{2+}$ and $[4Fe–4S]^{2+}$ states are EPR silent ($S=0$), as encountered, e.g., for most Fe/S proteins after purification or reconstitution. In contrast, [3Fe–4S] clusters, as found in damaged aconitase or in native succinate dehydrogenase, are paramagnetic in their oxidized +1 state ($S=1/2$, $g\sim 2.01$), as well as in the reduced, neutral state, but with integer spin $S=2$. Some reduced $[3Fe–4S]^0$ clusters show integer spin spectra with broad low-field signals around $g=8$–14 (Münck, Surerus, & Hendrich, 1993), depending on the zero-field splitting of the spin quintet.

Although reduced $[4Fe–4S]^+$ and $[2Fe–2S]^+$ clusters show rather similar spectra, temperature and power variations to some extent allow the discrimination of the Fe/S cluster types due to different spin relaxation behaviors. As a rule of thumb, most $[2Fe–2S]^+$ clusters can be readily detected up to 77 K or higher without any significant line broadening. The reason is a weak temperature dependence of spin relaxation (Fig. 4A). In contrast, most $[4Fe–4S]^+$ clusters cannot be detected above ~35 K (Fig. 4B) because the onset of

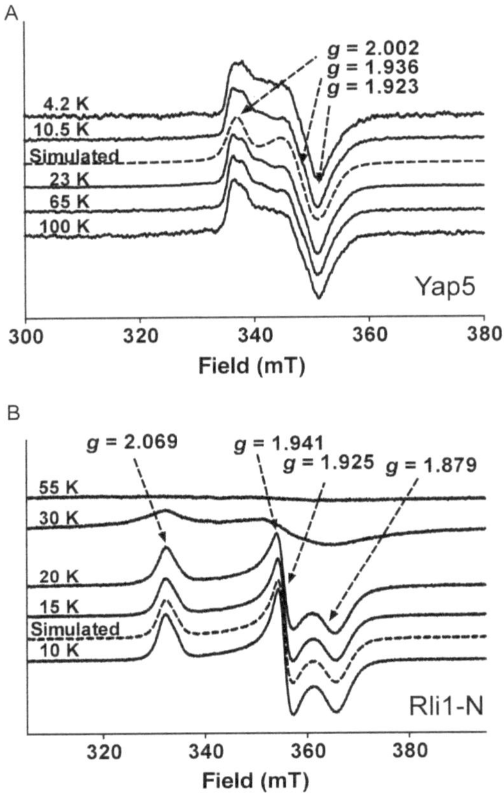

Fig. 4 EPR spectra of representative [2Fe–2S] and [4Fe–4S] proteins. Experimental EPR spectra (—) and simulation (– –) of dithionite-reduced reconstituted Yap5 (A) and the N-terminal domain (1–97) of Rli1 (B) from *S. cerevisiae* at different temperatures. *g*-values of the simulated spectrum of Yap5 (Microwave frequency, 9.458 GHz) and *g*-values of the experimental spectra of Rli1-N are indicated. *Panel (A) was taken with permission from Rietzschel, N., Pierik, A.J., Bill, E., Lill, R. & Mühlenhoff, U. (2015). The basic leucine zipper stress response regulator Yap5 senses high-iron conditions by coordination of [2Fe-2S] clusters.* Molecular and Cellular Biology, *35, 370–378. doi: 10.1128/MCB.01033-14.*

fast spin relaxation broadens the EPR lines beyond recognition. For the same reason, EPR signals of $[2Fe–2S]^+$ clusters can be easily saturated (attenuated) by high microwave power, whereas the fast relaxing $[4Fe–4S]^+$ clusters can cope with high power even at liquid-helium temperatures (not shown). These properties can help to unravel the superposition of $[4Fe–4S]^+$ and $[2Fe–2S]^+$ clusters: Appearance of novel peaks in the EPR spectrum of a reduced [2Fe–2S] protein at increased microwave power suggests the presence of a subpopulation of $[4Fe–4S]^+$ clusters.

4.3 Mössbauer Spectroscopy

Mössbauer spectroscopy is one of the most decisive techniques for the spectroscopic characterization of iron complexes in any oxidation or spin state. The method is routinely used for the identification of Fe/S clusters as well as for detailed electronic structure investigations. An overview can be found in several expert reviews (Beinert, 2000; Beinert et al., 1997; Pandelia, Lanz, Booker, & Krebs, 2014; Schünemann & Winkler, 2000). Mössbauer spectroscopy is based on resonant absorption of γ-rays from a radioactive "light"-source (^{57}Co) by ^{57}Fe Mössbauer nuclei in the sample, and it aims at the detection of hyperfine interaction of these nuclei with the electron shell of the atoms (Gütlich, Bill, & Trautwein, 2011). The Mössbauer isotope ^{57}Fe is stable and naturally abundant at ~2%. For Mössbauer spectroscopy, the (nonradioactive) ^{57}Fe isotope must be enriched in the Fe/S protein, e.g., by expression in *E. coli* (see below) or by standard reconstitution procedures using this isotope as the iron source (see Section 2). Because of the extremely high energy resolution of the nuclear levels, recoilless emission and absorption of the γ-rays is required (Mössbauer effect), which means that the sample has to be in solid state, like a frozen solution or a powder (Mössbauer, 2000).

Ground and excited states of the ^{57}Fe nucleus are spin manifolds with $I = 1/2$ and 3/2, respectively (Fig. 5A). Their total energy separation depends on the charge density at the nucleus, which gives rise to the so-called isomer shift (δ) of the Mössbauer transmission lines. In addition, the excited state has a quadrupole moment, which under the influence of a so-called electric field gradient (efg) from the electron shell leads to splitting (ΔE_Q) of the excited nuclear levels into two substates. In the measured spectrum, the two electric hyperfine interactions together give rise to a symmetric quadrupole doublet with isomer shift (δ) and electric quadrupole splitting (ΔE_Q).

The isomer shift δ is a key parameter for the assignment of iron oxidation states, due to the dependence of the charge density at the iron nucleus on the number of *d*-electrons. Interestingly, already in the early days of research on Fe/S clusters, a nice empirical correlation, $\delta(x) = [1.43–0.40x]$ mm/s, was found for δ and the oxidation number (x) of the quasi-tetrahedral high-spin [$Fe^{x+}S_4$] units of the clusters (Hoggins & Steinfink, 1976). The scale comprises Fe(II)–Fe(III) states and provides approximation for the distinct mixed-valence pattern arising from valence delocalization in $[3Fe–4S]^0$, $[4Fe–4S]^{1+/2+/3+}$, and other oligonuclear clusters (Bominaar, Hu, Münck, Girerd, & Borshch, 1995; Münck et al., 1993). Also, the electric

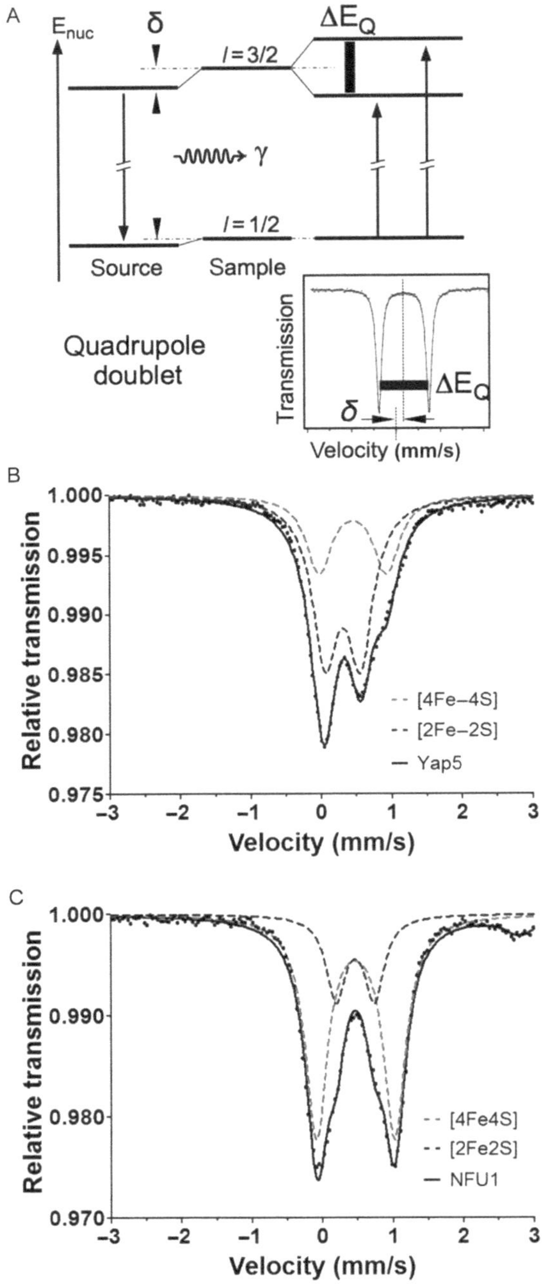

Fig. 5 See legend on opposite page.

quadrupole splitting ΔE_Q varies with the oxidation number because charge asymmetry of the valence electrons is a main source of the efg at the nucleus, but in addition ΔE_Q is also a sensitive indicator of the coordination environment.

Finally, if the Mössbauer atom is paramagnetic, or part of a paramagnetic cluster, internal magnetic fields at the Mössbauer nucleus from the unpaired spins in the electronic shell in conjunction with the electric interactions can cause full, complex splitting of the nuclear levels (not shown). Corresponding measurements at liquid helium temperatures (4 K) with applied magnetic field give rise to rich and wide-split magnetic Mössbauer spectra with a high content of chemical information but also high demand for interpretation. The magnetic hyperfine coupling parameters (A values) reveal details of the spin density distribution, and for the different Fe/S clusters disclose the typical spin coupling schemes and details of the cluster ligation.

Mössbauer spectroscopy, in contrast to EPR, detects all iron species of a sample, regardless of their oxidation and spin states. Therefore, even partial labeling with ^{57}Fe may be useful for some experiments. For the

Fig. 5 Mössbauer spectra of [2Fe–2S] and [4Fe–4S] proteins. (A) Nuclear level scheme of the Mössbauer isotope ^{57}Fe and the two-line Mössbauer transmission spectrum (quadrupole doublet) resulting from the electric hyperfine interactions, giving rise to isomer shift (δ) and quadrupole splitting (ΔE_Q). Since the energy modulation of the γ-rays is achieved by Doppler effect due to moving of the radioactive source during emission, the energy scale of Mössbauer spectra is given in units of the Doppler velocity (mm/s) (Gütlich et al., 2011). (B and C) Zero-field Mössbauer spectrum of the ^{57}Fe reconstituted [2Fe–2S] protein Yap5 from *S. cerevisiae* (B) and the human [4Fe–4S] protein NFU1 (C) recorded at 80 K. Fit (—) of the experimental data (●●) with two Lorentzian subspectra that can be assigned to [2Fe–2S] (– –) and [4Fe–4S] clusters (– –). The experimental data could be fitted with simulated spectra for [2Fe–2S] and [4Fe–4S] cofactors at a molar ratio of 4:1 (Yap5) and 1:4 (NFU1), respectively. In part C, the weak line in the experimental spectrum at ca. 2.9 mm/s was ignored. It indicates the presence of a further quadrupole doublet with high isomer shift and large quadrupole splitting due to a ferrous contamination (the corresponding left line must be hidden underneath the Fe/S cluster spectrum around −0.2 mm/s). The resultant large isomer shift $\delta > 1$ mm/s reveals high-spin Fe(II) with five or six hard N/O ligands, which hence does not belong to an Fe/S cluster site but may arise from traces of adventitiously bound iron. *Panel (A) was taken with permission from Rietzschel, N., Pierik, A.J., Bill, E., Lill, R. & Mühlenhoff, U. (2015). The basic leucine zipper stress response regulator Yap5 senses high-iron conditions by coordination of [2Fe-2S] clusters.* Molecular and Cellular Biology, 35, *370–378. doi: 10.1128/MCB.01033-14.*

identification of Fe/S clusters, often plain zero-field spectra recorded at 80–200 K provide already the data needed to this end (Fig. 5B and C) (Beinert et al., 1997; Bill, 2012). Under such conditions, not only the oxidized, diamagnetic $[2Fe–2S]^{2+}$ and $[4Fe–4S]^{2+}$ clusters exhibit simple quadrupole doublets but usually also the paramagnetic centers, because on the Mössbauer time scale of ca. 10^{-7} s, internal fields average to zero due to fast spin relaxation (Gütlich et al., 2011). Oxidized [2Fe–2S] clusters in their EPR-silent (2+) state exhibit a single, sometimes somewhat broadened quadrupole doublet with typical low δ and ΔE_Q values, revealing two similar Fe^{3+} sites. Since different iron sites yield superimposed subspectra, with intensity ratios proportional to their relative abundance, Mössbauer spectra of reduced $[2Fe–2S]^{1+}$ clusters show two quadrupole doublets with 1:1 intensity ratio and distinct parameters revealing valence-localized Fe^{3+} and Fe^{2+} sites. In contrast, reduction of the all-ferric $[3Fe–4S]^{+}$ clusters yields an Fe^{3+} subspectrum and a contribution from a mixed valence pair, $Fe^{2.5+}$, with an intensity ratio of 1:2. The cubane $[4Fe–4S]^{n+}$ clusters are valence-delocalized in all oxidation states, as was inferred from their Mössbauer signatures. Four similar $Fe^{2.5+}$ sites with usually unresolved spectral contributions are found in the oxidized (2+) state ($S=0$), and three similar $Fe^{2.25+}$ and one dissimilar Fe^{2+} sites occur after one-electron reduction ($S=1/2$). An interesting example is depicted in Fig. 5B and C, where the Fe/S reconstitution of yeast Yap5 and human NFU1 provided mixtures of oxidized [2Fe–2S] and [4Fe–4S] clusters, both in their EPR-silent (2+) states.

4.4 Equipment

- UV–vis spectrometer, 1 mL Fe/S protein at ~10 μ*M* concentration.
- CD spectrometer equipped with a stirring device; quartz cuvette with 2 mm of 5 mm path length; 300 μL Fe/S protein at >100 μ*M* concentration.
- EPR spectrometer with liquid nitrogen or helium cryostat; 300 μL Fe/S protein at >100 μ*M* concentration.
- Mössbauer spectrometer with liquid nitrogen or helium cryostat; superconducting magnet system for applied-field measurements and full analyses of paramagnetic spectra; 300–700 μL ^{57}Fe-enriched Fe/S protein at >200 μ*M* concentration for pure quadrupole spectra, 0.5 m*M* for applied-field spectra, and 1 m*M* for full analyses of wide split paramagnetic spectra, or for identification of individual subsites in Fe/S clusters.

4.5 Procedures

4.5.1 Sample Preparation for EPR

Dithionite reduction: Add 285 μL Fe/S protein to 15 μL of 40 m*M* Na-dithionite (2 m*M* final) in 0.1 *M* Tris, pH 8.3, mix fast, and incubate for 1–3 min to allow reduction. Some proteins may need longer reduction times, yet this increases the chance to damage the cluster. Transfer sample into an EPR tube and freeze the tube slowly (from bottom to top) in liquid nitrogen or cold ethanol. Dithionite results in a visible bleaching of Fe/S proteins that can be followed by UV/vis spectroscopy. Note that dithionite may destroy sensitive Fe/S clusters.

Preparation of oxidized [3Fe–4S]$^+$ proteins: Add 290 μL Fe/S protein to 10 μL of 30% H_2O_2, mix well, and transfer into an EPR tube. The sample is frozen slowly (from bottom to top) in liquid nitrogen.

4.5.2 Sample Preparation for Mössbauer

Preparation of ^{57}Fe-enriched Fe/S proteins: Stable Fe/S proteins are at best isolated in recombinant form from *E. coli* cultivated in standard rich medium supplemented with 25–50 μ*M* ^{57}Fe-FAC. To ensure proper Fe/S cluster insertion, decrease the temperature to 25–30°C following induction of expression and cultivate cells overnight. Removal of the natural isotope ^{56}Fe from the medium is not necessary as the degree of enrichment with ^{57}Fe is usually sufficient for Mössbauer analysis. Fe/S proteins that are isolated in their apoform can be chemically reconstituted in the presence of ^{57}Fe-FAC using the protocols described earlier. In this case, iron ions should not be added in large excess (e.g., no more than 4 Fe ions per [4Fe–4S] cluster), as iron ions tend to stick nonspecifically to proteins. Since Mössbauer spectroscopy detects all different iron sites, adventitiously bound iron may result in complicated spectra. The Fe/S cluster content of the proteins analyzed by Mössbauer spectroscopy should be verified by parallel EPR studies. Due to the high protein concentrations needed for Mössbauer spectroscopy, the reduction usually needs higher concentrations of Na-dithionite or longer reduction times.

Preparation of ^{57}Fe-FAC: ^{57}Fe is commercially available in form of chips of elemental iron. For the preparation of 10 mL of 50 m*M* ^{57}Fe-FAC, 33 mg of iron chips are dissolved completely in 1 mL 25% HCl (or 0.8 mL 32% HCl) under aerobic conditions for 1–3 days. The solution is diluted to 5 mL with water and 500 μL of 1 *M* citrate (1:1 with Fe) is added. The solution is neutralized slowly with 5% NH_3 (to pH 5–7) and finally filled up to 10 mL with water.

5. SUMMARY AND CONCLUSION

In this chapter, we provide protocols for the chemical and biochemical reconstitution of Fe/S proteins. We also introduce a strategy for developing an in vitro reconstitution assay for the initial step of mitochondrial Fe/S cluster synthesis. Since this assay system depends on all core ISC components defined to be needed in vivo, and since no artificial reductants such as DTT are required, we believe that the reaction closely mimics the physiological pathway. This conclusion has to be verified by further comparison of in vivo and in vitro test systems of this crucial step of Fe/S protein biogenesis. Finally, we discuss basic experimental details of several well-established spectroscopic techniques (UV–vis, CD, EPR, and Mössbauer) for the biophysical characterization of simple Fe/S proteins, and we provide hints for the interpretation of the various spectral results. These four spectroscopic techniques are complementary and powerful, especially when combined, and allow the identification of Fe/S cluster types and their coordination geometry. However, in vitro techniques alone cannot provide the definitive proof that the Fe/S protein under investigation is indeed an Fe/S protein in a living cell. Therefore, it is important that the results of these biochemical and spectroscopic techniques are complemented by in vivo approaches (see accompanying chapter "Biochemical analyses of human iron–sulfur protein biogenesis and of related diseases" by Stehling et al., this volume). Especially when Fe/S proteins need to be chemically reconstituted or when they are produced in recombinant form in a host such as *E. coli* in which they usually do not occur, there is a certain risk that the Fe/S cluster geometry may not be that of the natural protein or that the protein under investigation is not an Fe/S protein at all in its natural environment. Only when an Fe/S protein has been purified from its native host (best at physiological expression levels), when the biogenesis machinery-dependent Fe/S cluster binding in vivo has been demonstrated by, e.g., ^{55}Fe radiolabeling techniques (see accompanying chapter "Biochemical analyses of human iron–sulfur protein biogenesis and of related diseases" by Stehling et al., this volume), and when care has been taken that the protein of interest has not been modified or oxidized during purification, one can be reasonably confident that the isolated protein indeed closely resembles the one found in the cell.

The physiological pathways of Fe/S cluster synthesis, trafficking, and insertion into apoproteins in a cell differ substantially from the comparatively simple chemical Fe/S cluster reconstitution reactions. This is most evident from the comparatively complex biogenesis machinery needed for the individual steps of Fe/S protein assembly in a living cell. So far, the complete pathway involving all biogenesis factors required in vivo has not been reproduced in vitro in a faithful fashion in any of the known Fe/S protein biogenesis systems. However, partial steps of de novo Fe/S cluster synthesis, transfer, or target protein insertion have been successfully reconstituted and intensely analyzed (see above). Yet, many of these in vitro studies depend on the presence of artificial thiol-specific agents such as DTT and have to be interpreted cautiously with regard to their physiological relevance. Therefore, the development of assays that faithfully reconstitute the partial steps of the maturation of Fe/S proteins in the absence of artificial additives will be an important step toward a better understanding of the mechanism of Fe/S cluster formation in a living cell.

REFERENCES

Adam, A. C., Bornhövd, C., Prokisch, H., Neupert, W., & Hell, K. (2006). The Nfs1 interacting protein Isd11 has an essential role in Fe/S cluster biogenesis in mitochondria. *The EMBO Journal*, *25*, 174–183.

Agar, J. N., Zheng, L., Cash, V. L., Dean, D. R., & Johnson, M. K. (2000). Role of the IscU protein in iron-sulfur cluster biosynthesis: IscS-mediated assembly of a [Fe_2S_2] cluster in IscU. *Journal of the American Chemical Society*, *122*, 2136–2137.

Andreini, C., Rosato, A., & Banci, L. (2017). The relationship between environmental dioxygen and iron-sulfur proteins explored at the genome level. *PLoS One12*, e0171279. https://doi.org/10.1371/journal.pone.0171279.

Angerer, H. (2013). The superfamily of mitochondrial Complex1_LYR motif-containing (LYRM) proteins. *Biochemical Society Transactions*, *41*, 1335–1341. https://doi.org/10.1042/BST20130116.

Angerer, H., Radermacher, M., Mankowska, M., Steger, M., Zwicker, K., Heide, H., et al. (2014). The LYR protein subunit NB4M/NDUFA6 of mitochondrial complex I anchors an acyl carrier protein and is essential for catalytic activity. *Proceedings of the National Academy of Sciences of the United States of America*, *111*, 5207–5212. https://doi.org/10.1073/pnas.1322438111.

Atkinson, A., Smith, P., Fox, J. L., Cui, T. Z., Khalimonchuk, O., & Winge, D. R. (2011). The LYR protein Mzm1 functions in the insertion of the Rieske Fe/S protein in yeast mitochondria. *Molecular and Cellular Biology*, *31*, 3988–3996. https://doi.org/10.1128/MCB.05673-11.

Balk, J., & Schaedler, T. A. (2014). Iron cofactor assembly in plants. *Annual Review of Plant Biology*, *65*, 125–153. https://doi.org/10.1146/annurev-arplant-050213-035759.

Bandyopadhyay, S., Gama, F., Molina-Navarro, M. M., Gualberto, J. M., Claxton, R., Naik, S. G., et al. (2008). Chloroplast monothiol glutaredoxins as scaffold proteins for the assembly and delivery of [2Fe-2S] clusters. *The EMBO Journal*, *27*, 1122–1133.

Beinert, H. (2000). Iron-sulfur proteins: Ancient structures, still full of surprises. *Journal of Biological Inorganic Chemistry*, *5*, 2–15.

Beinert, H., Holm, R. H., & Münck, E. (1997). Iron-sulfur clusters: Nature's modular, multipurpose structures. *Science*, *277*, 653–659.

Beinert, H., & Sands, R. H. (1960). Studies on succinic and DPNH dehydrogenase preparations by paramagnetic resonance (EPR) spectroscopy. *Biochemical and Biophysical Research Communications*, *3*, 41–46.

Biederbick, A., Stehling, O., Rösser, R., Niggemeyer, B., Nakai, Y., Elsässer, H. P., et al. (2006). Role of human mitochondrial Nfs1 in cytosolic iron-sulfur protein biogenesis and iron regulation. *Molecular and Cellular Biology*, *26*, 5675–5687.

Bill, E. (2012). Iron-sulfur clusters—New features in enzymes and synthetic models. *Hyperfine Interactions*, *205*, 139–147.

Black, K. A., & Dos Santos, P. C. (2015). Shared-intermediates in the biosynthesis of thio-cofactors: Mechanism and functions of cysteine desulfurases and sulfur acceptors. *Biochimica et Biophysica Acta*, *1853*, 1470–1480. https://doi.org/10.1016/j.bbamcr.2014.10.018.

Blanc, B., Gerez, C., & Ollagnier de Choudens, S. (2015). Assembly of Fe/S proteins in bacterial systems: Biochemistry of the bacterial ISC system. *Biochimica et Biophysica Acta*, *1853*, 1436–1447. https://doi.org/10.1016/j.bbamcr.2014.12.009.

Bominaar, E. L., Hu, Z., Münck, E., Girerd, J.-J., & Borshch, S. (1995). Double exchange and vibronic coupling in mixed-valence systems. Electronic structure of exchange-coupled Siroheme-[Fe4S4]2+ chromophore in oxidized *E. coli* sulfite reductase. *Journal of the American Chemical Society*, *117*, 6976–6989.

Boniecki, M. T., Freibert, S. A., Mühlenhoff, U., Lill, R., & Cygler, M. (2017). Structure and functional dynamics of the mitochondrial Fe/S cluster synthesis complex. *Nature Communications*, *8*, 1287.

Bonomi, F., Iametti, S., Morleo, A., Ta, D., & Vickery, L. E. (2008). Studies on the mechanism of catalysis of iron-sulfur cluster transfer from IscU[2Fe2S] by HscA/HscB chaperones. *Biochemistry*, *47*, 12795–12801.

Brancaccio, D., Gallo, A., Mikolajczyk, M., Zovo, K., Palumaa, P., Novellino, E., et al. (2014). Formation of [4Fe-4S] clusters in the mitochondrial iron-sulfur cluster assembly machinery. *Journal of the American Chemical Society*, *136*, 16240–16250. https://doi.org/10.1021/ja507822j.

Braymer, J. J., & Lill, R. (2017). Iron-sulfur cluster biogenesis and trafficking in mitochondria. *The Journal of Biological Chemistry*, *292*, 12754–12763.

Bridwell-Rabb, J., Fox, N. G., Tsai, C. L., Winn, A. M., & Barondeau, D. P. (2014). Human frataxin activates Fe-S cluster biosynthesis by facilitating sulfur transfer chemistry. *Biochemistry*, *53*, 4904–4913. https://doi.org/10.1021/bi500532e.

Bych, K., Kerscher, S., Netz, D. J., Pierik, A. J., Zwicker, K., Huynen, M. A., et al. (2008). The iron-sulphur protein Ind1 is required for effective complex I assembly. *The EMBO Journal*, *27*, 1736–1746.

Cameron, J. M., Janer, A., Levandovskiy, V., Mackay, N., Rouault, T. A., Tong, W. H., et al. (2011). Mutations in iron-sulfur cluster scaffold genes NFU1 and BOLA3 cause a fatal deficiency of multiple respiratory chain and 2-oxoacid dehydrogenase enzymes. *American Journal of Human Genetics*, *89*, 486–495. https://doi.org/10.1016/j.ajhg.2011.08.011.

Cammack, R. (1993). Electron paramagnetic resonance spectroscopy of metalloproteins. *Methods in Molecular Biology*, *17*, 327–344. https://doi.org/10.1385/0-89603-215-9:327.

Cory, S. A., Van Vranken, J. G., Brignole, E. J., Patra, S., Winge, D. R., Drennan, C. L., et al. (2017). Structure of human Fe-S assembly subcomplex reveals unexpected cysteine desulfurase architecture and acyl-ACP-ISD11 interactions. *Proceedings of the National Academy of Sciences of the United States of America*, *114*, E5325–E5334. https://doi.org/10.1073/pnas.1702849114.

Craig, E. A., & Marszalek, J. (2017). How do J-proteins get Hsp70 to do so many different things? *Trends in Biochemical Sciences*, *42*, 355–368. https://doi.org/10.1016/j.tibs.2017.02.007.

Dailey, H. A., Finnegan, M. G., & Johnson, M. K. (1994). Human ferrochelatase is an iron-sulfur protein. *Biochemistry*, *33*, 403–407.

Emptage, M. H., Dreyers, J. L., Kennedy, M. C., & Beinert, H. (1983). Optical and EPR characterization of different species of active and inactive aconitase. *The Journal of Biological Chemistry*, *258*, 11106–11111.

Freibert, S. A., Goldberg, A. V., Hacker, C., Molik, S., Dean, P., Williams, T. A., et al. (2017). Evolutionary conservation and in vitro reconstitution of microsporidian iron-sulfur cluster biosynthesis. *Nature Communications*, *8*, 13932. https://doi.org/10.1038/ncomms13932.

Gao, H., Subramanian, S., Couturier, J., Naik, S. G., Kim, S. K., Leustek, T., et al. (2013). Arabidopsis thaliana Nfu2 accommodates [2Fe-2S] or [4Fe-4S] clusters and is competent for in vitro maturation of chloroplast [2Fe-2S] and [4Fe-4S] cluster-containing proteins. *Biochemistry*, *52*, 6633–6645. https://doi.org/10.1021/bi4007622.

Gelling, C., Dawes, I. W., Richhardt, N., Lill, R., & Mühlenhoff, U. (2008). Mitochondrial Iba57p is required for Fe/S cluster formation on aconitase and activation of radical SAM enzymes. *Molecular and Cellular Biology*, *28*, 1851–1861.

George, G. N., & Pickering, I. J. (2013). X-ray absorption spectroscopy of metals in biology. In G. C. K. Roberts (Ed.), *Encyclopedia of biophysics* (pp. 2762–2767). Berlin: Springer.

Goldberg, A. V., Molik, S., Tsaousis, A. D., Neumann, K., Kuhnke, G., Delbac, F., et al. (2008). Localization and functionality of microsporidian iron-sulphur cluster assembly proteins. *Nature*, *452*, 624–628.

Gütlich, P., Bill, E., & Trautwein, A. X. (2011). *Mössbauer spectroscopy and transition metal chemistry*. Berlin Heidelberg: Springer Verlag.

Hagen, W. R. (2006). EPR spectroscopy as a probe of metal centres in biological systems. *Dalton Transactions*, (37), 4415–4434. PMID: 16981015 [Epub 2006 Aug 11].

Hagen, W. R. (2009). *Biomolecular EPR spectroscopy*. Boca Raton: CRC Press.

Hoggins, J. T., & Steinfink, H. (1976). Empirical bonding relationships in metal-iron-sulfide compounds. *Inorganic Chemistry*, *15*, 1682–1685.

Hu, Y., & Ribbe, M. W. (2016). Biosynthesis of the metalloclusters of nitrogenases. *Annual Review of Biochemistry*, *85*, 455–483. https://doi.org/10.1146/annurev-biochem-060614-034108.

Kampinga, H. H., & Craig, E. A. (2010). The HSP70 chaperone machinery: J proteins as drivers of functional specificity. *Nature Reviews. Molecular Cell Biology*, *11*, 579–592. https://doi.org/10.1038/nrm2941.

Land, T., & Rouault, T. A. (1998). Targeting of a human iron-sulfur cluster assembly enzyme, nifs, to different subcellular compartments is regulated through alternative AUG utilization. *Molecular Cell*, *2*, 807–815.

Landgraf, B. J., McCarthy, E. L., & Booker, S. J. (2016). Radical S-adenosylmethionine enzymes in human health and disease. *Annual Review of Biochemistry*, *85*, 485–514. https://doi.org/10.1146/annurev-biochem-060713-035504.

Lange, H., Kaut, A., Kispal, G., & Lill, R. (2000). A mitochondrial ferredoxin is essential for biogenesis of cellular iron-sulfur proteins. *Proceedings of the National Academy of Sciences of the United States of America*, *97*, 1050–1055.

Li, J., Saxena, S., Pain, D., & Dancis, A. (2001). Adrenodoxin reductase homolog (Arh1p) of yeast mitochondria required for iron homeostasis. *The Journal of Biological Chemistry*, *276*, 1503–1509.

Li, K., Tong, W. H., Hughes, R. M., & Rouault, T. A. (2006). Roles of the mammalian cytosolic cysteine desulfurase, ISCS, and scaffold protein, ISCU in iron-sulfur cluster assembly. *The Journal of Biological Chemistry*, *281*, 12344–12351.

Lill, R. (2009). Function and biogenesis iron-sulphur proteins. *Nature*, *460*, 831–838.

Lill, R., Srinivasan, V., & Mühlenhoff, U. (2014). The role of mitochondria in cytosolic-nuclear iron-sulfur protein biogenesis and in cellular iron regulation. *Current Opinion in Microbiology*, *22*, 111–119. https://doi.org/10.1016/j.mib.2014.09.015.

Lovenberg, W. (1977). *Iron-sulfur proteins*. New York: Academic Press.
Malkin, R., & Rabinowitz, J. C. (1966). The reconstitution of clostridial ferredoxin. *Biochemical and Biophysical Research Communications*, *23*, 822–827.
Marelja, Z., Mullick Chowdhury, M., Dosche, C., Hille, C., Baumann, O., Lohmannsroben, H. G., et al. (2013). The L-cysteine desulfurase NFS1 is localized in the cytosol where it provides the sulfur for molybdenum cofactor biosynthesis in humans. *PLoS One 8*, e60869. https://doi.org/10.1371/journal.pone.0060869.
Melber, A., Na, U., Vashisht, A., Weiler, B. D., Lill, R., Wohlschlegel, J. A., et al. (2016). Role of Nfu1 and Bol3 in iron-sulfur cluster transfer to mitochondrial clients. *eLife 5*, e15991. https://doi.org/10.7554/eLife.15991.
Meyer, J., Moulis, J. M., & Lutz, M. (1986). High yield chemical assembly of [2Fe-2X] (X=S, Se) clusters into spinach apoferredoxin: Product characterization by resonance Raman spectroscopy. *Biochimica et Biophysica Acta*, *871*, 243–249.
Molik, S., Lill, R., & Mühlenhoff, U. (2007). Methods for studying iron metabolism in yeast mitochondria. *Methods in Cell Biology*, *80*, 261–280.
Mössbauer, R. L. (2000). The discovery of the Mössbauer effect. *Hyperfine Interactions*, *126*, 1–12.
Mühlenhoff, U., Balk, J., Richhardt, N., Kaiser, J. T., Sipos, K., Kispal, G., et al. (2004). Functional characterization of the eukaryotic cysteine desulfurase Nfs1p from *Saccharomyces cerevisiae*. *The Journal of Biological Chemistry*, *279*, 36906–36915.
Mühlenhoff, U., Richhardt, N., Gerber, J., & Lill, R. (2002). Characterization of iron-sulfur protein assembly in isolated mitochondria: A requirement for ATP, NADH and reduced iron. *The Journal of Biological Chemistry*, *277*, 29810–29816.
Mühlenhoff, U., Richter, N., Pines, O., Pierik, A. J., & Lill, R. (2011). Specialized function of yeast Isa1 and Isa2 proteins in the maturation of mitochondrial [4Fe-4S] proteins. *The Journal of Biological Chemistry*, *286*, 41205–41216. https://doi.org/10.1074/jbc.M111.296152.
Münck, E., Surerus, K. K., & Hendrich, M. P. (1993). Combining Mössbauer spectroscopy with integer spin electron paramagnetic resonance. *Methods in Enzymology*, *227*, 463–479.
Nakai, Y., Nakai, M., Hayashi, H., & Kagamiyama, H. (2001). Nuclear localization of yeast Nfs1p is required for cell survival. *The Journal of Biological Chemistry*, *276*, 8314–8320.
Navarro-Sastre, A., Tort, F., Stehling, O., Uzarska, M. A., Arranz, J. A., Del Toro, M., et al. (2011). A fatal mitochondrial disease is associated with defective NFU1 function in the maturation of a subset of mitochondrial Fe-S proteins. *American Journal of Human Genetics*, *89*, 656–667. https://doi.org/10.1016/j.ajhg.2011.10.005.
Netz, D. J., Mascarenhas, J., Stehling, O., Pierik, A. J., & Lill, R. (2014). Maturation of cytosolic and nuclear iron-sulfur proteins. *Trends in Cell Biology*, *24*, 303–312. https://doi.org/10.1016/j.tcb.2013.11.005.
Netz, D. J., Stith, C. M., Stumpfig, M., Kopf, G., Vogel, D., Genau, H. M., et al. (2012). Eukaryotic DNA polymerases require an iron-sulfur cluster for the formation of active complexes. *Nature Chemical Biology*, *8*, 125–132. https://doi.org/10.1038/nchembio.721.
Outten, F. W. (2015). Recent advances in the Suf Fe-S cluster biogenesis pathway: Beyond the proteobacteria. *Biochimica et Biophysica Acta*, *1853*, 1464–1469. https://doi.org/10.1016/j.bbamcr.2014.11.001.
Pandelia, M.-E., Lanz, N. D., Booker, S. J., & Krebs, C. (2014). Mössbauer spectroscopy of Fe/S proteins. *Biochimica et Biophysica Acta*, *1853*, 1395–1405. https://doi.org/10.1016/j.bbamcr.2014.12.005.
Parent, A., Elduque, X., Cornu, D., Belot, L., Le Caer, J. P., Grandas, A., et al. (2015). Mammalian frataxin directly enhances sulfur transfer of NFS1 persulfide to both ISCU and free thiols. *Nature Communications*, *6*, 5686. https://doi.org/10.1038/ncomms6686.

Pastore, A., & Puccio, H. (2013). Frataxin: A protein in search for a function. *Journal of Neurochemistry*, *126*(Suppl. 1), 43–52. https://doi.org/10.1111/jnc.12220.

Paul, V. D., & Lill, R. (2015). Biogenesis of cytosolic and nuclear iron-sulfur proteins and their role in genome stability. *Biochimica et Biophysica Acta*, *1853*, 1528–1539. https://doi.org/10.1016/j.bbamcr.2014.12.018.

Peters, J. W., Schut, G. J., Boyd, E. S., Mulder, D. W., Shepard, E. M., Broderick, J. B., et al. (2015). [FeFe]- and [NiFe]-hydrogenase diversity, mechanism, and maturation. *Biochimica et Biophysica Acta*, *1853*, 1350–1369. https://doi.org/10.1016/j.bbamcr.2014.11.021.

Pierik, A. J., Netz, D. J. A., & Lill, R. (2009). Analysis of iron-sulfur protein maturation in eukaryotes. *Nature Protocols*, *4*, 753–766.

Pukszta, S., Schilke, B., Dutkiewicz, R., Kominek, J., Moczulska, K., Stepien, B., et al. (2010). Co-evolution-driven switch of J-protein specificity towards an Hsp70 partner. *EMBO Reports*, *11*, 360–365. https://doi.org/10.1038/embor.2010.29.

Py, B., & Barras, F. (2015). Genetic approaches of the Fe-S cluster biogenesis process in bacteria: Historical account, methodological aspects and future challenges. *Biochimica et Biophysica Acta*, *1853*, 1429–1435. https://doi.org/10.1016/j.bbamcr.2014.12.024.

Py, B., Gerez, C., Angelini, S., Planel, R., Vinella, D., Loiseau, L., et al. (2012). Molecular organization, biochemical function, cellular role and evolution of NfuA, an atypical Fe-S carrier. *Molecular Microbiology*, *86*, 155–171. https://doi.org/10.1111/j.1365-2958.2012.08181.x.

Rabinowitz, J. C. (1978). Analysis of acid-labile sulfide and sulfhydryl groups. *Methods in Enzymology*, *53*, 275–277.

Ramelot, T. A., Cort, J. R., Goldsmith-Fischman, S., Kornhaber, G. J., Xiao, R., Shastry, R., et al. (2004). Solution NMR structure of the iron-sulfur cluster assembly protein U (IscU) with zinc bound at the active site. *Journal of Molecular Biology*, *344*, 567–583.

Rouault, T. A. (2015). Mammalian iron-sulphur proteins: Novel insights into biogenesis and function. *Nature Reviews. Molecular Cell Biology*, *16*, 45–55. https://doi.org/10.1038/nrm3909.

Rouault, T. A. (2017). *Iron-sulfur clusters in chemistry and biology*. Berlin: De Gruyter.

Rouhier, N., Couturier, J., Johnson, M. K., & Jacquot, J. P. (2010). Glutaredoxins: Roles in iron homeostasis. *Trends in Biochemical Sciences*, *35*, 43–52.

Schünemann, V., & Winkler, H. (2000). Structure and dynamics of biomolecules studied by Mössbauer spectroscopy. *Reports on Progress in Physics*, *63*, 263–353.

Shakamuri, P., Zhang, B., & Johnson, M. K. (2012). Monothiol glutaredoxins function in storing and transporting [Fe2S2] clusters assembled on IscU scaffold proteins. *Journal of the American Chemical Society*, *134*, 15213–15216. https://doi.org/10.1021/ja306061x.

Sharma, A. K., Pallesen, L. J., Spang, R. J., & Walden, W. E. (2010). Cytosolic iron-sulfur cluster assembly (CIA) system: Factors, mechanism, and relevance to cellular iron regulation. *The Journal of Biological Chemistry*, *285*, 26745–26751. https://doi.org/10.1074/jbc.R110.122218.

Sheftel, A. D., Stehling, O., Pierik, A. J., Elsässer, H. P., Mühlenhoff, U., Webert, H., et al. (2010). Humans possess two mitochondrial ferredoxins, Fdx1 and Fdx2, with distinct roles in steroidogenesis, heme, and Fe/S cluster biosynthesis. *Proceedings of the National Academy of Sciences of the United States of America*, *107*, 11775–11780. https://doi.org/10.1073/pnas.1004250107.

Sheftel, A. D., Stehling, O., Pierik, A. J., Netz, D. J., Kerscher, S., Elsässer, H. P., et al. (2009). Human Ind1, an iron-sulfur cluster assembly factor for respiratory complex I. *Molecular and Cellular Biology*, *29*, 6059–6073.

Sheftel, A. D., Wilbrecht, C., Stehling, O., Niggemeyer, B., Elsässer, H. P., Mühlenhoff, U., et al. (2012). The human mitochondrial ISCA1, ISCA2, and IBA57 proteins are required for [4Fe-4S] protein maturation. *Molecular Biology of the Cell*, *23*, 1157–1166. https://doi.org/10.1091/mbc.E11-09-0772.

Shi, Y., Ghosh, M., Kovtunovych, G., Crooks, D. R., & Rouault, T. A. (2012). Both human ferredoxins 1 and 2 and ferredoxin reductase are important for iron-sulfur cluster biogenesis. *Biochimica et Biophysica Acta*, *1823*, 484–492. https://doi.org/10.1016/j.bbamcr.2011.11.002.

Shi, Y., Ghosh, M. C., Tong, W. H., & Rouault, T. A. (2009). Human ISD11 is essential for both iron-sulfur cluster assembly and maintenance of normal cellular iron homeostasis. *Human Molecular Genetics*, *18*, 3014–3025.

Spiro, T. G. (1988). *Biological applications of Raman spectroscopy*. New York: John Wiley & Sons, Inc.

Stehling, O., Sheftel, A. D., & Lill, R. (2009). Chapter 12 Controlled expression of iron-sulfur cluster assembly components for respiratory chain complexes in mammalian cells. *Methods in Enzymology*, *456*, 209–231.

Stemmler, T. L., Lesuisse, E., Pain, D., & Dancis, A. (2010). Frataxin and mitochondrial FeS cluster biogenesis. *The Journal of Biological Chemistry*, *285*, 26737–26743. https://doi.org/10.1074/jbc.R110.118679.

Stephens, P. J., Thomson, A. J., Dunn, J. B., Keiderling, T. A., Rawlings, J., Rao, K. K., et al. (1978). Circular dichroism and magnetic circular dichroism of iron-sulfur proteins. *Biochemistry*, *17*, 4770–4778.

Sykes, A. G., & Cammack, R. (1999). *Iron-sulfur proteins*. New York: Academic Press.

Tovar, J., Leon-Avila, G., Sanchez, L. B., Sutak, R., Tachezy, J., Van Der Giezen, M., et al. (2003). Mitochondrial remnant organelles of *Giardia* function in iron-sulphur protein maturation. *Nature*, *426*, 172–176.

Uzarska, M. A., Dutkiewicz, R., Freibert, S. A., Lill, R., & Mühlenhoff, U. (2013). The mitochondrial Hsp70 chaperone Ssq1 facilitates Fe/S cluster transfer from Isu1 to Grx5 by complex formation. *Molecular Biology of the Cell*, *24*, 1830–1841. https://doi.org/10.1091/mbc.E12-09-0644.

Uzarska, M. A., Nasta, V., Weiler, B. D., Spantgar, F., Ciofi-Baffoni, S., Saviello, M. R., et al. (2016). Mitochondrial Bol1 and Bol3 function as assembly factors for specific iron-sulfur proteins. *eLife 5*, e16673. https://doi.org/10.7554/eLife.16673.

Van Vranken, J. G., Jeong, M. Y., Wei, P., Chen, Y. C., Gygi, S. P., Winge, D. R., et al. (2016). The mitochondrial acyl carrier protein (ACP) coordinates mitochondrial fatty acid synthesis with iron sulfur cluster biogenesis. *eLife 5*, e17828. https://doi.org/10.7554/eLife.17828.

Webert, H., Freibert, S. A., Gallo, A., Heidenreich, T., Linne, U., Amlacher, S., et al. (2014). Functional reconstitution of mitochondrial Fe/S cluster synthesis on Isu1 reveals the involvement of ferredoxin. *Nature Communications*, *5*, 5013.

Wiedemann, N., Urzica, E., Guiard, B., Müller, H., Lohaus, C., Meyer, H. E., et al. (2006). Essential role of Isd11 in mitochondrial iron-sulfur cluster synthesis on Isu scaffold proteins. *The EMBO Journal*, *25*, 184–195. https://doi.org/10.1038/sj.emboj.7600906.

Yan, R., Adinolfi, S., & Pastore, A. (2015). Ferredoxin, in conjunction with NADPH and ferredoxin-NADP reductase, transfers electrons to the IscS/IscU complex to promote iron-sulfur cluster assembly. *Biochimica et Biophysica Acta*, *1854*, 1113–1117. https://doi.org/10.1016/j.bbapap.2015.02.002.

Zheng, L., Cash, V. L., Flint, D. H., & Dean, D. R. (1998). Assembly of iron-sulfur clusters. Identification of an *iscSUA-hscBA-fdx* gene cluster from *Azotobacter vinelandii*. *The Journal of Biological Chemistry*, *273*, 13264–13272.

Zheng, L., White, R. H., Cash, V. L., & Dean, D. R. (1994). Mechanism for the desulfurization of L-cysteine catalyzed by the nifS gene product. *Biochemistry*, *33*, 4714–4720.

CHAPTER EIGHT

Biochemical Analyses of Human Iron–Sulfur Protein Biogenesis and of Related Diseases

Oliver Stehling*, Viktoria D. Paul*, Janina Bergmann*, Somsuvro Basu*, Roland Lill*,†,1

*Institut für Zytobiologie, Philipps-Universität Marburg, Marburg, Germany
†LOEWE Zentrum für Synthetische Mikrobiologie SynMikro, Marburg, Germany
[1]Corresponding author: e-mail address: lill@staff.uni-marburg.de

Contents

Abstract

Maturation of Fe/S proteins in mammals is an intricate process mediated by two assembly systems located in the mitochondrial and cytosolic–nuclear compartments. Malfunction particularly of the mitochondrial system gives rise to severe neurological, metabolic, or hematological disorders, often with fatal outcome. In this chapter,

Methods in Enzymology, Volume 599
ISSN 0076-6879
https://doi.org/10.1016/bs.mie.2017.11.004

we describe approaches for the differential biochemical investigation of cellular Fe/S protein maturation in mitochondria, cytosol, and nucleus. The analyses may also facilitate the identification of the affected Fe/S protein assembly step in diseased state. As Fe/S cluster insertion into target apoproteins is a frequent determinant of protein stability, examination of protein steady-state levels in biological samples frequently permits reliable first clues about the maturation process. In some specific cases, this approach allows the assessment of enzymatic or regulatory functions of Fe/S proteins, including the formation of lipoate cofactor by mitochondrial lipoic acid synthase or the posttranscriptional regulation of transferrin receptor and ferritin expression by the cytosolic iron regulatory proteins. More direct Fe/S protein maturation assays like enzymatic analyses may further validate the observed maturation defects. Here, we present a simple protocol for the determination of dihydropyrimidine dehydrogenase enzyme activity by thin-layer chromatography. In order to directly monitor Fe/S cluster insertion into target apoproteins, we have developed a ^{55}Fe radiolabeling technique tracing the in vivo Fe/S cofactor formation in mammalian tissue culture. The combination of the presented techniques represents a comprehensive strategy to assess the multiple facets of Fe/S protein assembly for both mechanistic analyses and for the elucidation of specific defects in Fe/S diseases.

1. INTRODUCTION

Iron–sulfur (Fe/S) clusters are simple and evolutionary ancient inorganic cofactors executing versatile biochemical functions in each realm of life, i.e., archaea, bacteria, and eukarya (Beinert, Holm, & Münck, 1997; Lill & Mühlenhoff, 2005; Py & Barras, 2010). In nongreen eukaryotes, Fe/S cluster-coordinating proteins are localized within mitochondria, cytosol, and nucleus. Unmatched to their structural simplicity, Fe/S clusters are assembled and inserted into apoproteins in an extraordinarily intricate biosynthetic process. According to present knowledge, two different biosynthetic systems are involved in Fe/S cluster synthesis and assembly, i.e., the mitochondrial ISC (iron–sulfur cluster) assembly system and the CIA (cytosolic iron–sulfur protein assembly) machinery, which are connected by the ISC export system (Lill, Srinivasan, & Mühlenhoff, 2014; Stehling & Lill, 2013). The latter involves the ABC transporter ABCB7 (yeast Atm1), which transfers an unknown, sulfur-containing compound (termed X–S) from the mitochondrial matrix to the cytosol. X–S is generated by the mitochondrial ISC system and is utilized by the CIA machinery for the maturation of both cytosolic and nuclear Fe/S proteins. The interdependence of these three parts of Fe/S protein biogenesis creates a branching pathway of biochemical events, with mitochondria playing the primary role in biogenesis. Numerous studies of the past two decades have provided a first mechanistic picture of eukaryotic Fe/S protein assembly which is remarkably complex and requires the participation of some 30 proteins to mature the more than 50 Fe/S proteins within a eukaryotic cell (Beilschmidt & Puccio, 2014; Braymer & Lill, 2017; Lill et al., 2015; Rouault, 2012).

The mitochondrial ISC system is highly conserved from yeast to man, and its action can be formally dissected into multiple steps (see Chapter "Biochemical reconstitution and spectroscopic analysis of iron–sulfur proteins" by Freibert et al., accompanying article in this issue). In an initial reaction, a [2Fe–2S] cluster is assembled de novo on the scaffold protein ISCU (in this chapter, we use the nomenclature for human proteins) (Agar et al., 2000; Mühlenhoff, Gerber, Richhardt, & Lill, 2003), harnessing sulfur provided by the cysteine desulfurase module NFS1-ISD11-ACP (Boniecki, Freibert, Mühlenhoff, Lill, & Cygler, 2017; Van Vranken et al., 2016), whose function is modulated by the iron-binding protein frataxin (Tsai & Barondeau, 2010). Fe/S cluster assembly further relies on the supply of electrons that are derived from NADPH via an electron transfer chain constituted by a ferredoxin reductase and a ferredoxin (FDX2 in humans) (Lange, Kaut, Kispal, & Lill, 2000; Sheftel et al., 2010; Shi, Ghosh, Kovtunovych, Crooks, & Rouault, 2012). This electron transfer chain is presumably needed to reduce the cysteine-derived sulfur to sulfide which is present in the cluster. A biochemical approach to reconstitute this initial part of the ISC system is described by Freibert et al. (2017) (accompanying article in this issue).

In the next step of mitochondrial ISC assembly, a dedicated chaperone system comprising Hsp70 and specialized J-type proteins facilitates the cluster transfer from ISCU to the monothiol glutaredoxin GLRX5, which then can mature target [2Fe–2S] proteins without further known factors (Uzarska, Dutkiewicz, Freibert, Lill, & Mühlenhoff, 2013). Conversion of the [2Fe–2S] to a [4Fe–4S] cluster requires the late-acting ISC-targeting factors IBA57, ISCA1, and ISCA2 (Gelling, Dawes, Richhardt, Lill, & Mühlenhoff, 2008; Mühlenhoff, Richter, Pines, Pierik, & Lill, 2011; Sheftel et al., 2012). Finally, dedicated ISC-targeting factors such as NFU1 and IND1 transiently bind the [4Fe–4S] cluster before mediating specific cluster insertion into the polypeptide chains of dedicated target apoproteins (Cameron et al., 2011; Navarro-Sastre et al., 2011; Sheftel, Stehling et al., 2009). The recently described ISC component BOLA3 (in yeast together with BOLA1) plays an important role in lipoic acid synthase (LIAS) maturation, yet the precise molecular role of the BOLA proteins is still unknown (Cameron et al., 2011; Uzarska et al., 2016).

The factors conducting de novo assembly of the [2Fe–2S] cluster on ISCU and the transfer to GLRX5 (collectively termed the "core ISC system") are also involved in the production of the aforementioned compound X–S that is exported into the cytosol by aid of the mitochondrial inner membrane ABC transporter ABCB7, a reaction requiring glutathione (Sipos et al., 2002). In the cytosol, X–S is used for de novo synthesis of a [4Fe–4S] cluster on the CIA scaffold proteins CFD1-NBP35 (Netz, Pierik, Stümpfig, Mühlenhoff, & Lill, 2007). They belong to the metal-binding subfamily of P-loop NTPases involved in the assembly of metal centers. Reminiscent of the mitochondrial pathway, an NADPH-dependent electron transfer chain constituted of NDOR1-CIAPIN1 (Tah18-Dre2 in yeast) is required for this assembly step (Netz et al., 2010). In the next phase of cytosolic–nuclear Fe/S protein biogenesis, the newly synthesized [4Fe–4S] cluster is released from the CFD1-NBP35 scaffold and transferred to target apoproteins. This step is assisted by the iron-only hydrogenase-like protein IOP1 (yeast Nar1) (Balk, Pierik, Netz, Mühlenhoff, & Lill, 2004; Song & Lee, 2008) and the CIA-targeting complex comprised of CIA1, CIA2B (Cia2 in yeast), and MMS19 (Gari et al., 2012; Stehling et al., 2013, 2012). These factors dynamically interact to form various subcomplexes that specifically assemble the majority of cytosolic and nuclear target Fe/S proteins (Paul & Lill, 2015). Conversely, the CIA2B homolog CIA2A entertains only one identified Fe/S target protein, the iron regulatory protein 1 (IRP1), a central element of intracellular iron homeostasis (Stehling et al., 2013). Additionally, CIA2A binds to IRP2, the non-Fe/S counterpart of IRP1, thus stabilizing it by complex formation (see below). Recently, we have identified a complex of ORAOV1 and YAE1D1 (yeast Lto1 and Yae1) that acts as an adaptor and specifically recruits the target Fe/S protein ABCE1 (yeast Rli1) to the CIA-targeting complex for maturation (Paul et al., 2015). To fulfill this adaptor function, yeast Lto1 utilizes a conserved C-terminal tryptophan residue to interact with the CIA-targeting complex, while yeast Yae1 establishes the contact to Rli1.

Owing to the intrinsic requirement of iron, Fe/S cluster assembly is tightly connected to cellular iron homeostasis. Mitochondria host the two major iron-consuming biosynthetic pathways of a cell, i.e., Fe/S cluster biogenesis and heme synthesis (Sheftel & Lill, 2009). Defects within the core ISC or the ISC export systems lead to the accumulation of iron within the organelle (Stehling, Wilbrecht, & Lill, 2014). The molecular basis of this phenomenon is related to two key cytosolic proteins regulating intracellular iron homeostasis, the above-mentioned IRP1 and IRP2 (Anderson, Shen,

Eisenstein, & Leibold, 2012; Kühn, 2015). IRP1 in its holoform is a [4Fe–4S] protein with aconitase activity, and in its apostate it binds to iron-responsive elements (IREs) of mRNAs encoding proteins engaged in trafficking (e.g., transferrin receptor, TFR1), storage (ferritin, FT), or utilization (mitochondrial aconitase, ACO2, δ-aminolevulinic acid synthase 2, ALAS2) of iron. Hence, Fe/S cluster assembly on IRP1, requiring both the core ISC and the CIA systems, is influencing intracellular iron handling by determining the equilibrium between apo- and holo-IRP1 (Sheftel & Lill, 2009). An additional critical level of regulation is introduced by the stabilization of IRP2 by the action of CIA2A, together leading to a tight coupling of Fe/S protein biogenesis and intracellular iron regulation in vertebrate cells (Johnson, Deck, Nizzi, & Eisenstein, 2017).

Cellular Fe/S protein assembly is an essential process (Lill, 2009; Lill, Fekete, Sipos, & Rotte, 2005; Lukes & Basu, 2015; Rouault, 2012), reflected by the fact that de(p)letion of numerous vertebrate ISC assembly, ISC export, and CIA components not only leads to cell death in human cell culture but is also lethal in zebrafish, mouse, and man during early embryonic development (Camaschella et al., 2007; Cossee et al., 2000; Nordin, Larsson, & Holmberg, 2012; Pondarre et al., 2006). To date, human disorders connected to defects in Fe/S protein biogenesis (Fe/S diseases) are solely linked to mutations in nuclear genes encoding mitochondrial ISC proteins. These maladies are associated with rather diverse clinical phenotypes, mainly showing neurological, hematological, and metabolic symptoms (Rouault, 2012; Stehling et al., 2014). For example, a missense mutation in the gene of ISCU leads to mRNA mis-splicing and low scaffold protein levels in muscle, giving rise to hereditary myopathy with lactic acidosis, a rare disease characterized by early onset of muscle weakness and exercise intolerance (Mochel et al., 2008; Olsson, Lind, Thornell, & Holmberg, 2008). Impaired expression of the core ISC component frataxin provokes the neurodegenerative disease Friedreich's ataxia (Schmucker & Puccio, 2010), whereas a splice mutation of the transfer factor GLRX5 triggers sideroblastic anemia with hypochromic erythrocytes as a result of ineffective heme formation (Camaschella et al., 2007). The so-called multiple mitochondrial dysfunctions syndromes (MMDS) are associated with defects in the ISC-targeting factors NFU1, BOLA3, IBA57, ISCA1, and ISCA2 (Ajit Bolar et al., 2013; Al-Hassnan et al., 2015; Cameron et al., 2011; Navarro-Sastre et al., 2011; Shukla et al., 2017). Individuals present with respiratory inadequacy, hypotonia, encephalopathy, deformities of brain, and neurological regression. At the cellular level, a respiratory deficiency due to defective [4Fe–4S] cluster-containing complexes I and II is observed. Further, hyperglycinemia, metabolic acidosis, and increased levels of lactate and ketoacids indicate the impaired activity of [4Fe–4S] enzymes of the citric acid cycle (i.e., aconitase and succinate dehydrogenase which also serves as respiratory chain complex II), and of LIAS-dependent enzymes such as pyruvate, 2-ketoglutarate, and branched-chain ketoacid dehydrogenases as well as the glycine cleavage system (GCS). These latter enzymes show a defect in lipoylation due to impaired activity of LIAS which carries two [4Fe–4S] clusters.

The cytosol and nucleus of eukaryotic cells harbor essential Fe/S proteins which participate in protein translation (ABCE1, yeast Rli1) and genome maintenance (e.g., CHLR1, DNA2, NTHL1, POLD1, PRIM2, and RTEL1) (Paul & Lill, 2015). Mutations in the Fe/S cluster-containing DNA helicases FANCJ and XPD are associated with Fanconi anemia, xeroderma pigmentosum, Cockayne syndrome, and trichothiodystrophy (Lehmann, 2003). Further, amino acid exchanges in the Fe/S cluster-containing glycosylase MUTYH near its cluster-binding site are associated with a colorectal cancer predisposition syndrome referred to as MUTYH-associated polyposis (MAP; Banda, Nunez, Burnside, Bradshaw, & David, 2017). The fundamental functions of many Fe/S proteins in central cellular pathways and physiology explain why Fe/S cluster assembly and thereby its key biosynthetic organelle, the mitochondrion, are indispensable for life (Lill, 2009).

In this chapter, we provide experimental approaches and protocols to assess the maturation of Fe/S proteins in mammalian cells including cell culture material and patient tissue samples. A detailed differential analysis of multiple cellular Fe/S proteins and their assisted processes may reveal critical information about the location of the defective biogenesis protein within the ISC or CIA pathways. First, we will highlight the significance of immunoblotting as a simple and robust method to study Fe/S protein maturation in human samples. Higher sensitivity and reliability is provided by Fe/S enzyme activity measurements, and some assays have previously been summarized (Molik, Lill, & Mühlenhoff, 2007; Stehling et al., 2007; Stehling, Sheftel, et al., 2009). Second, we provide a protocol to determine the activity of the clinically relevant Fe/S enzyme dihydropyrimidine dehydrogenase (DPYD) (Stehling et al., 2013, 2012). Third, the most direct way to assess Fe/S cluster insertion into target proteins is the assessment of Fe/S protein loading with radioactive ^{55}Fe which was originally described for yeast (Kispal, Csere, Prohl, & Lill, 1999; Pierik, Netz, & Lill, 2009), but for some

Fe/S proteins can also be applied for human cells. We will describe critical refinements of this technique allowing a more general use of the radiolabeling approach.

2. ANALYSIS OF Fe/S PROTEIN ASSEMBLY IN TISSUE MATERIAL BY IMMUNOBLOTTING

Immunoblotting is a versatile approach to determine the steady-state levels of Fe/S proteins in biological samples of various sources. Careful sample preparation is a particular prerequisite for the analysis of (primary) skin fibroblasts or of muscle biopsies from individuals showing Fe/S protein assembly defects.

2.1 Digitonin-Based Fractionation of Patient-Derived and Other Cultured Human Cells

Depending on the available material, antigen levels can be analyzed in total cell lysates as well as in cytosolic or mitochondria-enriched fractions. The latter samples usually produce stronger signals, facilitating the analysis of cytosolic or mitochondrial Fe/S proteins. Tissue culture material (including patient-derived skin fibroblasts or lymphoblasts) allows the separation of cytosolic and mitochondrial Fe/S proteins by digitonin-mediated plasma membrane permeabilization (Drapier & Hibbs, 1996). At low concentrations, the steroid glycoside acts as a cholesterol-chelating agent and thus interferes with plasma membrane integrity without impacting on other intracellular membranes. Following tabletop centrifugation, a supernatant fraction containing soluble cytosolic proteins can be separated from a "crude mitochondrial fraction" which additionally contains essentially all other membranous compartments including the plasma membrane.

2.1.1 Equipment

- Common tissue culture material
- Cell counting chamber
- Refrigerated tabletop microtube centrifuge

2.1.2 Buffers and Reagents

- Common tissue culture media and reagents
- Phenylmethanesulfonyl fluoride (PMSF, saturated 200–250 m*M* solution in ethanol, prepare always freshly)
- Digitonin (5% w/v in DMSO)

- Phosphate-buffered saline (PBS) without calcium and magnesium
- Mitobuffer: 5 m*M* Tris–hydrogen chloride (Tris–HCl) pH 7.4, 240 m*M* sucrose, 1.5 m*M* magnesium chloride ($MgCl_2$), 1 m*M* PMSF
- Digitonin buffer: 0.008% Digitonin in mitobuffer (prepared from stocks)

2.1.3 Procedure

1. Harvest cells by trypsination, wash twice with PBS, and determine cell number.
2. Resuspend cells in digitonin buffer (10 million cells per mL).
3. Incubate 10–15 min and remove up to one third of the total lysate for later analysis.
4. Centrifuge residual sample at 15,000 × *g* and 4°C for 10 min.
5. Save the supernatant as cytosolic fraction and separate into multiple aliquots.
6. Wash pellet by resuspending in 1 mL mitobuffer and centrifuging for 10 min at 15,000 × *g* and 4°C.
7. Remove supernatant and thoroughly resuspend the pellet in one volume of mitobuffer corresponding to the residual volume in step 4; separate sample into multiple aliquots.
8. Determine protein concentrations of all fractions.
9. Snap–freeze samples in liquid nitrogen and store aliquots at −80°C.

2.1.4 Notes

- Cell fractionation has to be strictly performed at 4°C using precooled buffers.
- Digitonin stock solution in DMSO can be stored at −20°C.

2.2 Preparation of Cell Extracts From Skeletal Muscle Biopsies

Skeletal muscle biopsies are particularly well suited for the analysis of mitochondrial Fe/S proteins because even small amounts of material may give rise to strong immunoblot signals. The tissue is sufficiently accessible, and, due to its high energy demand, is rich in mitochondria, amounting to 3%–20% of the cell volume (Reichmann, Hoppeler, Mathieu-Costello, von Bergen, & Pette, 1985). Only heart muscle tissue contains more mitochondria, constituting up to one third of the cell volume (Hom & Sheu, 2009). Other mitochondria-rich organs are liver and kidney (Fernandez-Vizarra, Enriquez, Perez-Martos, Montoya, & Fernandez-Silva, 2011), but biopsy samples are difficult to obtain. Proteins of interest can be successfully recovered from small biopsy samples (about 10 mg of wet tissue) by tissue homogenization followed by solubilization in RIPA (radioimmunoprecipitation assay) buffer (Harlow & Lane, 2006).

2.2.1 Equipment

- Tissue homogenizer (e.g., Ultra Turrax T25)
- Refrigerated tabletop microtube centrifuge

2.2.2 Buffers and Reagents

- Roche "cOmplete" protease inhibitor cocktail tablet dissolved in 2 mL of water
- Sigma Phosphatase Inhibitor Cocktail 2 solution
- "Benzonase" nuclease (25 U/μL)
- RIPA buffer: 50 m*M* Tris–HCl pH 8, 150 m*M* sodium chloride (NaCl), 1% (v/v) Triton X-100, 0.5% (w/v) sodium deoxycholate, 0.1% (w/v) sodium dodecyl sulfate (SDS); add protease inhibitor cocktail (1:25), phosphatase inhibitor (1:100), and benzonase (1:1000) always freshly

2.2.3 Procedure

1. Add 2.0 mL ice-cold RIPA buffer to 10 mg of tissue material.
2. Homogenize samples by five cycles of grinding for 5 s at a pestle speed of 13,500 rpm and 30 s of intermittent cooling.
3. Incubate samples with gentle agitation for another 2 h to allow for optimal solubilization.
4. Centrifuge for 20 min at 14,000 × *g* and 4°C to deplete insoluble material.
5. Save supernatant, determine protein concentration, and prepare aliquots.
6. Snap–freeze aliquots in liquid nitrogen and store at −80°C until use.

2.2.4 Notes

- Homogenization and sample preparation have to be performed at 4°C using precooled buffers.
- Avoid foam formation as well as possible.
- Be aware that ingredients of the RIPA solubilization buffer may interfere with protein determination assays.

2.3 TCA Precipitation of Proteins for Subsequent SDS-PAGE

Since protein concentrations in cell fractions (Section 2.1) and RIPA extracts (Section 2.2) are frequently low, we recommend trichloroacetic acid (TCA) precipitation in order to concentrate samples for SDS-polyacrylamide gel electrophoresis (PAGE). Moreover, TCA precipitation allows for the removal of membrane lipid contaminations, thereby improving sample quality for subsequent immunoblotting.

2.3.1 Equipment

- Refrigerated tabletop microtube centrifuge

2.3.2 Buffers and Reagents

- TCA (50% v/v)
- Acetone, ice-cold
- Laemmli sample buffer: 62.5 m*M* Tris–HCl pH 6.8, 10% (w/v) glycerol, 2% (w/v) SDS, 0.001% (w/v) bromophenol blue, 5% (v/v) 2-mercaptoethanol

2.3.3 Procedure

1. Determine concentrations of the respective protein solutions (see Sections 2.1.3 and 2.2.4).
2. Mix three volumes of protein solution with one volume of TCA solution.
3. Allow proteins to denature for 15 min on ice and precipitate sample at 14,000 × *g* and 4°C for 10 min.
4. Remove supernatant and wash pellet twice (14,000 × *g*, 4°C, 10 min) with 500 μL of ice-cold acetone.
5. Remove supernatant and let residual acetone evaporate at room temperature (the white pellet will become translucent).
6. Resuspend pellet in Laemmli sample buffer to a concentration of 1–2 μg/μL.
7. Heat samples to 95°C for 3 min while shaking thoroughly.
8. Precipitate nonsoluble material at 14,000 × *g* for 5 min prior to SDS-PAGE.

2.3.4 Notes

- Precipitation steps have to be performed at 4°C using precooled solutions.
- Avoid TCA precipitation of large protein amounts (>500 μg), as this will produce protein aggregates which are difficult to resolubilize.

2.4 Preparation of Gradient Gels for SDS-PAGE-Based Separation of Cellular Proteins

The degree by which proteins are sufficiently separated by linear SDS-PAGE is largely dependent on their size. Analysis of both small and large proteins in a given sample thus may require multiple gels with distinct acrylamide concentrations. In contrast to linear gels, continuous gradient gels allow the separation of small and large proteins in a single run, hence limiting the consumption of precious sample material, especially of patient-derived cells. We usually cast gradient gels ourselves by using a simple gravity-based mixing device, and thus are able to produce any appropriate acrylamide gradient. Precast gradient gels are commercially available.

2.4.1 Equipment

- Gradient mixer (reservoir volume at least 10 mL) with magnetic bar and outlet tap (Fig. 1)
- Magnetic stirrer
- Casting system for 12 × 15 cm gels of 1 mm thickness

2.4.2 Buffers and Reagents

- Acrylamide/bisacrylamide solution (30%, 37.5:1 ratio)
- 3 *M* Tris–HCl pH 8.8
- 0.6 *M* Tris–HCl pH 6.8
- SDS (10% w/v)
- 2 *M* Sucrose
- Bromophenol blue
- Ammonium persulfate (APS, 10% w/v)
- Tetramethylethylenediamine (TEMED)

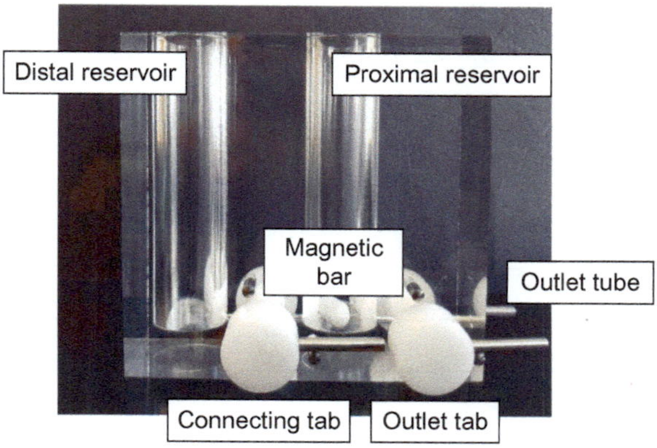

Fig. 1 Design of a mixing chamber for the preparation of gradient gels.

2.4.3 Procedure

1. Place the gradient mixer on a magnetic stirrer and apply a magnetic bar to the proximal reservoir connected to the outlet tab.
2. Connect one end of a silicone tubing with the outlet tap and the other end with the casting system; apply the tubing at the top center of the two plates framing the space for the gel.
3. Close the tap connecting the two mixer reservoirs, as well as the outlet tap.
4. For the separating gradient gel, prepare two acrylamide solutions with desired concentrations (examples are given in Table 1).
5. Transfer the blue-colored, sucrose-containing high-percentage acrylamide solution into the reservoir proximate to the mixer outlet, and the clear low-percentage acrylamide solution into the reservoir distal to the outlet.
6. Open the reservoir-connecting central tap and fill the reservoir-connecting tube in order to remove air bubbles that may clog the system (can be followed by the color of the acrylamide solution); immediately close the tap again.
7. Open the outlet tap and let the blue-colored, sucrose-containing acrylamide solution enter the gel casting system.
8. Open the tap connecting the two reservoirs of the mixing chamber and make sure that both reservoirs are emptied at equal rates.

Table 1 Composition of Acrylamide Solutions Suggested for Gradient Gel Preparation

	Separating Gel (2 × 7.5 mL)				
	Low-Percentage PAA Solution		High-Percentage PAA Solution		Stacking Gel (5 mL)
	6%	10%	16%	20%	6%
PAA (30%)	1.5 mL	2.5 mL	4 mL	5 mL	1 mL
Sucrose 2 *M*	—	—	1.5 mL	1.495 mL	—
Bromophenol blue	—	—	Grain	Grain	—
Tris/HCl	930 μL 3 *M* (pH 8.8)				1.04 mL 0.6 *M* (pH 6.8)
SDS (10%)	75 μL				50 μL
H_2O	Add to a total volume of 7.5 mL				2.9 mL
APS (10%)	22 μL				50 μL
TEMED	4 μL				5 μL

9. As soon as pouring of the gradient gel is completed, it has to be covered either by water or by any other appropriate liquid.
10. After polymerization of the gradient gel, a typical stacking gel may be poured on top of the gradient gel (see Table 1).

2.4.4 Notes

- Sucrose present in the high-percentage acrylamide solution increases density and ensures proper gradient formation.
- Bromophenol blue allows visualization of gradient formation.
- Make sure that the position of the mixing chamber is higher than that of the casting system in order to allow for proper gravity flow.
- Cooling of the acrylamide solutions may delay the polymerization process and hence prevent untimely polymerization and clogging of the mixing chamber.
- Immediately after the pouring process, rinse the mixing chamber thoroughly with water to remove residual acrylamide solution.

The gradient gel can be applied to conventional Tris–glycine gel electrophoresis. We usually perform the gel run at constant voltage of 110V for about 5h followed by semidry blotting onto nitrocellulose membranes for 2h at 350mA. Blots may be blocked by common procedures and cut horizontally in order to allow for simultaneous staining of multiple antigens of different molecular mass.

3. STRATEGIES TO ASSAY IRON–SULFUR PROTEIN ASSEMBLY DEFECTS BY IMMUNOBLOTTING

Fe/S clusters not only take part in protein function but also confer structural stabilization of the polypeptide backbone by connecting cognate amino acid ligands (Leal & Gomes, 2008; Walden et al., 2006). Disturbances in the Fe/S cluster assembly pathways, either experimentally or in disease states, may interfere with correct cluster ligation in a given Fe/S protein, and, as a consequence, may impact on the protein's structural integrity. This destabilization may lead to degradation of the improperly matured protein (Beilschmidt et al., 2017; Guillon et al., 2009). We have observed this consequence of impaired Fe/S cluster assembly in many cases, and a decline in the enzymatic activity of distinct Fe/S proteins was often tightly followed by a drop in their cellular steady-state levels (Biederbick et al., 2006; Navarro-Sastre et al., 2011; Sheftel et al., 2010, 2012; Sheftel, Stehling, et al., 2009; Stehling et al., 2013, 2008, 2012). Based on this correlation we frequently use common immunoblotting techniques to analyze total cell lysates for alterations in the levels of selected Fe/S reporter proteins. This approach is easy to apply, quick, allows the use of low amounts of both native and denatured material, and gives access to a multitude of different proteins of all cellular compartments in a single sample. Notably, simple immunoblotting against meaningful reporter proteins permits a rough estimation at which step Fe/S protein assembly might be corrupted in a certain condition. Since the core ISC assembly machinery is required for the assembly of both mitochondrial and extramitochondrial Fe/S proteins (Stehling & Lill, 2013), impaired function of one (or more) ISC core components will affect virtually all cellular Fe/S proteins, irrespective of cofactor type (i.e., [2Fe–2S] or [4Fe–4S]) and subcellular localization. In contrast, ISC-targeting factors are acting downstream of the ISC core system and are expected not to impact on mitochondrial [2Fe–2S] proteins or on cytosolic–nuclear Fe/S proteins. Maturation of the latter proteins is specifically affected by the core ISC assembly apparatus, the ISC export machinery, and the CIA system (Lill et al., 2014; Paul & Lill, 2015). Vice versa, impaired function of the latter two systems does usually not affect mitochondrial Fe/S protein assembly, even though in severe cases indirect effects cannot entirely be ruled out (Johnson et al., 2017; Kispal et al., 1999; Pondarre et al., 2006). According to the complex interdependencies of the multiple Fe/S protein assembly pathways, we strongly recommend to analyze the steady-state levels of various polypeptides. Table 2 provides an overview about expected changes in steady-state levels of reporter proteins and other selected parameters related to given Fe/S protein assembly defects.

3.1 Assessing Fe/S Protein Steady-State Levels

3.1.1 Analysis of Defects in Mitochondrial Core ISC Components

As the mitochondrial core ISC assembly system is key for the generation of [2Fe–2S] clusters, probing immunoblots for the [2Fe–2S] protein ferrochelatase (FECH) may provide a first clue about assembly defects within the early steps of Fe/S cofactor formation (Crooks, Ghosh, Haller, Tong, & Rouault, 2010; Sheftel et al., 2010). Further validation of a potential core ISC assembly defect requires subsequent analyses of mitochondrial [4Fe–4S] proteins including aconitase (ACO2) (Biederbick et al., 2006) or electron transfer flavoprotein-ubiquinone oxidoreductase (ETFDH; Watmough & Frerman, 2010). Moreover, we have observed that dysfunction of the core ISC system also affects extramitochondrial Fe/S proteins including IRP1, DPYD, the Nth-like DNA glycosylase 1/AP lyase (NTHL1), the catalytic subunit of

Table 2 Reference Chart for the Differential Analysis of Mammalian Fe–S Protein Assembly Defects

			ISC Assembly			CIA		
							CIA Targeting	
Compartment	Cluster Type	Reporter Protein[a]	Core ISC	ISC Targeting	ISC Export	Early CIA	CIA2A Branch	CIA1/CIA2B/ MMS19 Branch
Mitochondrion	[2Fe–2S]	FECH	↓					
		RCC-III (UQCRFS1, core 2)	↓	(↓)[b]	(↓)[b]			
	[4Fe–4S]	ACO2	↓	↓				
		LIAS (LA-DLAT, LA-DLST, LA-DBT, LA-GCS-H)	↓	↓				
		ETFDH	↓	↓				
	Multiple	RCC-I (NDUFS1, NDUFA13, NDUFB4, NDUFB6)	↓	↓	(↓)[b]			
		RCC-II (SDHB)	↓	↓	(↓)[b]			
Cytosol	[4Fe–4S]	IOP1	↓		↓	↓		
		IRP1	↓		↓	↓	↓	
		DPYD (protein and activity)	↓		↓	↓		↓
		NTHL1	↓		↓	↓		↓
		GPAT	↓		↓	↓		↓
		POLD1	↓		↓	↓		↓
		ABCE1 (only ^{55}Fe)	↓		↓	↘		↘
	None	IRP2	↑		↑	↘	↓	↑
		FT	↓		↓	↓	↘	↘
		TFR1	↑		↑	↑	↗	↗

[a]Steady-state protein level, unless stated otherwise.
[b]Indirect effects possible.
LA, lipoic acid; ↑, elevated; ↗, moderately elevated; ↓, decreased; ↘, moderately decreased.

DNA polymerase delta (POLD1), and the CIA component IOP1 (Biederbick et al., 2006; Stehling et al., 2012). Hence, inspection of the corresponding steady-state protein levels is useful to confirm an improper activity of the core ISC system.

3.1.2 Analysis of Defects in Late-Acting Mitochondrial ISC-Targeting Factors

In contrast to the core ISC system, deterioration of the mitochondrial ISC-targeting factors selectively affects the formation of mitochondrial [4Fe–4S] proteins (Stehling et al., 2014). Accordingly, the characterization of such defects may again involve the screening for mitochondrial [2Fe–2S] and of cytosolic Fe/S assembly defects as positive controls by differential immunoblot analyses. In the case of an ISC-targeting factor defect several representatives of the latter two Fe/S protein groups are hardly or only mildly affected in their steady-state levels, whereas the abundance of mitochondrial [4Fe–4S] proteins is substantially decreased (Ajit Bolar et al., 2013; Beilschmidt et al., 2017; Navarro-Sastre et al., 2011; Sheftel et al., 2012). Of particular relevance is the mitochondrial P-loop NTPase IND1 which has been implicated in cofactor delivery to Fe/S protein subunits of respiratory chain complex I (RCC-I) (Sheftel, Stehling, et al., 2009). Low levels of IND1 elicit a RCC-I deficiency phenotype reminiscent of ISC assembly or ISC-targeting deficiency as indicated by the destabilization of core and supernumerary subunits NDUFS3, NDUFA13, NDUFB4, and NDUFB6 (Ajit Bolar et al., 2013; Cameron et al., 2011; Sheftel et al., 2010, 2012).

Care has to be taken in the selection of suitable reporter polypeptides for the analysis of ISC-targeting factor defects. For example, the steady-state protein levels of NDUFS1 (a subunit of RCC-I) and of SDHB (succinate dehydrogenase subunit B) are highly sensitive to impairments in the mitochondrial ISC assembly system (Sheftel, Stehling, et al., 2009; Sheftel et al., 2012). However, both proteins harbor different types of Fe/S cofactors (Fiedorczuk et al., 2016; Sun et al., 2005), rendering the analysis of RCC-I and RCC-II generally not suitable for the discrimination of core ISC or ISC-targeting defects. Moreover, disturbed respiratory chain supercomplex formation in general may affect the abundance of RCC subunits. For example, deficiency of IND1 not only interferes with the assembly of RCC-I but also affects the mitochondrial amount of the RCC-III core 2 subunit (UQCRC2) (Sheftel, Stehling, et al., 2009). This interdependence of certain Fe/S proteins requires cautious interpretation of the RCC steady-state protein levels for the discrimination of ISC assembly defects.

3.1.3 Analysis of Defects in Cytosolic–Nuclear Fe/S Protein Assembly

Unlike the obvious mitochondrial phenotype caused by impaired function of the ISC assembly system, defects in the ISC export or in the CIA system are more difficult to identify. Though functional deficiency of the ISC export component ABCB7 (Atm1 in yeast) causes mitochondrial damage, it does not lead to a general inactivation of mitochondrial Fe/S enzymes (Cavadini et al., 2007; Kispal et al., 1999; Pondarre et al., 2006). Notably, FECH and ACO2 maintain their cofactors albeit respiratory chain components such as SDH may be affected, probably due to indirect effects of mitochondrial dysfunction. Instead of a substantial mitochondrial phenotype severe defects were observed in the maturation of the extramitochondrial Fe/S proteins including IRP1, demonstrating the critical function of ABCB7 in the biogenesis of cytosolic–nuclear Fe/S proteins. Accordingly, characterization of ABCB7 defects has to include immunoblotting for cytosolic–nuclear as well as for mitochondrial Fe/S proteins apart from RCC components.

Such a distinction between different mitochondrial Fe/S proteins is not required for the analysis of potential CIA defects. Experimental cell culture systems indicated no or only minor alterations in mitochondrial Fe/S protein formation upon interference with the cytosolic–nuclear Fe/S protein assembly pathway (Johnson et al., 2017; Stehling et al., 2013, 2008, 2012). However, due to the branching character of the CIA system, analysis of CIA defects not only requires immunoblotting for representative mitochondrial Fe/S proteins but also a comprehensive differential analysis of as many extramitochondrial Fe/S proteins as possible. For example, Fe/S cluster assembly on IRP1 is mainly dependent on CIA2A but not on the CIA-targeting complex CIA1-CIA2B-MMS19 (Stehling et al., 2013, 2012), whereas stability of DPYD or NTHL1 (O. Stehling, unpublished) requires the ternary CIA-targeting complex but not CIA2A. Notably, we found the stability of POLD1 to be mainly affected by CIA1 and MMS19, but hardly by CIA2B and CIA2A (Stehling et al., 2013, 2012). Recently, a binary CIA adaptor complex has been identified in yeast (Paul et al., 2015). Its constituents Yae1 and Lto1 (and their human counterparts; V.D. Paul, unpublished) appear to be solely dedicated to the maturation of the ribosomal recycling factor Rli1 (human ABCE1). Hence, a picture of a widely ramified CIA system emerges in which the unequivocal identification of cytosolic–nuclear Fe/S protein assembly defects requires a complex analytical approach.

Importantly, the stability of some Fe/S proteins is hardly affected by impaired Fe/S cluster maturation. Those include the P-loop NTPase-type assembly factors IND1, NBP35, and CFD1 of the ISC and CIA systems, as well as the Fe/S target protein ABCE1. Hence, in these cases the estimation of their steady-state levels by immunoblotting is not a suitable test of the efficiency of the maturation process.

3.2 Assessing Fe/S Protein Function

3.2.1 Indirect Analysis of Lipoic Acid Synthase Activity

LIAS is a mitochondrial Fe/S protein whose enzymatic function can be easily, yet indirectly, estimated by immunoblotting approaches. LIAS belongs to the radical *S*-adenosyl methionine (SAM) enzyme family and catalyzes the conversion of peptide-bound octanoate into the lipoic acid (LA) cofactor which is an essential part of the mitochondrial enzymes pyruvate dehydrogenase (PDH), alpha-ketoglutarate dehydrogenase (alpha-KGDH), branched-chain ketoacid dehydrogenase (BCKDH), the H protein of the GCS, and of 2-oxoadipate dehydrogenase (Mayr, Feichtinger, Tort, Ribes, & Sperl, 2014). LIAS requires two [4Fe–4S] clusters to conduct LA formation: An AdoMet radical cluster mediates the formation of a 5′-deoxyadenosyl radical, whereas a so-called auxiliary cluster is decomposed in order to provide two of its sulfur atoms for the attachment to the octanoate precursor (McLaughlin et al., 2016). Due to the critical relevance of each of the metalloclusters for LIAS function any disturbance of the ISC system is decreasing the amount of protein-attached LA (Ajit Bolar et al., 2013; Cameron et al., 2011; Navarro-Sastre et al., 2011; Sheftel et al., 2010). Using LA-specific antibodies, we usually determine the relative amount of LA attached to the E2 subunits of PDH (dihydrolipoamide *S*-acetyltransferase, DLAT) and alpha-KGDH (dihydrolipoamide *S*-succinyltransferase, DLST). Staining of the other LA-dependent enzymes is possible but due to their low abundance less straightforward. Since impaired lipoylation might interfere with E2 subunit stability alike in case of Fe/S proteins, we frequently measure the levels of the respective polypeptide chains by specific antibodies. Using this approach, it is possible to indirectly study the functionality of LIAS and hence the activity of the mitochondrial ISC system.

3.2.2 Assessing GPAT Autocatalytic Cleavage Activity

A reliable reporter enzyme for the analysis of extramitochondrial Fe/S protein assembly by immunoblotting is glutamine phosphoribosylpyrophosphate amidotransferase (GPAT, a.k.a. PPAT). GPAT catalyzes the first step of the de novo purine nucleotide synthesis pathway but is expressed as an inactive precursor (Grandoni, Switzer, Makaroff, & Zalkin, 1989; Zhou, Broyles, Dixon, & Zalkin, 1992). To become an active enzyme, GPAT requires both the incorporation of a [4Fe–4S] cluster and the Fe/S cluster-dependent subsequent autocatalytic removal of an N-terminal undecameric propeptide. Interference with Fe/S cofactor insertion due to impaired cluster assembly or experimental site-directed ligand mutagenesis prevents propeptide processing, and may allow the detection of the precursor protein by immunoblotting due to its size shift (Martelli et al., 2007). Notably, the crystal structure of *Bacillus subtilis* GPAT suggests that the presence of the propeptide may interfere with protein folding which in turn might contribute to the observed protein destabilization in case of an impaired Fe/S cluster insertion process (Martelli et al., 2007; Stehling et al., 2013, 2008). Accordingly, steady-state GPAT levels represent an integrated readout for cofactor-mediated protein stability in combination with Fe/S cluster-dependent autocatalytic enzyme activity. Thus, the levels of GPAT can be used as a direct measure of Fe/S protein assembly activities.

3.2.3 Probing for Alterations in Cellular Iron Homeostasis

Besides their enzymatic activity, Fe/S proteins may also exert regulatory activity on gene expression whose outcome is accessible by immunoblotting. A prototypical example is IRP1 which in its holoform exerts cytosolic aconitase activity. Yet, in its apoform IRP1 may bind to IREs of mRNAs encoding proteins involved in iron metabolism, thereby determining the translation efficiency of FT mRNA, and the mRNA stability of transferrin receptor (TFR1) (Anderson et al., 2012; Kühn, 2015). The equilibrium between apo- and holo-IRP1 is mainly determined by [4Fe–4S] cluster assembly and thus depends on both the mitochondrial ISC systems and the CIA machinery. Consequently, the functionality of the ISC and CIA systems is a critical determinant influencing FT and TFR1 steady-state protein levels. In turn, the cellular iron status is impacting on Fe/S protein assembly, particularly since the mitochondrial ISC system is a major iron-consuming pathway. Availability of iron outside mitochondria is sensed not only by the maturation status of IRP1 but also indirectly by the steady-state protein level of IRP2, a non-Fe/S homolog of IRP1 which constitutively binds to IREs. Regulation of IRP2 occurs through iron-dependent ubiquitinylation and proteasomal degradation involving the iron-sensing E3 ubiquitin ligase subunit FBXL5 (Ruiz & Bruick, 2014).

Despite the complexity of the iron regulatory network differential probing for its components provides clues about potential Fe/S protein assembly defects. Most importantly, impaired activities of the core ISC and the ISC export system give rise to profound iron accumulation within mitochondria at the expense of cytosolic iron pools (Stehling et al., 2014). Though the mechanistic basis of this phenomenon remains elusive so far, it not only leads to impaired assembly of Fe/S proteins including IRP1 (see Sections 3.1.1 and 3.1.3) but also to elevated steady-state levels of IRP2 (Sheftel et al., 2010) and TFR1, possibly due to an increased decay of FBXL5 under lower cytosolic iron concentrations (Cavadini et al., 2007; Johnson et al., 2017; Ruiz & Bruick, 2014). In contrast, dysfunction of late-acting ISC-targeting components affects iron regulatory proteins only mildly, if at all, facilitating the discrimination of defects in the early and late phases of mitochondrial Fe/S protein assembly (Ajit Bolar et al., 2013; Navarro-Sastre et al., 2011). Disturbances of the CIA system are again differentially reflected by the iron regulatory system. Impaired function of early-acting CIA components results in diminished maturation of IRP1 and thus elevation of TFR1 levels which in turn are likely to be related to a (usually moderate) FBXL5-mediated decrease of IRP2 (Johnson et al., 2017; Stehling et al., 2008). Inactivating the CIA2A branch of the CIA system in HeLa cells causes a nearly complete loss of IRP1 and IRP2 whereas failures of CIA2B and MMS19 hardly affect IRP1 maturation, yet cause a substantial increase in IRP2 steady-state level (Stehling et al., 2013, 2012). Importantly, alterations in TFR1 and FT steady-state protein levels have to be interpreted cautiously, because expression and turnover of the two proteins are regulated also by multiple other cues (reviewed by Arosio, Elia, & Poli, 2017).

4. ANALYSIS OF DPYD ENZYME ACTIVITY

The measurement of Fe/S enzyme activities is a direct way of assessing Fe/S protein assembly efficiencies. Several well-established protocols are available for Fe/S enzyme analyses, particularly for the determination of FECH, aconitase, and RCC function (Rustin et al., 1994; Stehling et al., 2007, 2009; Tong & Rouault, 2006; Trounce, Kim, Jun, & Wallace, 1996). Likewise, useful protocols exist for the analysis of some cytosolic Fe/S enzyme activities including those of holo-IRP1 (cytosolic aconitase), NTHL1, aldehyde oxidase, and xanthine dehydrogenase (Kurosaki, Demontis, Barzago, Garattini, & Terao, 1999; Martelli et al., 2007; Stehling et al., 2007; Tong & Rouault, 2006). Activity measurements of the clinically relevant Fe/S enzyme DPYD (Schnackerz, Dobritzsch, Lindqvist, & Cook, 2004), however, are challenging and often involve HPLC analyses. In the following, we will describe a rather simple thin-layer chromatography (TLC)-based approach to assess the functionality of DPYD.

DPYD is a homodimeric 220-kDa protein and catalyzes the reduction of pyrimidines in order to initiate their degradation (Schnackerz et al., 2004). Each DPYD monomer contains both an FAD and FMN moiety as well as four [4Fe–4S] clusters, the latter being arranged to form two entangled electron conduits that proceed across both subunits to connect the active sites of the enzyme. The substrate specificity of the enzyme is rather broad as DPYD is also acting on pyrimidine derivatives such as the anticancer drug 5-fluorouracil. We usually determine DPYD activity based on the conversion of [4-^{14}C]-thymine into [4-^{14}C]-dihydrothymine (Stehling et al., 2013, 2012). Substrate and product of the DPYD reaction are separated by TLC and quantitated by autoradiography. Representative results are shown in Fig. 2.

4.1 Setup of the DPYD Enzymatic Reaction

4.1.1 Equipment

- Microtube block incubator or heating cabinet

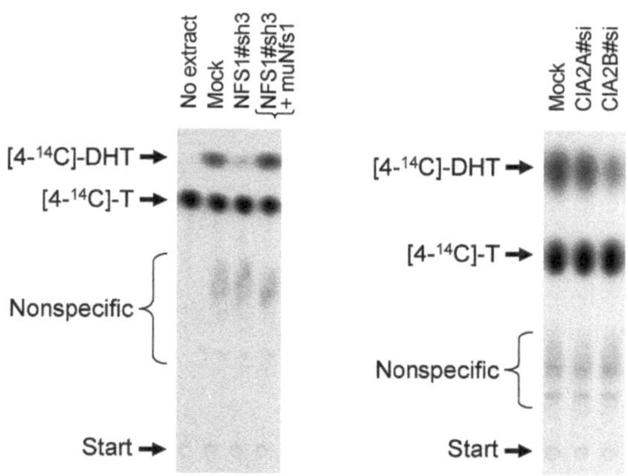

Fig. 2 Thin-layer chromatography analysis of DPYD enzyme activity. HeLa cells were depleted for NFS1 (*left panel*), CIA2A or CIA2B (*right panel*) by RNAi and analyzed for DPYD-dependent conversion of [4-^{14}C]-thymine ([4-^{14}C]-T) to [4-^{14}C]-dihydrothymine ([4-^{14}C]-DHT) by thin-layer chromatography and autoradiography (Stehling et al., 2013, 2012). The product of DPYD function, [4-^{14}C]-DHT, is decreased in samples of HeLa cells treated for 6 days with NFS1-directed shRNA (shNFS1#3) or CIA2B-directed siRNA (CIA2B#si). Formation of [4-^{14}C]-DHT was not affected in lysates of cells depleted for CIA2A (CIA2A#si) or in mock-transfected cells. Note the recovery of DPYD activity upon complementation of RNAi-mediated NFS1 deficiency by concomitant expression of RNAi-resistant murine Nfs1 (muNfs1, *left panel*). A reaction mixture devoid of any cell sample (no extract) was used to monitor [4-^{14}C]-T degradation. "Start" indicates the application sites of the DPYD product solution. "Nonspecific" reflects DPYD-independent substrate conversion.

4.1.2 Buffers and Reagents

- [4-^{14}C]-thymine (100 μCi/mL corresponding to 60 mCi/mmol exhibits sufficient activity)
- NADPH (375 m*M*, prepare freshly)
- Glucose-6-phosphate (G-6-P, 100 m*M*)
- Glucose-6-phosphate dehydrogenase (G6PDH, 1 kU/mL)
- Digitonin (5% w/v in DMSO)
- Perchloric acid (10% v/v)

4.1.3 Procedure

1. Prepare DPYD reaction buffer from stocks consisting of 25 m*M* Tris–HCl pH 7.5, 2.5 m*M* $MgCl_2$, 2 m*M* dithiothreitol (DTT, freshly prepared), 0.1% digitonin, 5 m*M* NADPH, 1 m*M* G-6-P, and 1 U/mL G6PDH.
2. Subject a whole-cell pellet of 125 μg total protein to one freeze–thaw cycle and suspend the sample in 25 μL of DPYD reaction buffer.
3. Add [4-^{14}C]-thymine to a final concentration of 30 μ*M* (1.8 μCi/mL) and mix well.
4. Incubate for up to 4 h at 32.5°C.
5. Stop the reaction by addition of 5 μL perchloric acid (10% v/v).
6. Centrifuge at 15,000 × *g* for 5 min at room temperature, save the supernatant (DPYD product solution), and discard the pellet.

4.1.4 Notes

- Do not use [2-^{14}C]-thymine in this assay since subsequent enzymatic reactions will lead to the production of volatile radioactive $^{14}CO_2$.
- Freeze–thawing of the cell sample supports cell opening.
- For samples with a high DPYD content like liver cells a total protein amount of 5–10 μg is sufficient for analysis.
- In our hands the rate of [4-^{14}C]-dihydrothymine formation remains constant for about 4 h.
- Ensure even tempering of the entire reaction vial in order to avoid evaporation from the reaction mixture.
- The DPYD product solution can be stored at −20°C for subsequent TLC analysis.

4.2 Thin-Layer Chromatography and Autoradiography

4.2.1 Equipment

- Plastic TLC sheets precoated with 0.1 mm cellulose (MN 300 or Cel 300)
- TLC developing chamber

- Filter paper sheets
- Separating funnel
- Autoradiography cassette
- Phosphor imager and screen, or X-ray film and darkroom

4.2.2 Buffers and Reagents

- *N*-Butanol
- Ammonia solution (25%)

4.2.3 Procedure

1. Pour 135 mL *N*-butanol, 67.5 mL water, and 22.5 mL ammonia solution (25%) corresponding to a ratio of 90:45:15 (v/v/v) into a separating funnel.
2. Stir thoroughly for 1 min, let the liquid phases separate for 5 min, and repeat this procedure four to five times.
3. Extract the organic phase by avoiding any aqueous contamination.
4. Line the inner walls of a TLC developing chamber with filter paper and fill in the funnel-extracted organic phase (representing the mobile phase in subsequent TLC).
5. Close the lid of the chamber and allow the mobile phase to spread throughout the entire filter paper.
6. Label both the intended starting front (about 1.5 cm distant from the sheet border) and the sample application sites (at least 1 cm apart from each other) at the TLC sheet using a soft pencil (take care not to damage the cellulose coat).
7. Place the labeled cellulose sheet into the TLC chamber and predevelop overnight.
8. Remove the TLC sheet from the chamber and let the mobile phase evaporate.
9. Spot a total of 6 μL of the DPYD product solution in 1 μL aliquots at the intended sample application sites at the TLC sheet with intermittent drying of the spots.
10. Place the TLC sheet into the TLC chamber, develop to the first third, remove the sheet from the chamber, and let the mobile phase evaporate.
11. Develop to the second third, let the mobile phase evaporate, and eventually develop the entire sheet.
12. Cover the dry TLC sheet with a transparent plastic bag, place it in an autoradiography cassette, and expose it either to a phosphor imager screen or to an X-ray film for several days.

4.2.4 Notes

- Use a fume hood for preparation and application of the TLC mobile phase.
- Make sure that the fluid level of the mobile phase within the TLC chamber does not cover the starting front at the TLC sheet.

5. DIRECT DETERMINATION OF Fe/S CLUSTER INSERTION INTO APOPROTEINS BY ^{55}Fe RADIOLABELING

The assembly of Fe/S proteins can be directly measured by the incorporation of radioactive iron (^{55}Fe) into the Fe/S cluster followed by immunoprecipitation of the protein and liquid scintillation counting. Whereas pulse radiolabeling of yeast cells with ^{55}Fe is a well-established method for the analysis of the de novo biogenesis of Fe/S cluster-containing proteins (Kispal et al., 1999; Pierik et al., 2009), radiolabeling in human tissue culture is hampered by several issues. For instance, the iron uptake by human cells cannot easily be increased by growth under iron-poor conditions as seen in yeast cells. Other critical aspects are the differences in the mechanisms of iron uptake and storage in yeast and human cells (Mühlenhoff et al., 2015). An additional element to consider is the limited plasmid-based overproduction of potential target Fe/S proteins in mammalian cells, rendering their detection difficult even upon cofactor radiolabeling. A particular problem is the high, nonspecific ^{55}Fe background signal during immuno- or affinity isolation of the Fe/S target proteins. In the following, we will describe an adapted Fe/S protein radiolabeling protocol for human tissue culture based on physiological ^{55}Fe uptake and insertion into apoproteins. ^{55}Fe/S cluster insertion is followed by specific Fe/S target protein release from the immunoprecipitate by TEV-protease cleavage, and direct target detection by the presence of a fluorescent tag (Fig. 3).

The principal route by which mammalian cells take up iron is by receptor-mediated endocytosis of transferrin (TF), a soluble bilobal iron-binding protein with two binding sites for ferric iron (Richardson & Ponka, 1997). Coupling of ^{55}Fe to transferrin (Ponka & Schulman, 1985) allows physiological cellular uptake and its incorporation into target Fe/S proteins. The dependence of ^{55}Fe incorporation into a given target apoprotein on established ISC or CIA factors provides evidence for the presence of a Fe/S cluster rather than ^{55}Fe only incorporation (Pierik et al., 2009; Upadhyay et al., 2014, 2017). Since the depletion of ISC or CIA factors in human cells may result in a destabilization of the Fe/S target proteins (see

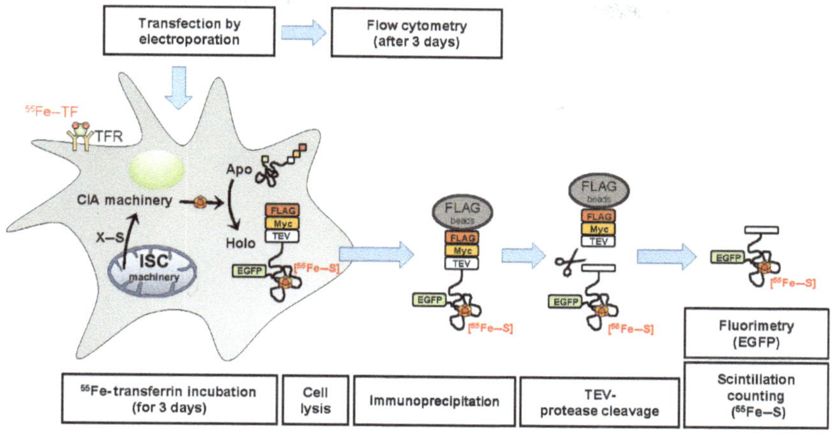

Fig. 3 Flow chart for the analysis of ^{55}Fe incorporation into EGFP-TEV-Myc-FLAG fusion proteins in mammalian cells in vivo. Plasmids encoding EGFP-TEV-Myc-FLAG-tagged Fe/S reporter proteins are transfected by electroporation into tissue culture cells. The expression efficiency of the EGFP fusion proteins may be analyzed in a nonradioactive aliquot of the transfection sample using flow cytometry. The remainder of transfected cells is cultured in medium supplemented with ^{55}Fe-loaded transferrin (^{55}Fe–TF), which is taken up by transferrin receptor (TFR1)-mediated endocytosis. After 3 days of tissue culture, cells are lysed and subjected to anti-FLAG (or anti-Myc) immunoprecipitation. For specific release of radiolabeled ^{55}Fe/S proteins the affinity resin is incubated with TEV protease. Then, the TEV supernatant is analyzed for both reporter protein-associated EGFP fluorescence (as a measure of protein amount) and radioactivity (as a measure of Fe/S cluster incorporation) by scintillation counting.

Section 3), the quantitative analysis of this Fe/S protein radiolabeling approach also requires the determination of the corresponding protein amount, in addition to the measurement of protein-bound ^{55}Fe.

5.1 ^{55}Fe Coupling to Transferrin

Human TF contains two ferric iron-binding sites per polypeptide and binds the metal preferentially in the presence of carbonate. The anions serve as bridging ligands between the metal ion and the protein, excluding water from the two coordination sites (Aisen & Listowsky, 1980; Richardson & Ponka, 1997). In our labeling protocol, we routinely use 1 mCi of $^{55}FeCl_3$ for coupling in twofold excess to human apotransferrin.

5.1.1 Equipment

- PD-10 desalting column
- 10-mL syringe and 0.22-μm filter

5.1.2 Buffers and Reagents

- Radioactive iron chloride ($^{55}FeCl_3$) in 0.5 N hydrogen chloride (HCl)
- 25% (w/v) sodium citrate
- 600 m*M* sodium hydrogen carbonate (bicarbonate, $NaHCO_3$)
- Human apotransferrin
- 20% bovine serum albumin (BSA)
- Hank's balanced salt solution (HBSS)

5.1.3 Procedure

1. Mix 1 mCi of $^{55}FeCl_3$ (with a specific activity of at least 50 mCi per mg corresponding to about 360 nmol ^{55}Fe per mCi) with a 100-fold molar excess of citrate. The color of the mixture will turn from clear to pale yellow.
2. Incubate 3–6 h at room temperature to allow for efficient ^{55}Fe-citrate formation.
3. Solve human apotransferrin in 600 m*M* $NaHCO_3$ at a molar amount corresponding to one half of the citrate-chelated ^{55}Fe (i.e., 180 nmol apo-TF).
4. Add the ^{55}Fe-citrate solution to the apotransferrin solution to a final volume of 2 mL. The color of the mixture will turn from yellow to salmon. The final molar ratio of transferrin to ^{55}Fe will be 1:2, complying with the stoichiometry of iron binding to transferrin.
5. Incubate at room temperature overnight.
6. Separate ^{55}Fe-loaded transferrin (^{55}Fe–TF) from the reaction mixture using a PD-10 gel filtration column and HBSS as buffer. Elution from the column can be followed by the salmon color of iron-loaded transferrin.
7. Dilute the eluate as desired (usually to 10 mL), add 1% BSA, filter sterile, and store at 4°C until use.

5.1.4 Notes

- The specific radioactivity of ^{55}Fe should be as high as possible. Reasonable results are obtained with a specific radioactivity of at least 100 GBq per mg (about 50 mCi per mg). A specific activity of less than 20 mCi per mg is not recommended.
- Carbonate has to be in high excess to hydrogen chloride (the typical solvent for $^{55}FeCl_3$) as a neutralizing agent. Moreover, iron loading onto transferrin requires bicarbonate anions.

5.2 ^{55}Fe Radiolabeling of Tissue Culture Cells

In order to analyze radioactive ^{55}Fe/S cluster insertion into apoproteins, we usually transfect cells with plasmids encoding reporter proteins carrying both an EGFP tag and a Myc-FLAG tandem tag, the latter being connected to the reporter protein by a TEV (Tobacco Etch Virus) cleavage site (Fig. 3). The tags allow efficient immunoprecipitation of the reporter protein by either Myc or FLAG affinity matrices, whereas the Fe/S protein can be subsequently released from the resin by addition of TEV protease. This step is critical in some cases, as it leads to the removal of adventitiously bound ^{55}Fe to the affinity matrix and a substantial decrease of the background signal. The EGFP tag remains fused to the reporter protein and can be used for quantitation of the precipitated material within the TEV eluate. Calculating the ratio of recovered radioactivity per recovered fluorescence represents a relative measure of the amount of mature Fe/S protein per total isolated reporter protein.

5.2.1 Equipment

- Transfection system for tissue culture cells (e.g., by electroporation (Stehling et al., 2009) or by a chemical transfection method)
- Common tissue culture equipment including 25-cm^2 culture flasks with filter caps and 6-well plates

5.2.2 Buffers and Reagents

- Common tissue culture media and reagents
- ^{55}Fe-loaded transferrin (see Section 5.1)

5.2.3 Procedure

1. Transfect cells with the reporter plasmid according to established protocols, and ensure the formation of a confluent cell layer within 3 days of tissue culture in a 25-cm^2 flask.
2. During these 3 days, grow cells in 4 mL of culture medium supplemented with ^{55}Fe-loaded transferrin at a final concentration of 2.5 μCi/mL.
3. Harvest cells by trypsination as usual and perform immunoprecipitation of the reporter protein (see Section 5.3).

5.2.4 Notes

- We usually perform plasmid transfection by electroporation (Stehling et al., 2009), but chemical transfection is also possible. For this purpose seed cells according to the transfection protocol of choice and add ^{55}Fe-transferrin when the transfection medium has to be exchanged.

- In our experience, cells have to be grown for at least 2 days in the presence of ^{55}Fe-loaded transferrin in order to achieve sufficient reporter protein labeling.
- Binding of TF to its cognate TFR1 receptor is highly species specific. Make sure that the TF employed corresponds to the respective cell type.
- Be aware that alterations in the Fe/S protein assembly processes as well as growth conditions including cell density effects are influencing TFR1 expression and hence ^{55}Fe–TF uptake (Ponka & Lok, 1999; Stehling et al., 2008).
- In order to monitor reporter protein expression it is sufficient to transfect cells for growth in 6-well plates. Culture for 3 days in the absence of ^{55}Fe-radiolabeling, and analyze fluorescence by flow cytometry or microscopy.

5.3 Immunoprecipitation of Tagged and Radiolabeled ^{55}Fe/S Proteins

5.3.1 Equipment

- Refrigerated tabletop microtube centrifuge
- Microplate reader device with fluorescence detector
- Liquid scintillation counter

5.3.2 Buffers and Reagents

- Bicinchoninic acid (BCA) assay reagents
- PMSF (saturated 200–250 m*M* solution in ethanol, prepare always freshly)
- Anti-FLAG M2 or Myc affinity resin
- TEV protease
- PBS without calcium and magnesium
- TNETG buffer: 10 m*M* Tris–HCl pH 7.4, 150 m*M* NaCl, 1 m*M* ethylenediaminetetraacetic acid (EDTA), 0.2% Triton X-100, 10% glycerol, 1 m*M* PMSF (add always freshly)
- TEV cleavage buffer: 10 m*M* Tris–HCl pH 7.4, 100 m*M* NaCl, 0.1% NP-40, 2 m*M* DTT (add always freshly)
- Liquid scintillation fluid

5.3.3 Procedure for Immunoprecipitation

1. Wash the required amount of affinity resin once with TNETG buffer (expect 15–25 μL resin per sample).
2. Resuspend resin in 100 μL TNETG buffer per sample.

3. Remove the ^{55}Fe–TF containing tissue culture medium from the confluent cell layer of a 25-cm^2 flask, flush cells carefully with PBS, and harvest cells by trypsination.
4. Wash cells twice with PBS and eventually remove buffer as much as possible.
5. Lyse the cell pellet in 425 μL TNETG buffer and incubate 5 min on ice.
6. Precipitate nonsoluble debris by centrifugation at 13000 rpm for 5 min at 4°C and collect the supernatant (clear cell extract).
7. Save three 3 μL aliquots of the cell extract for protein determination using the BCA assay.
8. Subject three 5 μL aliquots of the clear cell extract to scintillation counting in order to determine the total cellular iron uptake per mg of clear lysate protein.
9. Mix the clear cell extract with 100 μL of the prewashed affinity matrix.
10. Incubate for 2 h under gentle rotation.
11. Spin down resin at 3000 rpm for 5 min and remove the supernatant (non-attached material).
12. Wash the resin three times using 1 mL TNETG buffer.

5.3.4 Procedure for Reporter Protein Recovery

1. Wash the resin once with 500 μL TEV cleavage buffer and remove as much of the supernatant as possible.
2. Resuspend beads in 100-μL TEV cleavage buffer (make sure that DTT has been added freshly).
3. Add TEV protease at a final concentration of 50 μg/mL.
4. Incubate for 45 min at 16°C under gentle shaking in order to keep the resin suspended.
5. Add 150 μL TEV cleavage buffer, spin down resin, and transfer the supernatant (TEV eluate contains the released reporter protein) into new tubes.
6. Wash remaining resin three times with 500 μL of TEV cleavage buffer.
7. Subject resin to scintillation counting in order to monitor the efficiency of TEV-protease cleavage by analysis of residual-associated radioactivity.
8. Transfer 200 μL of the TEV eluate into one well of a 96-well plate.
9. Apply TEV cleavage buffer in another well (serves as blank).
10. Analyze EGFP fluorescence in a plate reader device.
11. Recover the TEV eluate (~180 μL) from the plate into the original tube.

12. Transfer 200 μL TEV eluate in 2-mL reaction tubes, add 1.8 mL scintillation fluid, and vortex thoroughly.
13. Determine radioactivity by liquid scintillation counting. The amount of $^{55}Fe/S$ cluster associated with the studied protein may be quantitated relative to the EGFP fluorescence within the TEV supernatants or to the protein content of the clear lysate applied to the immunoprecipitation.

5.3.5 Notes

- All precipitation steps have to be performed at 4°C. Buffers have to be kept on ice.
- Cutting pipette tips ensures proper aspiration of the affinity resin.
- Spin down resin not faster than 3000 rpm (tabletop centrifuge).
- Harvesting cells by trypsination removes surface-bound transferrin and allows the determination of total cellular ^{55}Fe uptake by scintillation counting.
- Avoid foam formation particularly during the immunoprecipitation and TEV cleavage procedures.
- TEV cleavage is particularly required, when mitochondrial iron accumulation occurs.
- Sample mixing with the liquid scintillation fluid by vortexing works well in 2-mL reaction tubes.
- The BCA assay is not compatible with DTT.

6. SUMMARY AND CONCLUSION

Fe/S cofactor formation and insertion into cellular target apoproteins is mediated by three subcellular systems hosted by mitochondria and the cytosolic–nuclear compartment. Inherited or spontaneous defects particularly of the mitochondrial ISC assembly and export systems give rise to severe diseases with characteristic biochemical presentation. Differential analysis of mitochondrial and cytosolic–nuclear Fe/S proteins as well as of Fe/S cluster-dependent biochemical and regulatory processes may help to narrow down the affected assembly step or even the assembly factor. Because the mitochondrial core ISC assembly system is required for the maturation of all cellular Fe/S proteins, its failure affects the insertion of all types of Fe/S clusters in each cellular compartment. Moreover, core ISC assembly defects result in severe disturbances of cellular iron homeostasis including mitochondrial iron accumulation. In contrast, the late-acting ISC-targeting factors are dedicated to the assembly of mitochondrial [4Fe–4S] proteins, and defects lead to the development of MMDS which do not or only indirectly affect cytosolic Fe/S protein maturation and cellular iron homeostasis. Mutations of the ABC transporter ABCB7, a key component of the mitochondrial ISC export system, do not generally affect mitochondrial Fe/S protein assembly, although they lead to a distinct type of mitochondrial iron accumulation in form of so-called ring sideroblasts (Sheftel, Richardson, et al., 2009). Instead, ABCB7 deficiency restricts the supply of the CIA system with a mitochondria-derived sulfur compound required for cytosolic–nuclear Fe/S protein assembly. The CIA machinery is a branching system with assembly factors dedicated to distinct Fe/S target proteins including those orchestrating cellular iron homeostasis.

Due to the complex interdependencies of the multiple Fe/S protein assembly pathways, the identification of disease-relevant alterations requires the analysis of multiple mitochondrial and cytosolic–nuclear Fe/S reporter proteins. As the stability of these proteins is frequently dependent on the maturation status, differential determination of steady-state protein levels in tissue material or cell culture samples by immunoblotting is a comprehensive and versatile method to identify compromised Fe/S protein assembly steps. Moreover, in some cases altered enzymatic or regulatory Fe/S enzyme activity can be addressed by immunoblotting, including the formation of lipoic acid on PDH-E2 or KGDH-E2 by LIAS or regulation of TFR1 and ferritin by iron regulatory proteins. Potential Fe/S protein assembly defects may then be further validated by enzymatic analyses such as thin-layer chromatography-based determination of DPYD enzyme activity. The most direct approach to study Fe/S protein maturation in tissue culture relies on metallocofactor radiolabeling by ^{55}Fe. Refinement of this technique for application in mammalian tissue culture provides a valuable tool to study the maturation of Fe/S proteins lacking detectable enzymatic activity in physiological samples. Combination of the presented techniques represents a comprehensive strategy to investigate the multiple facets of Fe/S protein assembly mechanisms and related Fe/S diseases.

REFERENCES

Agar, J. N., Krebs, C., Frazzon, J., Huynh, B. H., Dean, D. R., & Johnson, M. K. (2000). IscU as a scaffold for iron-sulfur cluster biosynthesis: Sequential assembly of [2Fe-2S] and [4Fe-4S] clusters in IscU. *Biochemistry*, *39*, 7856–7862.

Aisen, P., & Listowsky, I. (1980). Iron transport and storage proteins. *Annual Review of Biochemistry*, *49*, 357–393. https://doi.org/10.1146/annurev.bi.49.070180.002041.

Ajit Bolar, N., Vanlander, A. V., Wilbrecht, C., Van der Aa, N., Smet, J., De Paepe, B., et al. (2013). Mutation of the iron-sulfur cluster assembly gene IBA57 causes severe myopathy and encephalopathy. *Human Molecular Genetics*, *22*, 2590–2602. https://doi.org/10.1093/hmg/ddt107.

Al-Hassnan, Z. N., Al-Dosary, M., Alfadhel, M., Faqeih, E. A., Alsagob, M., Kenana, R., et al. (2015). ISCA2 mutation causes infantile neurodegenerative mitochondrial disorder. *Journal of Medical Genetics*, *52*, 186–194. https://doi.org/10.1136/jmedgenet-2014-102592.

Anderson, C. P., Shen, M., Eisenstein, R. S., & Leibold, E. A. (2012). Mammalian iron metabolism and its control by iron regulatory proteins. *Biochimica et Biophysica Acta*, *1823*, 1468–1483. https://doi.org/10.1016/j.bbamcr.2012.05.010.

Arosio, P., Elia, L., & Poli, M. (2017). Ferritin, cellular iron storage and regulation. *IUBMB Life*, *69*, 414–422. https://doi.org/10.1002/iub.1621.

Balk, J., Pierik, A. J., Netz, D. J., Mühlenhoff, U., & Lill, R. (2004). The hydrogenase-like Nar1p is essential for maturation of cytosolic and nuclear iron-sulphur proteins. *The EMBO Journal*, *23*, 2105–2115. https://doi.org/10.1038/sj.emboj.7600216.

Banda, D. M., Nunez, N. N., Burnside, M. A., Bradshaw, K. M., & David, S. S. (2017). Repair of 8-oxoG:A mismatches by the MUTYH glycosylase: Mechanism, metals and medicine. *Free Radical Biology & Medicine*, *107*, 202–215. https://doi.org/10.1016/j.freeradbiomed.2017.01.008.

Beilschmidt, L. K., Ollagnier de Choudens, S., Fournier, M., Sanakis, I., Hograindleur, M. A., Clemancey, M., et al. (2017). ISCA1 is essential for mitochondrial Fe4S4 biogenesis in vivo. *Nature Communications*, *8*, 15124. https://doi.org/10.1038/ncomms15124.

Beilschmidt, L. K., & Puccio, H. M. (2014). Mammalian Fe-S cluster biogenesis and its implication in disease. *Biochimie*, *100*, 48–60. https://doi.org/10.1016/j.biochi.2014.01.009.

Beinert, H., Holm, R. H., & Münck, E. (1997). Iron-sulfur clusters: Nature's modular, multipurpose structures. *Science*, *277*, 653–659.

Biederbick, A., Stehling, O., Rösser, R., Niggemeyer, B., Nakai, Y., Elsässer, H. P., et al. (2006). Role of human mitochondrial Nfs1 in cytosolic iron-sulfur protein biogenesis and iron regulation. *Molecular and Cellular Biology*, *26*, 5675–5687. https://doi.org/10.1128/MCB.00112-06.

Boniecki, M. T., Freibert, S. A., Mühlenhoff, U., Lill, R., & Cygler, M. (2017). Structure and functional dynamics of the mitochondrial Fe/S cluster synthesis complex. *Nature Communications*, *8*, 1287.

Braymer, J. J., & Lill, R. (2017). Iron-sulfur cluster biogenesis and trafficking in mitochondria. *The Journal of Biological Chemistry*, *292*, 12754–12763. https://doi.org/10.1074/jbc.R117.787101.

Camaschella, C., Campanella, A., De Falco, L., Boschetto, L., Merlini, R., Silvestri, L., et al. (2007). The human counterpart of zebrafish shiraz shows sideroblastic-like microcytic anemia and iron overload. *Blood*, *110*, 1353–1358.

Cameron, J. M., Janer, A., Levandovskiy, V., Mackay, N., Rouault, T. A., Tong, W. H., et al. (2011). Mutations in iron-sulfur cluster scaffold genes NFU1 and BOLA3 cause a fatal deficiency of multiple respiratory chain and 2-oxoacid dehydrogenase enzymes. *American Journal of Human Genetics*, *89*, 486–495. https://doi.org/10.1016/j.ajhg.2011.08.011.

Cavadini, P., Biasiotto, G., Poli, M., Levi, S., Verardi, R., Zanella, I., et al. (2007). RNA silencing of the mitochondrial ABCB7 transporter in HeLa cells causes an iron-deficient phenotype with mitochondrial iron overload. *Blood*, *109*, 3552–3559. https://doi.org/10.1182/blood-2006-08-041632.

Cossee, M., Puccio, H., Gansmüller, A., Koutnikova, H., Dierich, A., LeMeur, M., et al. (2000). Inactivation of the Friedreich ataxia mouse gene leads to early embryonic lethality without iron accumulation. *Human Molecular Genetics*, *9*, 1219–1226.

Crooks, D. R., Ghosh, M. C., Haller, R. G., Tong, W. H., & Rouault, T. A. (2010). Post-translational stability of the heme biosynthetic enzyme ferrochelatase is dependent on iron availability and intact iron-sulfur cluster assembly machinery. *Blood*, *115*, 860–869. https://doi.org/10.1182/blood-2009-09-243105.

Drapier, J. C., & Hibbs, J. B., Jr. (1996). Aconitases: A class of metalloproteins highly sensitive to nitric oxide synthesis. *Methods in Enzymology*, *269*, 26–36.

Fernandez-Vizarra, E., Enriquez, J. A., Perez-Martos, A., Montoya, J., & Fernandez-Silva, P. (2011). Tissue-specific differences in mitochondrial activity and biogenesis. *Mitochondrion*, *11*, 207–213. https://doi.org/10.1016/j.mito.2010.09.011.

Fiedorczuk, K., Letts, J. A., Degliesposti, G., Kaszuba, K., Skehel, M., & Sazanov, L. A. (2016). Atomic structure of the entire mammalian mitochondrial complex I. *Nature*, *538*, 406–410. https://doi.org/10.1038/nature19794.

Gari, K., Leon Ortiz, A. M., Borel, V., Flynn, H., Skehel, J. M., & Boulton, S. J. (2012). MMS19 links cytoplasmic iron-sulfur cluster assembly to DNA metabolism. *Science*, *337*, 243–245. https://doi.org/10.1126/science.1219664.

Gelling, C., Dawes, I. W., Richhardt, N., Lill, R., & Mühlenhoff, U. (2008). Mitochondrial Iba57p is required for Fe/S cluster formation on aconitase and activation of radical SAM enzymes. *Molecular and Cellular Biology*, *28*, 1851–1861. https://doi.org/10.1128/MCB.01963-07.

Grandoni, J. A., Switzer, R. L., Makaroff, C. A., & Zalkin, H. (1989). Evidence that the iron-sulfur cluster of Bacillus subtilis glutamine phosphoribosylpyrophosphate amidotransferase determines stability of the enzyme to degradation in vivo. *The Journal of Biological Chemistry*, *264*, 6058–6064.

Guillon, B., Bulteau, A. L., Wattenhofer-Donze, M., Schmucker, S., Friguet, B., Puccio, H., et al. (2009). Frataxin deficiency causes upregulation of mitochondrial Lon and ClpP proteases and severe loss of mitochondrial Fe-S proteins. *The FEBS Journal*, *276*, 1036–1047. https://doi.org/10.1111/j.1742-4658.2008.06847.x.

Harlow, E., & Lane, D. (2006). Lysing tissue-culture cells for immunoprecipitation. *Cold Spring Harbor Protocols 2006*, https://doi.org/10.1101/pdb.prot4531.

Hom, J., & Sheu, S. S. (2009). Morphological dynamics of mitochondria—A special emphasis on cardiac muscle cells. *Journal of Molecular and Cellular Cardiology*, *46*, 811–820. https://doi.org/10.1016/j.yjmcc.2009.02.023.

Johnson, N. B., Deck, K. M., Nizzi, C. P., & Eisenstein, R. S. (2017). A synergistic role of IRP1 and FBXL5 in coordinating iron metabolism during cell proliferation. *The Journal of Biological Chemistry*, *292*, 15976–15989. https://doi.org/10.1074/jbc.M117.785741.

Kispal, G., Csere, P., Prohl, C., & Lill, R. (1999). The mitochondrial proteins Atm1p and Nfs1p are essential for biogenesis of cytosolic Fe/S proteins. *The EMBO Journal*, *18*, 3981–3989. https://doi.org/10.1093/emboj/18.14.3981.

Kühn, L. C. (2015). Iron regulatory proteins and their role in controlling iron metabolism. *Metallomics*, 7, 232–243. https://doi.org/10.1039/c4mt00164h.

Kurosaki, M., Demontis, S., Barzago, M. M., Garattini, E., & Terao, M. (1999). Molecular cloning of the cDNA coding for mouse aldehyde oxidase: Tissue distribution and regulation in vivo by testosterone. *The Biochemical Journal*, *341*(Pt. 1), 71–80.

Lange, H., Kaut, A., Kispal, G., & Lill, R. (2000). A mitochondrial ferredoxin is essential for biogenesis of cellular iron-sulfur proteins. *Proceedings of the National Academy of Sciences of the United States of America*, *97*, 1050–1055.

Leal, S. S., & Gomes, C. M. (2008). On the relative contribution of ionic interactions over iron-sulfur clusters to ferredoxin stability. *Biochimica et Biophysica Acta*, *1784*, 1596–1600. https://doi.org/10.1016/j.bbapap.2008.05.001.

Lehmann, A. R. (2003). DNA repair-deficient diseases, xeroderma pigmentosum, Cockayne syndrome and trichothiodystrophy. *Biochimie*, *85*, 1101–1111.

Lill, R. (2009). Function and biogenesis iron-sulphur proteins. *Nature*, *460*, 831–838.

Lill, R., Dutkiewicz, R., Freibert, S. A., Heidenreich, T., Mascarenhas, J., Netz, D. J., et al. (2015). The role of mitochondria and the CIA machinery in the maturation of cytosolic and nuclear iron-sulfur proteins. *European Journal of Cell Biology*, *94*, 280–291. https://doi.org/10.1016/j.ejcb.2015.05.002.

Lill, R., Fekete, Z., Sipos, K., & Rotte, C. (2005). Is there an answer? Why are mitochondria essential for life? *IUBMB Life*, *57*, 701–703. https://doi.org/10.1080/15216540500305860.

Lill, R., & Mühlenhoff, U. (2005). Iron-sulfur-protein biogenesis in eukaryotes. *Trends in Biochemical Sciences*, *30*, 133–141. https://doi.org/10.1016/j.tibs.2005.01.006.

Lill, R., Srinivasan, V., & Mühlenhoff, U. (2014). The role of mitochondria in cytosolic-nuclear iron-sulfur protein biogenesis and in cellular iron regulation. *Current Opinion in Microbiology*, *22*, 111–119. https://doi.org/10.1016/j.mib.2014.09.015.

Lukes, J., & Basu, S. (2015). Fe/S protein biogenesis in trypanosomes—A review. *Biochimica et Biophysica Acta*, *1853*, 1481–1492. https://doi.org/10.1016/j.bbamcr.2014.08.015.

Martelli, A., Wattenhofer-Donze, M., Schmucker, S., Bouvet, S., Reutenauer, L., & Puccio, H. (2007). Frataxin is essential for extramitochondrial Fe-S cluster proteins in mammalian tissues. *Human Molecular Genetics*, *16*, 2651–2658.

Mayr, J. A., Feichtinger, R. G., Tort, F., Ribes, A., & Sperl, W. (2014). Lipoic acid biosynthesis defects. *Journal of Inherited Metabolic Disease*, *37*, 553–563. https://doi.org/10.1007/s10545-014-9705-8.

McLaughlin, M. I., Lanz, N. D., Goldman, P. J., Lee, K. H., Booker, S. J., & Drennan, C. L. (2016). Crystallographic snapshots of sulfur insertion by lipoyl synthase. *Proceedings of the National Academy of Sciences of the United States of America*, *113*, 9446–9450. https://doi.org/10.1073/pnas.1602486113.

Mochel, F., Knight, M. A., Tong, W. H., Hernandez, D., Ayyad, K., Taivassalo, T., et al. (2008). Splice mutation in the iron-sulfur cluster scaffold protein ISCU causes myopathy with exercise intolerance. *American Journal of Human Genetics*, *82*, 652–660. https://doi.org/10.1016/j.ajhg.2007.12.012.

Molik, S., Lill, R., & Mühlenhoff, U. (2007). Methods for studying iron metabolism in yeast mitochondria. *Methods in Cell Biology*, *80*, 261–280.

Mühlenhoff, U., Gerber, J., Richhardt, N., & Lill, R. (2003). Components involved in assembly and dislocation of iron-sulfur clusters on the scaffold protein Isu1p. *The EMBO Journal*, *22*, 4815–4825.

Mühlenhoff, U., Hoffmann, B., Richter, N., Rietzschel, N., Spantgar, F., Stehling, O., et al. (2015). Compartmentalization of iron between mitochondria and the cytosol and its regulation. *European Journal of Cell Biology*, *94*, 292–308. https://doi.org/10.1016/j.ejcb.2015.05.003.

Mühlenhoff, U., Richter, N., Pines, O., Pierik, A. J., & Lill, R. (2011). Specialized function of yeast Isa1 and Isa2 proteins in the maturation of mitochondrial [4Fe-4S] proteins. *The Journal of Biological Chemistry*, *286*, 41205–41216. https://doi.org/10.1074/jbc.M111.296152.

Navarro-Sastre, A., Tort, F., Stehling, O., Uzarska, M. A., Arranz, J. A., Del Toro, M., et al. (2011). A fatal mitochondrial disease is associated with defective NFU1 function in the maturation of a subset of mitochondrial Fe-S proteins. *American Journal of Human Genetics*, *89*, 656–667. https://doi.org/10.1016/j.ajhg.2011.10.005.

Netz, D. J., Pierik, A. J., Stümpfig, M., Mühlenhoff, U., & Lill, R. (2007). The Cfd1-Nbp35 complex acts as a scaffold for iron-sulfur protein assembly in the yeast cytosol. *Nature Chemical Biology*, *3*, 278–286. https://doi.org/10.1038/nchembio872.

Netz, D. J., Stümpfig, M., Dore, C., Mühlenhoff, U., Pierik, A. J., & Lill, R. (2010). Tah18 transfers electrons to Dre2 in cytosolic iron-sulfur protein biogenesis. *Nature Chemical Biology*, *6*, 758–765. https://doi.org/10.1038/nchembio.432.

Nordin, A., Larsson, E., & Holmberg, M. (2012). The defective splicing caused by the ISCU intron mutation in patients with myopathy with lactic acidosis is repressed by PTBP1 but can be derepressed by IGF2BP1. *Human Mutation, 33*, 467–470. https://doi.org/10.1002/humu.22002.

Olsson, A., Lind, L., Thornell, L. E., & Holmberg, M. (2008). Myopathy with lactic acidosis is linked to chromosome 12q23.3-24.11 and caused by an intron mutation in the ISCU gene resulting in a splicing defect. *Human Molecular Genetics, 17*, 1666–1672. https://doi.org/10.1093/hmg/ddn057.

Paul, V. D., & Lill, R. (2015). Biogenesis of cytosolic and nuclear iron-sulfur proteins and their role in genome stability. *Biochimica et Biophysica Acta, 1853*, 1528–1539. https://doi.org/10.1016/j.bbamcr.2014.12.018.

Paul, V. D., Mühlenhoff, U., Stümpfig, M., Seebacher, J., Kugler, K. G., Renicke, C., et al. (2015). The deca-GX3 proteins Yae1-Lto1 function as adaptors recruiting the ABC protein Rli1 for iron-sulfur cluster insertion. *eLife, 4*, e08231. https://doi.org/10.7554/eLife.08231.

Pierik, A. J., Netz, D. J., & Lill, R. (2009). Analysis of iron-sulfur protein maturation in eukaryotes. *Nature Protocols, 4*, 753–766. https://doi.org/10.1038/nprot.2009.39.

Pondarre, C., Antiochos, B. B., Campagna, D. R., Clarke, S. L., Greer, E. L., Deck, K. M., et al. (2006). The mitochondrial ATP-binding cassette transporter Abcb7 is essential in mice and participates in cytosolic iron-sulfur cluster biogenesis. *Human Molecular Genetics, 15*, 953–964. https://doi.org/10.1093/hmg/ddl012.

Ponka, P., & Lok, C. N. (1999). The transferrin receptor: Role in health and disease. *The International Journal of Biochemistry & Cell Biology, 31*, 1111–1137.

Ponka, P., & Schulman, H. M. (1985). Acquisition of iron from transferrin regulates reticulocyte heme synthesis. *The Journal of Biological Chemistry, 260*, 14717–14721.

Py, B., & Barras, F. (2010). Building Fe-S proteins: Bacterial strategies. *Nature Reviews Microbiology, 8*, 436–446.

Reichmann, H., Hoppeler, H., Mathieu-Costello, O., von Bergen, F., & Pette, D. (1985). Biochemical and ultrastructural changes of skeletal muscle mitochondria after chronic electrical stimulation in rabbits. *Pflügers Archiv, 404*, 1–9.

Richardson, D. R., & Ponka, P. (1997). The molecular mechanisms of the metabolism and transport of iron in normal and neoplastic cells. *Biochimica et Biophysica Acta, 1331*, 1–40.

Rouault, T. A. (2012). Biogenesis of iron-sulfur clusters in mammalian cells: New insights and relevance to human disease. *Disease Models & Mechanisms, 5*, 155–164. https://doi.org/10.1242/dmm.009019.

Ruiz, J. C., & Bruick, R. K. (2014). F-box and leucine-rich repeat protein 5 (FBXL5): Sensing intracellular iron and oxygen. *Journal of Inorganic Biochemistry, 133*, 73–77. https://doi.org/10.1016/j.jinorgbio.2014.01.015.

Rustin, P., Chretien, D., Bourgeron, T., Gerard, B., Rötig, A., Saudubray, J. M., et al. (1994). Biochemical and molecular investigations in respiratory chain deficiencies. *Clinica Chimica Acta, 228*, 35–51.

Schmucker, S., & Puccio, H. (2010). Understanding the molecular mechanisms of Friedreich's ataxia to develop therapeutic approaches. *Human Molecular Genetics, 19*, R103–110. https://doi.org/10.1093/hmg/ddq165.

Schnackerz, K. D., Dobritzsch, D., Lindqvist, Y., & Cook, P. F. (2004). Dihydropyrimidine dehydrogenase: A flavoprotein with four iron-sulfur clusters. *Biochimica et Biophysica Acta, 1701*, 61–74.

Sheftel, A. D., & Lill, R. (2009). The power plant of the cell is also a smithy: The emerging role of mitochondria in cellular iron homeostasis. *Annals of Medicine, 41*, 82–99. https://doi.org/10.1080/07853890802322229.

Sheftel, A. D., Richardson, D. R., Prchal, J., & Ponka, P. (2009). Mitochondrial iron metabolism and sideroblastic anemia. *Acta Haematologica, 122*, 120–133. https://doi.org/10.1159/000243796.

Sheftel, A. D., Stehling, O., Pierik, A. J., Elsässer, H. P., Mühlenhoff, U., Webert, H., et al. (2010). Humans possess two mitochondrial ferredoxins, Fdx1 and Fdx2, with distinct roles in steroidogenesis, heme, and Fe/S cluster biosynthesis. *Proceedings of the National Academy of Sciences of the United States of America*, *107*, 11775–11780. https://doi.org/10.1073/pnas.1004250107.

Sheftel, A. D., Stehling, O., Pierik, A. J., Netz, D. J., Kerscher, S., Elsässer, H. P., et al. (2009). Human ind1, an iron-sulfur cluster assembly factor for respiratory complex I. *Molecular and Cellular Biology*, *29*, 6059–6073. https://doi.org/10.1128/MCB.00817-09.

Sheftel, A. D., Wilbrecht, C., Stehling, O., Niggemeyer, B., Elsässer, H. P., Mühlenhoff, U., et al. (2012). The human mitochondrial ISCA1, ISCA2, and IBA57 proteins are required for [4Fe-4S] protein maturation. *Molecular Biology of the Cell*, *23*, 1157–1166. https://doi.org/10.1091/mbc.E11-09-0772.

Shi, Y., Ghosh, M., Kovtunovych, G., Crooks, D. R., & Rouault, T. A. (2012). Both human ferredoxins 1 and 2 and ferredoxin reductase are important for iron-sulfur cluster biogenesis. *Biochimica et Biophysica Acta*, *1823*, 484–492. https://doi.org/10.1016/j.bbamcr.2011.11.002.

Shukla, A., Hebbar, M., Srivastava, A., Kadavigere, R., Upadhyai, P., Kanthi, A., et al. (2017). Homozygous p.(Glu87Lys) variant in ISCA1 is associated with a multiple mitochondrial dysfunctions syndrome. *Journal of Human Genetics*, *62*, 723–727. https://doi.org/10.1038/jhg.2017.35.

Sipos, K., Lange, H., Fekete, Z., Ullmann, P., Lill, R., & Kispal, G. (2002). Maturation of cytosolic iron-sulfur proteins requires glutathione. *The Journal of Biological Chemistry*, *277*, 26944–26949. https://doi.org/10.1074/jbc.M200677200.

Song, D., & Lee, F. S. (2008). A role for iron-only hydrogenase like protein 1 in mammalian cytosolic iron-sulfur protein biogenesis. *The Journal of Biological Chemistry*, *283*, 9231–9238.

Stehling, O., & Lill, R. (2013). The role of mitochondria in cellular iron-sulfur protein biogenesis: Mechanisms, connected processes, and diseases. *Cold Spring Harbor Perspectives in Biology*, *5*, a011312. https://doi.org/10.1101/cshperspect.a011312.

Stehling, O., Mascarenhas, J., Vashisht, A. A., Sheftel, A. D., Niggemeyer, B., Rösser, R., et al. (2013). Human CIA2A-FAM96A and CIA2B-FAM96B integrate iron homeostasis and maturation of different subsets of cytosolic-nuclear iron-sulfur proteins. *Cell Metabolism*, *18*, 187–198. https://doi.org/10.1016/j.cmet.2013.06.015.

Stehling, O., Netz, D. J., Niggemeyer, B., Rösser, R., Eisenstein, R. S., Puccio, H., et al. (2008). Human Nbp35 is essential for both cytosolic iron-sulfur protein assembly and iron homeostasis. *Molecular and Cellular Biology*, *28*, 5517–5528. https://doi.org/10.1128/MCB.00545-08.

Stehling, O., Sheftel, A. D., & Lill, R. (2009). Controlled expression of iron-sulfur cluster assembly components for respiratory chain complexes in mammalian cells. *Methods in Enzymology*, *456*, 209–231. Chapter 12. https://doi.org/10.1016/s0076-6879(08)04412-1.

Stehling, O., Smith, P. M., Biederbick, A., Balk, J., Lill, R., & Mühlenhoff, U. (2007). Investigation of iron-sulfur protein maturation in eukaryotes. *Methods in Molecular Biology*, *372*, 325–342. https://doi.org/10.1007/978-1-59745-365-3_24.

Stehling, O., Vashisht, A. A., Mascarenhas, J., Jonsson, Z. O., Sharma, T., Netz, D. J., et al. (2012). MMS19 assembles iron-sulfur proteins required for DNA metabolism and genomic integrity. *Science*, *337*, 195–199. https://doi.org/10.1126/science.1219723.

Stehling, O., Wilbrecht, C., & Lill, R. (2014). Mitochondrial iron-sulfur protein biogenesis and human disease. *Biochimie*, *100*, 61–77. https://doi.org/10.1016/j.biochi.2014.01.010.

Sun, F., Huo, X., Zhai, Y., Wang, A., Xu, J., Su, D., et al. (2005). Crystal structure of mitochondrial respiratory membrane protein complex II. *Cell*, *121*, 1043–1057. https://doi.org/10.1016/j.cell.2005.05.025.

Tong, W. H., & Rouault, T. A. (2006). Functions of mitochondrial ISCU and cytosolic ISCU in mammalian iron-sulfur cluster biogenesis and iron homeostasis. *Cell Metabolism*, *3*, 199–210. https://doi.org/10.1016/j.cmet.2006.02.003.

Trounce, I. A., Kim, Y. L., Jun, A. S., & Wallace, D. C. (1996). Assessment of mitochondrial oxidative phosphorylation in patient muscle biopsies, lymphoblasts, and trans-mitochondrial cell lines. *Methods in Enzymology*, *264*, 484–509.

Tsai, C. L., & Barondeau, D. P. (2010). Human frataxin is an allosteric switch that activates the Fe-S cluster biosynthetic complex. *Biochemistry*, *49*, 9132–9139. https://doi.org/10.1021/bi1013062.

Upadhyay, A. S., Stehling, O., Panayiotou, C., Rösser, R., Lill, R., & Överby, A. K. (2017). Cellular requirements for iron-sulfur cluster insertion into the antiviral radical SAM protein viperin. *The Journal of Biological Chemistry*, *292*, 13879–13889. https://doi.org/10.1074/jbc.M117.780122.

Upadhyay, A. S., Vonderstein, K., Pichlmair, A., Stehling, O., Bennett, K. L., Dobler, G., et al. (2014). Viperin is an iron-sulfur protein that inhibits genome synthesis of tick-borne encephalitis virus via radical SAM domain activity. *Cellular Microbiology*, *16*, 834–848. https://doi.org/10.1111/cmi.12241.

Uzarska, M. A., Dutkiewicz, R., Freibert, S. A., Lill, R., & Mühlenhoff, U. (2013). The mitochondrial Hsp70 chaperone Ssq1 facilitates Fe/S cluster transfer from Isu1 to Grx5 by complex formation. *Molecular Biology of the Cell*, *24*, 1830–1841. https://doi.org/10.1091/mbc.E12-09-0644.

Uzarska, M. A., Nasta, V., Weiler, B. D., Spantgar, F., Ciofi-Baffoni, S., Saviello, M. R., et al. (2016). Mitochondrial Bol1 and Bol3 function as assembly factors for specific iron-sulfur proteins. *eLife*, *5*, pii: e16673. https://doi.org/10.7554/eLife.16673.

Van Vranken, J. G., Jeong, M. Y., Wei, P., Chen, Y. C., Gygi, S. P., Winge, D. R., et al. (2016). The mitochondrial acyl carrier protein (ACP) coordinates mitochondrial fatty acid synthesis with iron sulfur cluster biogenesis. *eLife*, *5*, pii: e17828. https://doi.org/10.7554/eLife.17828.

Walden, W. E., Selezneva, A. I., Dupuy, J., Volbeda, A., Fontecilla-Camps, J. C., Theil, E. C., et al. (2006). Structure of dual function iron regulatory protein 1 complexed with ferritin IRE-RNA. *Science*, *314*, 1903–1908. https://doi.org/10.1126/science.1133116.

Watmough, N. J., & Frerman, F. E. (2010). The electron transfer flavoprotein: Ubiquinone oxidoreductases. *Biochimica et Biophysica Acta*, *1797*, 1910–1916. https://doi.org/10.1016/j.bbabio.2010.10.007.

Zhou, G., Broyles, S. S., Dixon, J. E., & Zalkin, H. (1992). Avian glutamine phosphoribosylpyrophosphate amidotransferase propeptide processing and activity are dependent upon essential cysteine residues. *The Journal of Biological Chemistry*, *267*, 7936–7942.

CHAPTER NINE

Steps Toward Understanding Mitochondrial Fe/S Cluster Biogenesis

Andrew Melber, Dennis R. Winge[1]
University of Utah Health Sciences Center, Salt Lake City, Utah, United States
[1]Corresponding author: e-mail address: dennis.winge@hsc.utah.edu

Contents

Abstract

Iron–sulfur clusters (Fe/S clusters) are essential cofactors required throughout the clades of biology for performing a myriad of unique functions including nitrogen fixation, ribosome assembly, DNA repair, mitochondrial respiration, and metabolite catabolism. Although Fe/S clusters can be synthesized in vitro and transferred to a client protein without enzymatic assistance, biology has evolved intricate mechanisms to assemble and transfer Fe/S clusters within the cellular environment. In eukaryotes, the foundation of all cellular clusters starts within the mitochondria. The focus of this review is to detail the mitochondrial Fe/S biogenesis (ISC) pathway along with the Fe/S cluster transfer steps necessary to mature Fe/S proteins. New advances in our understanding of the mitochondrial Fe/S biogenesis machinery will be highlighted. Additionally, we will address various experimental approaches that have been successful in the identification and characterization of components of the ISC pathway.

Methods in Enzymology, Volume 599
ISSN 0076-6879
https://doi.org/10.1016/bs.mie.2017.09.004

1. INTRODUCTION

Iron–sulfur (Fe/S) clusters are redox-active cofactors that are essential for almost all life forms. Since Fe/S cluster proteins function in processes related to the central dogma of biology, Fe/S cluster availability is essential to cell viability. The breadth of metabolic pathways that utilize Fe/S cluster proteins/enzymes further highlights their importance and persistence in biology. Some novel uses of Fe/S clusters in biology include the following. Within the nucleus, Fe/S clusters bound to DNA repair enzymes can sense DNA damage (disruption of the nucleotide hydrogen bond pairing) by assessing the ability of an electron from a Fe/S cluster to diffuse along DNA (Arnold, Grodick, & Barton, 2016). Sensing of the important cellular nutrient, iron, occurs by a bifunctional protein in metazoans in which Fe-replete cells generate a Fe/S cluster-containing protein conformer with an aconitase catalytic function. However, this IRP1 Fe sensor in Fe-deficient cells is unpopulated with a Fe/S cluster and the apo-IRP1 is an active RNA-binding protein in which IRP1-RNA binding can either attenuate mRNA expression or stabilize a mRNA transcript (Anderson, Shen, Eisenstein, & Leibold, 2012; Kuhn, 2015). Within the mitochondria, the OXPHOS respiratory complexes I and II utilize multiple Fe/S clusters as an electron transfer wire for the reduction of ubiquinone by NADH or FADH, respectively. Clusters can perform a catalytic role in an array for nucleotide and amino acid biogenesis pathways throughout the cell. Radical *S*-adenosyl-L-methionine (SAM) protein lipoic acid synthase (LIAS) contains one Fe/S cluster that is consumed during its catalytic turnover to use the sulfide ions to generate lipoic acid (Cicchillo et al., 2004; Cronan, 2014; McLaughlin et al., 2016). The focus of this review will be on the mitochondrial iron–sulfur cluster biogenesis pathway (ISC) that generates [2Fe—2S], [4Fe—4S], and [3Fe—4S] clusters.

1.1 Mitochondrial Iron–Sulfur Clusters

Redox-active cofactors have the ability to donate or accept an electron to facilitate a unique set of chemistries that would otherwise be challenging for biology to perform (Beinert, 2000). Fe/S clusters of three nuclearities exist within the mitochondria. Proteins with [4Fe—4S] clusters often have four cysteinyl terminal ligands for the cluster in addition to the four bridging sulfide ions coordinating iron ions. One enzyme succinate dehydrogenase (SDH) contains a [3Fe—4S] cluster and this center lacks a fourth ligating cysteine residue. Replacing a cysteine at that sequence position in the bacterial SDH-related enzyme fumarate reductase converts that center into a [4Fe—4S] cluster (Hagerhall, Sled, Hederstedt, & Ohnishi, 1995; Manodori et al., 1992). This may imply that this center may be assembled as a [4Fe—4S] cofactor, but the fourth Fe lacking the terminal cysteine is unstable. Finally, multiple [2Fe—2S] centers exist within the mitochondrion.

Clusters with varying nuclearity also exhibit a wide spectrum of redox potentials ranging from −700 to +300 mV enabling diverse electron transfer steps (Johnson, Dean, Smith, & Johnson, 2005; Nanda et al., 2016). Two stable oxidation states often +2 or 1+ exist permitting one electron acceptance or donation (Johnson et al., 2005). Given this redox flexibility of Fe/S clusters, it is not surprising that they, along with heme and copper atoms, are found in abundance within the four OXPHOS respiratory chain complexes to transfer electrons through each complex. Respiratory complex I, NADH: ubiquinone oxidoreductase, contains eight Fe/S clusters, seven of which execute the transfer of electrons (Fiedorczuk et al., 2016; Gnandt, Dorner, Strampraad, de Vries, & Friedrich, 2016; Zhu, Vinothkumar, & Hirst, 2016). It is worth noting that the commonly used model organism *Saccharomyces cerevisiae* does not have a membrane-bound complex I; instead yeast have small soluble NADH:ubiquinone oxidoreductases that do not pump protons and lack Fe/S redox cofactors (Iwata et al., 2012). Complex II, SDH, contains three Fe/S clusters, a [2Fe—2S], a [4Fe—4S], and a [3Fe—4S] cluster, all within the Sdh2 (SdhB) subunit that transfer electrons to ubiquinone (Sun et al., 2005). Cytochrome bc_1, complex III, contains just a single [2Fe—2S] cluster that functions in concert with two heme cofactors to mediate electrons to cytochrome *c*1 for the reduction of diffusible cytochrome *c* (Iwata et al., 1998). Lastly, complex IV, cytochrome oxidase, although it lacks a Fe/S cluster for electron transfer (Tsukihara et al., 1995), requires the Fe/S protein ferredoxin to form the modified heme *a* moiety required for the reduction of molecular oxygen (Barros, Carlson, Glerum, & Tzagoloff, 2001; Barros & Nobrega, 1999; Barros, Nobrega, & Tzagoloff, 2002). Likewise, the diffusible electron carrier ubiquinone requires the function of the Fe/S cluster ferredoxin for its synthesis (Pierrel et al., 2010).

Fe/S cofactors are not only utilized by the electron transport chain in the mitochondria, many Fe/S clusters act in a catalytic capacity instead.

As mentioned, the SAM radical enzyme lipoic acid synthase (LIAS, Lip5 in *S. cerevisiae*) contains two [4Fe—4S] clusters, one of which is consumed during the catalytic cycle to liberate two sulfides used for the generation of a lipoate moiety generated on octanoyl-Acp1 (Cicchillo et al., 2004; Cronan, 2014; McLaughlin et al., 2016). Another mitochondrial radical SAM protein is biotin synthase, Bio2, which requires a [2Fe—2S] cluster that is consumed to provide the sulfurs to biotin and a stable [4Fe—4S] cluster (Berkovitch, Nicolet, Wan, Jarrett, & Drennan, 2004; Tse Sum Bui et al., 2003). A series of mitochondrial hydratases bind [4Fe—4S] clusters with only three thiolate ligands so that the remaining Fe(II) is available to bind substrates. The TCA cycle enzyme aconitase binds a [4Fe—4S] cluster with three thiolate ligands such that the open Fe(II) can coordinate the citrate substrate for isomerization (Robbins & Stout, 1989). In fungi, the [4Fe—4S] hydratases Aco2 and Lys4 preform steps in lysine biosynthesis with a similar substrate-binding Fe cluster ion mechanism as aconitase (Fazius, Shelest, Gebhardt, & Brock, 2012). Glutamate biosynthesis requires a Fe/S cluster, but it is unresolved whether the yeast glutamate synthase uses a [3Fe—4S] or a [4Fe—4S] cluster (Knaff, Hirasawa, Ameyibor, Fu, & Johnson, 1991; Miller & Stadtman, 1972). Mitochondrial β-oxidation of long-chain fatty acids involves the inner membrane electron transfer flavoprotein-ubiquinone oxidoreductase (ETF-QO) that mediates electron transfer from a series of FAD-containing acyl-CoA dehydrogenases to ubiquinone. The metazoan ETF-QO contains a [4Fe—4S] cluster in addition to FAD (Watmough & Frerman, 2010).

Proteins utilizing [2Fe—2S] clusters within the mitochondria include not only the mentioned SDH but also dihydroxyacid dehydratase, Ilv3, which uses its [2Fe—2S] cluster in the synthesis of branch chained amino acids (Muhlenhoff, Richter, Pines, Pierik, & Lill, 2011). In addition, the mitochondrial ferredoxin (Yah1 in yeast) requires a [2Fe—2S] cluster for its function in Fe/S, heme *a*, and coenzyme Q biosynthesis together with the NADH-Arh1 oxidoreductase (Barros et al., 2001; Manzella, Barros, & Nobrega, 1998; Pierrel et al., 2010). Ferredoxin donates electrons to the heme *a* biosynthesis pathways through Cox15, coenzyme Q biogenesis via Coq6, and Fe/S cluster biogenesis via the scaffold ISU complex, which will be discussed in the next section (Barros et al., 2002; Lange, Kaut, Kispal, & Lill, 2000; Li, Saxena, Pain, & Dancis, 2001; Pierrel et al., 2010). Given the array of pathways that ferredoxin is the functional reductase, dysfunctional [2Fe—2S] cluster biogenesis cause exacerbated defects in other mitochondrial biosynthesis pathways.

2. MITOCHONDRIAL IRON–SULFUR CLUSTER BIOGENESIS

Three bacterial pathways have independently arisen to generate Fe/S clusters for unique needs of the cell. These pathways are the ISC, SUF, and NIF systems. These systems all function by the same general mechanism: cysteine desulfurases produce sulfide ions, scaffold complexes assemble a de novo cluster, and a transfer system that delivers the clusters to client proteins (Roche et al., 2013). Eukaryotic mitochondria have adopted the ISC pathway from bacteria and integrated components of the NIF pathway for a single dedicated assembly pathway. Plant chloroplasts and the *Plasmodium* apicoplast retained the SUF pathway of Fe/S cluster biogenesis (Dellibovi-Ragheb, Gisselberg, & Prigge, 2013; Gisselberg, Dellibovi-Ragheb, Matthews, Bosch, & Prigge, 2013).

The mitochondrial Fe/S cluster biogenesis pathway (ISC) not only contributes functional clusters to the mitochondria but also provides an essential precursor to the cytosolic Fe/S assembly pathway (CIA) via the inner membrane Atm1 exporter (Kispal, Csere, Prohl, & Lill, 1999; Lill et al., 2015).

This review will discuss aspects of each component of the ISC pathway later, but first a general scheme of the pathway is illustrated in Fig. 1 and briefly described as follows. Within the ISU complex, iron and sulfide ions generated by a cysteine desulfurase enzyme are combined on a scaffold protein to form an initial [2Fe—2S] cluster. The preformed Fe/S cluster is then released from the scaffold by chaperones to a glutaredoxin that mediates delivery to [2Fe—2S] client proteins or to a second complex (ISA complex)

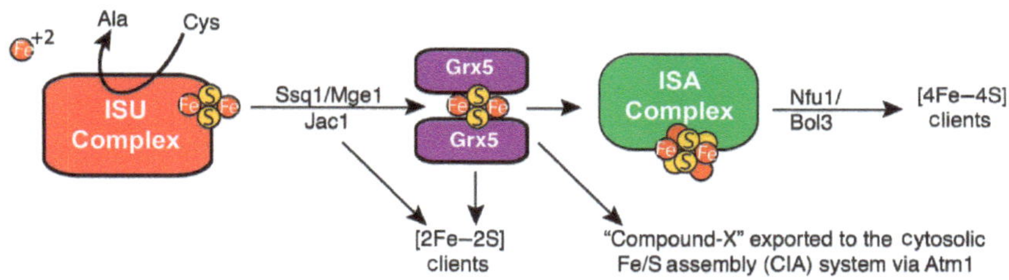

Fig. 1 General overview of mitochondrial Fe/S cluster biogenesis. All clusters in the cell require the initial biogenesis steps by the ISU complex that generates a [2Fe—2S] cluster from iron and cysteine before transferring it through Grx5 to the ISA complex. The ISA complex condenses two [2Fe–2S] clusters to create a single [4Fe—4S] cluster for mitochondrial client proteins. A summary of each component is described in Table 1 on the subsequent page.

Table 1 List of Mitochondrial Fe/S Assembly Components

Complex	Human	Yeast	Function
ISU complex	ISCUs	Isu1/ Isu2	Scaffold protein for the assembly of the initial [2Fe—2S] cluster
	NFS1	Nfs1	Cysteine desulfurase, provides a sulfide by reducing cysteine to alanine, requires a PLP cofactor
	ISD11	Isd11	Stabilizing binding partner of Nfs1 that activates persulfide formation
	ACPM	Acp1	Binds and stabilizes Isd11, potentially links Fe/S biogenesis to fatty acid biogenesis
	FDX2	Yah1	Ferredoxin, donates an electron to both Fe/S and heme biosynthesis pathways, contains a [2Fe—2S] cluster
	Frataxin	Yfh1	Stimulates [2Fe–2S] formation on ISCU by activating sulfur transfer from Nfs1 to ISCU
Cluster release complex	HSPA9	Ssq1	Hsp70 protein that binds Grx5 during the transfer of the [2Fe—2S] from ISU
	GRPE1	Mge1	Nucleotide exchange factor that cycles ATP on Ssq1 for cluster transfer
	HSC20	Jac1	DNAJ protein that facilitates transfer of the preformed [2Fe—2S] cluster from Isu1 to Grx5
	GLRX5	Grx5	Glutaredoxin that transfers performed [2Fe—2S] clusters to client proteins in the mitochondria
	BOLA1	Bol1	Assists Grx5 in [2Fe–2S] cluster delivery
	ABCB7	Atm1	Inner mitochondrial membrane transport that exports compound-X to the cytosolic Fe/S assembly pathway
ISA complex	ISCA1	Isa1	Binds a single [2Fe—2S] cluster to condense with the cluster from Isa2
	ISCA2	Isa2	Binds a single [2Fe—2S] cluster to condense with the cluster from Isa1
	IBA57	Iba57	Unknown function, but a component of the ISA complex required for [4Fe—4S] biogenesis
Late stage transfer	Nfu1	Nfu1	Transfers [4Fe–4S] clusters from the ISA complex to client proteins to protect clusters from oxidative damage
	BOLA3	Bol3	Assists Nfu1 in [4Fe–4S] transfer, exactly function is unknown

The table includes all known mitochondrial assembly proteins and their function.

in which two [2Fe—2S] clusters are condensed into a single [4Fe—4S] center, which is then transferred to mitochondrial [4Fe—4S] client proteins (Beilschmidt & Puccio, 2014; Stehling, Wilbrecht, & Lill, 2014). This pathway is dependent on ferrous iron, sulfur (from cysteine), the ferredoxin reductant, as well as NADH, ATP, and GTP (Amutha et al., 2008; Muhlenhoff, Richhardt, Gerber, & Lill, 2002; Pandey, Pain, Ghosh, Dancis, & Pain, 2015). A sulfur-containing adduct generated through the cysteine desulfurase, and the ISU scaffolding complex is extruded into the cytoplasm by the ABC transporter Atm1 (Kispal et al., 1999). The identity of this S-compound remains to be elucidated, although several compounds have been proposed (Schaedler et al., 2014; Srinivasan, Pierik, & Lill, 2014). The cytosolic iron–sulfur cluster biogenesis pathway (CIA) that utilizes this S-compound has been recently reviewed and will not be the focus of this review (Lill et al., 2015; Paul & Lill, 2015).

2.1 [2Fe—2S] Cluster Assembly by the ISU Complex

As mentioned, all cellular Fe/S clusters require the early steps of the ISC pathway within the mitochondria. This starts with the ISU complex. The ISU complex centers on the scaffolding protein Isu1 (a redundant scaffold protein exists in yeast, Isu2) on which the initial cluster is built. This ISU complex utilizes six different components. The scaffold receives ferrous iron from an unknown source, the sulfide comes from the cysteine desulfurase, Nfs1, and the electrons come from the Yah1 ferredoxin (Lange et al., 2000; Webert et al., 2014; Yan, Adinolfi, & Pastore, 2015). The other components that surround the scaffold include the Nfs1-stabilizing proteins Isd11 and Acp1 (Van Vranken et al., 2016; Wiedemann et al., 2006). Another protein important to Fe/S formation on the ISU scaffold protein is the Yfh1 frataxin, which also binds the Nfs1/Isd11 subcomplex (Schmucker et al., 2011; Shan, Napoli, & Cortopassi, 2007).

The ISU complex is dynamic in that different complexes form during cluster formation. This dynamic nature allows for different components to bind directly to the complex, such that the binding of ferredoxin to Nfs1 is displaced when Nfs1 associates with the scaffold or frataxin (Kim, Frederick, Reinen, Troupis, & Markley, 2013). The association of Yfh1 with Isu1 is mutually exclusive with Ssq1 binding due to overlapping binding site. Furthermore, the binding of Nfs1 to Isu1 is mutually exclusive with the binding ofJac1 due to the same interaction interface on Isu1 for Nfs1 and Jac1 (Majewska et al., 2013; Manicki et al., 2014). Nfs1 and Isd11 stably associate and Acp1 contributes to the stability of this subcomplex (Van Vranken et al., 2016). Isd11 is a member of the LYR protein family, and it is required for the structural stabilization and persulfide formation on Nfs1 (Adam, Bornhovd, Prokisch, Neupert, & Hell, 2006; Pandey, Golla, Yoon, Dancis, & Pain, 2012; Wiedemann et al., 2006). Isd11, as with all LYR proteins, is found associated with the acyl carrier protein Acp1 (Huttlin et al., 2015; Van Vranken et al., 2016). Acp1 functions in mitochondrial fatty acid synthesis yet has additional roles through its association with LYR proteins in the mitochondrial matrix. Fatty acid elongation occurs on a phosphopantetheine cofactor covalently bound to Acp1, and the interaction of Acp1 and two LYR proteins in complex I reveals the Acp1 acyl chain is embedded within the three-helical bundle of the LYR protein (Fiedorczuk et al., 2016; Wu, Gu, Guo, Huang, & Yang, 2016). Many of the other LYR proteins are assembly factors for OXPHOS respiratory complexes (Angerer et al., 2014; Atkinson et al., 2011; Floyd et al., 2016; Na et al., 2014). The association of Acp1 with the Nfs1/Isd11 subcomplex provides a potentially interesting regulatory mechanism that links the biogenesis of all cellular Fe/S clusters with the maturation of the respiratory complexes through fatty acid oxidation required to form the lipid that is covalently attached to Acp1.

Purification of the Nfs1/Isd11/Acp1 subcomplex reveals substoichiometric levels of Isu1, but this complex is not stable to electrophoresis on blue-native gel conditions (Van Vranken et al., 2016; Wiedemann et al., 2006). However, several groups have successfully purified the Yfh1/Isd11/Nfs1/Isu1 complex (Colin et al., 2013; Schmucker et al., 2011; Tsai & Barondeau, 2010) (Fig. 2A), although these groups did not evaluate whether Acp1 was also associated.

The Nfs1 cysteine desulfurase is a pyridoxal phosphate-dependent enzyme that liberates sulfane from cysteine releasing alanine and forming a persulfide intermediate on a conserved cysteine of Nfs1 (Kaiser et al., 2000; Mühlenhoff et al., 2004) (Fig. 2B). The persulfide sulfane is transferred to the scaffold Isu1 protein and reduction of the sulfane is required before its incorporation as sulfide ions with Fe(II) into the initial [2Fe—2S] cluster (Parent et al., 2015; Smith, Frazzon, Dean, & Johnson, 2005; Urbina, Silberg, Hoff, & Vickery, 2001) (Fig. 2C). This reduction is mediated by the mitochondrial ferredoxin, Yah1 (Cai, Tonelli, Frederick, & Markley, 2017; Lange et al., 2000; Webert et al., 2014) (Fig. 2A).

Transfer of the persulfide sulfur to the Isu1 scaffold appears to be mediated by frataxin (Bridwell-Rabb, Fox, Tsai, Winn, & Barondeau, 2014;

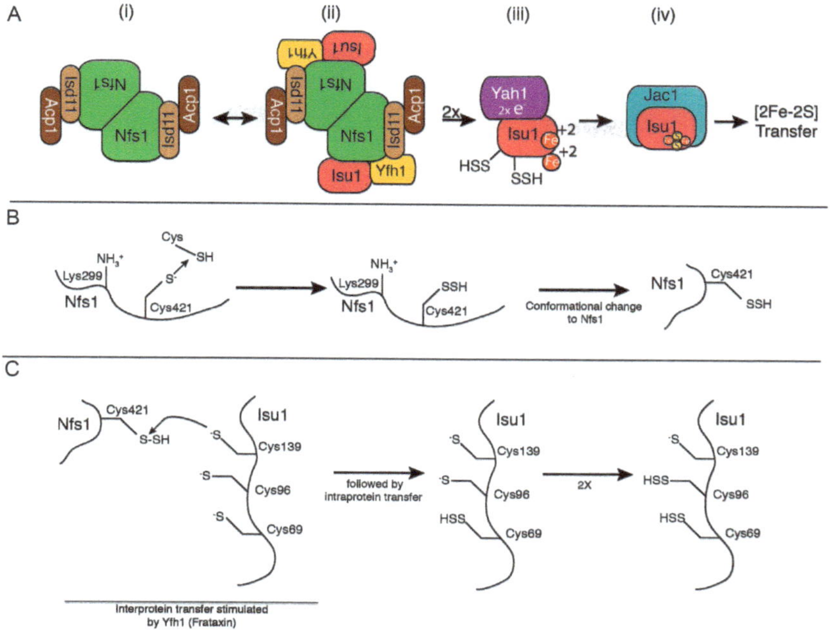

Fig. 2 Mitochondrial Fe/S cluster biogenesis. (A) Eukaryotic Nfs1 assembles with the LYR protein Isd11 and the acyl carrier protein Acp1. This heterotrimer exists in a complex with 2 of each subunit and is stable on BN-PAGE. The Isu1 scaffold protein associates to accept the persulfide ion after catalytic turnover of Nfs1 (shown in B). Frataxin Yfh1 appears to stimulate the transfer of the persulfide to Isu1 as shown in (C). The ferredoxin provides electrons likely for the reduction of the sulfane to sulfide ions for use as the bridging ligand. Whether ferredoxin associates with the entire Nfs1–Isu1 complex or not is to be resolved. Once the [2Fe—2S] cluster forms on the ISU scaffold, the binding of Jac1 initiates the dissociation of the complex and mediates cluster release.

Fox, Das, Chakrabarti, Lindahl, & Barondeau, 2015; Parent et al., 2015) (Fig. 2C). Frataxin was identified as the causative gene of Freidreich's Ataxia in patients, and the deletion of the homologue gene in yeast, *YFH1*, leads to excess mitochondrial iron accumulation (Babcock et al., 1997; Campuzano et al., 1996). Subsequently, frataxin was postulated to be the iron donor to the Isu1 scaffold in [2Fe—2S] biogenesis due of the protein's weak Fe(II)-binding ability (Adamec et al., 2000; Kondapalli, Kok, Dancis, & Stemmler, 2008; Yoon & Cowan, 2003). However, amino acid substitutions to the candidate carboxylates involved in Fe(II) binding led to only modest defects in Fe/S cluster formation (Correia, Wang, Craig, & Gomes, 2010; Foury, Pastore, & Trincal, 2007). A mutant allele of Isu1 was found to bypass the functional importance of Yfh1 in yeast (Yoon et al., 2015, 2014), questioning the Fe(II) donor function of Yfh1. Structural studies in bacterial systems revealed that the interaction of the bacterial frataxin, CyaY, with the ISU complex was independent of iron and oligomerization (a Fe-dependent phenomenon). The prokaryotic frataxin was found to directly bind the bacterial desulfurase, IscS, not the scaffold, IscU (Prischi et al., 2010). Taken together, the iron chaperone role of frataxin has fallen out of favor for other potential roles of frataxin.

Recent work on the role of frataxin suggested two other candidate functions. One report supported a role for frataxin in stimulating the binding of the cysteine substrate to Nfs1 (Pandey et al., 2013). An opposing model was later reported showing that frataxin stimulates sulfide transfer from Nfs1 to Isu1 (Bridwell-Rabb et al., 2014; Parent et al., 2015). The frataxin enhancement of sulfide transfer to Isu1 leads to formation of the [2Fe—2S] product on the Isu scaffold (Bridwell-Rabb et al., 2014; Fox et al., 2015). Further studies are needed to rigorously establish the role of frataxin.

The mechanism of Fe(II) delivery to the ISU scaffold remains unresolved. Despite numerous genetic screens to identify a putative Fe(II) chaperone, none have been identified. One appealing model for Fe(II) delivery is that Fe/S formation on the ISU scaffold could occur near the mitochondrial inner membrane juxtaposed to the Fe(II) transporters Mrs3 and Mrs4. Direct Fe(II) transfer to Isu1 may occur. Acp1 may be an attractive candidate to confer membrane association, since it has an acyl moiety bound to the Acp1 phosphopantetheine cofactor. However, the acyl moiety on Acp1 is not required for mitochondrial Fe/S biogenesis, although phosphopantetheine attachment is required (Van Vranken et al., 2016). Thus, Acp1 is unlikely to mediate membrane association.

2.2 Transfer of [2Fe—2S] Clusters From the Isu1 Scaffold

Once the ISU complex has completed synthesis of an initial [2Fe—2S] cluster, the cluster is released to the glutaredoxin, Grx5, by the DnaJ protein, Jac1, and the Hsp70 chaperone, Ssq1 (Fig. 1). This cycle is dependent on ATP hydrolysis by Ssq1 facilitated by the ATP exchange factor Mge1 (Dutkiewicz et al., 2003). The binding of Jac1 leads to the dissociation of the Nfs1/Isd11 complex (Majewska et al., 2013; Manicki et al., 2014) (Fig. 2A). The association of Grx5 and Ssq1 with the Jac1/Isu1 heterodimer stimulates ATP hydrolysis triggering a conformational change in Ssq1 that releases Jac1 and the [2Fe—2S] cluster to Grx5 (Kim, Tonelli, Frederick, Chow, & Markley, 2012; Uzarska, Dutkiewicz, Freibert, Lill, & Mühlenhoff, 2013). Cluster-bound Grx5 consists of a bridging [2Fe—2S] cluster between a Grx5 dimer (Banci et al., 2014) (Fig. 1). Each Grx5 molecule provides one cysteinyl ligand to the bridging [2Fe—2S] cluster, with a second thiolate ligand coming from a Grx5-bound glutathione that stably binds Grx5 (Banci et al., 2014; Johansson et al., 2011). Cells lacking Grx5 have defects in mitochondrial [4Fe—4S] and [2Fe—2S] centers as well as cytosolic Fe/S centers (Muhlenhoff, Gerber, Richhardt, & Lill, 2003; Rodriguez-Manzaneque, Tamarit, Belli, Ros, & Herrero, 2002; Uzarska et al., 2013). The functional importance of Grx5 is not entirely clear since all of these defects are modest relative to the essential nature of upstream components in the ISC pathway (Muhlenhoff et al., 2003). Depletion of downstream Fe/S biogenesis components in the ISA complex results in viable cells that are defective in respiratory growth due to the lack of mitochondrial [4Fe—4S] centers (Muhlenhoff et al., 2011).

[2Fe—2S] cluster transfer to client proteins is predicted to be mediated by the [2Fe—2S]-containing Grx5 homodimer. [2Fe—2S] transfer from a cytosolic GLRX3 glutaredoxin to cytosolic Fe/S client has been shown with GLRX3 to the client proteins Aft2 and anamorsin (Banci, Camponeschi, Ciofi-Baffoni, & Muzzioli, 2015). The homodimeric state of Grx5 with the bridging cluster protects the cluster prior to transfer (Banci et al., 2014).

While evidence supports a primary role for Grx5 as a [2Fe—2S] transfer agent in the mitochondria, an alternative model was proposed (Maio & Rouault, 2015). This alternative model posits that human mitochondrial [2Fe—2S] client proteins bind to the Jac1 (HSC20) associated with the ISU scaffold with direct cluster transfer stimulated by Ssq1 (HSPA9) (Maio et al., 2016, 2014; Maio, Kim, Singh, & Rouault, 2017). LYR sequence motifs either in the client protein or a LYR protein associated with a client protein associates with HSC20 to mediate the cluster transfer. The Fe/S-containing subunits respiratory complexes II (SDH) and III (Rieske Fe/S protein) were found to interact with HSC20 and the ISU scaffold (ISCU) in mammalian cells along with their respective LYR proteins. This group reported that Glrx5 (mammalian Grx5) was present with the chaperone/ISCU scaffold, but no mention was made of its relevance in cluster transfer. Unfortunately this model has not suggested an alternative function for Grx5, despite Grx5-deficient cells impairment in all mitochondrial and cytosolic Fe/S clusters tested to date.

Grx5 is implicated in presenting two [2Fe—2S] clusters to the ISA complex for formation of [4Fe—4S] clusters (Fig. 1). The [2Fe—2S] clusters are condensed into a single [4Fe—4S] cluster (Banci et al., 2014; Brancaccio et al., 2014; Rodriguez-Manzaneque et al., 2002; Shakamuri, Zhang, & Johnson, 2012).

BolA proteins have been reported to associate with glutaredoxin proteins in prokaryotes and eukaryotes (Li & Outten, 2012). In the eukaryotic cytosol, two glutaredoxins, Grx3 and Grx4, interact with the BolA protein Bol2 (Fra2) in a [2Fe—2S] cluster-bridged heterodimeric complex (Kumanovics et al., 2008; Li et al., 2009; Li, Mapolelo, Randeniya, Johnson, & Outten, 2012). Human homologues of this cytosolic glutaredoxin pair have been shown to mediate the in vitro transfer of [2Fe—2S] clusters to the client protein anamorsin and the [2Fe—2S]-linked human GLRX3–BOLA heterodimer is implicated in [2Fe—2S] cluster transfer in the cytoplasm (Banci et al., 2015; Frey, Palenchar, Wildemann, & Philpott, 2016). Mitochondria contain two BolA proteins, Bol1 and Bol3 (Uzarska et al., 2016). When the mutated mitochondrial BolA3 was identified in human patients with molecular defects in Fe/S centers, it was postulated that the BolA3 might function with Glrx5 (Cameron et al., 2011). However, recent studies revealed that the other mitochondrial BolA protein Bol1 interacted with Grx5 in yeast and in humans (Melber et al., 2016; Uzarska et al., 2016). Taken all together, a Grx5/Bol1 heterodimer bridged by a [2Fe—2S] cluster may be a specific transfer agent to an unidentified client protein within the mitochondria. How the Grx5/Bol1 and the Grx5/Grx5 dimers interplay with each other and with different fates of the bound [2Fe—2S] cluster (i.e., [2Fe—2S] clients or [4Fe—4S] biogenesis or export to CIA) is yet to be established.

2.3 [4Fe—4S] Cluster Formation by the ISA Complex

Within the cell, [4Fe—4S] clusters are the most commonly used clusters; however, distinct machinery is used to generate these clusters in the cytoplasm vs the mitochondrial matrix. As mentioned, within the mitochondria, conversion of [2Fe—2S] centers into [4Fe—4S] clusters requires the ISA complex in the mitochondrial matrix (Muhlenhoff et al., 2011). In contrast, formation of nuclear and cytoplasmic [4Fe—4S] clusters requires a distinct cytosolic iron sulfur assembly pathway (CIA) that utilizes the substrate exported from the matrix by Atm1 (Kispal et al., 1999; Lill et al., 2015).

Given that the ISA complex only produces [4Fe—4S] clusters for the mitochondria, cells lacking the ISA complex are only impaired in mitochondrial function, but not in cellular viability as these cells are able to export the sulfur-containing component required for cytoplasmic Fe/S biogenesis. The

ISA complex is a heterotrimeric complex comprised of Isa1, Isa2, and Iba57 (Gelling, Dawes, Richhardt, Lill, & Muhlenhoff, 2008; Muhlenhoff et al., 2011; Sheftel et al., 2012).

Following successful transfer of two [2Fe—2S] clusters by Grx5, one to Isa1 and the other to Isa2, the clusters are condensed to a single [4Fe—4S] cluster (Banci et al., 2014; Brancaccio et al., 2014). A remaining challenge in mitochondrial [4Fe—4S] cluster biogenesis is to understand the role of Iba57 in the trimeric complex. Patients of multiple mitochondrial dysfunction syndromes (MMDSs) present with functional defects in many mitochondrial Fe/S centers; recently, causative mutations were identified in the genes coding for Iba57 and Isca2 of the ISA complex (Ajit Bolar et al., 2013; Al-Hassnan et al., 2015; Lossos et al., 2015; Torraco et al., 2017).

2.4 [4Fe—4S] Cluster Transfer

MMDS patients have also been identified with mutations in the gene for NFU1 (Ahting et al., 2015; Cameron et al., 2011; Invernizzi et al., 2014; Navarro-Sastre et al., 2011; Nizon et al., 2014; Tonduti et al., 2015). Nfu1 has been characterized to bind [4Fe—4S] clusters formed on the ISA complex and to transfer them to a host of client proteins present from yeast to human, including aconitase, Sdh2 (the Fe/S subunit SDH), and LIAS (Cai et al., 2016; Melber et al., 2016). While patients present in a similar fashion whether the causative mutation is in genes coding the ISA complex or in the transfer proteins, the transfer step is only conditionally important to yeast. This is seen in the mild growth defects of Nfu1-deficient cells cultured in anoxic conditions relative to aerobic conditions. The anoxic cells do not display the same defects in [4Fe—4S] enzymes, suggesting that transfer systems are in place to protect preformed clusters from oxidative damage en route to client proteins (Melber et al., 2016).

Some species, e.g., *Trypanosoma brucei* contain multiple Nfu gene products that localize to the mitochondria (Benz, Kovarova, Kralova-Hromadova, Pierik, & Lukes, 2016). Two distinct Nfu species Nfu2 and Nfu3 are individually essential for *Trypanosome* growth in bloodstream and procyclic stages and may exhibit different client selectivity for [4Fe—4S] cluster transfer.

The other gene identified with causative mutations for MMDS was BOLA3, a family member of the BOLA family whose Bol1 family member functions upstream with Grx5 as discussed earlier (Baker et al., 2014; Cameron et al., 2011; Haack et al., 2013). Similar to the proposed [2Fe—2S] transfer system of Grx and BolA proteins, Nfu1 functions with Bol3, but the role of Bol3 in late stage cluster transfer with Nfu1 remains unresolved (Melber et al., 2016; Uzarska et al., 2016). Potential functions could include facilitating release of the stable [4Fe—4S] cluster bound to Nfu1 or stabilizing the interaction with client proteins.

Another late step cluster transfer protein is Ind1, which was implicated in respiratory complex I assembly (Bych et al., 2008; Sheftel et al., 2009). Human Ind1 is a P-loop NTPase resembling the Nbp35 P-loop protein involved in cytoplasmic Fe/S formation. Depletion of Ind1 in mammalian cells leads to a marked attenuation in complex I activity and diminution in the abundance of several complex I subunits (Sheftel et al., 2009). Ind1 has a potential Fe/S-binding motif and is capable of binding a Fe/S cluster, although the role of the cluster in Ind1 remains unresolved. Since Ind1 has only been shown to be associated with complex I function, it is no surprise that patient mutations in human IND1 present differently that of MMDS patients (Calvo et al., 2010; Kevelam et al., 2013). It is not known whether Nfu1 collaborates with Ind1 in the assembly of complex I, but some of the MMDS patients with mutations in the NFU1 gene do present with defects in complex I (Ahting et al., 2015; Cameron et al., 2011; Navarro-Sastre et al., 2011).

2.5 Unresolved Questions of Mitochondrial Fe/S Biogenesis

In recent years, significant advancement in the mitochondrial Fe/S biogenesis pathway has been made; however, many questions remain unanswered. Numerous discrepancies exist in our understanding of the role of frataxin in Fe/S biogenesis, and further studies are needed to definitively establish its metabolic role. It is unlikely that frataxin is the key Fe(II) donor to the ISU complex. As mentioned, an attractive postulate is that the mitochondrial Fe(II) transporters Mrs3 and Mrs4 may have a direct role in Fe(II) delivery to the ISU scaffold (Zhang et al., 2006). The role of GTP in Fe/S cluster biogenesis has not been resolved, and no GTPase has been implicated in the process. Within the ISA complex, no function has yet to be assigned to Iba57. Brancaccio et al. showed that [4Fe—4S] cluster condensation could occur with just Isa1 and Isa2 (Brancaccio et al., 2014). The condensation step is a reductive coupling, so the reductant for this reaction is not known. Iba57 was suggested to participate in this reductive coupling on the ISA complex (Brancaccio et al., 2014). This step requires additional characterization.

The controversy surrounding the [2Fe—2S] cluster transfer system needs resolution concerning whether Jac1 (HSC20) mediates direct donation of

[2Fe—2S] clusters to clients proteins in the absence of Grx5. An important issue remaining is the role of Bol1 in the Grx5 transfer steps. Clearly, in vitro, a [2Fe—2S]-bridged heterodimer of Bol1/Grx5 forms, however, does this cluster-loaded species exist stably within the mitochondrial matrix, and what purpose does it serve? Additionally, the function of the other mitochondrial BolA protein, Bol3, with Nfu1 in [4Fe—4S] transfer remains in question. Recent in vitro studies of the transfer between Nfu1 and client proteins are being expanded to include Bol3 to develop a mechanistic understanding. With further insight to the BolA proteins, we will hopefully understand why the phenotypes of the BolA deletion strains are so mild in yeast compared to all other ISC components.

Additionally, entire branches of Fe/S cluster biology potentially still exist awaiting discovery. While Atm1 has been characterized to export an undefined sulfide-containing substrate for the CIA pathway, no additional protein partners have been described to mediate the transition of a [2Fe—2S] cluster to the exported "sulfide" compound or escorting the "sulfide" complex out through the mitochondrial outer membrane pores and be presented to the cytoplasmic CIA machinery. Within the cytoplasm, the monothiol glutaredoxins may participate in this CIA presentation if the exported "sulfur" compound is a [2Fe—2S] cluster (Frey et al., 2016). While an understanding of the basic biosynthesis for [2Fe—2S] and [4Fe—4S] clusters exists, no biosynthetic machinery has been described for [3Fe—4S] clusters. [3Fe—4S] clusters may spontaneously arise in client proteins lacking one terminal cysteinyl ligand. Alternatively, endogenous oxidants may participate. Certainly in times of high oxidative stress, the [4Fe—4S] cluster of aconitase (having only three terminal cysteinyl ligands) is converted by oxidants to a [3Fe—4S] cluster (Flint, Tuminello, & Emptage, 1993). Might this be the mechanism by which SdhB creates a [3Fe—4S] cluster? Lastly, given that Fe/S clusters can be damaged by oxidation and an undefined repair pathway was reported in *Escherichia coli*, what surveillance mechanisms exist in the matrix to identify damaged clusters and does machinery exist that can fix oxidized clusters (Djaman, Outten, & Imlay, 2004)? Recently, it was shown that impairment of the matrix Fe/S ISU complex results in polythiol modifications of Fe/S apoprotein cysteinyl ligands (Christ, Leichert, Willms, Lill, & Muhlenhoff, 2016). Cysteine S-thiolation may be modification that enables these proteins to be readily reutilized for Fe/S cluster binding upon reduction of the labile polysulfide linkage. The radical SAM Fe/S enzyme LIAS, which consumes a [4Fe—4S] cluster during its catalytic turnover, must have a facile way for the cluster to be restored for subsequent rounds of catalysis. It is unknown whether ISC components moonlight to repair damaged clusters?

3. DEFINING NEW COMPONENTS OF Fe/S CLUSTER BIOGENESIS

Here we will highlight methods to identify novel components and potential directions and approaches for the characterization of Fe/S cluster biogenesis components. Numerous approaches have been used to identify and properly place novel components into mitochondrial iron–sulfur biogenesis. The first eukaryotic Fe/S assembly proteins identification came simply from identifying eukaryotic homologues of bacterial Fe/S assembly proteins (Garland, Hoff, Vickery, & Culotta, 1999; Kispal et al., 1999). Proteomics and genetic suppressor screens have been informative for identification of many components (Rodriguez-Manzaneque et al., 2002; Stehling et al., 2012) (Fig. 3). Additionally, identification of gene mutations in patients with mitochondrial disorders affecting Fe/S enzymes has successfully introduced new proteins involved in Fe/S cluster biogenesis. Two striking examples of this include frataxin as the gene associated with Friedreich's Ataxia and *BOLA3* first identified in MMDS (Cameron et al., 2011; Campuzano et al., 1996). The overlap of these approaches has led to rich insights into deciphering a protein's mechanism of action. In our study of Nfu1 the [4Fe—4S] transfer protein, physical interactions of Nfu1 identified by proteomics with a host of client proteins and ISA biogenesis components combined with genetic studies showed genetic interactions with the ISA complex, and finally NFU1 patient mutants were informative in establishing the role of Nfu1 in [4Fe—4S] cluster transfer (Melber et al., 2016). These approaches are useful in the identification of novel components, and also informative in the functional characterization of the protein.

As mentioned earlier, Fe/S centers are found throughout the cell, requiring bifurcations within the Fe/S cluster biogenesis pathways. Utilizing this, one can quickly assess where a novel component is acting by quantifying Fe/S enzymatic activities of enzymes in various cellular compartments. Additionally, some Fe/S cluster clients are destabilized without their cluster (such as the mitochondrial proteins Sdh2 [2Fe—2S], [4Fe—4S], [3Fe—4S], or Rip1 [2Fe—2S]), and in cases like lipoic acid, antibodies have been developed to observe the loss of products produced by Fe/S-containing enzymes; thus immunoblotting can be utilized to determine loss of protein stability or

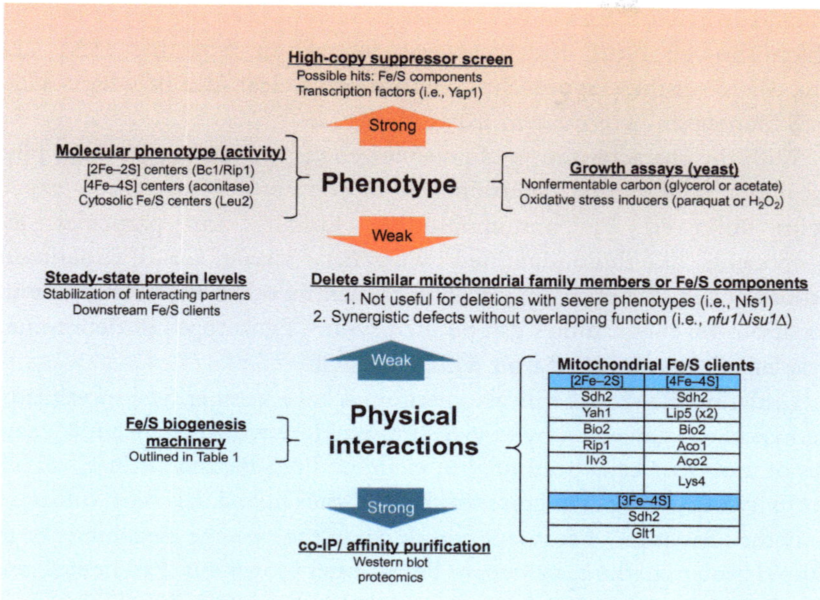

Fig. 3 Potential workflow to determine the function of a novel Fe/S component. The workflow separates biochemical and genetic approaches for molecular characterization of novel proteins. Biochemical approaches require an approach for purification, most likely by affinity tag or immunoprecipitation. Genetic approaches are based on the phenotype presented by deletion of the protein, weak growth phenotypes allow for determining synergistic interactions with other deletions, but do not serve useful for suppresser screens. In the case of a strong growth phenotype the opposite is true. Many of these approaches will inform one another, for example, often high-copy suppressors turn out to be interacting partners. This overlap is important for verifying the true interactions and function of a protein.

activity due to failure to produce Fe/S clusters (Fig. 3). If all cellular clusters are affected, it is likely an early factor working with the ISU complex, if defects are only found in mitochondrial [4Fe—4S] clusters one can focus studies on the ISA complex, and if only cytosolic clusters are perturbed, the focus would be on the CIA pathway.

These molecular phenotypes are often presented as more general growth defects on particular medias. In the case of the ISU complex and the CIA pathway, many loss of function mutations or deletions to biogenesis machinery results in inviable yeast, so many of these genes must put under a repressible promoter to thoroughly characterize these Fe/S components. Once a phenotype is established, it becomes an excellent tool to screen for suppressors, either through mutagenesis, as has been an important step in the studies of frataxin or with high-copy suppressor screens, like our study of Nfu1, which showed Nfu1 had a genetic interaction with the ISA2 gene (Fig. 3). All of these approaches will provide a clear idea of where within Fe/S biogenesis a component may act.

With the characterization of possible genetic interacting proteins, physical interactions can be determined with coimmunoprecipitation experiments followed by immunoblotting looking for particular ISC components, like those identified by a genetic screen. Co-IP experiments can also be used for proteomic analysis to identify other interacting proteins. Variations on these studies can be used with ^{55}Fe isotopes to determine if these novel proteins bind iron within the cell.

Lastly, in vitro biochemical experiments have been critical to validating the hypothesis generated by in vivo studies. Here we will give a few examples of tools that have been used to examine Fe/S biogenesis in vitro. The reactivity of cysteine, the common Fe/S cluster ligand, has been utilized to study the formation of covalent persulfide on the cysteine desulfurase or the scaffold protein in the early steps of Fe/S cluster biogenesis. Parent et al. used the covalent attachment of a maleimide peptide to visualize discrete band shifts on SDS-PAGE gels to determine the number of reactive cysteine present on a protein (Parent et al., 2015). To examine cluster transfer steps to client proteins the activation of Fe/S cluster client enzymes provide a sensitive readout. Alternatively, HSQC NMR studies have elegantly been implemented to study Fe/S cluster transfer (Banci et al., 2014; Brancaccio et al., 2014). This approach has been used to show the transfer of a [4Fe—4S] cluster from Nfu1 to aconitase (Cai et al., 2016). More detailed approaches for in vitro characterization of the Fe/S cluster proteins are described throughout this issue of *Methods in Enzymology*.

3.1 Lessons From Past Fe/S Studies

Of course all individual results need to be taken with caution. Mitochondrial defects can often propagate into loss of mitochondrial DNA, which can result in pleiotropic defects that may confuse the interpretation. An example of this phenomenon is that the deletion of the gene encoding the yeast ISA component Iba57 leads to instability in the mitochondrial DNA leading to a petite phenotype (Gelling et al., 2008). Loss of mtDNA would impair complex III biogenesis leading to the absence of the Rieske Fe/S subunit (Rip1). The absence of Rip1 arises from loss of mtDNA and not loss of the ISA complex.

Growth phenotypes can also create confusion; in the case of *nfu1Δ*, it contains a synthetic defect with *isu1Δ*, *nfu1Δisu1Δ*, leading to an early hypothesis that Nfu1 acts as an alternative scaffold (Schilke, Voisine, Beinert, & Craig, 1999). However, upon our studies, we believe this synthetic defect arises from an accumulation of defects in completely separate parts of Fe/S biogenesis opposed to reflecting Isu1 and Nfu1 performing similar functions. It is important to note that many proteomic studies looking for copurified binding partners often identify components of the Fe/S biogenesis, and that it is imperative to resolve the specific interactions with follow-up studies. Another curious challenge with proteomic studies is the identification of potential novel components. In our proteomic studies of Nfu1 and the mitochondrial BolA proteins, Bol1 and Bol3, the matrix peroxiredoxin Prx1 reproducibly and strongly interacted with Bol1. Since Grx5 also strongly interacted with Prx1, the potential of a glutaredoxin interaction with a peroxiredoxin was suggested. However, cells depleted of Prx1 had no Fe/S cluster defect, so the significance of this putative Prx1–Grx5 interaction is questioned.

An understudied aspect of mitochondrial Fe/S biogenesis concerns its regulation. Limited information is available on what mechanisms exist to modulate the flux of Fe/S cluster synthesis to accommodate cellular demands for the cofactor. When Fe/S cluster formation is impaired, it is known that Isu1 in yeast accumulates (Andrew, Song, Schilke, & Craig, 2008). This process arises from an increased stability of the Isu1 protein due to stable interactions with either Nfs1 or Jac1 (Song, Marszalek, & Craig, 2012). Under steady-state conditions in wild-type cells, Isu1 is susceptible to degradation by the matrix Lon protease during its transition from cluster formation to cluster transfer (Ciesielski, Schilke, Marszalek, & Craig, 2016). This highlights one candidate mechanism to alter the cellular abundance of one key ISU biogenesis components. Fe/S cluster regulation is likely to become a significant future topic (Mettert & Kiley, 2015). The pathway may be modulated by reversible posttranslational modifications including phosphorylation or acetylation.

4. CONCLUDING REMARKS

While all data can be informative it is important to use multiple approaches for validating the function of Fe/S biogenesis components. In our studies of Nfu1, we never showed the physical transfer of a [4Fe—4S] cluster from Nfu1 to one of the mitochondrial clients; hence, it was very satisfying when Cai et al. published the observation of an activation of apo-aconitase by incubation of holo-Nfu1 (Cai et al., 2016). After many years of intense debate, the many studies of frataxin are finally coalescing to understand a mechanistic function. These are excellent examples of how two separate studies complemented each to further strengthen a hypothesis. Additional studies will be required to reconcile the two opposing models of [2Fe—2S] cluster transfer (either by Jac1 or Grx5), while Grx5 has been shown to transfer to clients in vitro the in vivo validation of this function has not thoroughly described the interaction with client proteins. Likewise, Jac1 has been shown to be coimmunoprecipitated with many [2Fe—2S] client proteins in vivo; however, the demonstration of [2Fe—2S] cluster transfer to client proteins with isolated proteins in vitro with and without Grx5 is necessary to resolve the controversy. Thus, the field of mitochondrial Fe/S cluster biogenesis has matured significantly from the late 1990s. However, unresolved questions persist in our complete understanding of the mechanisms of cluster transfer, Fe/S cluster repair processes, and cellular regulation of the pathway. The field remains an inviting opportunity for new investigators to focus their studies.

REFERENCES

Adam, A. C., Bornhovd, C., Prokisch, H., Neupert, W., & Hell, K. (2006). The Nfs1 interacting protein Isd11 has an essential role in Fe/S cluster biogenesis in mitochondria. *The EMBO Journal*, *25*, 174–183.

Adamec, J., Rusnak, F., Owen, W. G., Naylor, S., Benson, L. M., Gacy, A. M., et al. (2000). Iron-dependent self-assembly of recombinant yeast frataxin: Implications for Friedreich ataxia. *American Journal of Human Genetics*, *67*, 549–562.

Ahting, U., Mayr, J. A., Vanlander, A. V., Hardy, S. A., Santra, S., Makowski, C., et al. (2015). Clinical, biochemical, and genetic spectrum of seven patients with NFU1 deficiency. *Frontiers in Genetics*, *6*, 123.

Ajit Bolar, N., Vanlander, A. V., Wilbrecht, C., Van der Aa, N., Smet, J., De Paepe, B., et al. (2013). Mutation of the iron-sulfur cluster assembly gene IBA57 causes severe myopathy and encephalopathy. *Human Molecular Genetics*, *22*, 2590–2602.

Al-Hassnan, Z. N., Al-Dosary, M., Alfadhel, M., Faqeih, E. A., Alsagob, M., Kenana, R., et al. (2015). ISCA2 mutation causes infantile neurodegenerative mitochondrial disorder. *Journal of Medical Genetics*, *52*, 186–194.

Amutha, B., Gordon, D. M., Gu, Y., Lyver, E. R., Dancis, A., & Pain, D. (2008). GTP is required for iron-sulfur cluster biogenesis in mitochondria. *The Journal of Biological Chemistry*, *283*, 1362–1371.

Anderson, C. P., Shen, M., Eisenstein, R. S., & Leibold, E. A. (2012). Mammalian iron metabolism and its control by iron regulatory proteins. *Biochimica et Biophysica Acta*, *1823*, 1468–1483.

Andrew, A. J., Song, J. Y., Schilke, B., & Craig, E. A. (2008). Posttranslational regulation of the scaffold for Fe-S cluster biogenesis, Isu. *Molecular Biology of the Cell*, *19*, 5259–5266.

Angerer, H., Radermacher, M., Mankowska, M., Steger, M., Zwicker, K., Heide, H., et al. (2014). The LYR protein subunit NB4M/NDUFA6 of mitochondrial complex I anchors an acyl carrier protein and is essential for catalytic activity. *Proceedings of the National Academy of Sciences of the United States of America*, *111*, 5207–5212.

Arnold, A. R., Grodick, M. A., & Barton, J. K. (2016). DNA charge transport: From chemical principles to the cell. *Cell Chemical Biology*, *23*, 183–197.

Atkinson, A., Smith, P., Fox, J. L., Cui, T. Z., Khalimonchuk, O., & Winge, D. R. (2011). The LYR protein Mzm1 functions in the insertion of the rieske Fe/S protein in yeast mitochondria. *Molecular and Cellular Biology*, *31*, 3988–3996.

Babcock, M., de Silva, D., Oaks, R., Davis-Kaplan, S., Jiralerspong, S., Montermini, L., et al. (1997). Regulation of mitochondrial iron accumulation by Yfh1p, a putative homolog of frataxin. *Science*, *276*, 1709–1712.

Baker, P. R., 2nd, Friederich, M. W., Swanson, M. A., Shaikh, T., Bhattacharya, K., Scharer, G. H., et al. (2014). Variant non ketotic hyperglycinemia is caused by mutations in LIAS, BOLA3 and the novel gene GLRX5. *Brain*, *137*, 366–379.

Banci, L., Brancaccio, D., Ciofi-Baffoni, S., Del Conte, R., Gadepalli, R., Mikolajczyk, M., et al. (2014). [2Fe-2S] cluster transfer in iron-sulfur protein biogenesis. *Proceedings of the National Academy of Sciences of the United States of America*, *111*, 6203–6208.

Banci, L., Camponeschi, F., Ciofi-Baffoni, S., & Muzzioli, R. (2015). Elucidating the molecular function of human BOLA2 in GRX3-dependent anamorsin maturation pathway. *Journal of the American Chemical Society*, *137*, 16133–16143.

Barros, M. H., Carlson, C. G., Glerum, D. M., & Tzagoloff, A. (2001). Involvement of mitochondrial ferredoxin and Cox15p in hydroxylation of heme O. *FEBS Letters*, *492*, 133–138.

Barros, M. H., & Nobrega, F. G. (1999). YAH1 of *Saccharomyces cerevisiae*: A new essential gene that codes for a protein homologous to human adrenodoxin. *Gene*, *233*, 197–203.

Barros, M. H., Nobrega, F. G., & Tzagoloff, A. (2002). Mitochondrial ferredoxin is required for heme A synthesis in *Saccharomyces cerevisiae*. *The Journal of Biological Chemistry*, *277*, 9997–10002.

Beilschmidt, L. K., & Puccio, H. M. (2014). Mammalian Fe-S cluster biogenesis and its implication in disease. *Biochimie*, *100*, 48–60.

Beinert, H. (2000). Iron-sulfur proteins: Ancient structures, still full of surprises. *Journal of Biological Inorganic Chemistry*, *5*, 2–15.

Benz, C., Kovarova, J., Kralova-Hromadova, I., Pierik, A. J., & Lukes, J. (2016). Roles of the Nfu Fe-S targeting factors in the trypanosome mitochondrion. *International Journal for Parasitology*, *46*, 641–651.

Berkovitch, F., Nicolet, Y., Wan, J. T., Jarrett, J. T., & Drennan, C. L. (2004). Crystal structure of biotin synthase, an S-adenosylmethionine-dependent radical enzyme. *Science*, *303*, 76–79.

Brancaccio, D., Gallo, A., Mikolajczyk, M., Zovo, K., Palumaa, P., Novellino, E., et al. (2014). Formation of [4Fe-4S] clusters in the mitochondrial iron-sulfur cluster assembly machinery. *Journal of the American Chemical Society*, *136*, 16240–16250.

Bridwell-Rabb, J., Fox, N. G., Tsai, C. L., Winn, A. M., & Barondeau, D. P. (2014). Human frataxin activates Fe-S cluster biosynthesis by facilitating sulfur transfer chemistry. *Biochemistry*, *53*, 4904–4913.

Bych, K., Kerscher, S., Netz, D. J., Pierik, A. J., Zwicker, K., Huynen, M. A., et al. (2008). The iron-sulphur protein Ind1 is required for effective complex I assembly. *The EMBO Journal*, *27*, 1736–1746.

Cai, K., Liu, G., Frederick, R. O., Xiao, R., Montelione, G. T., & Markley, J. L. (2016). Structural/functional properties of human NFU1, an intermediate [4Fe- 4S] carrier in human mitochondrial iron-sulfur cluster biogenesis. *Structure*, *24*, 2080–2091.

Cai, K., Tonelli, M., Frederick, R. O., & Markley, J. L. (2017). Human mitochondrial ferredoxin 1 (FDX1) and ferredoxin 2 (FDX2) both bind cysteine desulfurase and donate electrons for iron-sulfur cluster biosynthesis. *Biochemistry*, *56*, 487–499.

Calvo, S. E., Tucker, E. J., Compton, A. G., Kirby, D. M., Crawford, G., Burtt, N. P., et al. (2010). High-throughput, pooled sequencing identifies mutations in NUBPL and FOXRED1 in human complex I deficiency. *Nature Genetics*, *42*, 851–858.

Cameron, J. M., Janer, A., Levandovskiy, V., Mackay, N., Rouault, T. A., Tong, W.-H., et al. (2011). Mutations in iron-sulfur cluster scaffold genes NFU1 and BOLA3 cause a fatal deficiency of multiple respiratory chain and 2-oxoacid dehydrogenase enzymes. *American Journal of Human Genetics*, *89*, 486–495.

Campuzano, V., Montermini, L., Molto, M. D., Pianese, L., Cossee, M., Cavalcanti, F., et al. (1996). Friedreich's ataxia: Autosomal recessive disease caused by an intronic GAA triplet repeat expansion. *Science*, *271*, 1423–1427.

Christ, S., Leichert, L. I., Willms, A., Lill, R., & Muhlenhoff, U. (2016). Defects in mitochondrial iron-sulfur cluster assembly induce cysteine S-polythiolation on iron-sulfur apoproteins. *Antioxidants & Redox Signaling*, *25*, 28–40.

Cicchillo, R. M., Lee, K. H., Baleanu-Gogonea, C., Nesbitt, N. M., Krebs, C., & Booker, S. J. (2004). Escherichia coli lipoyl synthase binds two distinct [4Fe-4S] clusters per polypeptide. *Biochemistry*, *43*, 11770–11781.

Ciesielski, S. J., Schilke, B., Marszalek, J., & Craig, E. A. (2016). Protection of scaffold protein Isu from degradation by the Lon protease Pim1 as a component of Fe-S cluster biogenesis regulation. *Molecular Biology of the Cell*, *27*, 1060–1068.

Colin, F., Martelli, A., Clemancey, M., Latour, J. M., Gambarelli, S., Zeppieri, L., et al. (2013). Mammalian frataxin controls sulfur production and iron entry during de novo Fe4S4 cluster assembly. *Journal of the American Chemical Society*, *135*, 733–740.

Correia, A. R., Wang, T., Craig, E. A., & Gomes, C. M. (2010). Iron-binding activity in yeast frataxin entails a trade off with stability in the alpha1/beta1 acidic ridge region. *The Biochemical Journal*, *426*, 197–203.

Cronan, J. E. (2014). The structure of lipoyl synthase, a remarkable enzyme that performs the last step of an extraordinary biosynthetic pathway. *The Biochemical Journal*, *464*, e1–3.

Dellibovi-Ragheb, T. A., Gisselberg, J. E., & Prigge, S. T. (2013). Parasites FeS up: Iron-sulfur cluster biogenesis in eukaryotic pathogens. *PLoS Pathogens*, *9*, e1003227.

Djaman, O., Outten, F. W., & Imlay, J. A. (2004). Repair of oxidized iron-sulfur clusters in Escherichia coli. *The Journal of Biological Chemistry*, *279*, 44590–44599.

Dutkiewicz, R., Schilke, B., Knieszner, H., Walter, W., Craig, E. A., & Marszalek, J. (2003). Ssq1, a mitochondrial Hsp70 involved in iron-sulfur (Fe/S) center biogenesis. Similarities to and differences from its bacterial counterpart. *The Journal of Biological Chemistry*, *278*, 29719–29727.

Fazius, F., Shelest, E., Gebhardt, P., & Brock, M. (2012). The fungal alpha-aminoadipate pathway for lysine biosynthesis requires two enzymes of the aconitase family for the isomerization of homocitrate to homoisocitrate. *Molecular Microbiology*, *86*, 1508–1530.

Fiedorczuk, K., Letts, J. A., Degliesposti, G., Kaszuba, K., Skehel, M., & Sazanov, L. A. (2016). Atomic structure of the entire mammalian mitochondrial complex I. *Nature*, *538*, 406–410.

Flint, D. H., Tuminello, J. F., & Emptage, M. H. (1993). The inactivation of Fe-S cluster containing hydro-lyases by superoxide. *The Journal of Biological Chemistry*, *268*, 22369–22376.

Floyd, B. J., Wilkerson, E. M., Veling, M. T., Minogue, C. E., Xia, C., Beebe, E. T., et al. (2016). Mitochondrial protein interaction mapping identifies regulators of respiratory chain function. *Molecular Cell*, *63*, 621–632.

Foury, F., Pastore, A., & Trincal, M. (2007). Acidic residues of yeast frataxin have an essential role in Fe-S cluster assembly. *EMBO Reports*, *8*, 194–199.

Fox, N. G., Das, D., Chakrabarti, M., Lindahl, P. A., & Barondeau, D. P. (2015). Frataxin accelerates [2Fe-2S] cluster formation on the human Fe-S assembly complex. *Biochemistry*, *54*, 3880–3889.

Frey, A. G., Palenchar, D. J., Wildemann, J. D., & Philpott, C. C. (2016). A glutaredoxin-BolA complex serves as an iron-sulfur cluster chaperone for the cytosolic cluster assembly machinery. *The Journal of Biological Chemistry, 291*, 22344–22356.

Garland, S. A., Hoff, K., Vickery, L. E., & Culotta, V. C. (1999). Saccharomyces cerevisiae ISU1 and ISU2: Members of a well-conserved gene family for iron-sulfur cluster assembly. *Journal of Molecular Biology, 294*, 897–907.

Gelling, C., Dawes, I. W., Richhardt, N., Lill, R., & Muhlenhoff, U. (2008). Mitochondrial Iba57p is required for Fe/S cluster formation on aconitase and activation of radical SAM enzymes. *Molecular and Cellular Biology, 28*, 1851–1861.

Gisselberg, J. E., Dellibovi-Ragheb, T. A., Matthews, K. A., Bosch, G., & Prigge, S. T. (2013). The suf iron-sulfur cluster synthesis pathway is required for apicoplast maintenance in malaria parasites. *PLoS Pathogens, 9*, e1003655.

Gnandt, E., Dorner, K., Strampraad, M. F., de Vries, S., & Friedrich, T. (2016). The multitude of iron-sulfur clusters in respiratory complex I. *Biochimica et Biophysica Acta, 1857*, 1068–1072.

Haack, T. B., Rolinski, B., Haberberger, B., Zimmermann, F., Schum, J., Strecker, V., et al. (2013). Homozygous missense mutation in BOLA3 causes multiple mitochondrial dysfunctions syndrome in two siblings. *Journal of Inherited Metabolic Disease, 36*, 55–62.

Hagerhall, C., Sled, V., Hederstedt, L., & Ohnishi, T. (1995). The trinuclear iron-sulfur cluster S3 in Bacillus subtilis succinate: Menaquinone reductase; effects of a mutation in the putative cluster ligation motif on enzyme activity and EPR properties. *Biochimica et Biophysica Acta, 1229*, 356–362.

Huttlin, E. L., Ting, L., Bruckner, R. J., Gebreab, F., Gygi, M. P., Szpyt, J., et al. (2015). The BioPlex network: A systematic exploration of the human interactome. *Cell, 162*, 425–440.

Invernizzi, F., Ardissone, A., Lamantea, E., Garavaglia, B., Zeviani, M., Farina, L., et al. (2014). Cavitating leukoencephalopathy with multiple mitochondrial dysfunction syndrome and NFU1 mutations. *Frontiers in Genetics, 5*, 412.

Iwata, S., Lee, J. W., Okada, K., Lee, J. K., Iwata, M., Rasmussen, B., et al. (1998). Complete structure of the 11-subunit bovine mitochondrial cytochrome bc1 complex. *Science, 281*, 64–71.

Iwata, M., Lee, Y., Yamashita, T., Yagi, T., Iwata, S., Cameron, A. D., et al. (2012). The structure of the yeast NADH dehydrogenase (Ndi1) reveals overlapping binding sites for water- and lipid-soluble substrates. *Proceedings of the National Academy of Sciences of the United States of America, 109*, 15247–15252.

Johansson, C., Roos, A. K., Montano, S. J., Sengupta, R., Filippakopoulos, P., Guo, K., et al. (2011). The crystal structure of human GLRX5: Iron-sulfur cluster co-ordination, tetrameric assembly and monomer activity. *The Biochemical Journal, 433*, 303–311.

Johnson, D. C., Dean, D. R., Smith, A. D., & Johnson, M. K. (2005). Structure, function, and formation of biological iron-sulfur clusters. *Annual Review of Biochemistry, 74*, 247–281.

Kaiser, J. T., Clausen, T., Bourenkow, G. P., Bartunik, H.-D., Steinbacher, S., & Huber, R. (2000). Crystal structure of a NifS-like protein from Thermotoga maritima: Implications for iron sulphur cluster assembly. *Journal of Molecular Biology, 297*, 451–464.

Kevelam, S. H., Rodenburg, R. J., Wolf, N. I., Ferreira, P., Lunsing, R. J., Nijtmans, L. G., et al. (2013). NUBPL mutations in patients with complex I deficiency and a distinct MRI pattern. *Neurology, 80*, 1577–1583.

Kim, J. H., Frederick, R. O., Reinen, N. M., Troupis, A. T., & Markley, J. L. (2013). [2Fe-2S]-ferredoxin binds directly to cysteine desulfurase and supplies an electron for iron-sulfur cluster assembly but is displaced by the scaffold protein or bacterial frataxin. *Journal of the American Chemical Society, 135*, 8117–8120.

Kim, J. H., Tonelli, M., Frederick, R. O., Chow, D. C., & Markley, J. L. (2012). Specialized Hsp70 chaperone (HscA) binds preferentially to the disordered form, whereas J-protein (HscB) binds preferentially to the structured form of the iron-sulfur cluster scaffold protein (IscU). *The Journal of Biological Chemistry*, *287*, 31406–31413.

Kispal, G., Csere, P., Prohl, C., & Lill, R. (1999). The mitochondrial proteins Atm1p and Nfs1p are essential for biogenesis of cytosolic Fe/S proteins. *The EMBO Journal*, *18*, 3981–3989.

Knaff, D. B., Hirasawa, M., Ameyibor, E., Fu, W., & Johnson, M. K. (1991). Spectroscopic evidence for a [3Fe-4S] cluster in spinach glutamate synthase. *The Journal of Biological Chemistry*, *266*, 15080–15084.

Kondapalli, K. C., Kok, N. M., Dancis, A., & Stemmler, T. L. (2008). Drosophila frataxin: An iron chaperone during cellular Fe-S cluster bioassembly. *Biochemistry*, *47*, 6917–6927.

Kuhn, L. C. (2015). Iron regulatory proteins and their role in controlling iron metabolism. *Metallomics*, 7, 232–243.

Kumanovics, A., Chen, O. S., Li, L., Bagley, D., Adkins, E. M., Lin, H., et al. (2008). Identification of FRA1 and FRA2 as genes involved in regulating the yeast iron regulon in response to decreased mitochondrial iron-sulfur cluster synthesis. *The Journal of Biological Chemistry*, *283*, 10276–10286.

Lange, H., Kaut, A., Kispal, G., & Lill, R. (2000). A mitochondrial ferredoxin is essential for biogenesis of cellular iron-sulfur proteins. *Proceedings of the National Academy of Sciences of the United States of America*, *97*, 1050–1055.

Li, H., Mapolelo, D. T., Dingra, N. N., Naik, S. G., Lees, N. S., Hoffman, B. M., et al. (2009). The yeast iron regulatory proteins Grx3/4 and Fra2 form heterodimeric complexes containing a [2Fe-2S] cluster with cysteinyl and histidyl ligation. *Biochemistry*, *48*, 9569–9581.

Li, H., Mapolelo, D. T., Randeniya, S., Johnson, M. K., & Outten, C. E. (2012). Human glutaredoxin 3 forms [2Fe-2S]-bridged complexes with human BolA2. *Biochemistry*, *51*, 1687–1696.

Li, H., & Outten, C. E. (2012). Monothiol CGFS glutaredoxins and BolA-like proteins: [2Fe-2S] binding partners in iron homeostasis. *Biochemistry*, *51*, 4377–4389.

Li, J., Saxena, S., Pain, D., & Dancis, A. (2001). Adrenodoxin reductase homolog (Arh1p) of yeast mitochondria required for iron homeostasis. *The Journal of Biological Chemistry*, *276*, 1503–1509.

Lill, R., Dutkiewicz, R., Freibert, S. A., Heidenreich, T., Mascarenhas, J., Netz, D. J., et al. (2015). The role of mitochondria and the CIA machinery in the maturation of cytosolic and nuclear iron-sulfur proteins. *European Journal of Cell Biology*, *94*, 280–291.

Lossos, A., Stumpfig, C., Stevanin, G., Gaussen, M., Zimmerman, B. E., Mundwiller, E., et al. (2015). Fe/S protein assembly gene IBA57 mutation causes hereditary spastic paraplegia. *Neurology*, *84*, 659–667.

Maio, N., Ghezzi, D., Verrigni, D., Rizza, T., Bertini, E., Martinelli, D., et al. (2016). Disease-causing SDHAF1 mutations impair transfer of Fe-S clusters to SDHB. *Cell Metabolism*, *23*, 292–302.

Maio, N., Kim, K. S., Singh, A., & Rouault, T. A. (2017). A single adaptable Cochaperone-Scaffold complex delivers nascent iron-sulfur clusters to mammalian respiratory chain complexes I-III. *Cell Metabolism*, *25*, 945–953, e946.

Maio, N., & Rouault, T. A. (2015). Iron-sulfur cluster biogenesis in mammalian cells: New insights into the molecular mechanisms of cluster delivery. *Biochimica et Biophysica Acta*, *1853*, 1493–1512.

Maio, N., Singh, A., Uhrigshardt, H., Saxena, N., Tong, W. H., & Rouault, T. A. (2014). Cochaperone binding to LYR motifs confers specificity of iron sulfur cluster delivery. *Cell Metabolism*, *19*, 445–457.

Majewska, J., Ciesielski, S. J., Schilke, B., Kominek, J., Blenska, A., Delewski, W., et al. (2013). Binding of the chaperone Jac1 protein and cysteine desulfurase Nfs1 to the iron-sulfur cluster scaffold Isu protein is mutually exclusive. *The Journal of Biological Chemistry, 288*, 29134–29142.

Manicki, M., Majewska, J., Ciesielski, S., Schilke, B., Blenska, A., Kominek, J., et al. (2014). Overlapping binding sites of the frataxin homologue assembly factor and the heat shock protein 70 transfer factor on the Isu iron-sulfur cluster scaffold protein. *The Journal of Biological Chemistry, 289*, 30268–30278.

Manodori, A., Cecchini, G., Schroder, I., Gunsalus, R. P., Werth, M. T., & Johnson, M. K. (1992). [3Fe-4S] to [4Fe-4S] cluster conversion in Escherichia coli fumarate reductase by site-directed mutagenesis. *Biochemistry, 31*, 2703–2712.

Manzella, L., Barros, M. H., & Nobrega, F. G. (1998). ARH1 of Saccharomyces cerevisiae: A new essential gene that codes for a protein homologous to the human adrenodoxin reductase. *Yeast, 14*, 839–846.

McLaughlin, M. I., Lanz, N. D., Goldman, P. J., Lee, K. H., Booker, S. J., & Drennan, C. L. (2016). Crystallographic snapshots of sulfur insertion by lipoyl synthase. *Proceedings of the National Academy of Sciences of the United States of America, 113*, 9446–9450.

Melber, A., Na, U., Vashisht, A., Weiler, B. D., Lill, R., Wohlschlegel, J. A., et al. (2016). Role of Nfu1 and Bol3 in iron-sulfur cluster transfer to mitochondrial clients. *eLife, 5*, e15991.

Mettert, E. L., & Kiley, P. J. (2015). How is Fe-S cluster formation regulated? *Annual Review of Microbiology, 69*, 505–526.

Miller, R. E., & Stadtman, E. R. (1972). Glutamate synthase from Escherichia coli. An iron-sulfide flavoprotein. *The Journal of Biological Chemistry, 247*, 7407–7419.

Mühlenhoff, U., Balk, J., Richhardt, N., Kaiser, J. T., Sipos, K., Kispal, G., et al. (2004). Functional characterization of the eukaryotic cysteine desulfurase Nfs1 from *Saccharomyces cerevisiae*. *The Journal of Biological Chemistry, 279*, 36906–36915.

Muhlenhoff, U., Gerber, J., Richhardt, N., & Lill, R. (2003). Components involved in assembly and dislocation of iron-sulfur clusters on the scaffold protein Isu1p. *The EMBO Journal, 22*, 4815–4825.

Muhlenhoff, U., Richhardt, N., Gerber, J., & Lill, R. (2002). Characterization of iron-sulfur protein assembly in isolated mitochondria. A requirement for ATP, NADH, and reduced iron. *The Journal of Biological Chemistry, 277*, 29810–29816.

Muhlenhoff, U., Richter, N., Pines, O., Pierik, A. J., & Lill, R. (2011). Specialized function of yeast Isa1 and Isa2 proteins in the maturation of mitochondrial [4Fe-4S] proteins. *The Journal of Biological Chemistry, 286*, 41205–41216.

Na, U., Yu, W., Cox, J., Bricker, D. K., Brockmann, K., Rutter, J., et al. (2014). The LYR factors SDHAF1 and SDHAF3 mediate maturation of the iron-sulfur subunit of succinate dehydrogenase. *Cell Metabolism, 20*, 253–266.

Nanda, V., Senn, S., Pike, D. H., Rodriguez-Granillo, A., Hansen, W. A., Khare, S. D., et al. (2016). Structural principles for computational and de novo design of 4Fe-4S metalloproteins. *Biochimica et Biophysica Acta, 1857*, 531–538.

Navarro-Sastre, A., Tort, F., Stehling, O., Uzarska, M. A., Arranz, J. A., Del Toro, M., et al. (2011). A fatal mitochondrial disease is associated with defective NFU1 function in the maturation of a subset of mitochondrial Fe-S proteins. *American Journal of Human Genetics, 89*, 656–667.

Nizon, M., Boutron, A., Boddaert, N., Slama, A., Delpech, H., Sardet, C., et al. (2014). Leukoencephalopathy with cysts and hyperglycinemia may result from NFU1 deficiency. *Mitochondrion, 15*, 59–64.

Pandey, A., Golla, R., Yoon, H., Dancis, A., & Pain, D. (2012). Persulfide formation on mitochondrial cysteine desulfurase: Enzyme activation by a eukaryote-specific interacting protein and Fe-S cluster synthesis. *The Biochemical Journal, 448*, 171–187.

Pandey, A., Gordon, D. M., Pain, J., Stemmler, T. L., Dancis, A., & Pain, D. (2013). Frataxin directly stimulates mitochondrial cysteine desulfurase by exposing substrate-binding sites, and a mutant Fe-S cluster scaffold protein with frataxin-bypassing ability acts similarly. *The Journal of Biological Chemistry*, *288*, 36773–36786.

Pandey, A., Pain, J., Ghosh, A. K., Dancis, A., & Pain, D. (2015). Fe-S cluster biogenesis in isolated mammalian mitochondria: Coordinated use of persulfide sulfur and iron and requirements for GTP, NADH, and ATP. *The Journal of Biological Chemistry*, *290*, 640–657.

Parent, A., Elduque, X., Cornu, D., Belot, L., Le Caer, J. P., Grandas, A., et al. (2015). Mammalian frataxin directly enhances sulfur transfer of NFS1 persulfide to both ISCU and free thiols. *Nature Communications*, *6*, 5686.

Paul, V. D., & Lill, R. (2015). Biogenesis of cytosolic and nuclear iron-sulfur proteins and their role in genome stability. *Biochimica et Biophysica Acta*, *1853*, 1528–1539.

Pierrel, F., Hamelin, O., Douki, T., Kieffer-Jaquinod, S., Muhlenhoff, U., Ozeir, M., et al. (2010). Involvement of mitochondrial ferredoxin and para-aminobenzoic acid in yeast coenzyme Q biosynthesis. *Chemistry & Biology*, *17*, 449–459.

Prischi, F., Konarev, P. V., Iannuzzi, C., Pastore, C., Adinolfi, S., Martin, S. R., et al. (2010). Structural bases for the interaction of frataxin with the central components of iron-sulphur cluster assembly. *Nature Communications*, *1*, 95.

Robbins, A. H., & Stout, C. D. (1989). Structure of activated aconitase: Formation of the [4Fe-4S] cluster in the crystal. *Proceedings of the National Academy of Sciences of the United States of America*, *86*, 3639–3643.

Roche, B., Aussel, L., Ezraty, B., Mandin, P., Py, B., & Barras, F. (2013). Iron/sulfur proteins biogenesis in prokaryotes: Formation, regulation and diversity. *Biochimica et Biophysica Acta*, *1827*, 455–469.

Rodriguez-Manzaneque, M. T., Tamarit, J., Belli, G., Ros, J., & Herrero, E. (2002). Grx5 is a mitochondrial glutaredoxin required for the activity of iron/sulfur enzymes. *Molecular Biology of the Cell*, *13*, 1109–1121.

Schaedler, T. A., Thornton, J. D., Kruse, I., Schwarzlander, M., Meyer, A. J., van Veen, H. W., et al. (2014). A conserved mitochondrial ATP-binding cassette transporter exports glutathione polysulfide for cytosolic metal cofactor assembly. *The Journal of Biological Chemistry*, *289*, 23264–23274.

Schilke, B., Voisine, C., Beinert, H., & Craig, E. (1999). Evidence for a conserved system for iron metabolism in the mitochondria of Saccharomyces cerevisiae. *Proceedings of the National Academy of Sciences of the United States of America*, *96*, 10206–10211.

Schmucker, S., Martelli, A., Colin, F., Page, A., Wattenhofer-Donze, M., Reutenauer, L., et al. (2011). Mammalian frataxin: An essential function for cellular viability through an interaction with a preformed ISCU/NFS1/ISD11 iron-sulfur assembly complex. *PLoS One*, *6*, e16199.

Shakamuri, P., Zhang, B., & Johnson, M. K. (2012). Monothiol glutaredoxins function in storing and transporting [Fe2S2] clusters assembled on IscU scaffold proteins. *Journal of the American Chemical Society*, *134*, 15213–15216.

Shan, Y., Napoli, E., & Cortopassi, G. (2007). Mitochondrial frataxin interacts with ISD11 of the NFS1/ISCU complex and multiple mitochondrial chaperones. *Human Molecular Genetics*, *16*, 929–941.

Sheftel, A. D., Stehling, O., Pierik, A. J., Netz, D. J., Kerscher, S., Elsasser, H. P., et al. (2009). Human ind1, an iron-sulfur cluster assembly factor for respiratory complex I. *Molecular and Cellular Biology*, *29*, 6059–6073.

Sheftel, A. D., Wilbrecht, C., Stehling, O., Niggemeyer, B., Elsasser, H. P., Muhlenhoff, U., et al. (2012). The human mitochondrial ISCA1, ISCA2, and IBA57 proteins are required for [4Fe-4S] protein maturation. *Molecular Biology of the Cell*, *23*, 1157–1166.

Smith, A. D., Frazzon, J., Dean, D. R., & Johnson, M. K. (2005). Role of conserved cysteines in mediating sulfur transfer from IscS to IscU. *FEBS Letters, 579*, 5236–5240.

Song, J. Y., Marszalek, J., & Craig, E. A. (2012). Cysteine desulfurase Nfs1 and Pim1 protease control levels of Isu, the Fe-S cluster biogenesis scaffold. *Proceedings of the National Academy of Sciences of the United States of America, 109*, 10370–10375.

Srinivasan, V., Pierik, A. J., & Lill, R. (2014). Crystal structures of nucleotide-free and glutathione-bound mitochondrial ABC transporter Atm1. *Science, 343*, 1137–1140.

Stehling, O., Vashisht, A. A., Mascarenhas, J., Jonsson, Z. O., Sharma, T., Netz, D. J., et al. (2012). MMS19 assembles iron-sulfur proteins required for DNA metabolism and genomic integrity. *Science, 337*, 195–199.

Stehling, O., Wilbrecht, C., & Lill, R. (2014). Mitochondrial iron-sulfur protein biogenesis and human disease. *Biochimie, 100*, 61–77.

Sun, F., Huo, X., Zhai, Y., Wang, A., Xu, J., Su, D., et al. (2005). Crystal structure of mitochondrial respiratory membrane protein complex II. *Cell, 121*, 1043–1057.

Tonduti, D., Dorboz, I., Imbard, A., Slama, A., Boutron, A., Pichard, S., et al. (2015). New spastic paraplegia phenotype associated to mutation of NFU1. *Orphanet Journal of Rare Diseases, 10*, 13.

Torraco, A., Ardissone, A., Invernizzi, F., Rizza, T., Fiermonte, G., Niceta, M., et al. (2017). Novel mutations in IBA57 are associated with leukodystrophy and variable clinical phenotypes. *Journal of Neurology, 264*, 102–111.

Tsai, C. L., & Barondeau, D. P. (2010). Human frataxin is an allosteric switch that activates the Fe-S cluster biosynthetic complex. *Biochemistry, 49*, 9132–9139.

Tse Sum Bui, B., Benda, R., Schunemann, V., Florentin, D., Trautwein, A. X., & Marquet, A. (2003). Fate of the (2Fe-2S)(2+) cluster of *Escherichia coli* biotin synthase during reaction: A Mossbauer characterization. *Biochemistry, 42*, 8791–8798.

Tsukihara, T., Aoyama, H., Yamashita, E., Tomizaki, T., Yamaguchi, H., Shinzawa-Itoh, K., et al. (1995). Structures of metal sites of oxidized bovine heart cytochrome c oxidase at 2.8 Å. *Science, 269*, 1069–1074.

Urbina, H. D., Silberg, J. J., Hoff, K. G., & Vickery, L. E. (2001). Transfer of sulfur from IscS to IscU during Fe/S cluster assembly. *The Journal of Biological Chemistry, 276*, 44521–44526.

Uzarska, M. A., Dutkiewicz, R., Freibert, S.-A., Lill, R., & Mühlenhoff, U. (2013). The mitochondrial Hsp70 chaperone Ssq1 facilitates Fe/S cluster transfer from Isu1 to Grx5 by complex formation. *Molecular Biology of the Cell, 24*, 1830–1841.

Uzarska, M. A., Nasta, V., Weiler, B. D., Spantgar, F., Ciofi-Baffoni, S., Saviello, M. R., et al. (2016). Mitochondrial Bol1 and Bol3 function as assembly factors for specific iron-sulfur proteins. *eLife, 5*, e16673.

Van Vranken, J. G., Jeong, M. Y., Wei, P., Chen, Y. C., Gygi, S. P., Winge, D. R., et al. (2016). The mitochondrial acyl carrier protein (ACP) coordinates mitochondrial fatty acid synthesis with iron sulfur cluster biogenesis. *eLife, 5*, e17828.

Watmough, N. J., & Frerman, F. E. (2010). The electron transfer flavoprotein:Ubiquinone oxidoreductases. *Biochimica et Biophysica Acta, 1797*, 1910–1916.

Webert, H., Freibert, S. A., Gallo, A., Heidenreich, T., Linne, U., Amlacher, S., et al. (2014). Functional reconstitution of mitochondrial Fe/S cluster synthesis on Isu1 reveals the involvement of ferredoxin. *Nature Communications, 5*, 5013.

Wiedemann, N., Urzica, E., Guiard, B., Muller, H., Lohaus, C., Meyer, H. E., et al. (2006). Essential role of Isd11 in mitochondrial iron-sulfur cluster synthesis on Isu scaffold proteins. *The EMBO Journal, 25*, 184–195.

Wu, M., Gu, J., Guo, R., Huang, Y., & Yang, M. (2016). Structure of mammalian respiratory supercomplex I1III2IV1. *Cell, 167*, 1598–1609, e1510.

Yan, R., Adinolfi, S., & Pastore, A. (2015). Ferredoxin, in conjunction with NADPH and ferredoxin-NADP reductase, transfers electrons to the IscS/IscU complex to promote iron-sulfur cluster assembly. *Biochimica et Biophysica Acta*, *1854*, 1113–1117.

Yoon, T., & Cowan, J. A. (2003). Iron-sulfur cluster biosynthesis. Characterization of frataxin as an iron donor for assembly of [2Fe-2S] clusters in ISU-type proteins. *Journal of the American Chemical Society*, *125*, 6078–6084.

Yoon, H., Knight, S. A., Pandey, A., Pain, J., Turkarslan, S., Pain, D., et al. (2015). Turning Saccharomyces cerevisiae into a Frataxin-independent organism. *PLoS Genetics*, *11*, e1005135.

Yoon, H., Knight, S. A., Pandey, A., Pain, J., Zhang, Y., Pain, D., et al. (2014). Frataxin-bypassing Isu1: Characterization of the bypass activity in cells and mitochondria. *The Biochemical Journal*, *459*, 71–81.

Zhang, Y., Lyver, E. R., Knight, S. A., Pain, D., Lesuisse, E., & Dancis, A. (2006). Mrs3p, Mrs4p, and frataxin provide iron for Fe-S cluster synthesis in mitochondria. *The Journal of Biological Chemistry*, *281*, 22493–22502.

Zhu, J., Vinothkumar, K. R., & Hirst, J. (2016). Structure of mammalian respiratory complex I. *Nature*, *536*, 354–358.

CHAPTER TEN

Approaches to Interrogate the Role of Nucleotide Hydrolysis by Metal Trafficking NTPases: The Nbp35–Cfd1 Iron–Sulfur Cluster Scaffold as a Case Study

John D. Grossman, Eric J. Camire, Deborah L. Perlstein[1]
Boston University, Boston, MA, United States
[1]Corresponding author: e-mail address: dperl@bu.edu

Contents

Abstract

Nucleotide hydrolases play integral yet poorly understood roles in several metallocluster biosynthetic pathways. For example, the cytosolic iron–sulfur cluster assembly (CIA) is initiated by the CIA scaffold, an ATPase which builds new iron–sulfur clusters for proteins

Methods in Enzymology, Volume 599
ISSN 0076-6879
https://doi.org/10.1016/bs.mie.2017.11.005

localized to the cytosol and the nucleus in eukaryotic organisms. While in vivo studies have demonstrated the scaffold's nucleotide hydrolase domain is vital for its function, in vitro approaches have not revealed tight allosteric coupling between the cluster scaffolding site and the ATPase site. Thus, the role of ATP hydrolysis has been hard to pinpoint. Herein, we describe methods to probe the nucleotide affinity and hydrolysis activity of the CIA scaffold from yeast, which is comprised of two homologous polypeptides called Nbp35 and Cfd1. In particular, we report two different equilibrium binding assays that make use of commercially available fluorescent nucleotide analogs. Importantly, these assays can be applied to probe nucleotide affinity of both the apo- and holo-forms of the CIA scaffold. Generally, these fluorescent nucleotide analogs have been underutilized to probe metal trafficking NTPase because one of the most commonly used probes, mantATP, which is labeled with the methylanthraniloyl probe via the 2′ or 3′ sugar hydroxyls, has an absorption which overlaps with the UV–Vis features of many metal-binding proteins. However, by exploiting analogs like BODIPY-FL and trinitrophenyl-labeled nucleotides which have better photophysical properties for metalloprotein applications, these approaches have the potential to reveal the mechanistic underpinnings of NTPases required for metallocluster biosynthesis.

1. INTRODUCTION

1.1 Introduction to Cytosolic Iron–Sulfur Cluster Assembly and the Nbp35–Cfd1 Scaffold

Iron–sulfur (FeS) clusters are required for countless biochemical pathways including several of the most fundamental processes of life, such as respiration, nitrogen fixation, and photosynthesis (Johnson, Dean, Smith, & Johnson, 2005). Since the assembly and insertion of these vital metalloclusters requires a dedicated biosynthetic pathway, cluster biogenesis proteins are essential and conserved from bacteria to humans (Johnson et al., 2005; Lill, 2009). Recent work has shown that defects in FeS biogenesis can directly result in nuclear genome instability (Veatch, McMurray, Nelson, & Gottschling, 2009). This linkage between FeS cluster homeostasis and DNA integrity is likely mediated by the growing number of DNA replication and repair enzymes that require FeS cofactors for their activity (Arnold, Grodick, & Barton, 2016; Banda, Nunez, Burnside, Bradshaw, & David, 2017; Fuss, Tsai, Ishida, & Tainer, 2015; Netz, Stith, et al., 2012). Thus, there is great interest in elucidating the mechanism by which [Fe_4S_4] proteins residing outside of the mitochondria acquire their cofactors. This process is carried out by the cytosolic iron–sulfur cluster assembly (CIA) system (Paul & Lill, 2015).

While the mitochondrial FeS biogenesis machinery is homologous to the well-studied bacterial *ISC* system, many of the CIA factors are unique to eukaryotes or, if they do have bacterial homologs, the bacterial enzymes have not yet been extensively investigated. For example, the CIA pathway begins with a cluster scaffolding complex that assembles nascent clusters (Fig. 1). Although the CIA scaffold is distinct in sequence and structure from canonical U-type scaffolds found in the mitochondria, both types of scaffolds have cysteine motifs for building and binding newly forged clusters (Blanc, Gerez, & Ollagnier de Choudens, 2015; Netz, Pierik, et al., 2012). However, U-type scaffolds do not have NTPase domains nor do they require nucleotide hydrolysis to assemble clusters. In contrast, CIA scaffold harbors an ATPase domain that is essential for cluster biogenesis in the cytosol (Camire, Grossman, Thole, Fleischman, & Perlstein, 2015; Netz, Pierik, et al., 2012; Roy, Solodovnikova, Nicholson, Antholine, & Walden, 2003; Vitale, Fabre, & Hurt, 1996). Recent work has shown that mutation of the ATPase site results in a defect in cluster assembly on the scaffold in vivo (Netz, Pierik, et al., 2012; Vitale et al., 1996). However, it has been challenging to dissect the molecular details underlying how the nucleotide hydrolase domain influences the cluster scaffolding function with in vitro approaches. For example, it was reported by the Lill lab that the rate of

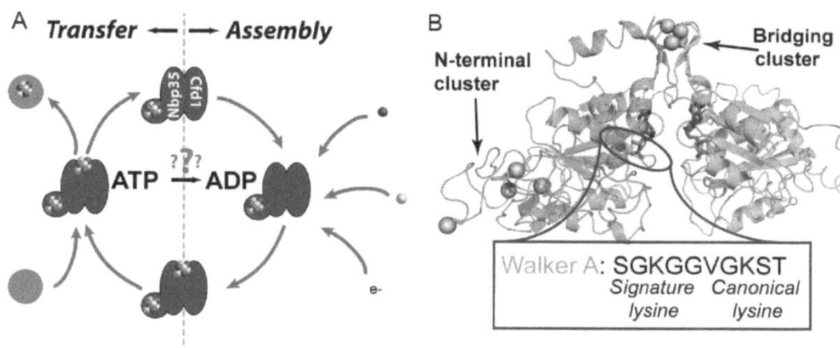

Fig. 1 (A) As the CIA scaffold, the yeast Nbp35–Cfd1 complex must assemble a nascent cluster and transfer this cluster to a downstream recipient. This cluster scaffolding activity requires nucleotide hydrolysis in vivo, but the role of the ATPase activity is not well understood. Since the CIA scaffold and its homologs belong to the deviant Walker A family of nucleotide hydrolases, we hypothesize that nucleotide binding and hydrolysis help the scaffold to switch between the cluster assembly phase (*right*) and cluster transfer phase (*left*) of its reaction cycle. (B) A threading model of the Nbp35–Cfd1 complex generated by the Phyre2 server based on the unpublished structure of an archaeal homolog (PDB-ID: 3KB1) (Kelley, Mezulis, Yates, Wass, & Sternberg, 2015). Both the ATPase site and the cluster scaffolding site are situated at the dimer interface. The P-loop or Walker A motif of Nbp35 and Cfd1 contains two lysines, the canonical lysine and the signature lysine of the deviant Walker A family of NTPases (Leipe, Wolf, Koonin, & Aravind, 2002).

cluster transfer from scaffold to target is unaffected by the inclusion of nucleotide in the assay mixture (Netz, Pierik, Stümpfig, Mühlenhoff, & Lill, 2007). A similar finding was reported for a bacterial homolog derived from *Salmonella enterica* called ApbC (Boyd, Pierik, Netz, Lill, & Downs, 2008). Additionally, the ATP hydrolysis kinetics appear to be unaffected by the presence of cluster in the scaffolding site as both apo- and holo-scaffold, where the FeS clusters have been chemically reconstituted, exhibit similar ATP hydrolysis kinetics (Boyd, Sondelski, & Downs, 2009; Camire et al., 2015). Finally, mutation to some residues of ApbC results in a 30%–35% drop in the rate of cluster transfer from the scaffold, whereas other mutations lead to complete inhibition of nucleotide hydrolysis but they do not produce any measurable effect on the rate of cluster transfer in vitro (Boyd, Sondelski, et al., 2009). Thus, the role of nucleotide hydrolysis has been challenging to unravel.

1.2 The CIA Scaffold and Its Homologs Are Members of the Deviant Walker A Family of NTPases

One hint as to the possible function of nucleotide binding and hydrolysis might play in the assembly and transfer of a nascent cluster from the scaffold can be gained from the realization that the CIA scaffold and its homologs belong to a large and functionally diverse family of NTPases called the "deviant Walker A NTPases" (Leipe et al., 2002). The deviant Walker A NTPase family is comprised of the MRP/Nbp35 family of cluster scaffolding ATPases and many other NTPase families, playing roles in several biochemical processes including heavy metal resistance, nitrogen reduction, bacterial plasmid partitioning, regulation of bacterial cell division, and membrane protein insertion (Bange & Sinning, 2013; Bhattacharjee et al., 2000; Lahiri, Pulakat, & Gavini, 2008; Lutkenhaus, 2012; Lutkenhaus & Sundaramoorthy, 2003; Shan, 2016). The hallmark of the family is the presence of two lysines in the P-loop or Walker A motif, the canonical Walker A lysine and an additional "signature" lysine unique to the deviant Walker A family (Fig. 1B) (Leipe et al., 2002). Although the functions of proteins within this family are quite diverse, they share the common theme that nucleotide binding and hydrolysis drive conformational changes mediated by the signature lysine (Lahiri et al., 2008; Lutkenhaus & Sundaramoorthy, 2003; Mateja et al., 2009; Schindelin, Kisker, Schlessman, Howard, & Rees, 1997; Zhou, Radaev, Rosen, & Gatti, 2000). This nucleotide-driven conformational change regulates dynamic interactions with upstream and/or downstream protein(s) in the pathway (Bange & Sinning, 2013; Lutkenhaus, 2012;

Schindelin et al., 1997; Shan, 2016). In turn, the protein–protein interaction partner or, in some cases, a small molecule substrate or macromolecular cargos affect nucleotide affinity and/or hydrolysis activity of the NTPase. This commonality in function across the deviant Walker A family suggests that the CIA scaffold and its homologs utilize nucleotide to regulate interaction(s) with other CIA factors, such as those required to assemble a nascent cluster on the scaffold or to mobilize this cluster to a downstream recipient.

Herein, we describe how kinetic assays and fluorescent nucleotide analogs can be exploited to probe the role of nucleotide hydrolysis by the yeast CIA scaffold. Notably, Nbp35, Cfd1, and their homologs are not the only metal-binding NTPases required for metallocofactor maturation. In fact, several other metal trafficking and metallocluster biosynthetic pathways also contain NTPases (Table 1). Yet, the role of the nucleotide in many of these other systems is also poorly understood. We believe wider adoption of the fluorescence-based approaches described herein has the potential to reveal the mechanistic underpinnings of several metallocluster biosynthetic enzymes.

2. CLONING, EXPRESSION, AND PURIFICATION OF THE CIA SCAFFOLD

In yeast, the CIA scaffold is comprised of two homologous polypeptides called Nbp35 and Cfd1 (Netz et al., 2007; Roy et al., 2003; Vitale et al., 1996). Both Nbp35 and Cfd1 are prototypical members of the MRP/Nbp35 family of cluster scaffolding ATPases (Leipe et al., 2002). Both proteins are highly conserved across all eukaryotes, except Archaeplastida (red algae, green algae, and land plants) and the SAR supergroup (Stramenopiles, Alveolates, and Rhizaria) are missing the Cfd1 ortholog (Bych, Netz, et al., 2008; Tsaousis, Gentekaki, Eme, Gaston, & Roger, 2014). Some eukaryotes encode additional paralogs housed within organelles. The mitochondrial Ind1, also known as NUBPL, appears to play a role in Complex I biosynthesis, and the HCF101 homolog found within the chloroplast is important for assembly of Photosystem I (Bych, Kerscher, et al., 2008; Schwenkert et al., 2010). In addition, MRP/Nbp35 family is widely distributed in bacteria and archaea (Leipe et al., 2002). However, the role(s) of these proteins is not well understood as they have not been extensively investigated, with the exception of the ApbC analog from *S. enterica* (Boyd, Drevland, Downs, & Graham, 2009; Boyd et al., 2008; Boyd, Sondelski, et al., 2009). In this section, we describe the strategy to clone,

Table 1 A Select List of NTPases Required for Metallocofactor Maturation

Protein	Pathway or Enzyme	Nucleotide	Metal or Metallocofactor	Refs[a]
ArsA[b]	Detoxification of heavy metals	ATP	AsO_3^{3-}, SbO_3^{3-}	Walmsley, Zhou, Borges-Walmsley, and Rosen (2001), Karkaria and Rosen (1991)
NifH[b]	Mo-dependent nitrogenase maturation	ATP	4Fe–4S, MoO_4^{2-}	Hu, Corbett, et al. (2006), Hu, Corbettt, et al. (2006)
HydF	[FeFe] Hydrogenase maturation	GTP	2Fe–2S, 4Fe–4S	Shepard et al. (2010)
HypB	[NiFe] Hydrogenase maturation	GTP	Ni^{2+}	Gasper, Scrima, and Wittinghofer (2006), Lacasse and Zamble (2016), Sydor, Lebrette, Ariyakumaran, Cavazza, and Zamble (2014)
SufC	*SUF* system FeS cluster assembly	ATP	4Fe–4S	Hirabayashi et al. (2015), Eccleston, Petrovic, Davis, Rangachari, and Wilson (2006), Petrovic et al. (2008)
CooC[b]	Carbon monoxide dehydrogenase maturation	ATP	Ni^{2+}	Jeoung, Giese, Grunwald, and Dobbek (2009, 2010)
UreG	Urease maturation	GTP	Ni^{2+}	Fong et al. (2013), Soriano and Hausinger (1999)
MeaB	Methylmalonyl-CoA mutase maturation	GTP	Adenosyl-cobalamin	Padovani and Banerjee (2009), Jost, Cracan, Hubbard, Banerjee, and Drennan (2015)

[a]Representative publications investigating the role of nucleotide in metal trafficking or metallocofactor assembly and/or employing the use of fluorescent nucleotide probes.
[b]Also members of the deviant Walker A family of NTPases.

express, and purify Nbp35 and Cfd1 (Camire et al., 2015; Netz et al., 2007). We expect analogous approaches could be and have been successful for accessing other proteins in the MRP/Nbp35 family.

2.1 Cloning Strategy

Nbp35 and Cfd1 can form homocomplexes and heterocomplexes (Camire et al., 2015; Netz et al., 2007). Our lab has observed dimeric complexes between Nbp35 and Cfd1 (Camire et al., 2015), and the Lill lab has reported a heterotetrameric complex of Nbp35 and Cfd1 (Netz et al., 2007). In total, three different quaternary structures of the CIA scaffold have been investigated, an Nbp35 homocomplex, a Cfd1 homocomplex, and an Nbp35–Cfd1 heterocomplex. Curiously, these quaternary states appear to be quite stable and do not readily reorganize in extracts of *Escherichia coli* or yeast, or following mixing of the purified proteins in vitro (Camire et al., 2015; Pallesen, Solodovnikova, Sharma, & Walden, 2013). Therefore, to obtain the heterocomplex, Nbp35 and Cfd1 must be cotranslated.

For heterologous overexpression in *E. coli*, simultaneous expression of Nbp35 and Cfd1 is achieved via the pET-Duet series of vectors (EMD Millipore) (Camire et al., 2015; Netz et al., 2007). The pRSF-Duet vector houses a kanamycin resistance gene for selection, a RSF1030 replicon, and two multiple cloning sites, MCS-1 and MCS-2, each preceded by an IPTG-inducible T7 promoter, *lac* operator, and ribosome-binding site. This plasmid is compatible with most expression strains derived from BL21 (DE3); however, their use in pLysS strains can lead to diminished expression due to the presence of a plasmid bearing the P15A (pACYC184) replicon in these strains. A His-tag precedes MCS-1 and an S-tag is present after MCS-2. For Nbp35 and Cfd1, we have found that N-terminal tags yield the highest overexpression as compared to C-terminal affinity tags.

For purification of the heterodimer, we have found that tandem affinity purification is required to isolate the heterodimer away from the homodimeric complexes which have different nucleotide affinities and hydrolysis activities (Camire et al., 2015). We and others have employed the His- and Strep-II tags (IBA Lifesciences) appended to Cfd1 and Nbp35, respectively (Camire et al., 2015; Netz et al., 2007). The His-tag is introduced by cloning Cfd1 in-frame with the six consecutive histidine residues encoded immediately before MCS-1n in the pRSF-Duet vector. This allows for purification of complexes containing His-Cfd1, the Cfd1 homodimer and the Nbp35–Cfd1 heterodimer, via immobilized metal affinity chromatography (IMAC). The Strep-II peptide tag (WSHPQFEKSA) is appended to the N-terminus of Nbp35. This tag allows for purification of complexes containing Nbp35, the Nbp35 homodimer and the Nbp35–Cfd1 heterodimer, via streptactin chromatography. The streptactin resin features a modified streptavidin engineered to bind to the eight residue Strep-II tag. The desired protein is then eluted from the streptactin resin with desthiobiotin or biotin. Since the Strep-II tag is not present in the pET-Duet series of vectors, we have introduced the tag during the PCR amplification of Nbp35 before its insertion into MCS-2 of pRSF-Duet. While the Strep-tagged Nbp35 homodimer binds well to the column, the binding of the Nbp35–Cfd1 heterocomplex can be weak, leading to loss of protein during washing. To address this problem, we have also recently appended a Twin-Strep tag to the N-terminus of Nbp35 (Schmidt et al., 2013). This tag (WSHPQFEKGGGSGGGSGGSAWSHPQFEK) encodes two Strep tags separated by a spacer that allows for a tighter interaction with the affinity resin and increased recovery of the Nbp35–Cfd1 complex following the streptactin affinity purification step.

2.2 Protein Overproduction in *E. coli*

We express the CIA scaffold proteins in the standard BL21(DE3) *E. coli* strain. We have found that the plasmids for expression of the CIA scaffold proteins must be freshly transformed into the expression strain for good overexpression. Overall, Cfd1 expresses quite well, whereas the expression level of Nbp35 can be variable. Therefore, we have found it necessary to analyze samples of the cultures before and after the addition of isopropyl-β-D-1-thiogalactopyranoside (IPTG) to assess overexpression before continuing on with purification of the protein. It is particularly important to compare cells derived from cultures before and after IPTG addition since uninduced cells have bands of molecular weights similar to those of Nbp35 and Cfd1 (Fig. 2A). Overexpression is achieved by growing the cells in standard LB media with the appropriate antibiotics to maintain plasmid selection. The cells are grown to log phase, then inducing protein production by the addition of IPTG (1 m*M*). Cells are harvested by centrifugation and stored at −80°C.

2.3 Purification of the CIA Scaffold

The Nbp35 and Cfd1 homodimers are obtained via a standard His-tag purification using IMAC. The only complicating factor is Nbp35's low expression, which can result in lower purity of the Nbp35 homodimer compared

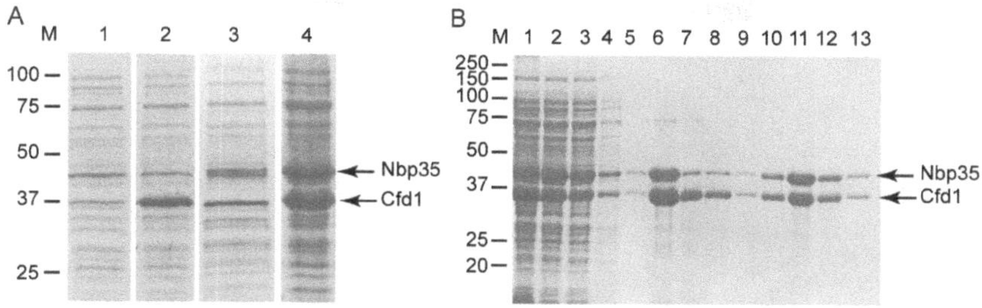

Fig. 2 SDS-PAGE analysis of CIA scaffold expression and purification. (A) Expression of the CIA scaffold proteins. Lane 1: Preinduction; Lane 2: Cfd1 induction; Lane 3: Nbp35 induction; Lane 4: Nbp35–Cfd1 complex induction. (B) Purification of the Nbp35–Cfd1 complex. Lane 1: Whole-cell extract; Lane 2: Clarified lysate; Lane 3: His column flow-through; Lanes 4–5: His column washes; Lane 6: Pooled His column elution; Lane 7: Streptactin column flow-through; Lanes 8–9: Streptactin column washes; Lanes 10–13: Streptactin column elution fractions.

to the heterodimer or the Cfd1 homodimer. The major challenge in the purification of Nbp35–Cfd1 heterocomplex is isolating the 1:1 heterodimer from the homodimers. This is achieved through the use of a two-step affinity chromatography (IMAC followed by streptactin) purification, taking advantage of the His-tag on Cfd1 and the Strep-tag on Nbp35. When the expression of Nbp35 and Cfd1 is balanced, it often appears as if the proteins are 1:1 after only the IMAC step. However, the IMAC eluate will contain a mixture of Cfd1 homodimer and the Nbp35–Cfd1 heterodimer. Since these different quaternary states exhibit very different nucleotide hydrolysis kinetics (Camire et al., 2015), we have found that it is absolutely essential to complete the tandem affinity purification before in vitro analysis of nucleotide affinity or hydrolysis activity. Although one could exploit the two chromatography steps any order, we have found that purifying the protein via IMAC followed by streptactin is advantageous since it can prolong the life of the more expensive and less robust streptactin resin. Purification steps are carried out stepwise in 1 day with chilled buffers and the following protocol described below.

2.3.1 Equipment

- Nickel–NTA agarose resin (GoldBio) or a similar resin from other vendors
- Streptactin resin (IBA Life Sciences)
- Amicon Ultra-15 centrifugal concentrator (EMD Millipore) or similar device
- PD10 desalting column (GE Healthcare) or similar gel filtration columns
- Standard SDS-PAGE gel equipment

2.3.2 Buffers and Reagents

- Lysis buffer: 50 m*M* Tris (hydroxymethyl)-aminomethane hydrochloride (Tris–HCl; pH 8.0), 300 m*M* sodium chloride (NaCl), 5 m*M* betamercaptoethanol (β-ME), 10 (v/v)% glycerol. The β-ME is added immediately before use
- HisWash buffer: 50 m*M* Tris–HCl (pH 8.0), 300 m*M* NaCl, 5 m*M* β-ME, 10 (v/v)% glycerol, 5 m*M* imidazole
- HisElution buffer: 50 m*M* Tris–HCl (pH 8.0), 300 m*M* NaCl, 5 m*M* β-ME, 10 (v/v)% glycerol, 250 m*M* imidazole
- StrepWash buffer: 50 m*M* Tris–HCl (pH 8.0), 300 m*M* NaCl, 5 m*M* 1,4-dithiothreitol (DTT) 10 (v/v)% glycerol. Add DTT immediately before use from a 1 *M* stock
- StrepElution buffer: 50 m*M* Tris–HCl (pH 8.0), 300 m*M* NaCl, 5 m*M* DTT, 10 (v/v)% glycerol, 2.5 m*M* D-desthiobiotin
- Storage buffer: 50 m*M* Tris–HCl (pH 8.0), 100 m*M* NaCl, 5 m*M* DTT, 10 (v/v)% glycerol
- PMSF: 100 m*M* phenylmethylsulfonyl fluoride (PMSF) in ethanol
- Pierce universal nuclease, 250 U/μL (ThermoFisher)
- Lysozyme from hen egg white (GoldBio)

2.3.3 Procedure for Purification of the Nbp35–Cfd1 Complex

1. Prepare the buffers at room temperature and then chill them to 4°C. Add all reducing agents (β-ME or DTT) to buffers immediately before use.
2. Remove frozen cell pellet (Section 2.2) to Erlenmeyer flask and record mass. Add 10 mL lysis buffer per gram cell pellet. Typically, we purify from ~14 to 16 g cell paste. Stir with spatula and/or stir bar until the mixture is homogeneous and the cells are completely resuspended.
3. Add imidazole (5 m*M* final concentration), PMSF (1 m*M* final concentration), lysozyme (1 mg/mL final concentration), nuclease (5 U/mL) and stir for 1 h on ice.
4. Place sample in an ice water bath inside the sonicator and sonicate for six cycles of 20 s on and 40 s off at 65% amplitude. Care should be taken to ensure that temperature does not increase and frothing is kept to a minimum.
5. Remove a small sample for SDS-PAGE analysis and centrifuge remaining lysate at 48,000 × *g* for 45 min. While centrifuging, equilibrate Ni–NTA resin with 10 column volumes (CV) of HisWash buffer. Typically, we use 2 mL of resin for every 10–15 g of wet cell paste.

6. Flow the clarified lysate over the Ni–NTA resin followed by 15 CV of HisWash buffer. Samples of the clarified lysate, flow-through, and wash fractions should be reserved for SDS-PAGE analysis.
7. Elute protein with 10 CV of HisElution buffer and pool fractions containing protein.
8. Pass pooled His elution over 2 mL Streptactin resin preequilibrated with 10 CV StrepWash buffer. Wash with 5 CV StrepWash buffer, followed by elution with StrepElution buffer. Remove small aliquots for SDS-PAGE. Pool fractions containing protein.
9. Concentrate pooled Strep elution fractions with an Amicon centrifugal concentrator to a final volume of no more than 1.5–2 mL.
10. While protein is concentrating, equilibrate a PD10 desalting column with storage buffer. Once concentrated, buffer exchange the protein over the PD10 and pool protein-containing fractions.
11. Determine protein concentration by established methods (i.e., Bradford), then aliquot, and flash freeze in liquid nitrogen before storing the protein at −80°C.
12. Analyze fractions reserved in steps 5, 6, and 8 via SDS-PAGE. Expected results shown in Fig. 2.

2.4 Chemical Reconstitution of the FeS Clusters of the Nbp35–Cfd1 Heterodimer

As a scaffold, Nbp35 and Cfd1 can rapidly transfer the scaffold-assembled cluster to an apo-recipient (Fig. 1A) (Netz et al., 2007). The number of [Fe_4S_4] clusters bound by the CIA scaffold depends on the quaternary state of the protein (Netz, Pierik, et al., 2012). As prototypical members of the MRP/Nbp35 family of scaffolds, Nbp35 and Cfd1 each contain a CxxC motif close to their C-termini that is used to ligate an [Fe_4S_4] cluster which bridges the protein–protein interaction interface (Fig. 3) (Netz, Pierik, et al., 2012). Nbp35 has an additional domain at its N-terminus which binds an [Fe_4S_4] cluster (Fig. 3). Thus, the Cfd1 homodimer can bind one [Fe_4S_4] cluster, the Nbp35–Cfd1 heterodimer can bind two [Fe_4S_4] clusters, and the Nbp35 homodimer can bind three [Fe_4S_4] clusters (Fig. 3B). Although the precise role of each FeS cluster is not well understood, the current working model is that Nbp35's N-terminal cluster is assembled first and remains stably bound to the protein, whereas the bridging cluster is the site of nascent cluster assembly and transfer (Fig. 1) (Netz, Pierik, et al., 2012; Netz et al., 2007; Pallesen et al., 2013). Since the bridging cluster binding site is likely to be surface exposed (Fig. 1B), the bridging cluster is prone to oxidative

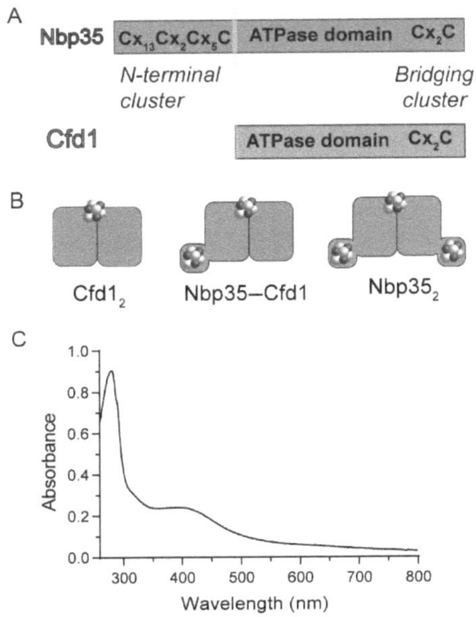

Fig. 3 (A) The primary structure of Nbp35 and Cfd1 highlighting the relative locations of the cluster-ligating cysteine motifs and ATPase domain. (B) Each quaternary state of the CIA scaffold binds a different number of FeS clusters. (C) UV–Vis spectrum of the Nbp35–Cfd1 heterodimer following reconstitution of its FeS clusters.

degradation. Therefore, aerobic purification of the CIA scaffold largely results in purification of the apo-form of the protein. To study the cluster bound proteins, the holo-form must be generated via chemical reconstitution of the [Fe_4S_4] cluster(s). Generally, we have found that solutions with 1.0–1.5 mg/mL of the protein with a 12-fold molar excess of ferrous iron and sulfide over the dimeric concentration of the protein can successfully achieve stoichiometric reconstitution of the two [Fe_4S_4] clusters per heterodimer. Our protocol for anaerobic reconstitution of the FeS clusters is described below.

2.4.1 Equipment

- Anaerobic chamber (COY Lab Products)
- Coolbox (BioCision) or Lab Armor beads
- Sephadex G-25 column or equivalent desalting column within the anaerobic chamber
- Microcentrifuge within the anaerobic chamber
- Anaerobic pipette tips and microcentrifuge tubes

2.4.2 Buffers and Reagents

- Iron(II) ammonium sulfate hexahydrate: 100 m*M* $Fe(NH_4)_4(SO_4)_2$ in anaerobic water
- Sodium sulfide nonahydrate: 100 m*M* Na_2S in anaerobic water
- Magnesium chloride hexahydrate: 1 *M* $MgCl_2$ in anaerobic water
- Dithiothreitol (DTT): 1 *M* DTT in anaerobic water
- Anaerobic storage buffer (Section 2.3.2)

2.4.3 Procedure

1. Anaerobic water, glycerol, and consumables are prepared at least 72 h before beginning the reconstitution as described previously in this series (Lanz et al., 2012). The night before reconstitution, bring the following solid reagents into the box: 19 mg $Fe(NH_4)_4(SO_4)_2$, 12 mg $Na_2S{\cdot}9H_2O$, 77 mg DTT, and 102 mg $MgCl_2{\cdot}6H_2O$. The following day, add 0.5 mL anaerobic water to make the following solutions: 100 m*M* $Fe(NH_4)_4(SO_4)_2$, 100 m*M* Na_2S, 1 *M* DTT, 1 *M* $MgCl_2$.
2. To degas and prereduce the protein, bring in 1 mL aliquots of 1.0–1.5 mg/mL protein in a 1.5-mL tube into the box. Add DTT from the anaerobic stock to a final concentration of 5 m*M*. Slowly stir with a micro stir bar for 2 h at room temperature with the cap open. For temperature-sensitive proteins, degassing can be carried out on a cold block or Armor beads for 3–4 h.
3. Make a 1:10 dilution of the 100 m*M* iron and sulfide stock solutions to make a 10 m*M* working stock with anaerobic water. Add in a 12-fold molar excess of iron over the dimeric concentration of the Nbp35–Cfd1 complex from the 10 m*M* $Fe(NH_4)_4(SO_4)_2$ solution. Add iron slowly in small aliquots with stirring to limit high localized concentrations of the stock. Incubate at room temperature for 10 min with gentle stirring.
4. Add a 12-fold molar excess of sulfide from the 10 m*M* Na_2S solution in the same manner as described for the iron in step 3. Move the tubes to a cold block, cap, and incubate with gentle stirring for 2 h. During this time, the solution should noticeably change color to a yellow-brown to brown color.
5. Centrifuge the sample at maximum speed for 10 min to pellet any precipitated protein.
6. Apply the supernatant to a 20 mL G-25 desalting column equilibrated with anaerobic storage buffer without DTT.
7. Collect elution fractions and concentrate protein-containing fractions with an Amicon micro concentrator.

2.5 Quantification of FeS Cluster Insertion

Immediately following reconstitution, it is important to quantify the amount of iron and sulfide bound to assess the extent of cluster loading. A quick way to monitor the extent of the reconstitution is to take a UV–Vis spectrum of the sample using an anaerobically sealed quartz cuvette. The [Fe_4S_4] clusters give a characteristic, broad peak at 410 nm (Fig. 3C). To quantitatively determine the amount of iron and sulfide bound, we have adapted the use of a ferrozine assay and a sulfide assay for the determination of iron and sulfide, respectively (Beinert, 1978, 1983; Fish, 1988; Stookey, 1970). When one is characterizing a new protein, it is advantageous to characterize the FeS clusters via electron paramagnetic resonance spectroscopy and Mossbauer spectroscopy to correctly assign the number and type of FeS cluster(s) (Netz, Pierik, et al., 2012; Netz et al., 2007).

The ferrozine method has been described in numerous variations in the literature. In essence, iron is removed from the protein and solubilized by treatment with acid. The precipitated protein is separated by centrifugation. The supernatant containing the liberated iron is treated with sodium ascorbate to reduce the iron and ferrozine (3-(2-pyridyl)-5,6-bis(4-phenylsulfonic acid)-l,2,4-triazine) is added to yield a purple species. This colored complex has a maximum absorbance at 562 nm with an extinction coefficient of 27.9 mM^{-1} cm^{-1}. The absorption at 562 nm is used to calculate the concentration of iron in the sample. In our case, we would expect eight iron atoms per heterodimeric Nbp35–Cfd1 complex (Netz, Pierik, et al., 2012).

The sulfide assay is adapted from the method of Beinert (Beinert, 1983) and has been recently described by others in this series (Yan & Kiley, 2009). Briefly, the sulfide reacts with *N*,*N*-dimethyl-*p*-phenylenediamine (DMPD) to form the methylene blue product which has a known extinction coefficient of 34.5 mM^{-1} cm^{-1} at 670 nm. However, special care must be taken to avoid loss of the sulfide as the H_2S gas (Yan & Kiley, 2009). Since this assay can be technically challenging to perform for novices, we encourage new researchers in our group to practice the assay with standard solutions of sodium sulfide before using the assay to determine the amount of sulfide associated with FeS-binding proteins.

2.5.1 Equipment

- Stoppered quartz cuvette
- Micro magnetic stir bars

- UV–Vis spectrophotometer
- Amicon Ultra-0.5 mL centrifugal concentrator (EMD Millipore) or similar
- Microcentrifuge

2.5.2 Buffers and Reagents

- Storage buffer: described in Section 2.4.2
- Hydrochloric acid: 2 *M* HCl
- Ferrozine (3-(2-pyridyl)-5,6-diphenyl-1,2,4-triazine-4′,4″-disulfonic acid sodium salt): 10 m*M* in water, make fresh before beginning the assay
- Sodium ascorbate: 75 m*M* in water, make fresh before beginning the assay
- Saturated ammonium acetate ($NH_4CO_2CH_3$) solution (~19 *M*) in water
- Zinc acetate ($Zn(OAc)_2$) dihydrate: 1% solution in water, make fresh before beginning the assay
- Sodium hydroxide (NaOH): 12% NaOH
- DMPD: 0.1% DMPD in 5 *M* HCl
- Ferric chloride ($FeCl_3$): 23 m*M* $FeCl_3$ in 1.2 *M* HCl

2.5.3 Ferrozine Assay Procedure

1. Remove three, 30 μL aliquots of the reconstituted protein (Section 2.4.3) from the Coy chamber in 1.5-mL Eppendorf tubes. Also remove a 30 μL aliquot of the buffer used to elute the G25 column for the background determination.
2. Add 70 μL of 2 *M* HCl and incubate for 5 min on a heat block set to 95°C.
3. After incubation, centrifuge at 16,000 × *g* for 10–15 min to pellet precipitated protein. Transfer supernatants to fresh tubes and reserve.
4. Carefully resuspend the pellet in 100 μL of 2 *M* HCl and incubate at 95°C for 15 min. Centrifuge again as in step 3.
5. Combine the supernatants from steps 3 and 4 for a total volume of 200 μL and add 360 μL water.
6. Add 40 μL 75 m*M* sodium ascorbate and 200 μL 10 m*M* ferrozine to each tube and vortex briefly to mix.
7. Add 200 μL of saturated ammonium acetate (~19 *M*) and vortex to mix. Take care when pipetting as the saturated solution is very viscous. The final volume of the tubes should be 1 mL and purple in color.
8. Record the UV–Vis spectrum of the samples from 400 to 700 nm. Samples containing iron will display a peak at 562 nm. The absorbance value of this peak, minus the absorbance value obtained from the buffer without any protein, can be used to calculate the amount of iron associated with the protein using the extinction coefficient (27.9 mM^{-1} cm^{-1}).

9. To determine the ratio of iron to protein, determine the protein concentration of the reconstituted sample by the method of Bradford using BSA as the standard. The concentration of the iron in solution divided by the concentration of the dimeric CIA scaffold yields the number of iron atoms bound per dimer.

2.5.4 Sulfide Assay Procedure

1. Prepare three, 200 μL aliquots of the protein from the Coy chamber in 1.5-mL tubes. These tubes should contain approximately 20 μ*M* of sulfide, calculated from the protein concentration and the expected number of sulfur atoms bound. Also remove a 200 μL aliquot of the buffer.
2. Add a micro stir bar and stir gently. Load a pipette with 600 μL fresh 1% $Zn(OAc)_2$ and another pipette with 30 μL 12% NaOH. While stirring gently, add the zinc acetate followed immediately by the sodium hydroxide. Cap tubes and stir gently at ambient temperature for 45 min. The solution will turn cloudy.
3. After the incubation period, turn off the stir plate and underlay the cloudy solution with 150 μL fresh 0.1% DMPD. Then, immediately after, inject 30 μL of 23 m*M* $FeCl_3$ just under the surface of the solution. Cap tubes and resume with stirring until homogeneous, more vigorous stirring may be required to fully homogenize the solution. Continue stirring for 15 min.
4. Remove stir bars and centrifuge the tubes for 20 min at room temperature to pellet the precipitated protein. Remove the supernatant for analysis.
5. Take a UV–Vis scan of the samples from 550 to 800 nm. Samples will be blue and spectra will display a characteristic spectrum with a valley at 710 nm and two peaks at 670 and 750 nm. The absorbance value of the sample's 670 nm peak, minus the absorbance value of the buffer-only control, can be used to calculate the amount of sulfide in the sample using the extinction coefficient ($34.5\ mM^{-1}\ cm^{-1}$). As with the ferrozine assay, the concentration of sulfide divided by the concentration of the protein dimer yields the number of sulfide bound per dimer.

3. KINETIC ASSAYS OF NUCLEOTIDE HYDROLYSIS

The ATP hydrolysis kinetics of MRP/Nbp35 cluster scaffolds can be monitored via continuous or discontinuous assays. For the discontinuous assay, we have found that direct observation of conversion of the NTP to

NDP via reverse-phase HPLC to be the most efficient approach (Camire et al., 2015). However, we note that Downs and coworkers have also applied the molybdate-based malachite green assay to detect the release of inorganic phosphate by ApbC (Boyd, Sondelski, et al., 2009). Notably, the reagents to carry out this malachite green assay are commercially available in kit format from several vendors.

For the continuous assay, we and others have employed a standard coupled assay in which ATP hydrolysis is detected via the pyruvate kinase (PK)/lactate dehydrogenase (LDH) coupling system, which couples ATP hydrolysis to the reduction of NADH that can be followed at 340 nm (Boyd, Sondelski, et al., 2009; Camire et al., 2015). As with any coupled assay, care must be taken to ensure neither of the coupling enzyme steps are rate limiting so that the oxidation of NADH is an accurate readout for the rate of ATP hydrolysis. Since the ADP is consumed by the coupling system, this assay is convenient for NTPases that exhibit product inhibition, such as the CIA scaffold and its bacterial homolog ApbC (Boyd, Sondelski, et al., 2009; Camire et al., 2015).

To obtain the steady-state kinetic parameters k_{cat} and V_{max}, the initial velocity is monitored as a function of varying ATP concentration. The data are fit either with the standard Michaelis–Menten equation for classic hyperbolic kinetics or the Hill equation for enzymes with sigmoidal kinetics (Boyd, Sondelski, et al., 2009; Camire et al., 2015). Since the MRP/Nbp35 family of ATPases are generally slow ATPases, with reported k_{cat} values in the range of 2–4 min^{-1}, we have found it to be necessary to evaluate several site-directed mutants of the enzyme to ensure that the hydrolysis activity observed via the coupled assay is not due to a contaminating ATPase, or another process unrelated to enzymatic activity, such as the nonenzymatic oxidation of NADH (Boyd, Sondelski, et al., 2009; Camire et al., 2015).

3.1 Discontinuous Assay for Nucleotide Hydrolysis

To determine if the CIA scaffold hydrolyzed ATP or GTP, we made use of a discontinuous HPLC assay (Camire et al., 2015). Briefly, the CIA scaffold is incubated with nucleotide and samples are removed over time, quenched with acid, reneutralized, and analyzed by reverse-phase HPLC (Prodigy 5 μm ODS-2 150 × 4.6 mm column, Phenomenex). Separation of ATP from ADP is accomplished via 0%–10% linear gradient of methanol in a 100 m*M* potassium phosphate buffer (pH 6) over 12 min. The amount of nucleotide formed is determined by comparison to a standard curve generated with known amounts of ADP (4.4 min) and ATP (5.0 min) standards. To evaluate whether GTP is a substrate, a similar approach can be taken where GTP is separated from GDP on the same column. The column is equilibrated in solvent A (50 m*M* sodium phosphate (pH 6.5) and 2 m*M* tetrabutylammonium bromide with 3% acetonitrile) and eluted with a linear gradient from 3% to 40% acetonitrile over 15 min. GTP elutes at 7.4 min and GDP elutes at 4.4 min.

3.2 Continuous Assay for Nucleotide Hydrolysis

To determine of the steady-state kinetic parameters, k_{cat} and K_M, we have made use of standard PK/LDH coupling system using the following protocol.

3.2.1 Equipment

- Spectrophotometer such as CaryBio-100, equipped with a multicell changer and peltier temperature control
- Quartz cuvettes

3.2.2 Buffers and Reagents

- 1 *M* Tris–HCl, pH 8.0
- ATP (disodium salt): 150 m*M* in 1 *M* Tris–HCl (pH 8.0)
- 5 × ATPase assay buffer: 250 m*M* Tris–HCl (pH 8.0), 1 *M* KCl, and 50 m*M* $MgCl_2$
- Phosphoenolpyruvate (PEP) potassium salt: 100 m*M* in 1 × ATPase assay buffer
- NADH, sodium salt: 10 m*M* in 100 m*M* Tris–HCl (pH 8.0)
- PK/LDH-coupled enzyme mix in 50% glycerol (Sigma-Aldrich)

3.2.3 Coupled Enzyme Assay Procedure

1. Prepare dilutions of ATP in 1 *M* Tris–HCl (pH 8.0) varying in concentration from 0 to 150 m*M* to make 10 × working stock solutions.
2. Prepare the assay master mix containing PEP, NADH, DTT, and the assay buffer. The final volume of each assay is 500 μL, and 380 μL of this master mix is used for each assay. Every 380 μL of master mix contains 10 μL 10 m*M* NADH, 20 μL 100 m*M* PEP, 100 μL 5 × ATPase buffer, 0.5 μL 1 *M* DTT, and 249.5 μL water. The amount of master mix made depends on the number of assays required. For example, if 10 assays will be completed, then 3.8 mL of master mix will be needed. Generally, it is

advantageous to make an excess of master mix in case a few points need to be repeated or to account for pipetting error.

3. Set the temperature control on the UV–Vis spectrophotometer to 37°C and blank the instrument with 1 × assay buffer.
4. Add 380 μL master mix, 20 μL PK/LDH coupling enzyme mix, and 50 μL of the ATP solution (concentration varies) to the cuvette. Place cuvette cell holder and incubate for 5 min at 37°C. This incubation brings the assay mixture to temperature and it serves to convert any contaminating ADP back to ATP. Also preincubate an aliquot of the CIA scaffold (1 mg/mL) at 37°C.
5. Initiate reaction by the addition of CIA scaffold (50 μL). Cover cuvette with parafilm and carefully mix assay components by inversion. Place cuvette into spectrophotometer at 37°C and record the absorbance at 340 nm over time, ≤10 min. The absorbance at 340 nm should linearly decrease as NADH is consumed. For each concentration of ATP, a separate cuvette should be run without the protein to act as a control for background change in A_{340nm} unrelated to ATPase activity.
6. Determine the initial rate by determining the slope of the linear portion of the plot of A_{340nm} vs time (min). Subtract any background rate of NADH consumption observed when no CIA scaffold is added to obtain the ΔA_{340}. The initial rate of ATP hydrolysis (v_0) is then determined by applying Eq. (1)

$$v_0 = \frac{\Delta A_{340} \cdot volume \cdot 10^9}{l \cdot \varepsilon \cdot mg} \tag{1}$$

where v_0 is in the units of nmol min^{-1} mg^{-1}, ΔA_{340} is the background subtracted change in absorbance at 340 nm per min, *volume* is the assay volume in mL (0.5 mL), l is the pathlength of the cuvette in cm (1 cm), ε is the extinction coefficient of NADH (6.22 mM^{-1} cm^{-1}), and *mg* is the total mass of the protein in the cuvette (~0.05 mg in a typical assay).

7. Plotting of the initial rate vs ATP concentration (mM) and subsequent fitting to the Michaelis–Menten equation (Eq. 2) allows for determination of the steady-state kinetic parameters

$$v_0 = \frac{V_{max} \cdot [ATP]}{(K_M + [ATP])} \tag{2}$$

where v_o is the initial rate (determined for each [ATP] in step 6), V_{max} is the maximum rate (nmol min^{-1} mg^{-1}), K_M is the Michaelis constant (units, m*M*), and *[ATP]* is the m*M* concentration of ATP. The turnover number, k_{cat}, in units of inverse time (min^{-1}) can be obtained by dividing the V_{max} by the nmol of enzyme in each 0.5 mL assay (~1.5 nmol for assays run with ~1 mg/mL CIA stock solution).

4. NUCLEOTIDE AFFINITY MONITORED VIA FLUORESCENCE SPECTROSCOPY

Fluorescent nucleotide analogs are well-known tools for probing NTPases (Bagshaw, 2001; Cremo, 2003; Hiratsuka, 2003; Jameson & Eccleston, 1997). Generally, nucleotides can be labeled through the base, the sugar, or the phosphate. Many varieties of fluorescently labeled nucleotide analogs are commercially available from vendors such as Jena Biosciences, ThermoFisher, and Sigma-Aldrich. While the best analog for a particular application must be determined empirically, information about the structure of the NTPase and/or information about what analogs bind to NTPases within the same family can be helpful for selection of an appropriate probe. For example, the MRP/Nbp35 family of cluster scaffolds represent a large family within the "deviant Walker A" branch of the SIMIBI (*Si*gnal recognition particle, *Mi*nD and *Bi*oD) superfamily of NTPases (Leipe et al., 2002). The mantATP analog, which contains the *N′*-methylanthraniloyl probe to either the 2′ or 3′ hydroxyl of the sugar, binds to several deviant Walker A NTPases including MinD (Okuno et al., 2010), ArsA (Walmsley et al., 2001), Get3 (Rome, Rao, Clemons, & Shan, 2013), and ParA (Dobruk-Serkowska et al., 2012). Structural studies have revealed that nucleotide at the dimer interface in a head-to-head orientation with the sugar hydroxyls pointing out toward the dimer interface, thus explaining why this family of enzymes appears to readily bind nucleotides modified through the sugar (Jeoung et al., 2010; Mateja et al., 2009; Schindelin et al., 1997; Wu, Park, Holyoak, & Lutkenhaus, 2011; Zhou et al., 2000).

In this section, we describe methods we have developed to probe the nucleotide affinity of the yeast CIA scaffold to begin elucidating how nucleotide affinity is affected by the presence of an FeS cluster on the scaffold. We note that while fluorescent nucleotide analogs have been widely used to probe the mechanism of numerous ATPases and GTPases, they have not been frequently employed to understand NTPases involved in metal trafficking or metallocluster biosynthesis. For example, Mo-dependent nitrogenase, [NiFe]-hydrogenase, [FeFe]-hydrogenase, carbon monoxide dehydrogenase, and many other metalloenzymes all depend on at least one NTPase for assembly of their complex metalloclusters (Table 1). Yet, few studies of metal trafficking NTPases have employed fluorescent nucleotides to gain mechanistic insight (Eccleston et al., 2006; Karkaria & Rosen, 1991; Petrovic et al., 2008; Sydor et al., 2014; Walmsley et al., 2001). Thus, we think an opportunity exists for the community to begin using these well-established tools and approaches that have shown to be successful for mechanistic investigations of other NTPases to answer long-standing questions about the role of nucleotide hydrolysis in metal trafficking and metallocofactor biosynthesis.

4.1 Nucleotide Affinity Evaluated via Fluorescence Anisotropy

The binding of a mantATP to the CIA scaffold can be monitored via a fluorescence anisotropy (FA) assay (Camire et al., 2015; Pollard, 2010; Rossi & Taylor, 2011). In this experiment, mantATP is excited with plane-polarized light. When free in solution, the nucleotide rapidly rotates, leading to depolarization of the emitted light or low anisotropy. As increasing concentrations of protein are titrated into the solution, the nucleotide binds to the protein. Formation of the protein–ligand complex slows the fluorophore's rotational diffusion leading to greater retention of polarization in the emitted light or high anisotropy (Fig. 4). Thus, the degree of binding between mantATP and the scaffold can be determined by monitoring the anisotropy of the emitted light. The execution of FA assays requires access to a fluorimeter or, in our case, a fluorescence plate reader, capable of excitation with plane-polarized light and detection of the fluorescence emission in planes both parallel to and perpendicular to the plane of the polarized excitation. Anisotropy (A) is simply the difference between the fluorescence in the parallel and perpendicular planes divided by the total fluorescence (see Eq. 3 in Section 4.1.3 for the specifics of how anisotropy is calculated).

While these FA assays are relatively straightforward to perform, careful considerations must be made in their design and analysis. We refer novice users to several instructive reviews (Pollard, 2010; Rossi & Taylor, 2011), and would like to point out a few critical design and analysis considerations. First, the concentration of fluorophore used in the assay must be optimized to ensure it is high enough to yield a robust signal while also ensuring the concentration of the fluorescent probe is well below the expected K_D for the protein–ligand interaction for evaluation of affinity via the hyperbolic binding equation as described in Section 4.1.3. Additionally, several validation experiments must also be performed to ensure fluorophore specifically

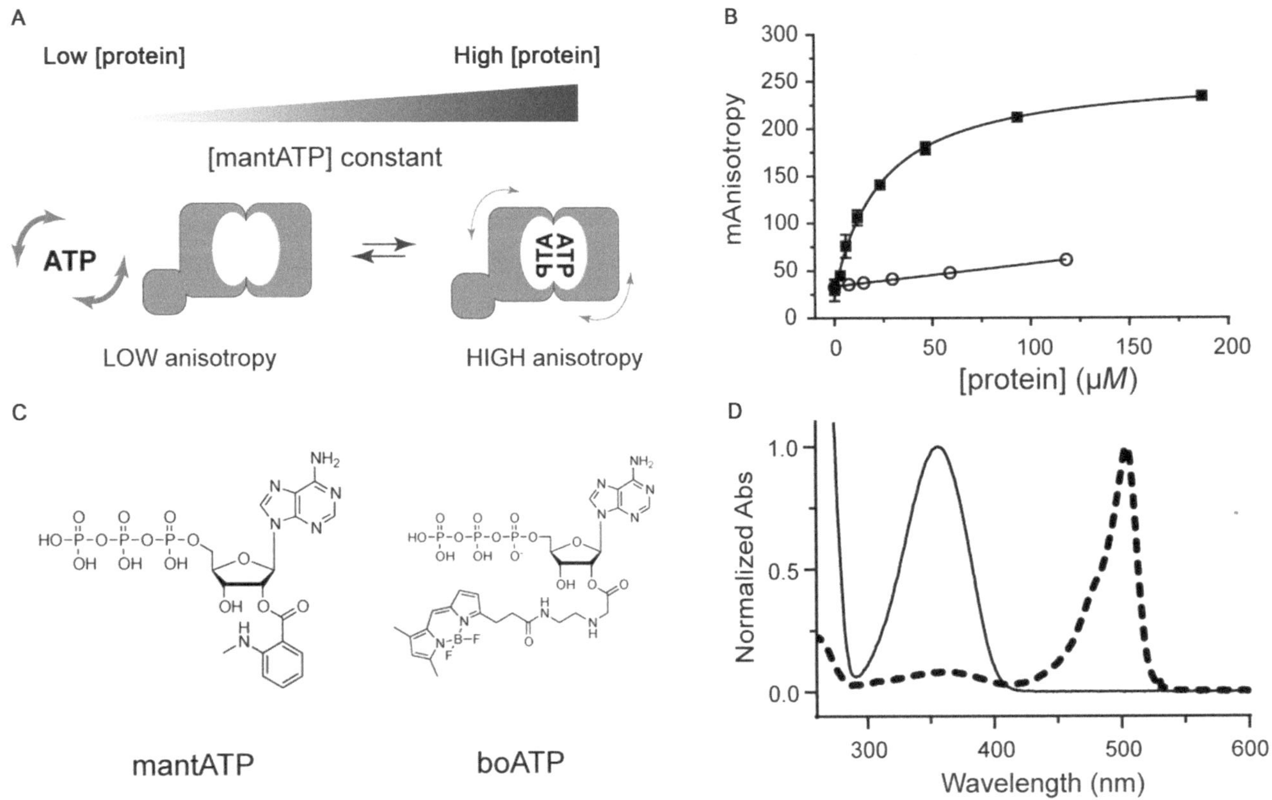

Fig. 4 See legend on opposite page.

binds to the protein without significant nonspecific interactions. For example, during design of the FA assay for the CIA scaffold (Camire et al., 2015), we evaluated the stoichiometry of the mantATP·CIA scaffold interaction and the magnesium ion dependence of the interaction (Fig. 4B). We additionally demonstrated that ATPase site mutants impact nucleotide affinity.

To monitor affinity of the holo-scaffold for nucleotide, the mant-labeled analog cannot be used because the absorption envelope of the FeS cluster overlaps with that of mantATP leading to inner filter effect problems as holo-scaffold is titrated into solution (Figs. 3 and 4). This limitation could explain why few metal-binding NTPases have been interrogated with the mant-labeled probes as many metallocofactors also exhibit absorption features that overlap with that of the mant label. To get around this limitation, we have made use of boATP (2′/3′-*O*-(*N*-(2-aminoethyl)urethane BODIPY-FL ATP, Fig. 4C) to monitor affinity of the holo-scaffold (Grossman, Camire, & Perlstein, unpublished results). The BODIPY-FL probe has an absorption peak that is shifted away from the FeS cluster, relieving the inner filter effect problems that prevent the use of mant-labeled probe. Notably, a wide range of fluorescent nucleotide analogs are available from Jena Biosciences that feature the same linker as boATP. Alternatively, custom analogs can be synthesized to access an analog with photophysical properties suited for a particular system (Cremo, 2003; Jameson & Eccleston, 1997; Wilke, Francis, & Carlson, 2015). However, the boATP probe features a longer linker between the sugar and the fluorophore than that of mantATP (Fig. 4C). While this longer linker could be advantageous for some proteins with deeply buried active sites, it can be problematic for FA assays as it can allow for rotational motion of the fluorescent probe while bound to the protein. If this occurs, then anisotropy cannot be exploited to

Fig. 4 Introduction to fluorescence anisotropy (FA) assays. (A) In a FA assay, the protein is titrated into a solution containing a low and constant amount of the fluorescent nucleotide, such as mantATP. As the protein concentration is increased, the mantATP binds to the protein. The resulting change in anisotropy, stemming from restriction of the fluorophore's rotational diffusion (*arrows*), reports on formation of the protein–ligand interaction. (B) The binding of mantATP (0.5 μ*M*) to the Cfd1 homodimer in the presence (■) and absence (○) of 10 m*M* $MgCl_2$ as monitored by FA. In this example, the *Y*-axis is mAnisotropy, the anisotropy multiplied by 1000. (C) Chemical structure of mantATP (*left*) and boATP (*right*). Although the structures are shown as being labeled via the 2′-OH of the ribose for clarity, both analogs exist as an equilibrium mixture of 2′- and 3′-labeled isomers. (D) Normalized absorbance spectra of mantATP (*solid line*) and boATP (*dashed line*) showing absorbance maxima at 355 and 448 nm for mantATP and boATP, respectively.

monitor the binding event and another property, such as a change in fluorescence, must be used to instead (Section 4.2) (Sydor et al., 2014). The utility of these analogs for a particular application must be carefully validated for each individual system as described for mantATP FA assays with the CIA scaffold. In the following sections, we describe our standard FA assay.

4.1.1 Equipment

- SpectraMax5 plate reader, or equivalent, equipped for fluorescence polarization measurements
- Black, 96-well, round-bottom plate

4.1.2 Buffers and Reagents

- 4 mg/mL bovine serum albumin (BSA) in storage buffer (Section 2.3.2)
- 200 m*M* $MgCl_2$ in water
- 20 μ*M* mantATP (Jena Bioscience) in storage buffer

4.1.3 Procedure for FA mantATP Binding Experiment

1. Prepare twofold serial dilutions of the CIA scaffold in the storage buffer. Ideally the dilutions should cover concentrations ranging from 0.1 to 10 × the expected K_D.
2. To each well, add 180 μL of the diluted protein, 10 μL of 200 m*M* $MgCl_2$, and 5 μL of 4 mg/mL BSA. Carefully pipette the contents up and down to thoroughly mix, paying special attention to avoid formation of bubbles. Any bubbles that do form can be removed by popping them with a small pipette tip or small gauge needle.
3. Using a plate reader, read the background fluorescence polarization (excitation, 355 nm; emission 448 nm for mant-labeled nucleotide) of each well to obtain the background fluorescence (both parallel and perpendicular modes) in the absence of the probe.
4. Add the fluorescent probe to each well. The final concentration of the probe should be at least an order of magnitude below the expected K_D. For our system, we add 5 μL of 20 μ*M* mantATP for a final concentration of 0.5 μ*M* in a 200 μL sample. Well contents should be mixed again as described in step 2.
5. After a 1 h incubation at 25°C to ensure equilibration, the fluorescence polarization is read again at the same excitation and emission wavelengths. Evaluation of mantATP hydrolysis via the discontinuous assay (Section 3.1) revealed no significant hydrolysis to mantADP occurs over

the 1 h incubation. After background subtraction, the anisotropy in each well can then be calculated using Eq. (3)

$$r = \frac{I_{para} - I_{perp}}{I_{para} + 2I_{perp}} \tag{3}$$

where r is the anisotropy, I_{para} is the intensity of the emitted light in the parallel plane, and I_{perp} is the intensity of the emitted light in the perpendicular plane. Background subtraction is achieved by subtracting the average I_{perp} and I_{para} measurements of the wells without mantATP (step 3) from the measurements following addition of mantATP (step 4).

6. Plot the anisotropy (calculated via Eq. 3) vs the concentration of protein and determine the equilibrium dissociation constant, K_D, using the hyperbolic binding equation (Eq. 4)

$$r = (r_{max} - r_{min})\left(\frac{x}{x + K_D}\right) + r_{min} \tag{4}$$

where r is the anisotropy, r_{max} is the maximum anisotropy, r_{min} is the minimum anisotropy, x is the concentration of the protein, and K_D is the equilibrium dissociation constant. Fitting to a hyperbolic binding curve is appropriate when the concentration of mantATP, the fixed component, is an order of magnitude below the K_D. If instead, the fixed component is on the order of the K_D the data should be fit with the quadratic binding equation as described for TNP-ATP equilibrium binding (Section 4.2.3). Additionally, the variable component (the protein) is titrated over a range of concentrations that span an order of magnitude above and below the K_D.

4.2 Equilibrium Binding Assays With Trinitrophenyl-ATP

Since the mant-labeled probes cannot be used to monitor binding to the holo-scaffold, we have also developed a nucleotide binding assay based on the fluorescent analog 2′(3′)-*O*-(2,4,6-trinitrophenyl)-ATP (TNP-ATP, Fig. 5) (Hiratsuka, 2003). Like the boATP analog, the TNP-ATP's absorption and emission are shifted out of the FeS cluster envelope, making it useful in assays with either the apo- or holo-form of the CIA scaffold (Fig. 5). However, unlike the FA assays that exploit mantATP or boATP, the environmental sensitivity of the TNP fluorescence provides the basis for detection of its binding. When free in solution, TNP-ATP has relatively

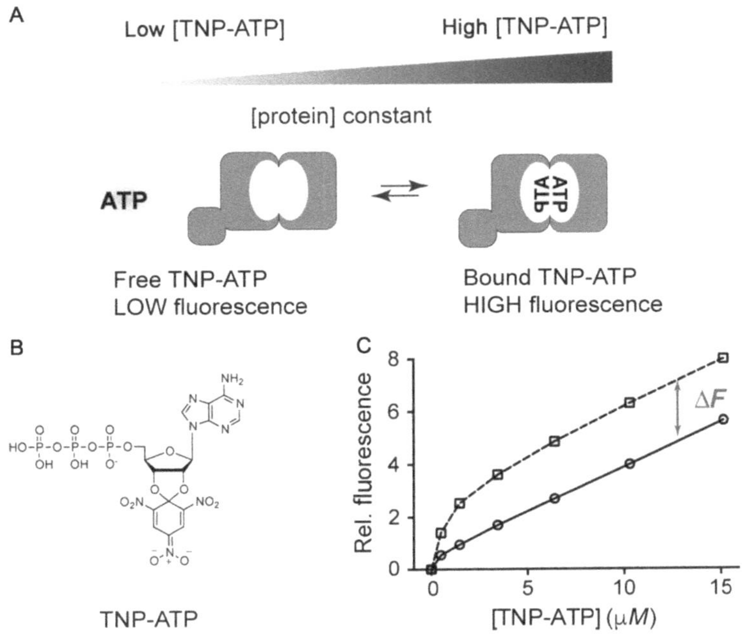

Fig. 5 Equilibrium binding assays with TNP-ATP. (A) The binding of TNP-ATP to the CIA scaffold is monitored by titrating the nucleotide into a solution containing a low and constant amount of protein. As the concentration of TNP-ATP is increased, it binds to the protein. The resulting enhancement of the TNP-ATP fluorescence provides the basis for detection of formation of the protein–ligand complex. (B) The structure of TNP-ATP. (C) Plot of relative fluorescence vs [TNP-ATP] in the presence (□) and absence (○) of Cfd1 (1 μ*M*). Subtraction of the fluorescence intensity in the presence and absence of the protein yields the ΔF values required to generate the binding isotherm.

low fluorescence. Upon binding to the protein, the TNP probe is placed in a more hydrophobic environment, leading to enhancement of the fluorescence (Fig. 5). The assay is performed by titrating increasing amounts of TNP-ATP into a solution in the absence and presence of a fixed concentration of the protein. The protein-dependent difference in the fluorescence (ΔF) is monitored and plotted against the concentration of fluorescent nucleotide. The K_D of the interaction can then be calculated by fitting the data to an appropriate binding equation (Pollard, 2010).

Since a low and constant concentration of receptor (the CIA scaffold) is titrated with increasing amounts of the ligand, the assay setup is more similar to traditional binding assay as compared to the FA assays which feature a reverse titration. Notably, this difference also means that the FA is performed under single-site conditions, where the concentration of nucleotide is much lower than the concentration of the CIA scaffold at the end of the titration (Fig. 4). However, the TNP-ATP assays are performed under multisite conditions, where the concentration of nucleotide is much higher than the concentration of the CIA scaffold at the end of the titration (Fig. 5). As with the FA assay, the assay conditions (buffer components and concentrations of nucleotide and protein) and data analysis specifics for each system must be optimized. For the CIA scaffolding proteins, we have found a protein concentration of 1 μ*M* yields a sufficient ΔF signal. With this concentration of protein, the quadratic binding equation (Eq. 7) is most appropriate for fitting the data as the K_D for the interaction is close to the concentration of protein used in the assay (1 μ*M*). A detailed protocol for monitoring TNP-ATP binding to the CIA scaffold is described below.

4.2.1 Equipment

- Horiba Jobin Yvon FluoroMax3 fluorimeter or similar
- Stoppered fluorescence cuvette fitted with gastight septum cap
- Gastight Hamilton syringe, capable of small (5 μL) additions

4.2.2 Buffers and Reagents

- Assay buffer: 50 m*M* Tris–HCl (pH 8.0), 100 m*M* NaCl, 10 m*M* $MgCl_2$, 10 (v/v)% glycerol
- TNP-ATP (ThermoFisher): supplied by vendor as a 5–10 m*M* solution

4.2.3 Procedure for TNP-ATP Binding Assay

1. Place the assay solution, either the buffer alone or buffer with the CIA scaffold, into the fluorescence cuvette. Our TNP-ATP binding assay is carried out in a volume of 1.5 mL. The CIA scaffold is held constant at a concentration of 1 μ*M* dimer. For the holo-scaffold, the solutions of TNP-ATP and the holo-scaffold must be prepared within in the anaerobic chamber using anaerobic solutions as described in Section 2.4.3. The titrations are then carried out by injection of the nucleotide solution housed within a gastight syringe through the cuvette septum.
2. Prepare dilutions of TNP-ATP in assay buffer. Dilutions should be designed such that the total volume of TNP-ATP added to the assay does not exceed 4% of the original assay volume ($\leq$60 μL). For example, we make dilutions that range from 30 to 1600 μ*M* TNP-ATP and add up to twelve 5 μL additions to cover a range of TNP-ATP concentrations, from 0 to 20 μ*M* in the assay.

3. For fluorimeter setup, the temperature controller should be set to 25° C, the excitation wavelength set to 469 nm, and emission is monitored from 500 to 600 nm.
4. Place the cuvette prepared in step 1 in the fluorimeter and allow both to equilibrate to 25°C. When the instrument and cuvette have reached a steady temperature, blank the instrument.
5. Add the first aliquot (5 μL) of TNP-ATP into the solution containing the CIA scaffold. For aerobic assays, cover with parafilm and invert gently several times to mix. For anaerobic titrations, TNP-ATP is added through the stopper with a gastight syringe and mixed by inversion. Return the cuvette to the fluorimeter and wait 5 min to reach thermal equilibrium.
6. Collect fluorescence spectrum of the solution. It is important to make sure binding equilibrium has been reached and the fluorescence intensity is not changing with time. In our experiments, the CIA scaffold reaches binding equilibrium within a few minutes of TNP-ATP addition.
7. Repeat steps 5 and 6 for all TNP-ATP aliquots.
8. Repeat steps 3 through 6 with 1 × assay buffer in place of the protein solution.
9. For data analysis, the observed fluorescence must be corrected for the inner filter effect (Eq. 5)

$$F_{corr} = F_{obs}{}^{*}10^{(OD_{ex} + OD_{em})/2} \tag{5}$$

where F_{corr} is the fluorescence corrected for the inner filter effect, F_{obs} is the observed fluorescence (step 6), OD_{ex} is the absorbance of the solution at the excitation wavelength, and OD_{em} is the absorbance of the solution at the emission wavelength.
10. Once the data collected both in the presence and absence of protein have been processed to account for inner filter effect, difference in fluorescence intensity (ΔF) between the protein-containing sample and the buffer control can be determined. This is done by subtracting the values of excitation wavelength maximum of the buffer-only control from its corresponding protein-containing sample (Eq. 6)

$$\Delta F = F_{corr,\,Protein} - F_{corr,\,Buffer} \tag{6}$$

where ΔF is the change in fluorescence, $F_{corr,Protein}$ is the inner filter effect-corrected fluorescence of the TNP-ATP in the presence of protein, and $F_{corr,Buffer}$ is the inner filter effect-corrected fluorescence of the TNP-ATP in the absence of protein (buffer only).

11. Plot ΔF vs the concentration of TNP-ATP to obtain the binding isotherm. The equilibrium dissociation constant of the interaction, K_D, can be determined by fitting the data to the quadratic equation (Eq. 7)

$$\Delta F = (\Delta F_{max} - \Delta F_{min})\frac{(R_T + x + K_D) - \sqrt{(R_T + x + K_D)^2 - 4xR_T}}{2R_T} + \Delta F_{min} \tag{7}$$

where ΔF is the change in fluorescence, ΔF_{min} is the minimum ΔF signal, ΔF_{max} is the maximum ΔF signal, x is the concentration of TNP-ATP, R_T is the concentration of protein, and K_D is the dissociation constant.

ACKNOWLEDGMENTS

This work was supported by a grant from the National Science Foundation (CHE-1555295) to D.L.P.

REFERENCES

Arnold, A. R., Grodick, M. A., & Barton, J. K. (2016). DNA charge transport: From chemical principles to the cell. *Cell Chemical Biology*, *23*(1), 183–197. https://doi.org/10.1016/j.chembiol.2015.11.010.

Bagshaw, C. (2001). ATP analogues at a glance. *Journal of Cell Science*, *114*(Pt. 3), 459–460.

Banda, D. M., Nunez, N. N., Burnside, M. A., Bradshaw, K. M., & David, S. S. (2017). Repair of 8-oxoG: A mismatches by the MUTYH glycosylase: Mechanism, metals and medicine. *Free Radical Biology and Medicine*, *107*, 202–215. https://doi.org/10.1016/j.freeradbiomed.2017.01.008.

Bange, G., & Sinning, I. (2013). SIMIBI twins in protein targeting and localization. *Nature Structural & Molecular Biology*, *20*(7), 776–780. https://doi.org/10.1038/nsmb.2605.

Beinert, H. (1978). Micro methods for the quantitative determination of iron and copper in biological material. *Methods in Enzymology*, *54*, 435–445.

Beinert, H. (1983). Semi-micro methods for analysis of labile sulfide and of labile sulfide plus sulfane sulfur in unusually stable iron-sulfur proteins. *Analytical Biochemistry*, *131*(2), 373–378.

Bhattacharjee, H., Zhou, T., Li, J., Gatti, D. L., Walmsley, A. R., & Rosen, B. P. (2000). Structure-function relationships in an anion-translocating ATPase. *Biochemical Society Transactions*, *28*(4), 520–526.

Blanc, B., Gerez, C., & Ollagnier de Choudens, S. (2015). Assembly of Fe/S proteins in bacterial systems: Biochemistry of the bacterial ISC system. *Biochimica et Biophysica Acta*, *1853*(6), 1436–1447. https://doi.org/10.1016/j.bbamcr.2014.12.009.

Boyd, J. M., Drevland, R. M., Downs, D. M., & Graham, D. E. (2009). Archaeal ApbC/Nbp35 homologs function as iron-sulfur cluster carrier proteins. *Journal of Bacteriology*, *191*(5), 1490–1497. https://doi.org/10.1128/JB.01469-08.

Boyd, J. M., Pierik, A. J., Netz, D. J. A., Lill, R., & Downs, D. M. (2008). Bacterial ApbC can bind and effectively transfer iron-sulfur clusters. *Biochemistry*, *47*(31), 8195–8202. https://doi.org/10.1021/bi800551y.

Boyd, J. M., Sondelski, J. L., & Downs, D. M. (2009). Bacterial ApbC protein has two biochemical activities that are required for in vivo function. *Journal of Biological Chemistry, 284*(1), 110–118. https://doi.org/10.1074/jbc.M807003200.

Bych, K., Kerscher, S., Netz, D. J. A., Pierik, A. J., Zwicker, K., Huynen, M. A., et al. (2008). The iron-sulphur protein Ind1 is required for effective complex I assembly. *EMBO Journal, 27*(12), 1736–1746. https://doi.org/10.1038/emboj.2008.98.

Bych, K., Netz, D. J. A., Vigani, G., Bill, E., Lill, R., Pierik, A. J., et al. (2008). The essential cytosolic iron-sulfur protein Nbp35 acts without Cfd1 partner in the green lineage. *Journal of Biological Chemistry, 283*(51), 35797–35804. https://doi.org/10.1074/jbc.M807303200.

Camire, E. J., Grossman, J. D., Thole, G. J., Fleischman, N. M., & Perlstein, D. L. (2015). The yeast Nbp35-Cfd1 cytosolic iron-sulfur cluster scaffold is an ATPase. *Journal of Biological Chemistry, 290*(39), 23793–23802. https://doi.org/10.1074/jbc.M115.667022.

Cremo, C. R. (2003). Fluorescent nucleotides: Synthesis and characterization. *Methods in Enzymology, 360*, 128–177.

Dobruk-Serkowska, A., Caccamo, M., Rodriguez-Castaneda, F., Wu, M., Bryce, K., Ng, I., et al. (2012). Uncoupling of nucleotide hydrolysis and polymerization in the ParA protein superfamily disrupts DNA segregation dynamics. *Journal of Biological Chemistry, 287*(51), 42545–42553. https://doi.org/10.1074/jbc.M112.410324.

Eccleston, J. F., Petrovic, A., Davis, C. T., Rangachari, K., & Wilson, R. J. (2006). The kinetic mechanism of the SufC ATPase: The cleavage step is accelerated by SufB. *Journal of Biological Chemistry, 281*(13), 8371–8378. https://doi.org/10.1074/jbc.M513455200.

Fish, W. W. (1988). Rapid colorimetric micromethod for the quantitation of complexed iron in biological samples. *Methods in Enzymology, 158*, 357–364.

Fong, Y. H., Wong, H. C., Yuen, M. H., Lau, P. H., Chen, Y. W., & Wong, K. B. (2013). Structure of UreG/UreF/UreH complex reveals how urease accessory proteins facilitate maturation of Helicobacter pylori urease. *PLoS Biology, 11*(10), e1001678. https://doi.org/10.1371/journal.pbio.1001678.

Fuss, J. O., Tsai, C. L., Ishida, J. P., & Tainer, J. A. (2015). Emerging critical roles of Fe-S clusters in DNA replication and repair. *Biochimica et Biophysica Acta, 1853*(6), 1253–1271. https://doi.org/10.1016/j.bbamcr.2015.01.018.

Gasper, R., Scrima, A., & Wittinghofer, A. (2006). Structural insights into HypB, a GTP-binding protein that regulates metal binding. *Journal of Biological Chemistry, 281*(37), 27492–27502. https://doi.org/10.1074/jbc.M600809200.

Hirabayashi, K., Yuda, E., Tanaka, N., Katayama, S., Iwasaki, K., Matsumoto, T., et al. (2015). Functional dynamics revealed by the structure of the SufBCD complex, a novel ATP-binding cassette (ABC) protein that serves as a scaffold for iron-sulfur cluster biogenesis. *Journal of Biological Chemistry, 290*(50), 29717–29731. https://doi.org/10.1074/jbc.M115.680934.

Hiratsuka, T. (2003). Fluorescent and colored trinitrophenylated analogs of ATP and GTP. *European Journal of Biochemistry, 270*(17), 3479–3485.

Hu, Y. L., Corbett, M. C., Fay, A. W., Webber, J. A., Hodgson, K. O., Hedman, B., et al. (2006). FeMo cofactor maturation on NifEN. *Proceedings of the National Academy of Sciences of the United States of America, 103*(46), 17119–17124. https://doi.org/10.1073/pnas.0602647103.

Hu, Y. L., Corbettt, M. C., Fay, A. W., Webber, J. A., Hodgson, K. O., Hedman, B., et al. (2006). Nitrogenase Fe protein: A molybdate/homocitrate insertase. *Proceedings of the National Academy of Sciences of the United States of America, 103*(46), 17125–17130. https://doi.org/10.1073/pnas.0602651103.

Jameson, D. M., & Eccleston, J. F. (1997). Fluorescent nucleotide analogs: Synthesis and applications. *Methods in Enzymology, 278*, 363–390.

Jeoung, J. H., Giese, T., Grunwald, M., & Dobbek, H. (2009). CooC1 from Carboxydothermus hydrogenoformans is a nickel-binding ATPase. *Biochemistry, 48*(48), 11505–11513. https://doi.org/10.1021/bi901443z.

Jeoung, J. H., Giese, T., Grunwald, M., & Dobbek, H. (2010). Crystal structure of the ATP-dependent maturation factor of Ni,Fe-containing carbon monoxide dehydrogenases. *Journal of Molecular Biology, 396*(4), 1165–1179. https://doi.org/10.1016/j.jmb.2009.12.062.

Johnson, D. C., Dean, D. R., Smith, A. D., & Johnson, M. K. (2005). Structure, function, and formation of biological iron-sulfur clusters. *Annual Review of Biochemistry, 74*, 247–281. https://doi.org/10.1146/annurev.biochem.74.082803.133518.

Jost, M., Cracan, V., Hubbard, P. A., Banerjee, R., & Drennan, C. L. (2015). Visualization of a radical B12 enzyme with its G-protein chaperone. *Proceedings of the National Academy of Sciences of the United States of America, 112*(8), 2419–2424. https://doi.org/10.1073/pnas.1419582112.

Karkaria, C. E., & Rosen, B. P. (1991). Trinitrophenyl-ATP binding to the ArsA protein: The catalytic subunit of an anion pump. *Archives of Biochemistry and Biophysics, 288*(1), 107–111.

Kelley, L. A., Mezulis, S., Yates, C. M., Wass, M. N., & Sternberg, M. J. (2015). The Phyre2 web portal for protein modeling, prediction and analysis. *Nature Protocols, 10*(6), 845–858. https://doi.org/10.1038/nprot.2015.053.

Lacasse, M. J., & Zamble, D. B. (2016). [NiFe]-hydrogenase maturation. *Biochemistry, 55*(12), 1689–1701. https://doi.org/10.1021/acs.biochem.5b01328.

Lahiri, S., Pulakat, L., & Gavini, N. (2008). NifH: Structural and mechanistic similarities with proteins involved in diverse biological processes. *American Journal of Biochemistry and Biotechnology, 4*(3), 304–316.

Lanz, N. D., Grove, T. L., Gogonea, C. B., Lee, K. H., Krebs, C., & Booker, S. J. (2012). RlmN and AtsB as models for the overproduction and characterization of radical SAM proteins. *Methods in Enzymology, 516*, 125–152. https://doi.org/10.1016/b978-0-12-394291-3.00030-7.

Leipe, D. D., Wolf, Y. I., Koonin, E. V., & Aravind, L. (2002). Classification and evolution of P-loop GTPases and related ATPases. *Journal of Molecular Biology, 317*(1), 41–72. https://doi.org/10.1006/jmbi.2001.5378.

Lill, R. (2009). Function and biogenesis of iron-sulphur proteins. *Nature, 460*(7257), 831–838. https://doi.org/10.1038/nature08301.

Lutkenhaus, J. (2012). The ParA/MinD family puts things in their place. *Trends in Microbiology, 20*(9), 411–418.

Lutkenhaus, J., & Sundaramoorthy, M. (2003). MinD and role of the deviant Walker A motif, dimerization and membrane binding in oscillation. *Molecular Microbiology, 48*(2), 295–303.

Mateja, A., Szlachcic, A., Downing, M. E., Dobosz, M., Mariappan, M., Hegde, R. S., et al. (2009). The structural basis of tail-anchored membrane protein recognition by Get3. *Nature, 461*(7262), 361–366. https://doi.org/10.1038/nature08319.

Netz, D. J. A., Pierik, A. J., Stumpfig, M., Bill, E., Sharma, A. K., Pallesen, L. J., et al. (2012). A bridging [4Fe-4S] cluster and nucleotide binding are essential for function of the Cfd1-Nbp35 complex as a scaffold in iron-sulfur protein maturation. *Journal of Biological Chemistry, 287*(15), 12365–12378. https://doi.org/10.1074/jbc.M111.328914.

Netz, D. J. A., Pierik, A. J., Stümpfig, M., Mühlenhoff, U., & Lill, R. (2007). The Cfd1-Nbp35 complex acts as a scaffold for iron-sulfur protein assembly in the yeast cytosol. *Nature Chemical Biology, 3*(5), 278–286. https://doi.org/10.1038/nchembio872.

Netz, D. J. A., Stith, C. M., Stümpfig, M., Köpf, G., Vogel, D., Genau, H. M., et al. (2012). Eukaryotic DNA polymerases require an iron-sulfur cluster for the formation of active complexes. *Nature Chemical Biology, 8*(1), 125–132. https://doi.org/10.1038/nchembio.721.

Okuno, T., Ohgita, T., Sasa, T., Nonaka, A., Funasaki, N., & Kogure, K. (2010). Fluorescence polarization analysis for revealing molecular mechanism of nucleotide-dependent phospholipid membrane binding of MinD adenosine 5′-triphosphate, adenosine triphosphatase. *Biological and Pharmaceutical Bulletin*, *33*(10), 1746–1750.

Padovani, D., & Banerjee, R. (2009). A G-protein editor gates coenzyme B-12 loading and is corrupted in methylmalonic aciduria. *Proceedings of the National Academy of Sciences of the United States of America*, *106*(51), 21567–21572. https://doi.org/10.1073/pnas.0908106106.

Pallesen, L. J., Solodovnikova, N., Sharma, A. K., & Walden, W. E. (2013). Interaction with Cfd1 increases the kinetic lability of FeS on the Nbp35 scaffold. *Journal of Biological Chemistry*, *288*(32), 23358–23367. https://doi.org/10.1074/jbc.M113.486878.

Paul, V. D., & Lill, R. (2015). Biogenesis of cytosolic and nuclear iron-sulfur proteins and their role in genome stability. *Biochimica et Biophysica Acta*, *1853*, 1528–1539. https://doi.org/10.1016/j.bbamcr.2014.12.018.

Petrovic, A., Davis, C. T., Rangachari, K., Clough, B., Wilson, R. J. M., & Eccleston, J. F. (2008). Hydrodynamic characterization of the SufBC and SufCD complexes and their interaction with fluorescent adenosine nucleotides. *Protein Science*, *17*(7), 1264–1274. https://doi.org/10.1110/ps.034652.108.

Pollard, T. D. (2010). A guide to simple and informative binding assays. *Molecular Biology of the Cell*, *21*(23), 4061–4067. https://doi.org/10.1091/mbc.E10-08-0683.

Rome, M. E., Rao, M., Clemons, W. M., & Shan, S. O. (2013). Precise timing of ATPase activation drives targeting of tail-anchored proteins. *Proceedings of the National Academy of Sciences of the United States of America*, *110*(19), 7666–7671. https://doi.org/10.1073/pnas.1222054110.

Rossi, A. M., & Taylor, C. W. (2011). Analysis of protein-ligand interactions by fluorescence polarization. *Nature Protocols*, *6*(3), 365–387. https://doi.org/10.1038/nprot.2011.305.

Roy, A., Solodovnikova, N., Nicholson, T., Antholine, W., & Walden, W. E. (2003). A novel eukaryotic factor for cytosolic Fe-S cluster assembly. *EMBO Journal*, *22*(18), 4826–4835. https://doi.org/10.1093/emboj/cdg455.

Schindelin, H., Kisker, C., Schlessman, J. L., Howard, J. B., & Rees, D. C. (1997). Structure of ADP·AIF4$^-$ stabilized nitrogenase complex and its implications for signal transduction. *Nature*, *387*(6631), 370–376. https://doi.org/10.1038/387370a0.

Schmidt, T. G., Batz, L., Bonet, L., Carl, U., Holzapfel, G., Kiem, K., et al. (2013). Development of the Twin-Strep-tag(R) and its application for purification of recombinant proteins from cell culture supernatants. *Protein Expression and Purification*, *92*(1), 54–61. https://doi.org/10.1016/j.pep.2013.08.021.

Schwenkert, S., Netz, D. J. A., Frazzon, J., Pierik, A. J., Bill, E., Gross, J., et al. (2010). Chloroplast HCF101 is a scaffold protein for [4Fe-4S] cluster assembly. *Biochemical Journal*, *425*(1), 207–214. https://doi.org/10.1042/BJ20091290.

Shan, S. O. (2016). ATPase and GTPase tangos drive intracellular protein transport. *Trends in Biochemical Sciences*, *41*(12), 1050–1060. https://doi.org/10.1016/j.tibs.2016.08.012.

Shepard, E. M., McGlynn, S. E., Bueling, A. L., Grady-Smith, C. S., George, S. J., Winslow, M. A., et al. (2010). Synthesis of the 2Fe subcluster of the [FeFe]-hydrogenase H cluster on the HydF scaffold. *Proceedings of the National Academy of Sciences of the United States of America*, *107*(23), 10448–10453. https://doi.org/10.1073/pnas.1001937107.

Soriano, A., & Hausinger, R. P. (1999). GTP-dependent activation of urease apoprotein in complex with the UreD, UreF, and UreG accessory proteins. *Proceedings of the National Academy of Sciences of the United States of America*, *96*(20), 11140–11144. https://doi.org/10.1073/pnas.96.20.11140.

Stookey, L. L. (1970). Ferrozine—A new spectrophotometric reagent for iron. *Analytical Chemistry*, *42*(7), 779–781. https://doi.org/10.1021/ac60289a016.

Sydor, A. M., Lebrette, H., Ariyakumaran, R., Cavazza, C., & Zamble, D. B. (2014). Relationship between Ni(II) and Zn(II) coordination and nucleotide binding by the Helicobacter pylori [NiFe]-hydrogenase and urease maturation factor HypB. *Journal of Biological Chemistry*, *289*(7), 3828–3841. https://doi.org/10.1074/jbc.M113.502781.

Tsaousis, A. D., Gentekaki, E., Eme, L., Gaston, D., & Roger, A. J. (2014). Evolution of the cytosolic iron-sulfur cluster assembly machinery in Blastocystis species and other microbial eukaryotes. *Eukaryotic Cell*, *13*(1), 143–153. https://doi.org/10.1128/ec.00158-13.

Veatch, J. R., McMurray, M. A., Nelson, Z. W., & Gottschling, D. E. (2009). Mitochondrial dysfunction leads to nuclear genome instability via an iron-sulfur cluster defect. *Cell*, *137*(7), 1247–1258. https://doi.org/10.1016/j.cell.2009.04.014.

Vitale, G., Fabre, E., & Hurt, E. C. (1996). NBP35 encodes an essential and evolutionary conserved protein in Saccharomyces cerevisiae with homology to a superfamily of bacterial ATPases. *Gene*, *178*(1–2), 97–106.

Walmsley, A. R., Zhou, T., Borges-Walmsley, M. I., & Rosen, B. P. (2001). A kinetic model for the action of a resistance efflux pump. *Journal of Biological Chemistry*, *276*(9), 6378–6391. https://doi.org/10.1074/jbc.M008105200.

Wilke, K. E., Francis, S., & Carlson, E. E. (2015). Inactivation of multiple bacterial histidine kinases by targeting the ATP-binding domain. *ACS Chemical Biology*, *10*(1), 328–335. https://doi.org/10.1021/cb5008019.

Wu, W., Park, K. T., Holyoak, T., & Lutkenhaus, J. (2011). Determination of the structure of the MinD-ATP complex reveals the orientation of MinD on the membrane and the relative location of the binding sites for MinE and MinC. *Molecular Microbiology*, *79*(6), 1515–1528. https://doi.org/10.1111/j.1365-2958.2010.07536.x.

Yan, A., & Kiley, P. J. (2009). Techniques to isolate O2-sensitive proteins: [4Fe-4S]-FNR as an example. *Methods in Enzymology*, *463*, 787–805. https://doi.org/10.1016/s0076-6879(09)63042-1.

Zhou, T. Q., Radaev, S., Rosen, B. P., & Gatti, D. L. (2000). Structure of the ArsA ATPase: The catalytic subunit of a heavy metal resistance pump. *EMBO Journal*, *19*(17), 4838–4845. https://doi.org/10.1093/emboj/19.17.4838.

CHAPTER ELEVEN

Characterization of Glutaredoxin Fe–S Cluster-Binding Interactions Using Circular Dichroism Spectroscopy

Angela-Nadia Albetel, Caryn E. Outten[1]
University of South Carolina, Columbia, SC, United States
[1]Corresponding author: e-mail address: outten@sc.edu

Contents

Abstract

Monothiol glutaredoxins (Grxs) with a conserved Cys-Gly-Phe-Ser (CGFS) active site are iron–sulfur (Fe–S) cluster-binding proteins that interact with a variety of partner proteins and perform crucial roles in iron metabolism including Fe–S cluster transfer, Fe–S cluster repair, and iron signaling. Various analytical and spectroscopic methods are currently being used to monitor and characterize glutaredoxin Fe–S cluster-dependent

Methods in Enzymology, Volume 599
ISSN 0076-6879
https://doi.org/10.1016/bs.mie.2017.11.003

interactions at the molecular level. The electronic, magnetic, and vibrational properties of the protein-bound Fe–S cluster provide a convenient handle to probe the structure, function, and coordination chemistry of Grx complexes. However, some limitations arise from sample preparation requirements, complexity of individual techniques, or the necessity for combining multiple methods in order to achieve a complete investigation. In this chapter, we focus on the use of UV–visible circular dichroism spectroscopy as a fast and simple initial approach for investigating glutaredoxin Fe–S cluster-dependent interactions.

1. INTRODUCTION

The ubiquitous monothiol CGFS glutaredoxin (Grx) proteins play important roles in iron metabolism by assisting in cellular iron sensing, Fe–S cluster biosynthesis, transfer, repair, and storage (Couturier, Przybyla-Toscano, Roret, Didierjean, & Rouhier, 2015). A wide variety of CGFS glutaredoxin isoforms have been identified in both prokaryotes and multiple compartments of eukaryotes, all with the ability to either assemble or acquire an Fe–S cluster, which is subsequently stored or transferred to target proteins. CGFS Grxs with a single GRX domain are found in prokaryotes and the mitochondria of eukaryotes, while multidomain Grxs with one to three GRX domains linked to an N-terminal thioredoxin (TRX)-like domain are found exclusively in eukaryotes (Couturier, Jacquot, & Rouhier, 2009). In the absence of a binding partner, CGFS Grxs primarily form homodimers with a bridging [2Fe–2S] cluster ligated by the conserved cysteine in the CGFS motif from each monomer and two glutathione (GSH) molecules stabilized by the GSH-binding pocket of each Grx monomer (Iwema et al., 2009). Cellular processes involving CGFS Grxs often proceed as Fe–S cluster-dependent protein–protein interactions via a mechanism that employs structural rearrangements and ligand exchange. Two primary types of Grx Fe–S cluster-dependent interactions have been identified thus far in which Grxs interact with one or two binding partners to form (1) a heterocomplex with the Fe–S cluster ligated at the interface of the Grx domain(s) and the interacting partner protein as illustrated by the Grx–BolA complexes (Banci, Camponeschi, Ciofi-Baffoni, & Muzzioli, 2015; Dlouhy et al., 2016; Li et al., 2009; Li, Mapolelo, Randeniya, Johnson, & Outten, 2012; Nasta, Giachetti, Ciofi-Baffoni, & Banci, 2017; Roret et al., 2014; Uzarska et al., 2016; Yeung et al., 2011) or (2) the formation of an entirely new Fe–S cluster coordinated protein complex obtained via complete and intact transfer of the Fe–S cluster from Grx to the interacting partner (Banci et al., 2014; Banci, Camponeschi, et al., 2015; Banci, Ciofi-Baffoni, et al., 2015; Bandyopadhyay et al., 2008; Fidai, Wachnowsky, & Cowan, 2016; Mapolelo et al., 2013; Poor et al., 2014; Qi & Cowan, 2011; Shakamuri, Zhang, & Johnson, 2012; Vranish, Das, & Barondeau, 2016; Vranish et al., 2015; Xia et al., 2015; Yeung et al., 2011; Zhang et al., 2013).

Biological Fe–S clusters are commonly coordinated by cysteines via the sulfur atom of the protein thiol moiety. However, sometimes noncysteinyl ligands such as histidine, serine, or water are used as ligands. Owing to the fact that the structural, electronic, and vibrational properties of Fe–S clusters in metalloenzymes depend on the type of ligands and protein conformation in the vicinity of the Fe–S cofactor, spectroscopic methods such as UV–visible absorption, circular dichroism (CD), electron paramagnetic resonance (EPR), resonance Raman, Mossbauer, and NMR are commonly employed when characterizing Fe–S containing proteins and/or Fe–S cluster-dependent interactions in vitro. Taken together and in association with analytical methods and site-directed mutagenesis, these techniques provide information on the number, oxidation state, and structure type of the cluster involved, as well as the amino acid residues coordinating the cluster in both the initial and final complexes. However, no one spectroscopic probe available can accomplish all these criteria at the present moment and complementary techniques often need to be employed.

Most Grxs identified thus far have been shown to ligate a $[2Fe–2S]^{2+}$ cluster that can be observed easily using UV–visible absorption and CD spectroscopies. Although $[3Fe–4S]^{+}$ and $[4Fe–4S]^{2+}$ cluster-coordinating Grxs have also been reported (Zhang et al., 2013), for simplicity these will not be discussed here. Solutions of oxidized $[2Fe–2S]^{2+}$ cluster-containing Grxs absorb light strongly, exhibiting a reddish-brown color and broad UV–visible absorption spectra consisting of multiple overlapping absorption bands through the visible and near UV regions. These bands reflect many unresolved S to Fe^{3+} charge transfer electronic transitions, with greater intensity closer to the near UV region (Fig. 1A). Owing to the high similarities and lack of resolved features in the absorption spectra of $[2Fe–2S]^{2+}$ cluster-containing proteins, UV–visible absorption spectroscopy is generally not the method of choice when monitoring and/or characterizing Grx Fe–S cluster-dependent interactions, as it cannot distinguish between Fe–S clusters with subtle changes in the protein surroundings.

On the other hand, the interaction of the achiral Fe–S cluster with the chiral protein moiety generates a suitable handle for characterization via CD absorption spectroscopy (Stephens et al., 1978). This method measures the

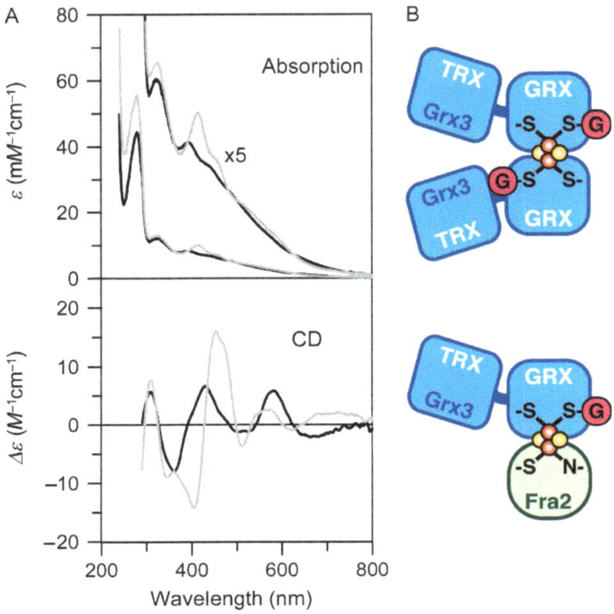

Fig. 1 (A) Comparison of the UV–visible absorption (*top*) and CD spectra (*bottom*) of as-purified [2Fe–2S] Grx3 (*gray line*) and [2Fe–2S] Grx3–Fra2 (*black line*). (B) Models for Grx3 homodimer and Grx3–Fra2 heterodimer. *Panel (A): Adapted with permission from Li, H., Mapolelo, D. T., Dingra, N. N., Naik, S. G., Lees, N. S., Hoffman, B. M., et al. (2009). The yeast iron regulatory proteins Grx3/4 and Fra2 form heterodimeric complexes containing a [2Fe–2S] cluster with cysteinyl and histidyl ligation.* Biochemistry, 48*(40), 9569–9581. https://doi.org/10.1021/bi901182w. Copyright (2009) American Chemical Society.*

difference in absorption of left- and right-polarized light as a function of wavelength for a given chiral chromophore. CD absorption is observed continuously for [2Fe–2S] cluster proteins, and the spectral range spans the near IR, the visible, and near UV. In the UV range, CD reports on protein secondary structure, whereas in the visible and near UV range, CD reports on the electronics of the Fe–S cluster. In contrast to their corresponding absorption spectra, the CD spectra of [2Fe–2S] cluster-containing proteins are significantly more structured due to both positive and negative bands with well-defined peaks accompanied by shoulders and inflections. Thus, overlapping bands in the absorption spectrum can be clearly resolved by CD if their signs are different. Owing to its ability to report on subtle changes in cluster coordination environment, CD is an elegant diagnostic tool for investigating Fe–S cluster structural rearrangements orchestrated by protein–protein interactions and/or ligand exchanges. In addition, unlike other spectroscopic methods which require concentrated protein solutions, denaturing solvents, and/or low temperature conditions; CD absorption spectroscopy provides the advantage of measurements on dilute protein solutions under physiological conditions.

Given these advantages, UV–visible CD spectroscopy has been routinely and extensively employed as a method to assess Fe–S cluster-dependent interactions in biological systems. Due to its reliability, simplicity, and unique sensitivity to Fe–S cluster coordination changes, this technique has been our preferred approach to investigate the interaction between CGFS-type glutaredoxins and BolA-like proteins from both prokaryotes (Dlouhy et al., 2016; Mapolelo et al., 2013) and eukaryotes (Li et al., 2011, 2009, 2012; Poor et al., 2014). In this chapter, we will outline several CD-focused approaches for characterizing and probing different Fe–S-dependent interactions in which CGFS Grxs participate.

2. GENERAL EQUIPMENT

- Anaerobic chamber (glove box)
- 1-cm/0.5-mL quartz spectrophotometer cell (semi-microcuvette) sealed with a rubber septum to maintain anaerobicity when needed (Starna Cells, Inc.)
- Pipettes and pipette tips
- CD spectropolarimeter (Jasco) equipped with a temperature controlling (Peltier) device to maintain the temperature below 10°C when needed
- Spectra Manager™ (Jasco) for instrument control, data acquisition, and data processing
- GraphPad Prism or similar graphing software for graphing, spectral analysis, and curve fitting via nonlinear regression

3. GENERAL NOTES

1. CD spectra are measured over the visible and near UV range (290–700 nm).
2. Typical CD parameters include: sensitivity, 100 mdeg; data pitch, 1 nm; scanning mode, continuous; scanning speed, 200 nm min^{-1}; response, 2 s; and bandwidth, 10 nm.
3. Semi-microquartz cuvettes are used to minimize the volume of protein solution required.

4. In the case of samples requiring anaerobic conditions, solutions should be manipulated and cells should be filled inside an anaerobic chamber. Cells should be sealed with a septum as shown in Fig. 2B.
5. The concentration of the [2Fe–2S] cluster in each sample is obtained by dividing the concentration of total iron in the sample by two. The iron concentration is typically determined using the ferrozine assay (Riemer, Hoepken, Czerwinska, Robinson, & Dringen, 2004). In addition, the [2Fe–2S] concentration can be estimated with UV–visible absorption spectroscopy from the dominant band in the 390–430 nm range. The extinction coefficient for this peak is typically in the range of 8–11 mM^{-1} cm^{-1} per [2Fe–2S] cluster for Grx homodimers and 5–9 mM^{-1} cm^{-1} per [2Fe–2S] cluster for Grx heterocomplexes (Bandyopadhyay et al., 2008; Dlouhy et al., 2016; Li et al., 2011, 2009, 2012; Nasta et al., 2017; Roret et al., 2014; Zhang et al., 2013), similar to other $[2Fe–2S]^{2+}$ cluster proteins (Dailey, Finnegan, & Johnson, 1994). Fe–S cluster loading and protein purity can thus be assessed from the A_{390}:A_{280} ratio calculated from the UV–visible absorption spectrum.
6. When using a semi-microcuvette (1 cm path length), Fe–S cluster concentrations should be adjusted to the optimum range (±100 mdeg) in the UV and visible region (typically 30–100 μM Fe–S cluster).
7. If protein stability is affected by temperature, the samples should be kept on ice at all times, and the temperature should be maintained below 10°C while scanning the CD spectrum using the Peltier device.

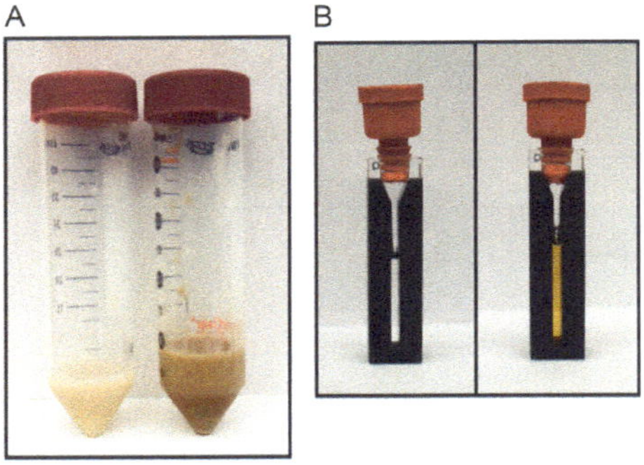

Fig. 2 Overexpression and purified samples of Fra2 and Grx3–Fra2 heterodimer. (A) Cell pellets obtained upon expression of recombinant *S. cerevisiae* Fra2 (*left*) and coexpression of Grx3 with Fra2 (*right*) in *E. coli*. (B) UV–visible absorption/CD samples of apo, purified Fra2 (*left*) and [2Fe–2S]-bridged Grx3–Fra2 (*right*) in a semi-microcuvette.

8. The same cuvette should be used for recording all of the spectra whenever possible to avoid reading errors arising from intrinsic cuvette absorption when subtracting the blank.
9. Handle cuvettes with gloves and wipe with a disposable wiper (e.g., Kimwipes®) before recording each spectrum in order to avoid introducing errors from fingerprint marks.
10. 50 m*M* Tris-MES (pH 8.0) or 50–100 m*M* Tris–HCl (pH 7.4–8.0) buffer solutions are generally used in our group to purify and to characterize CGFS Grxs and their complexes. GSH (1–5 m*M*) is sometimes added to the buffer in order to improve Fe–S cluster stability for Grx homodimers (Bandyopadhyay et al., 2008; Dlouhy et al., 2016; Poor et al., 2014). However, the effect of GSH on Grx heterocomplexes varies depending on the specific proteins. For example, GSH does not appear to affect the stability of the $[2Fe–2S]^{2+}$ cluster in *Saccharomyces cerevisiae* Grx3–Fra2 or Grx4–Fra2 heterodimers (Li et al., 2011, 2009), but destabilizes the $[2Fe–2S]^{2+}$ cluster in *Escherichia coli* Grx4–IbaG heterodimers (Dlouhy et al., 2016).

4. Fe–S CLUSTER-DEPENDENT COMPLEX FORMATION

The results for our studies on the *S. cerevisiae* Grx3/4–Fra2 interaction are used in this section to illustrate the use of UV–visible CD spectroscopy to rapidly and efficiently probe Fe–S cluster-dependent complex formation. By comparing the CD spectra of the [2Fe–2S] centers of the as-purified [2Fe–2S] Grx3/4 homodimers and the complex obtained upon coexpressing Grx3/4–Fra2 or titrating [2Fe–2S] Grx3/4 homodimer with apo Fra2, we were able to distinguish and characterize the two Fe–S-containing complexes and ultimately conclude that Grx3/4 and Fra2 interact by forming a [2Fe–2S] cluster-bridged heterodimer (Li et al., 2009). Furthermore, when coupled with site-directed mutagenesis, these CD-monitored interactions can be used to explore the role of specific amino acid residues in Fe–S cluster-dependent interactions (Li et al., 2011).

4.1 Characterization of Fe–S Cluster Binding in a Copurified Protein Complex

Genetic and biochemical studies indicated that *S. cerevisiae* Grx3/4 and Fra2 interact in vivo and are required for regulation of iron homeostasis (Kumanovics et al., 2008; Lesuisse et al., 2005; Ojeda et al., 2006; Pujol-Carrion, Belli, Herrero, Nogues, & de la Torre-Ruiz, 2006). To characterize the iron-dependent interactions between Grx3/4 and Fra2 at the molecular level, Grx3 or Grx4 was coexpressed with Fra2 in *E. coli* for overexpression and purification, and the purified complexes were subjected to a detailed biochemical characterization. *E. coli* cells coexpressing recombinant Grx3 and Fra2 or Grx4 and Fra2 exhibited a reddish-brown color (Fig. 2A) that was maintained throughout the purification procedure (Fig. 2B). This color is a strong indication of the presence of a Fe–S cluster-binding protein. For comparison, *E. coli* cells expressing Fra2 alone, which does not bind an Fe–S cluster, is shown in Fig. 2A (left panel), resulting in a colorless purified protein solution (Fig. 2B, left panel). The SDS–PAGE gel of the colored fractions indicated that Grx3/4 and Fra2 coeluted as a complex (Li et al., 2009). The shift in the intensity and position of the CD absorption bands of the as-purified heterocomplexes vs the as-purified [2Fe–2S]–Grx3/4 homodimers demonstrated that the Grx3/4–Fra2 interaction included an [2Fe–2S] center with a distinctly different coordination environment (Fig. 1A). However, comprehensive analytical and spectroscopic characterization of the purified Grx3–Fra2 and Grx4–Fra2 complexes was necessary to further characterize the protein–protein and metal–protein interactions (Li et al., 2009). These results confirmed the presence of [2Fe–2S] cluster-bridged Grx3–Fra2 and Grx4–Fra2 heterodimers, in which the Fe–S centers were ligated by one active site cysteine contributed by Grx3 or Grx4, one glutathione (GSH) molecule, and two ligands (one histidine and one cysteine) contributed by the Fra2 partner (Fig. 1B) (Li et al., 2011, 2009).

4.1.1 Buffers and Reagents

- Purified proteins with known concentrations (both protein and Fe–S cluster concentration)
- *Sample buffers*: Same as the buffers used for concentrating the protein complexes after the final purification step

4.1.2 Procedure

1. Scan and record the UV–visible CD absorption spectrum of the sample buffer for the baseline normalization of all spectra in the 1-cm cuvette. This spectrum is called the blank hereafter.
2. Clean the quartz cuvette after each sample by rinsing with water to avoid contamination. Flush cuvette with ethanol and blow nitrogen to completely dry the cuvette between different samples.

3. For each Fe–S-containing protein sample, prepare ~350 μL of 30–100 μ*M* final Fe–S cluster concentration in the cuvette and record the UV–visible CD absorption spectrum.
4. Subtract the corresponding blank from the spectrum of each sample.

4.1.3 Data Analysis

1. Export files and convert the observed CD ellipticity units (θ_{obs} in millidegrees) to the differential molar extinction coefficient ($\Delta\varepsilon$ in M^{-1} cm^{-1}) using Eq. (1). $\Delta\varepsilon$, also known as the molar CD, is the difference in molar extinction coefficients for left vs right polarized light ($\Delta\varepsilon = \varepsilon_L - \varepsilon_R$). In Eq. (1), 32,980 is the ellipticity to absorbance conversion factor, *l* is the cuvette path length in cm, and [Fe–S] is the molar concentration of the [2Fe–2S] cluster in the sample:

$$\Delta\varepsilon = \frac{\theta_{obs} \times 10^6}{32,980 \times l \times [Fe - S]} \quad (1)$$

2. Use GraphPad Prism or similar graphing software to plot the UV–visible CD spectra for each sample.
3. Overlap all spectra together for comparison and analysis (Fig. 1A). Differences in the intensity and/or position of CD absorption bands are indicative of changes in the Fe–S cluster coordination environments.

4.2 Characterization of Fe–S Cluster Complex Formation via Titration

As another approach, Fe–S cluster-dependent binding interactions can be investigated by separately expressing and purifying the potential partner proteins, followed by CD-monitored titration of the holo protein with the apo partner. For instance, CD-monitored titration of purified [2Fe–2S] Grx3 with apo Fra2 has been performed to probe the stoichiometry, binding affinity, and cluster coordination changes that occur upon interaction of these partner proteins (Li et al., 2011). A typical CD-monitored titration reaction is performed by incubating a fixed amount of holo protein with varying amounts of its apo partner and scanning the UV–visible CD absorption spectra (Fig. 3).

4.2.1 Buffers and Reagents

- Refer to Section 4.1.1.

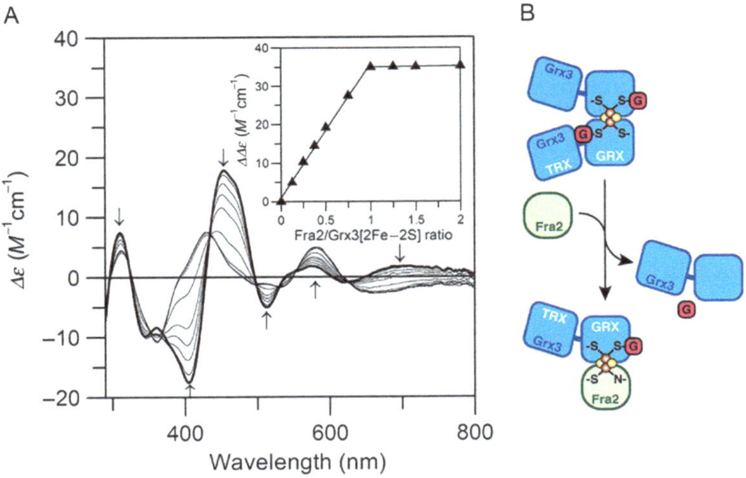

Fig. 3 (A) Titration of apo Fra2 into the [2Fe–2S] Grx3 homodimer. As-purified [2Fe–2S] Grx3 is shown as a *thick black line*. *Arrows* indicate the direction of changes in CD intensity with increasing apo Fra2 concentration. The *inset* shows the maximum difference in CD intensity (at 463 and 403 nm) as a function of the Fra2:Grx3 $[2Fe–2S]^{2+}$ cluster ratio. (B) Model for conversion of [2Fe–2S] Grx3 homodimer to [2Fe–2S] Grx3–Fra2 heterodimer. *Panel (A): This research was originally published in Li, H., Mapolelo, D. T., Dingra, N. N., Keller, G., Riggs-Gelasco, P. J., Winge, D. R., et al. (2011). Histidine 103 in Fra2 is an iron–sulfur cluster ligand in the [2Fe–2S] Fra2–Grx3 complex and is required for in vivo iron signaling in yeast.* The Journal of Biological Chemistry, 286*(1), 867–876. https://doi.org/10.1074/jbc.M110.184176. Copyright © The American Society for Biochemistry and Molecular Biology.*

4.2.2 Procedure

1. Follow steps 1 and 2 in Section 4.1.2 to record the blank.
2. Record the UV–visible CD absorption spectrum of the Fe–S cluster-containing holo protein.
3. Prepare ~350 μL reactions with 30–100 μ*M* fixed final cluster concentration of holo protein and different concentrations (0.125–5-fold excess) of apo protein binding partner in 1.5-mL microcentrifuge tubes.
4. Allow reactions to reach equilibrium by incubating for 5–15 min at room temperature, with occasional stirring.
5. Record the UV–visible CD absorption spectra for each reaction.

4.2.3 Data Analysis

1. Export files and convert the observed CD units (θ_{obs} in millidegrees) to molar CD units ($\Delta\varepsilon$ in M^{-1} cm^{-1}) using Eq. (1) (see Section 4.1.3).
2. Use GraphPad Prism or similar graphing software to plot the UV–visible CD spectra for each sample.

3. Overlap all spectra together for comparison and analysis (Fig. 3A).
4. In order to determine the stoichiometry of the binding interaction between partner proteins, plot the difference in the $\Delta\varepsilon_{obs}$ values at two different wavelengths that exhibit the maximum changes as a function of [2Fe–2S]:apo protein ratio (Fig. 3A, *inset*). The shape of this binding curve provides information about the nature of the binding interaction. In the case of the Grx3–Fra2 interaction, the initial binding curve is a straight line, indicating that Fra2 binding to [2Fe–2S] Grx3 is stoichiometric under these conditions. In other words, the K_d for this interaction is below the concentration of [2Fe–2S] Grx3 homodimer used in the assay (50 μ*M*) and therefore too low be directly measured under these conditions. However, the intercept of the initial binding curve and the saturation binding plateau provides a measure of the binding stoichiometry.

4.2.4 Notes

1. For a comprehensive and accurate analysis of the ligand exchange reaction, the products of the final titration should be subsequently isolated using appropriate chromatography methods, and subjected to analytical, biochemical, and spectroscopic analyses in order to assess the oligomeric state and Fe:S:protein ratio of each product. These additional steps were used to confirm formation of the [2Fe–2S]-bridged Grx3–Fra2 heterodimer and release of the apo Grx3 monomer (Fig. 3B) (Li et al., 2011).

5. THERMODYNAMIC AND KINETIC CHARACTERIZATION OF Fe–S CLUSTER TRANSFER REACTIONS

CGFS glutaredoxins, either alone or in concert with BolA proteins, have been shown to accept, store, and deliver Fe–S clusters to a wide variety of acceptor proteins (Banci et al., 2014; Banci, Camponeschi, et al., 2015; Banci, Ciofi-Baffoni, et al., 2015; Bandyopadhyay et al., 2008; Boutigny et al., 2013; Frey, Palenchar, Wildemann, & Philpott, 2016; Hoffmann et al., 2011; Iñigo et al., 2016; Kim, Chung, Kim, Lee, & Roe, 2010; Mapolelo et al., 2013; Melber et al., 2016; Mühlenhoff et al., 2010; Poor et al., 2014; Shakamuri et al., 2012; Ströher et al., 2016; Uzarska et al., 2016; Vranish et al., 2016; Xia et al., 2015; Zhang et al., 2013). In *S. cerevisiae*, transfer of a [2Fe–2S] cluster from Grx3/4–Fra2 to the transcription factor Aft1/2 is proposed to be the key mechanism for regulating Aft1/2 activity in response to iron (Poor et al., 2014). We monitored the thermodynamics of Fe–S cluster transfer from Grx3–Fra2 to Aft2 by incubating [2Fe–2S] Grx3–Fra2 with Aft2 at different ratios and measuring the UV–visible CD spectra, following a procedure similar to the one described in Section 4.2.

In order to determine whether the rate of Fe–S cluster transfer is physiologically relevant, CD-monitored Fe–S cluster transfer reactions can be also followed kinetically either by scanning the full UV–visible CD absorption spectrum of the two interacting proteins with time (full wavelength scan) or by monitoring the changes in CD at a single fixed wavelength with time (fixed wavelength scan). Although both methods can be used for the kinetic evaluation of the reaction, in the case of a very fast Fe–S cluster transfer, the latter would be better for capturing the initial changes in the CD spectrum due to the shorter time between scans, therefore ensuring a more accurate determination of the kinetic rate constant. We describe here the procedure for each of these methods.

5.1 CD-Monitored Thermodynamic Analysis of Fe–S Cluster Transfer Reactions

Similar to the Fe–S cluster complex formation titrations described in Section 4.2, by varying the amount of acceptor protein added to the [2Fe–2S] cluster donor, thermodynamic analysis of transfer of an intact Fe–S cluster from an acceptor to a donor can be assessed. Similar equipment, procedures, data analysis, and notes described for the Grx–Fra2 Fe–S complex titrations (see Section 4.2) are employed when determining the stoichiometry and binding ratio of the intact cluster transfer from a donor to acceptor (Fig. 4). Additional equipment and procedures are described later.

5.1.1 Additional Equipment

- Concentrator (e.g., Amicon® stirred cell or centrifugal concentrator).

5.1.2 Buffers and Reagents

- Purified proteins with known Fe–S cluster concentrations for the donor and protein concentration for the acceptor
- *Interaction buffer*: Same as the buffer used for concentrating the protein samples after the final purification step. If different buffers are used when purifying the two proteins to be used in the study, choose the buffer suitable to the least stable protein. For the [2Fe–2S]–Grx3–Fra2 to Aft2 cluster transfer reaction, 50 mM Tris–HCl pH 8.0, 500 mM NaCl buffer

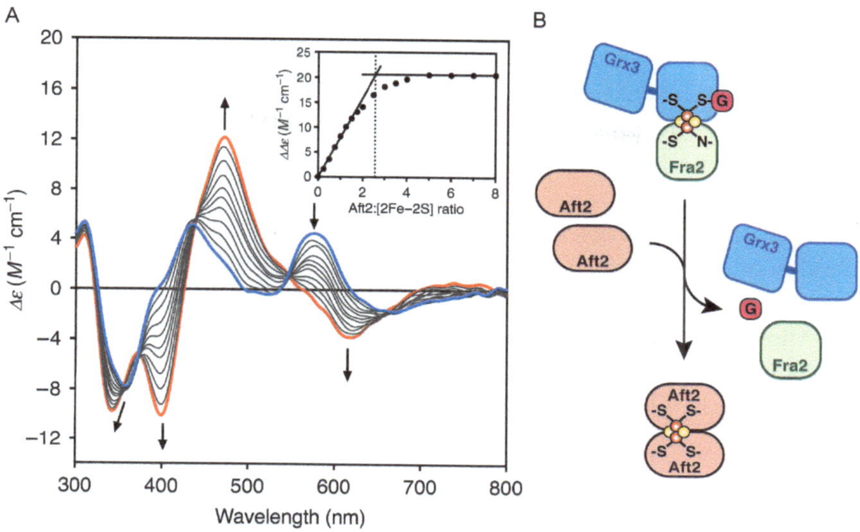

Fig. 4 Thermodynamic analysis of the Fe–S cluster transfer from *S. cerevisiae* [2Fe–2S] Grx3–Fra2 to Aft2 monitored by UV–visible CD spectroscopy. (A) Different ratios of acceptor:donor proteins are incubated and allowed to reach equilibrium. The CD spectrum of the reaction at each ratio is recorded, then the difference in CD intensity between two different wavelengths exhibiting the largest variations (400 and 472 nm) is plotted as a function of the Aft2:[2Fe–2S] ratio (*inset*). *Blue line* is [2Fe–2S]–Fra2–Grx3 without addition of Aft2, *black lines* indicate titration with 0.25–3.5 excess Aft2, and *red line* is 5:1 [Aft2]:[2Fe–2S] ratio. *Arrows* indicate the direction of intensity changes with increasing [Aft2]. (B) Model of the Fe–S cluster transfer reaction from [2Fe–2S]–Fra2–Grx3 to Aft2. *Panel (A): Taken with permission from Poor, C. B., Wegner, S. V., Li, H., Dlouhy, A. C., Schuermann, J. P., Sanishvili, R., et al. (2014). Molecular mechanism and structure of the* Saccharomyces cerevisiae *iron regulator Aft2.* Proceedings of the National Academy of Sciences of the United States of America, 111*(11), 4043–4048. https://doi.org/10.1073/pnas.1318869111. Copyright © 2012, National Academy of Sciences.*

solution was used considering that Aft2 stability decreases with decreasing salt concentration

- Reductant (i.e., dithiothreitol (DTT) or Tris-(2-carboxyethyl)-phosphine (TCEP))

5.1.3 Procedure

1. Follow steps 1 and 2 in Section 4.1.2 to obtain the CD spectrum of the blank solution.
2. Prereduce the acceptor protein by incubating on ice with reductant (5–10 m*M* DTT or 40 m*M* TCEP) for approximately 30 min. Stir occasionally. This step is necessary in order to break any disulfide bonds and ensures that all thiol moieties involved in ligating the Fe–S cluster are reduced and available.

3. Buffer exchange to remove excess reductant by repeated dilution–concentration cycles, using either a stirred cell or centrifugal concentrator inside the anaerobic chamber.
4. Determine the protein concentration using the Bradford assay (BioRad).
5. Record the UV–visible CD spectrum of the Fe–S cluster-containing holo protein.
6. Prepare ~350 μL reactions with 30–100 μ*M* fixed final cluster concentration of holo protein and different concentrations (0.125–5-fold excess) of prereduced apo protein to be titrated in 1.5-mL microcentrifuge tubes.
7. Allow reactions to reach equilibrium by incubating 5–15 min at room temperature, with occasional stirring.
8. Record the UV–visible CD absorption spectra for each reaction.

5.1.4 Data Analysis

1. See Section 4.2.3.

5.1.5 Notes

1. Control experiments lacking the acceptor protein should be carried out in order to demonstrate that the changes in CD occur as a result of the interaction rather than Fe–S cluster instability of the holo donor protein.
2. As a control for cluster disassembly/reassembly, the Fe–S donor can be replaced with equivalent amounts of Fe^{2+} and S^{2-} and the cluster transfer reaction repeated. Lack of a distinct [2Fe–2S] cluster signal for this control reaction indicates that the Fe–S cluster is specifically transferred as an intact unit from the donor to the acceptor protein. Intact cluster transfer can also be confirmed by including EDTA (~0.1–1 m*M*) in the transfer mixture to chelate free iron.
3. Excess reductant should be removed prior to the Fe–S cluster transfer reaction to avoid any interference in the transfer reaction. In particular, the strong reductant DTT has been shown to accelerate Fe–S cluster transfer reactions in a nonphysiological manner (Vranish et al., 2016). If required, GSH (1–10 m*M*) can be included in the Fe–S transfer reaction instead to better mimic intracellular conditions.
4. The Fe–S cluster transfer can be theoretically simulated by adding the CD spectra of the holo donor protein and the CD spectrum of acceptor protein in 10% increments (Bandyopadhyay et al., 2008). A close correlation between the experimental and simulated data indicates that an intact and direct cluster transfer reaction occurred.

5.2 CD-Monitored Kinetic Analysis of Fe–S Cluster Transfer Reactions Based on Full Wavelength Scans

Changes in the intensity and position of peaks that occur with time in the UV–visible region of the CD spectra upon mixing a Fe–S cluster donor and its potential acceptor partner are readily observed in vitro and are a definitive indication of changes in the Fe–S cluster coordination. Similar to the Fe–S cluster-dependent interactions described in Section 4, when coupled with site-directed mutagenesis, CD-monitored Fe–S cluster transfer experiments offer information not only on the presence of a Fe–S cluster transfer, but also on the molecular details of the mechanism by which the reaction occurs.

5.2.1 Additional Equipment

- Air-tight Hamilton syringes and rubber stopper
- Concentrator (e.g., Amicon® stirred cell or centrifugal concentrator)
- Timer

5.2.2 Buffers and Reagents

- See Section 5.1.2.

5.2.3 Procedure

1. All samples are to be prepared in an anaerobic chamber.
2. Select "spectrum measurement" mode in the CD software and select the range of wavelength to be measured.
3. Follow steps 1 and 2 in Section 4.1.2 to obtain the CD spectrum of the blank solution.
4. Transfer donor protein into the cleaned and dried cuvette, seal with a rubber septum to maintain anaerobicity (Fig. 2B), and take outside of the anaerobic chamber.
5. Scan and record the CD spectrum of the donor protein.
6. Bring cuvette back into the anaerobic chamber and transfer donor protein to a microcentrifuge tube.
7. Rinse cuvette thoroughly with distilled water and dry as described in Section 4.1.2, step 2.
8. Prereduce the acceptor protein and buffer exchange to remove excess reductant as described in Section 5.1.3, steps 2–4.
9. Calculate the volume of donor and acceptor proteins needed to prepare ~350 μL reaction having a 1:2 [2Fe–2S]:acceptor protein ratio, and a final [2Fe–2S] cluster concentration of 30–100 μ*M*.

10. Prepare the acceptor protein by diluting into the interaction buffer to the concentration and volume calculated in step 9, and place into the 1-cm semi-microcuvette sealed with a rubber septum.
11. Record the UV–visible CD absorption spectrum of the acceptor protein.
12. Using a Hamilton airtight syringe with a needle, obtain the calculated amount of Fe–S cluster donor protein in the glove box, cap the needle with a rubber stopper (Fig. 5, *left*), then take it outside to the CD spectrophotometer.

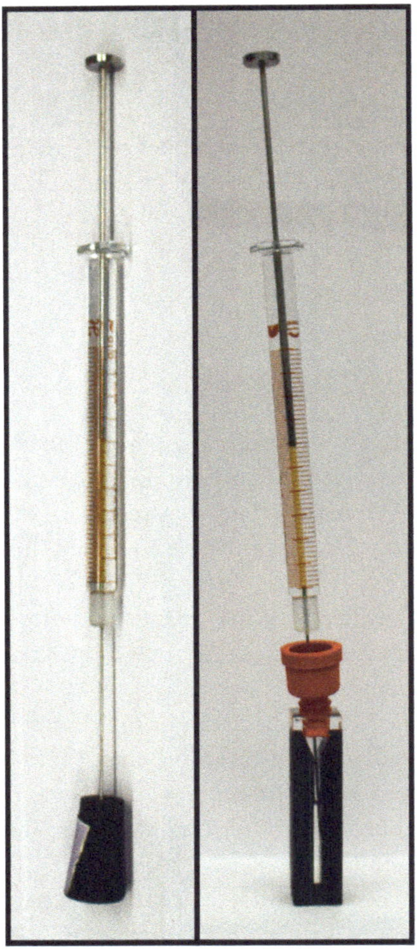

Fig. 5 Set up for Fe–S cluster transfer reactions. (*Left*) Hamilton syringe filled inside anaerobic chamber with [2Fe–2S]–Grx3–Fra2 and sealed with a rubber stopper to maintain anaerobicity. (*Right*) Hamilton syringe inserted via the rubber septum into the cuvette containing prereduced anaerobic Aft2.

13. Remove stopper from the syringe needle and quickly insert the needle in the cuvette containing the acceptor protein through the rubber stopper used to seal the cuvette (Fig. 5, *right*).
14. Inject donor protein into the acceptor protein.
15. Start timer simultaneously with injecting the donor protein.
16. Mix quickly by inverting the cuvette 3–4 times.
17. Place cuvette inside the cuvette compartment, close the instrument lid, and start scanning the CD spectrum.
18. Record the time that passes from the moment the reaction is initiated (donor protein is injected) until the scan is started. This will be called the "lag time" hereafter.
19. When scan is done, start scanning again, and record the time each scan is started.
20. Continue scanning spectra and recording the time until no changes in CD are observed.

5.2.4 Data Analysis

1. Export files and convert the observed CD units (θ_{obs}) to $\Delta\varepsilon$ (M^{-1} cm^{-1}) using Eq. (1). For the [Fe–S] term in this equation, use the protein concentration for the acceptor protein, and the Fe–S cluster concentration for the donor protein and the cluster transfer reaction data points.
2. Use graphing software (Microsoft Excel, GraphPad Prism, etc.) to plot the CD spectra collected for the donor protein, and for each of the time points recorded during the Fe–S cluster transfer reaction.
3. Overlap all spectra together for comparison and analysis (Fig. 6A).
4. Plot the change in CD values ($\Delta\varepsilon_{obs}$) at a wavelength position exhibiting the largest spectral changes, as a function of time. Add the "lag time" to each of the time values recorded during data collecting.
5. Use kinetics simulation software (e.g., Chemical Kinetics Simulator or COPASI) to fit the data to a kinetic model.

5.2.5 Notes

1. In order to minimize the amount of time that passes from the moment the donor protein is injected into the cuvette containing the acceptor protein, until the CD spectrum is scanned, have the software and computer set up and ready for scanning prior to injection.
2. The Jasco CD spectrophotometer also has an "interval scan measurement" scanning mode that can be used for scanning the spectrum at designated time intervals.

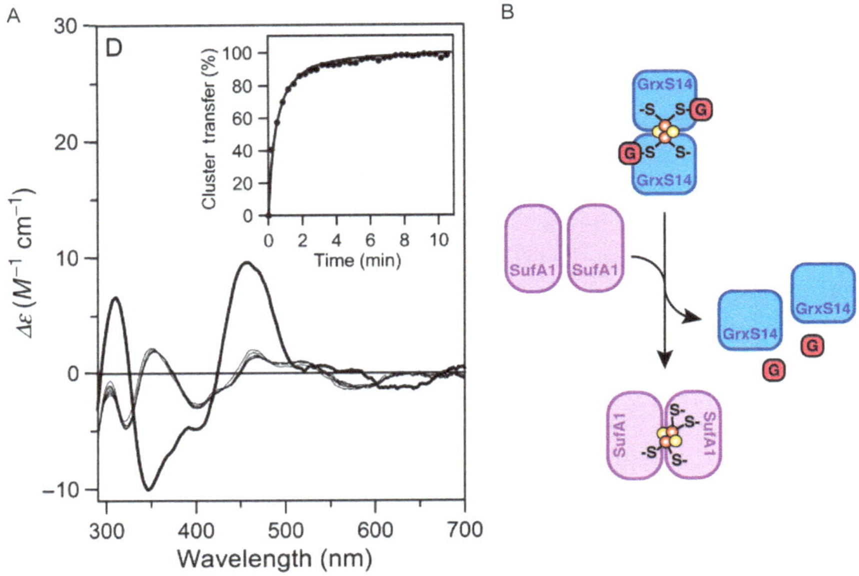

Fig. 6 CD-monitored wavelength scan and fixed wavelength kinetic analysis (*inset*) for the Fe–S cluster transfer reaction (*thin gray lines*) from *Arabidopsis thaliana* [2Fe–2S]–GrxS14 (*thick black line*) to *A. thaliana* SufA1. *Adapted with permission from Mapolelo, D. T., Zhang, B., Randeniya, S., Albetel, A. N., Li, H., Couturier, J., et al. (2013). Monothiol glutaredoxins and A-type proteins: Partners in Fe–S cluster trafficking.* Dalton Transactions, *42(9), 3107–3115. https://doi.org/10.1039/c2dt32263c. Copyright © 2012, Royal Society of Chemistry.*

5.3 CD-Monitored Kinetic Analysis of Fe–S Cluster Transfer Reactions at a Single Wavelength

A more direct kinetic measurement for the rate of Fe–S cluster transfer reactions is obtained when the changes in CD are monitored in real time. The same equipment, buffers, reagents, and notes described for the "full wavelength scan" method (see Section 5.2) are also needed when employing the single wavelength time course measurement for monitoring Fe–S cluster transfers. The procedure is similar, except changes in CD at only one wavelength are monitored. However, a wavelength scan of the interaction is necessary prior to employing the single wavelength method in order to be able to select the wavelength where the most dramatic changes in the CD values occur during the Fe–S cluster transfer reaction.

5.3.1 Additional Equipment

- Refer to Section 5.2.1.

5.3.2 Procedure

1. Use the results obtained from the "wavelength scan" method described earlier to select the wavelength at which the changes in CD values with time are the greatest.
2. Select "Time course measurement" in the CD software and use the wavelength selected in step 1 to set up the instrument.
3. Prepare samples in the anaerobic chamber as described in Section 5.2.3, steps 3–9.
4. Transfer the calculated amount of acceptor protein to the cuvette, seal cuvette using a rubber septum (Fig. 2B), and bring it outside of the anaerobic chamber.
5. Place cuvette into the cuvette compartment and close lid. Record CD value reading.
6. Using a Hamilton airtight syringe with a needle, obtain the calculated amount of Fe–S cluster donor protein in the glove box, cap the needle with a rubber stopper (Fig. 5, *left*), then take it outside to the CD spectrophotometer.
7. Remove stopper from the syringe needle and quickly insert the needle in the cuvette containing the acceptor protein through the rubber stopper used to seal the cuvette (Fig. 5, *right*).
8. Inject donor protein into the acceptor protein.
9. Simultaneously start the timer.
10. Mix quickly by inverting the cuvette 3–4 times.
11. Place cuvette inside the cuvette compartment, close the instrument lid, and start the program.
12. Record the time that passes from the moment the donor protein is injected into the cuvette until the scan is started.
13. Continue recording until no changes in CD value occur and a plateau is reached.

5.3.3 Data Analysis

1. Export the file and use software suitable for curve fitting to plot the change in CD values ($\Delta \varepsilon_{obs}$) as a function of time. Subtract the CD value recorded in Section 5.3.1, step 5, from each of the CD values on the y-axis.
2. Using a kinetics simulation software, estimate the rate constant and fit the data to a kinetic curve.

5.3.4 Notes

1. Some cluster transfer reactions occur extremely fast, making it difficult to determine the rate constant accurately. In such cases, a CD spectrometer

equipped with a rapid mixer such as the stopped-flow accessories available from Jasco or Applied Photophysics (e.g., RX2000 stopped-flow accessory) is needed. These systems are capable of observing reactions on the 2.1 and 10 ms or higher scale, respectively.

6. CD-MONITORED PH TITRATIONS OF Fe–S CLUSTER COMPLEXES

CD spectroscopy is also an effective tool for characterizing the effects of pH on the coordination environment of a [2Fe–2S] cluster. If an amino acid residue that either directly coordinates or is located sufficiently close to the [2Fe–2S] center has a dissociable proton that is solvent accessible, the spectroscopic features of the Fe–S center may reflect measurable changes as a function of pH. Such behavior has been reported using CD, UV–visible absorption, NMR, and resonance Raman spectroscopy for Rieske-type [2Fe–2S] cluster proteins (Cys_2His_2 coordination) (Iwasaki, Imai, Urushiyama, & Oshima, 1996; Konkle et al., 2009; Kuila et al., 1992; Lin et al., 2006), and using resonance Raman spectroscopy for the mitochondrial membrane [2Fe–2S] mitoNEET protein family (Cys_3His_1 coordination) (Tirrell et al., 2009). Subtle changes in the CD spectra of *Azotobacter vinelandii* [3Fe–4S] cluster ferredoxin (Cys_4 coordination) as a function of pH have also been reported due to protonation of a histidine introduced in place of a phenylalanine in the vicinity of the cluster (Chen et al., 2002). We have used this approach to study the interaction between CGFS Grxs and BolA proteins in *E. coli*. Specifically, we have compared the pH dependence of the CD spectra of the *E. coli* [2Fe–2S]–Grx4 homodimer to the [2Fe–2S]–Grx4–IbaG heterodimer (IbaG is a BolA homolog). This approach confirmed histidine ligation to the Fe–S cluster in the Grx4–IbaG complex and allowed calculation of the pK_a of the dissociable proton (Dlouhy et al., 2016). Protonation/deprotonation of a nitrogen atom in the histidine imidazole ring is proposed to induce subtle changes in the protein environment surrounding the cluster in the [2Fe–2S]–Grx4–IbaG complex, leading to marked differences in the spectroscopic properties of the isolated complex as a function of pH (Fig. 7). In comparison, the *E. coli* [2Fe–2S]–Grx4 homodimer with all-Cys ligation does not exhibit these pH-dependent changes (Dlouhy et al., 2016).

6.1 Additional Equipment

- Micro pH meter to measure pH of low volume samples (<500 μL).

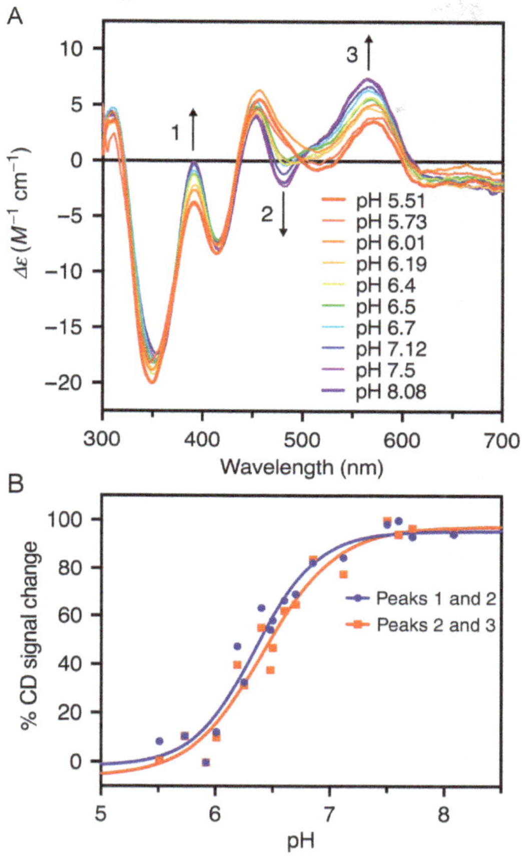

Fig. 7 CD-monitored pH titration studies on *E. coli* $[2Fe–2S]^{2+}$ Grx4–IbaG. (A) CD spectra of the protein complex equilibrated in buffer at the indicated pH values. (B) The difference in CD intensity at two wavelengths indicated by the *arrows* plotted as a function of pH was used to determine the pK_a of the dissociable proton near the Fe–S cluster. *Taken with permission from Dlouhy, A. C., Li, H., Albetel, A. N., Zhang, B., Mapolelo, D. T., Randeniya, S., et al. (2016). The* Escherichia coli *BolA protein IbaG forms a histidine-ligated [2Fe–2S]-bridged complex with Grx4.* Biochemistry, *55(49), 6869–6879. https://doi.org/10.1021/acs.biochem.6b00812. Copyright © (2016) American Chemical Society.*

6.2 Buffers and Reagents

- 50 m*M* Tris-MES, 150 m*M* NaCl with pH ranging from 5.5 to 9.0 in 0.1 pH unit increments.

6.3 Procedure

1. Mix 40–100 μ*M* Fe–S cluster-containing protein with sample buffer at different pH values in separate chilled microcentrifuge tubes. The final

dilution of concentrated protein to buffer should be at least 1:10 to obtain a final pH close to the respective buffer pH.

2. Allow the protein to reach conformational equilibrium by incubating for at least 30 min, with occasional stirring, either at room temperature or on ice, depending on thermodynamic stability.
3. Using a micro pH meter, read and record pH of each sample before scanning.
4. Transfer to 1-cm quartz semi-microcuvette and scan CD absorption spectrum.
5. Rinse cuvette with distilled water and dry after scanning the spectrum of each buffer and/or protein sample as described in Section 4.1.2, step 2 to avoid contaminating the samples or diluting the protein concentration.
6. Acquire and subtract the CD spectrum of the sample buffer from the CD spectrum of each sample.

6.4 Data Processing and pK_a Determination

1. Export files and use GraphPad Prism or similar curve fitting software to plot, overlap, and compare the CD spectra obtained at different pH values. Additionally, plot the change in CD values ($\Delta\theta_{obs}$ or $\Delta\Delta\varepsilon_{obs}$) at a wavelength exhibiting the greatest spectral shift as a function of pH. Often the changes in the CD spectral features may be subtle due to rising peaks overlapping disappearing peaks and/or slight conformational changes around the Fe–S center. In such cases, the difference in the CD values at two different wavelengths should be plotted as a function of pH (Fig. 7B).
2. Determine the pK_a and the number of protons involved in the conversion from the conformation observed at low pH to the one observed at high pH, by fitting the data to an ideal titration curve obtained using Hill's equation defined in Eq. (2). This is a traditional model used to measure cooperativity in protein–ligand interactions in which the observed CD intensity (CD_{obs}) is fit to a sigmoidal curve simulated by the Hill's coefficient (n), the pK_a, and the CD intensity of the protonated (CD_P) and deprotonated (CD_D) forms of the protein complex.

$$CD_{obs} = CD_D + (CD_P - CD_D)\left[\frac{(10^{pH})^n}{(10^{pH})^n + (10^{pK_a})^n}\right]$$
$$\text{or } CD_{obs} = CD_D + \frac{(CD_P - CD_D)}{1 + 10^{n(pK_a - pH)}} \quad (2)$$

In this equation, the Hill coefficient (n) represents the cooperativity of the proton dissociation. When $n=1$, the equation is a standard Henderson–Hasselbalch protonation function.

6.5 Notes

- For recording the blank, scanning one buffer solution at any pH value should suffice because the pH of the buffer has insignificant effects on the CD spectrum.
- Owing to pH variations as a function of temperature, all samples need to be handled at the same or very close temperature during preparation and scanning.
- In some cases, protein degradation can occur at pH values below 6 or above 8.5. If the degradation is not severe, centrifuge the sample for 1 min at 7000–8000 rpm, and then transfer the supernatant to a new microcentrifuge tube before reading the pH and scanning the CD spectrum.
- The existence of two separate protonation events as observed for Rieske [2Fe–2S] proteins with two His ligands to the Fe–S cluster requires fitting the titration curve to the sum of two Henderson–Hasselbalch equations (Konkle et al., 2009).

7. SUMMARY AND CONCLUSIONS

CD spectroscopy has emerged as a useful and preferred tool for studying Fe–S cluster-dependent interactions due to its unique ability to capture slight changes in the conformational properties of and around the Fe–S centers in a protein environment. Unlike other spectroscopic techniques currently available, CD spectroscopy requires small protein sample volumes at low concentrations, with measurements performed under physiological conditions. These advantages are particularly important since some Fe–S clusters—containing proteins are difficult to purify in large quantities with an intact Fe–S cluster due to Fe–S cluster instability and sensitivity to oxygen. Various types of Fe–S cluster-dependent interactions can be investigated, and a wide range of information can be gleaned using CD spectroscopy ranging from primary protein–protein interaction data to detailed information on Fe–S cluster transfer mechanisms at the molecular level. However, in spite of the valuable information acquired using CD spectroscopy, further studies are often necessary to complement this approach and answer questions regarding the fine details of Fe–S cluster-dependent interactions.

The versatility of CGFS Grxs in carrying out different types of Fe–S cluster-dependent interactions and the compilation of data available to date, renders them an ideal model for illustrating the use of CD spectroscopy in studying similar cellular processes.

ACKNOWLEDGMENTS

Research reported in this publication was supported by the National Institute of General Medical Sciences of the National Institutes of Health under award numbers R01GM100069 and R35GM118164 (to C.E.O.).

REFERENCES

Banci, L., Brancaccio, D., Ciofi-Baffoni, S., Del Conte, R., Gadepalli, R., Mikolajczyk, M., et al. (2014). [2Fe–2S] cluster transfer in iron–sulfur protein biogenesis. *Proceedings of the National Academy of Sciences of the United States of America*, *111*(17), 6203–6208. https://doi.org/10.1073/pnas.1400102111.

Banci, L., Camponeschi, F., Ciofi-Baffoni, S., & Muzzioli, R. (2015). Elucidating the molecular function of human BOLA2 in GRX3-dependent anamorsin maturation pathway. *Journal of the American Chemical Society*, *137*(51), 16133–16143. https://doi.org/10.1021/jacs.5b10592.

Banci, L., Ciofi-Baffoni, S., Gajda, K., Muzzioli, R., Peruzzini, R., & Winkelmann, J. (2015). N-terminal domains mediate [2Fe–2S] cluster transfer from glutaredoxin-3 to anamorsin. *Nature Chemical Biology*, *11*(10), 772–778. https://doi.org/10.1038/nchembio.1892.

Bandyopadhyay, S., Gama, F., Molina-Navarro, M. M., Gualberto, J. M., Claxton, R., Naik, S. G., et al. (2008). Chloroplast monothiol glutaredoxins as scaffold proteins for the assembly and delivery of [2Fe–2S] clusters. *The EMBO Journal*, *27*(7), 1122–1133. https://doi.org/10.1038/emboj.2008.50.

Boutigny, S., Saini, A., Baidoo, E. E., Yeung, N., Keasling, J. D., & Butland, G. (2013). Physical and functional interactions of a monothiol glutaredoxin and an iron sulfur cluster carrier protein with the sulfur-donating radical S-adenosyl-L-methionine enzyme MiaB. *The Journal of Biological Chemistry*, *288*(20), 14200–14211. https://doi.org/10.1074/jbc.M113.460360.

Chen, K., Bonagura, C. A., Tilley, G. J., McEvoy, J. P., Jung, Y. S., Armstrong, F. A., et al. (2002). Crystal structures of ferredoxin variants exhibiting large changes in [Fe–S] reduction potential. *Nature Structural Biology*, *9*(3), 188–192. https://doi.org/10.1038/nsb751.

Couturier, J., Jacquot, J. P., & Rouhier, N. (2009). Evolution and diversity of glutaredoxins in photosynthetic organisms. *Cellular and Molecular Life Sciences*, *66*(15), 2539–2557. https://doi.org/10.1007/s00018-009-0054-y.

Couturier, J., Przybyla-Toscano, J., Roret, T., Didierjean, C., & Rouhier, N. (2015). The roles of glutaredoxins ligating Fe–S clusters: Sensing, transfer or repair functions? *Biochimica et Biophysica Acta*, *1853*(6), 1513–1527. https://doi.org/10.1016/j.bbamcr.2014.09.018.

Dailey, H. A., Finnegan, M. G., & Johnson, M. K. (1994). Human ferrochelatase is an iron–sulfur protein. *Biochemistry*, *33*(2), 403–407.

Dlouhy, A. C., Li, H., Albetel, A. N., Zhang, B., Mapolelo, D. T., Randeniya, S., et al. (2016). The *Escherichia coli* BolA protein IbaG forms a histidine-ligated [2Fe–2S]-bridged complex with Grx4. *Biochemistry*, *55*(49), 6869–6879. https://doi.org/10.1021/acs.biochem.6b00812.

Fidai, I., Wachnowsky, C., & Cowan, J. A. (2016). Mapping cellular Fe–S cluster uptake and exchange reactions—Divergent pathways for iron–sulfur cluster delivery to human ferredoxins. *Metallomics*, *8*(12), 1283–1293. https://doi.org/10.1039/c6mt00193a.

Frey, A. G., Palenchar, D. J., Wildemann, J. D., & Philpott, C. C. (2016). A glutaredoxin–BolA complex serves as an iron–sulfur cluster chaperone for the cytosolic cluster assembly machinery. *The Journal of Biological Chemistry*, *291*, 22344–22356. https://doi.org/10.1074/jbc.M116.744946.

Hoffmann, B., Uzarska, M. A., Berndt, C., Godoy, J. R., Haunhorst, P., Lillig, C. H., et al. (2011). The multidomain thioredoxin-monothiol glutaredoxins represent a distinct functional group. *Antioxidants & Redox Signaling*, *15*(1), 19–30. https://doi.org/10.1089/ars.2010.3811.

Iñigo, S., Durand, A. N., Ritter, A., Le Gall, S., Termathe, M., Klassen, R., et al. (2016). Glutaredoxin GRXS17 associates with the cytosolic iron–sulfur cluster assembly pathway. *Plant Physiology*, *172*(2), 858–873. https://doi.org/10.1104/pp.16.00261.

Iwasaki, T., Imai, T., Urushiyama, A., & Oshima, T. (1996). Redox-linked ionization of sulredoxin, an archaeal Rieske-type [2Fe–2S] protein from Sulfolobus sp. strain 7. *The Journal of Biological Chemistry*, *271*(44), 27659–27663.

Iwema, T., Picciocchi, A., Traore, D. A., Ferrer, J. L., Chauvat, F., & Jacquamet, L. (2009). Structural basis for delivery of the intact [Fe2S2] cluster by monothiol glutaredoxin. *Biochemistry*, *48*(26), 6041–6043. https://doi.org/10.1021/bi900440m.

Kim, K. D., Chung, W. H., Kim, H. J., Lee, K. C., & Roe, J. H. (2010). Monothiol glutaredoxin Grx5 interacts with Fe–S scaffold proteins Isa1 and Isa2 and supports Fe–S assembly and DNA integrity in mitochondria of fission yeast. *Biochemical and Biophysical Research Communications*, *392*(3), 467–472. https://doi.org/10.1016/j.bbrc.2010.01.051.

Konkle, M. E., Muellner, S. K., Schwander, A. L., Dicus, M. M., Pokhrel, R., Britt, R. D., et al. (2009). Effects of pH on the Rieske protein from Thermus thermophilus: A spectroscopic and structural analysis. *Biochemistry*, *48*(41), 9848–9857. https://doi.org/10.1021/bi901126u.

Kuila, D., Schoonover, J. R., Dyer, R. B., Batie, C. J., Ballou, D. P., Fee, J. A., et al. (1992). Resonance Raman studies of Rieske-type proteins. *Biochimica et Biophysica Acta*, *1140*(2), 175–183.

Kumanovics, A., Chen, O. S., Li, L., Bagley, D., Adkins, E. M., Lin, H., et al. (2008). Identification of FRA1 and FRA2 as genes involved in regulating the yeast iron regulon in response to decreased mitochondrial iron–sulfur cluster synthesis. *The Journal of Biological Chemistry*, *283*(16), 10276–10286. https://doi.org/10.1074/jbc.M801160200.

Lesuisse, E., Knight, S. A., Courel, M., Santos, R., Camadro, J. M., & Dancis, A. (2005). Genome-wide screen for genes with effects on distinct iron uptake activities in Saccharomyces cerevisiae. *Genetics*, *169*(1), 107–122. https://doi.org/10.1534/genetics.104.035873.

Li, H., Mapolelo, D. T., Dingra, N. N., Keller, G., Riggs-Gelasco, P. J., Winge, D. R., et al. (2011). Histidine 103 in Fra2 is an iron–sulfur cluster ligand in the [2Fe–2S] Fra2–Grx3 complex and is required for in vivo iron signaling in yeast. *The Journal of Biological Chemistry*, *286*(1), 867–876. https://doi.org/10.1074/jbc.M110.184176.

Li, H., Mapolelo, D. T., Dingra, N. N., Naik, S. G., Lees, N. S., Hoffman, B. M., et al. (2009). The yeast iron regulatory proteins Grx3/4 and Fra2 form heterodimeric complexes containing a [2Fe–2S] cluster with cysteinyl and histidyl ligation. *Biochemistry*, *48*(40), 9569–9581. https://doi.org/10.1021/bi901182w.

Li, H., Mapolelo, D. T., Randeniya, S., Johnson, M. K., & Outten, C. E. (2012). Human glutaredoxin 3 forms [2Fe–2S]-bridged complexes with human BolA2. *Biochemistry*, *51*(8), 1687–1696. https://doi.org/10.1021/bi2019089.

Lin, I. J., Chen, Y., Fee, J. A., Song, J., Westler, W. M., & Markley, J. L. (2006). Rieske protein from Thermus thermophilus: 15N NMR titration study demonstrates the role of iron-ligated histidines in the pH dependence of the reduction potential. *Journal of the American Chemical Society*, *128*(33), 10672–10673. https://doi.org/10.1021/ja0627388.

Mapolelo, D. T., Zhang, B., Randeniya, S., Albetel, A. N., Li, H., Couturier, J., et al. (2013). Monothiol glutaredoxins and A-type proteins: Partners in Fe–S cluster trafficking. *Dalton Transactions*, *42*(9), 3107–3115. https://doi.org/10.1039/c2dt32263c.

Melber, A., Na, U., Vashisht, A., Weiler, B. D., Lill, R., Wohlschlegel, J. A., et al. (2016). Role of Nfu1 and Bol3 in iron–sulfur cluster transfer to mitochondrial clients. *eLife*, *5*, e15991, https://doi.org/10.7554/eLife.15991.

Mühlenhoff, U., Molik, S., Godoy, J. R., Uzarska, M. A., Richter, N., Seubert, A., et al. (2010). Cytosolic monothiol glutaredoxins function in intracellular iron sensing and trafficking via their bound iron–sulfur cluster. *Cell Metabolism*, *12*(4), 373–385. https://doi.org/10.1016/j.cmet.2010.08.001.

Nasta, V., Giachetti, A., Ciofi-Baffoni, S., & Banci, L. (2017). Structural insights into the molecular function of human [2Fe–2S] BOLA1–GRX5 and [2Fe–2S] BOLA3–GRX5 complexes. *Biochimica et Biophysica Acta*, *1861*, 2119–2131. https://doi.org/10.1016/j.bbagen.2017.05.005.

Ojeda, L., Keller, G., Mühlenhoff, U., Rutherford, J. C., Lill, R., & Winge, D. R. (2006). Role of glutaredoxin-3 and glutaredoxin-4 in the iron regulation of the Aft1 transcriptional activator in Saccharomyces cerevisiae. *The Journal of Biological Chemistry*, *281*(26), 17661–17669. https://doi.org/10.1074/jbc.M602165200.

Poor, C. B., Wegner, S. V., Li, H., Dlouhy, A. C., Schuermann, J. P., Sanishvili, R., et al. (2014). Molecular mechanism and structure of the *Saccharomyces cerevisiae* iron regulator Aft2. *Proceedings of the National Academy of Sciences of the United States of America*, *111*(11), 4043–4048. https://doi.org/10.1073/pnas.1318869111.

Pujol-Carrion, N., Belli, G., Herrero, E., Nogues, A., & de la Torre-Ruiz, M. A. (2006). Glutaredoxins Grx3 and Grx4 regulate nuclear localisation of Aft1 and the oxidative stress response in Saccharomyces cerevisiae. *Journal of Cell Science*, *119*(Pt. 21), 4554–4564. https://doi.org/10.1242/jcs.03229.

Qi, W., & Cowan, J. A. (2011). Mechanism of glutaredoxin-ISU [2Fe-2S] cluster exchange. *Chemical Communications (Cambridge, England)*, *47*(17), 4989–4991. https://doi.org/10.1039/c0cc05079b.

Riemer, J., Hoepken, H. H., Czerwinska, H., Robinson, S. R., & Dringen, R. (2004). Colorimetric ferrozine-based assay for the quantitation of iron in cultured cells. *Analytical Biochemistry*, *331*(2), 370–375. https://doi.org/10.1016/j.ab.2004.03.049.

Roret, T., Tsan, P., Couturier, J., Zhang, B., Johnson, M. K., Rouhier, N., et al. (2014). Structural and spectroscopic insights into BolA–glutaredoxin complexes. *The Journal of Biological Chemistry*, *289*(35), 24588–24598. https://doi.org/10.1074/jbc.M114.572701.

Shakamuri, P., Zhang, B., & Johnson, M. K. (2012). Monothiol glutaredoxins function in storing and transporting [Fe2S2] clusters assembled on IscU scaffold proteins. *Journal of the American Chemical Society*, *134*(37), 15213–15216. https://doi.org/10.1021/ja306061x.

Stephens, P. J., Thomson, A. J., Dunn, J. B., Keiderling, T. A., Rawlings, J., Rao, K. K., et al. (1978). Circular dichroism and magnetic circular dichroism of iron–sulfur proteins. *Biochemistry*, *17*(22), 4770–4778.

Ströher, E., Grassl, J., Carrie, C., Fenske, R., Whelan, J., & Millar, A. H. (2016). Glutaredoxin S15 is involved in Fe–S cluster transfer in mitochondria influencing lipoic acid-dependent enzymes, plant growth, and arsenic tolerance in Arabidopsis. *Plant Physiology*, *170*(3), 1284–1299. https://doi.org/10.1104/pp.15.01308.

Tirrell, T. F., Paddock, M. L., Conlan, A. R., Smoll, E. J., Jr., Nechushtai, R., Jennings, P. A., et al. (2009). Resonance Raman studies of the (His)(Cys)3 2Fe–2S cluster of MitoNEET: Comparison to the (Cys)4 mutant and implications of the effects of pH on the labile metal center. *Biochemistry*, *48*(22), 4747–4752. https://doi.org/10.1021/bi900028r.

Uzarska, M. A., Nasta, V., Weiler, B. D., Spantgar, F., Ciofi-Baffoni, S., Saviello, M. R., et al. (2016). Mitochondrial Bol1 and Bol3 function as assembly factors for specific iron–sulfur proteins. *eLife*, *5*, e16673. https://doi.org/10.7554/eLife.16673.

Vranish, J. N., Das, D., & Barondeau, D. P. (2016). Real-time kinetic probes support monothiol glutaredoxins as intermediate carriers in Fe–S cluster biosynthetic pathways. *ACS Chemical Biology*, *11*(11), 3114–3121. https://doi.org/10.1021/acschembio.6b00632.

Vranish, J. N., Russell, W. K., Yu, L. E., Cox, R. M., Russell, D. H., & Barondeau, D. P. (2015). Fluorescent probes for tracking the transfer of iron–sulfur cluster and other metal cofactors in biosynthetic reaction pathways. *Journal of the American Chemical Society*, *137*(1), 390–398. https://doi.org/10.1021/ja510998s.

Xia, H., Li, B., Zhang, Z., Wang, Q., Qiao, T., & Li, K. (2015). Human glutaredoxin 3 can bind and effectively transfer [4Fe–4S] cluster to apo-iron regulatory protein 1. *Biochemical and Biophysical Research Communications*, *465*(3), 620–624. https://doi.org/10.1016/j.bbrc.2015.08.073.

Yeung, N., Gold, B., Liu, N. L., Prathapam, R., Sterling, H. J., Willams, E. R., et al. (2011). The E. coli monothiol glutaredoxin GrxD forms homodimeric and heterodimeric FeS cluster containing complexes. *Biochemistry*, *50*(41), 8957–8969. https://doi.org/10.1021/bi2008883.

Zhang, B., Bandyopadhyay, S., Shakamuri, P., Naik, S. G., Huynh, B. H., Couturier, J., et al. (2013). Monothiol glutaredoxins can bind linear [Fe3S4]+ and [Fe4S4]2+ clusters in addition to [Fe2S2]2+ clusters: Spectroscopic characterization and functional implications. *Journal of the American Chemical Society*, *135*(40), 15153–15164. https://doi.org/10.1021/ja407059n.

CHAPTER TWELVE

Conformationally Gated Electron Transfer in Nitrogenase. Isolation, Purification, and Characterization of Nitrogenase From *Gluconacetobacter diazotrophicus*

Cedric P. Owens*[,1], Faik A. Tezcan†[,1]
*Schmid College of Science and Technology, Chapman University, Orange, CA, United States
†University of California, San Diego, La Jolla, CA, United States
[1]Corresponding authors: e-mail address: cpowens@chapman.edu; tezcan@ucsd.edu

Contents

Methods in Enzymology, Volume 599
ISSN 0076-6879
https://doi.org/10.1016/bs.mie.2017.09.007

Abstract

Nitrogenase is a complex, bacterial enzyme that catalyzes the ATP-dependent reduction of dinitrogen (N_2) to ammonia (NH_3). In its most prevalent form, it consists of two proteins, the catalytic molybdenum-iron protein (MoFeP) and its specific reductase, the iron protein (FeP). A defining feature of nitrogenase is that electron and proton transfer processes linked to substrate reduction are synchronized by conformational changes driven by ATP-dependent FeP–MoFeP interactions. Yet, despite extensive crystallographic, spectroscopic, and biochemical information on nitrogenase, the structural basis of the ATP-dependent synchronization mechanism is not understood in detail. In this chapter, we summarize some of our efforts toward obtaining such an understanding.

Experimental investigations of the structure–function relationships in nitrogenase are challenged by the fact that it cannot be readily expressed heterologously in non-diazotrophic bacteria, and the purification protocols for nitrogenase are only known for a small number of diazotrophic organisms. Here, we present methods for purifying and characterizing nitrogenase from a new model organism, *Gluconacetobacter diazotrophicus*. We also describe procedures for observing redox-dependent conformational changes in *G. diazotrophicus* nitrogenase by X-ray crystallography and electron paramagnetic resonance spectroscopy, which have provided new insights into the redox-dependent conformational gating processes in nitrogenase.

Graphical Abstract

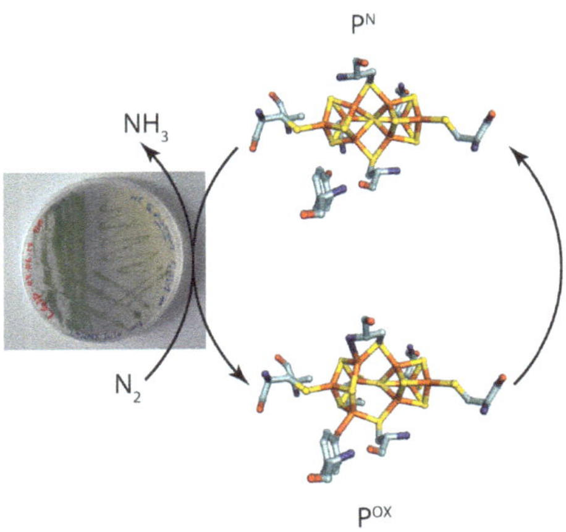

ABBREVIATIONS

ATP adenosine triphosphate
Av *Azotobacter vinelandii*
Cp *Clostridium pasteurianum*
DG docking geometry
DT sodium dithionite

EPR	electron paramagnetic resonance
ET	electron transfer
FeP	iron protein
FPLC	fast protein liquid chromatography
Gd	*Gluconacetobacter diazotrophicus*
IDS	indigo carmine (5,5′-indigodisulfonic acid)
Kp	*Klebsiella pneumoniae*
MoFeP	molybdenum-iron protein
MPD	2-methyl-2,4-pentanediol
RMSD	root mean square deviation
SDS-PAGE	sodium dodecyl sulfate polyacrylamide gel electrophoresis
TRIS	2-amino-2-hydroxymethylpropane-1,3-diol

Highlights

- Nitrogenase from *G. diazotrophicus* has been isolated and structurally characterized for the first time.
- The P-cluster of *G. diazotrophicus* MoFeP undergoes redox-dependent structural changes that are distinct from those of hitherto-characterized MoFeP variants from other organisms.
- The reversible ligation of the *G. diazotrophicus* P-cluster to a Tyr ligand strongly suggests the importance of hard, O-based ligands in mediating conformationally gated electron transfer (ET) in nitrogenase.

1. INTRODUCTION

1.1 Biological and Abiological Nitrogen Fixation

Conversion of atmospheric N_2 into biologically usable NH_3 is necessary for life on earth. Developing methods for enhancing plant NH_3 supply is critical for agriculture since lack of fixed nitrogen, such as NH_3, often limits crop growth (Oldroyd & Dixon, 2014; Rogers & Oldroyd, 2014). The two most important sources of NH_3 are biological nitrogen fixation by the enzyme nitrogenase and synthetic nitrogen fixation via the Haber–Bosch process (Hirsch & Mauchline, 2015).

In the Haber–Bosch process, N_2 is converted to NH_3 according to Eq. (1). This process requires high temperatures (350–550 K) to enable the activation of the inert N_2 triple bond, while high pressures (>300 atm) are concomitantly needed to drive this overall exergonic reaction forward.

$$N_2 + 3H_2 \rightarrow 2NH_3 \tag{1}$$

In contrast to the extreme conditions required in the Haber–Bosch process, biological nitrogen fixation takes place under ambient conditions, albeit at the expense of 16 ATP molecules in each catalytic cycle (Eq. 2).

$$N_2 + 8H^+ + 8e^- + 16ATP \rightarrow 2NH_3 + H_2 + 16ADP + 16P_i \quad (2)$$

To overcome the high kinetic and thermodynamic stability of N_2, nitrogenase employs an exquisitely complex, but incompletely understood choreography of electron transfer (ET) and proton transfer (PT) processes.

Mo-containing nitrogenase is the most widely distributed form of this enzyme (Fig. 1A) (Dos Santos, Fang, Mason, Setubal, & Dixon, 2012). It consists of two protein components, the molybdenum-iron protein (MoFeP) and the iron protein (FeP). MoFeP is an $(\alpha\beta)_2$ dimer of dimers and contains two unique metal clusters, the catalytic iron-molybdenum cofactor FeMoco, a [7Fe:9S:1Mo:1C] cluster, and the P-cluster with the composition of [8Fe:7S]. FeP is a γ_2 homodimer harboring a [4Fe:4S] cluster at the dimer interface and two ATP-binding sites (Rees et al., 2005; Seefeldt, Hoffman, & Dean, 2009) (Fig. 1).

As shown in Eq. (2), each enzymatic turnover reaction by nitrogenase is an $8e^-$ process. These electrons are delivered one at a time to MoFeP by FeP, with two ATP molecules consumed by FeP for each electron transferred (Katz, Owens, & Tezcan, 2016; Seefeldt et al., 2009). FeP is the only known biological agent that can reduce MoFeP and activate it for catalysis.

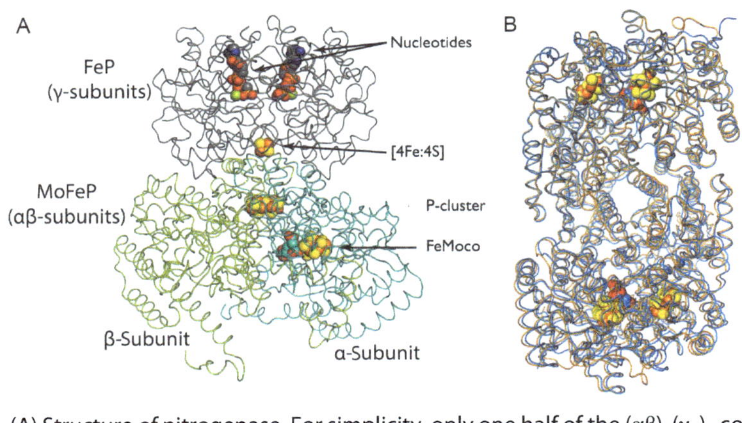

Fig. 1 (A) Structure of nitrogenase. For simplicity, only one half of the $(\alpha\beta)_2(\gamma_2)_2$ complex is shown. (B) Structural alignment of the entire $(\alpha\beta)_2$ MoFeP of *A. vinelandii* (*orange*) with that of *G. diazotrophicus* (*blue*), illustrating their structural similarity. Metal clusters and bound nucleotides are shown as *spheres*.

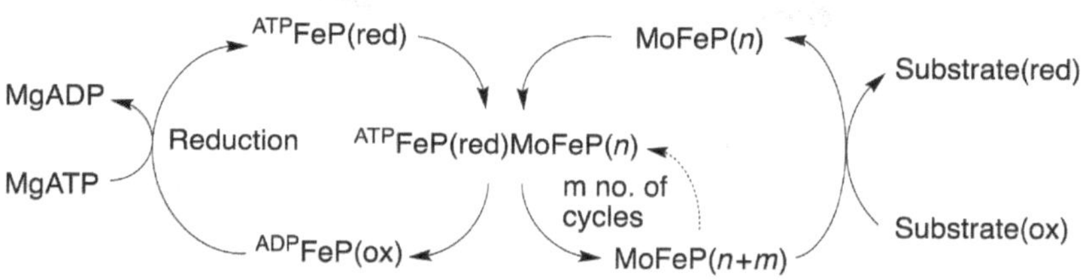

Fig. 2 Model for electron transfer and ATP hydrolysis during nitrogenase turnover. The number of cycles depends on the substrate, where $m = 8$ for N_2 and $m = 2$ for C_2H_2 and H^+. The value (n) refers to the number of electrons accumulated on FeMoco.

In turn, MoFeP is the only protein that can support ATP hydrolysis by FeP. These observations underscore the importance of specific ATP-dependent interactions between these two proteins and imply the presence of conformationally gated redox events (see Sections 1.2 and 1.3). Electrons donated by FeP are first received by the P-cluster and then relayed to FeMoco, which reduces substrates in a stepwise fashion (Fig. 2) (Hoffman, Lukoyanov, Yang, Dean, & Seefeldt, 2014; Katz et al., 2016; Lowe & Thorneley, 1984; Seefeldt et al., 2009; Wilson, Nyborg, & Watt, 2001).

1.2 Evidence for Gated-ET in Nitrogenase

Alongside the enigmatic process of N_2 reduction by FeMoco, the mechanism of ATP-driven ET events within the FeP–MoFeP complex remains one of the key unanswered questions in biological nitrogen fixation. While the structural basis for how MoFeP docking activates ATP hydrolysis by FeP is relatively well understood (Rees et al., 2005; Schindelin, Kisker, Schlessman, Howard, & Rees, 1997; Tezcan et al., 2005) (Section 1.3), the mechanism of how FeP induces conformational changes within in MoFeP is not (Section 2). Compared to other biological redox systems, FeP and MoFeP possess several distinguishing features. First, FeP differs from common electron shuttles such as flavodoxins, ferredoxins, and cytochromes by its large size and in its requirement to bind and hydrolyze ATP during ET. ATP-dependent docking interactions between FeP and MoFeP involve the formation of unusually large interfaces (Rees et al., 2005; Tezcan et al., 2005), which are more typical of permanent rather than transient protein complexes. Because of these extensive interactions, FeP dissociates from MoFeP very slowly, with an estimated rate of 6–10 s^{-1} (Lowe & Thorneley, 1984; Owens, Katz, Carter, Luca, & Tezcan, 2015; Wilson et al., 2001). It is postulated that the unusually long lifetime of the FeP–MoFeP complex may be necessary for the orchestration of gated redox events within MoFeP

(Tezcan, Kaiser, Howard, & Rees, 2015). Several other lines of evidence support conformationally controlled ET within the nitrogenase complex (Hoffman et al., 2014; Katz et al., 2016; Seefeldt et al., 2009):

(1) The rate of ET between FeP and MoFeP is dependent on the osmolarity of the solution, implying a conformational rearrangement of the FeP–MoFeP complex is required prior to interprotein ET (Danyal, Mayweather, Dean, Seefeldt, & Hoffman, 2010).

(2) The order of ET appears to be (i) slow transfer between P-cluster and FeMoco and (ii) rapid "backfill" from the P-cluster to the FeP [4Fe:4S] (Danyal, Dean, Hoffman, & Seefeldt, 2011), suggesting that the conformational gating likely involves ET from P-cluster to FeMoco.

(3) Potent, low-potential photosensitizers, placed in close proximity to P-cluster, are not able to support efficient catalysis when illuminated with light, likely due to the fact that MoFeP conformations populated during ET events in catalysis cannot be accessed in absence of FeP (Roth, Nguyen, & Tezcan, 2010; Roth & Tezcan, 2012).

These combined findings paint a picture in which reduced, ATP-bound FeP binds MoFeP, and induces conformational changes within MoFeP to activate ET between P-cluster and FeMoco (Hoffman et al., 2014; Katz et al., 2016).

1.3 Structural Dynamics of the FeP–MoFeP Complex

The crystal structures of several FeP–MoFeP complexes (from *Azotobacter vinelandii*) have been obtained under conditions relevant for ATP hydrolysis and catalytic turnover (Fig. 3) (Rees et al., 2005; Tezcan et al., 2005). These structures reveal that FeP adopts at least three distinct, nucleotide-state-dependent docking geometries (DG) (Fig. 3). In DG1, FeP largely docks with the β-subunit surface of MoFeP ("Encounter complex" in Fig. 3) (Owens et al., 2015; Willing, Georgiadis, Rees, & Howard, 1989; Willing & Howard, 1990), forming a complex that is stabilized predominantly by electrostatic interactions between negatively charged FeP residues γ110–112, and positively charged MoFeP residues β399–401 (Owens et al., 2015; Tezcan et al., 2005). In this geometry, the [4Fe:4S] of FeP is positioned >20 Å away from the P-cluster, suggesting that this conformation is not active for interprotein ET. At the same time, several lines of evidence suggest that the DG1 complex is a functionally relevant initial encounter complex whose mechanistic role is to increase the rate of association between FeP and MoFeP during turnover, especially when FeP/MoFeP

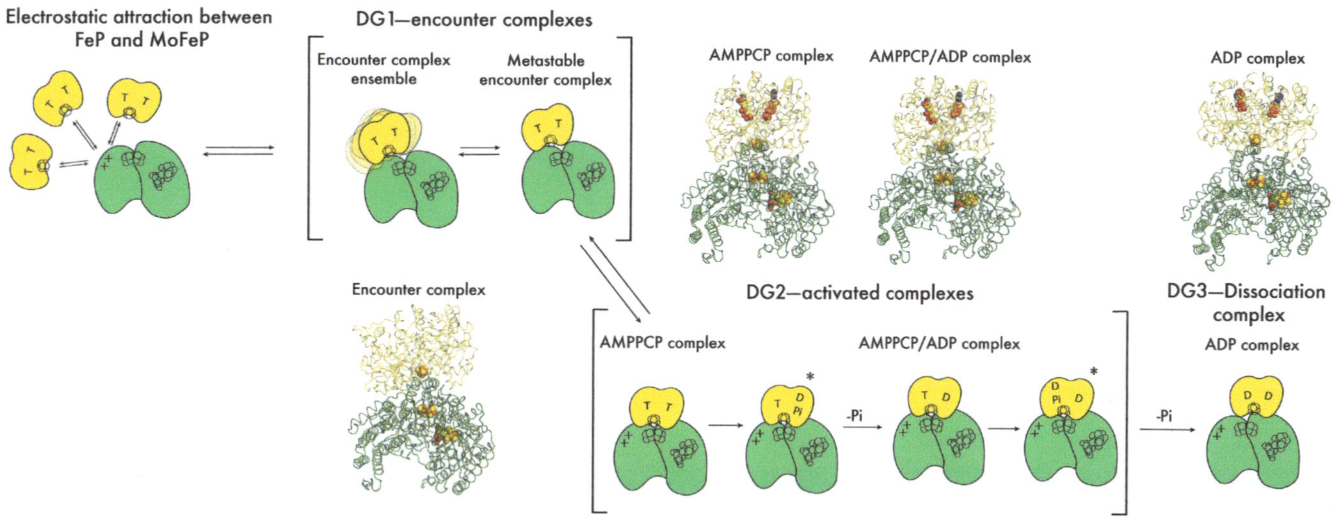

Fig. 3 Proposed scheme for nitrogenase turnover depicting ATP-dependent FeP–MoFeP interactions populated during turnover, where *T* represents ATP and D represents ADP. Conformations marked with an * are likely substrates, but have not yet been experimentally observed. The structures of known turnover intermediates are depicted as *cartoons*.

ratios are low (Owens et al., 2015; Ubbink, 2009). Indeed, elimination of favorable H-bonding and electrostatic interactions between FeP and MoFeP through point mutations such as βLys400Glu and βArg401Glu leads to considerable reduction of the catalytic activity of MoFeP (Owens et al., 2015).

DG2 represents the ATP-activated state of the nitrogenase complex and has been obtained in the presence of the nonhydrolyzable ATP analog AMPPCP or ADP·AlF_4^-, a transition-state analog for ATP hydrolysis (Fig. 3). In DG2, FeP is centrally located atop the αβ-interface surface of MoFeP, allowing the [4Fe:4S] to approach the P-cluster as closely as physically possible, yielding an ET-conducive distance of ~15 Å (Chiu et al., 2001; Schindelin et al., 1997; Tezcan et al., 2005). At the same time, FeP adopts a compact structure (particularly in the ADP·AlF_4^--bound form) that enables residues on each FeP subunit to extend across the dimer interface to participate in the hydrolysis of the bound nucleotide in the adjacent subunit. In DG2, the interface between FeP and MoFeP is particularly large (>3700 $Å^2$), which allows the FeP to undergo nucleotide-state-dependent conformational changes while retaining its central position on the MoFeP surface. Interestingly, one of the crystallographically characterized DG2 complexes is obtained in the presence of equimolar AMPPCP and ADP, with the two nucleotides bound asymmetrically by the FeP subunits ("AMPPCP/ADP complex" in Fig. 3) (Tezcan et al., 2015). This binding mode suggests that ATP hydrolysis and phosphate release within DG2 may proceed by a stepwise, rather than a concerted mechanism. Such a stepwise mechanism could be expected to prolong the lifetime of the activated FeP–MoFeP complex and, in turn, could orchestrate the sequence of intracomplex ET required for substrate reduction through the associated Fe-protein conformational changes (Spatzal, Perez, Einsle, Howard, & Rees, 2014).

Finally, DG3 is obtained in the presence of ADP and may represent a FeP–MoFeP conformation that is populated after ET and ATP hydrolysis have taken place within DG2 (ADP complex in Fig. 3) (Tezcan et al., 2005). In DG3, the ADP-bound FeP is found in a relaxed conformation typical of uncomplexed FeP variants and interacts primarily with the α-subunit surface of MoFeP. Like in the DG1 complex, the distance of >20 Å between the [4Fe:4S] cluster and the P-cluster is prohibitively long for ET. The mechanistic relevance of DG3 has yet to be probed experimentally through structure/function studies.

Taken together, these crystallographic snapshots suggest that FeP explores an unusually large conformational space along the MoFeP surface during ATP hydrolysis, whereby distinct docking geometries are selectively populated, depending on the FeP nucleotide state. This allows the distance between the [4Fe:4S] and the P-clusters to be directly coupled to the nucleotide state, providing a way for actively modulating electron flow in nitrogenase. Interestingly, the correlation between the nucleotide state and the docking positions of FeP suggests that irreversible ATP hydrolysis may serve as a ratchet that provides directionality to protein–protein interactions, much in the same way that ATP hydrolysis ensures directed motion in proteins such as AAA^+ domains (Hanson & Whiteheart, 2005; Rappas, Bose, & Zhang, 2007) and helicase (Enemark & Joshua-Tor, 2008). Yet, none of the FeP–MoFeP complex structures offer any clues about the nature of the proposed redox-gating mechanism *within* MoFeP that could control ET between the P-cluster and FeMoco: in these complexes, MoFeP displays no structural changes near the P-cluster, FeMoco, or the intervening region between these clusters suggestive of a gating process.

1.4 Redox-Dependent Structural Changes in the P-Cluster of *A. vinelandii* MoFe Protein

A survey of all available MoFeP crystal structures (uncomplexed or complexed with FeP) reveals that the only notable conformational change observed on the ET pathway to FeMoco is on the P-cluster itself. It was established that the dithionite (DT)-reduced, all-ferrous P-cluster (P^N) could be reversibly oxidized by one and two electrons to P^{+1} and P^{OX} states, respectively (Chan, Christiansen, Dean, & Seefeldt, 1999; Igarashi & Seefeldt, 2003; Seefeldt et al., 2009). As demonstrated by a crystal structure of the *A. vinelandii* MoFeP (*Av*-MoFeP) (Peters et al., 1997), the oxidation of the P-cluster leads to a dramatic structural rearrangement: in the P^N state, the P-cluster can be viewed as a fusion of two closed [4Fe:4S] cubane architectures joined by a shared hexacoordinate sulfide, S1 (Fig. 4). In the P^{OX} state, one of the cubane units opens up, whereby Fe5 bonds to the backbone amide N of *Av*-αCys88, Fe6 ligates to *Av*-βSer188, and Fe5 and Fe6 dissociate from S1. Since both the α-Cys88 backbone N and the β-Ser188 hydroxyl group likely have to be deprotonated to coordinate Fe ions, they would stabilize the P-cluster in an oxidized state, thus lowering its reduction potential and enabling it to deliver electrons to FeMoco. Therefore, any structural perturbation induced by FeP–MoFeP association that favors ligation by *Av*-βSer188 and the *Av*-αCys88 backbone N would provide a basis for the conformational redox gating in nitrogenase.

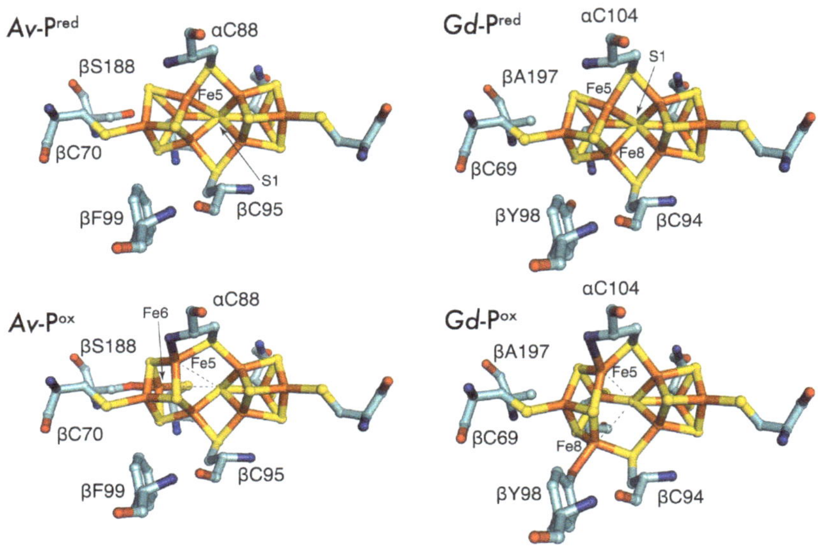

Fig. 4 Redox-dependent structural changes in *Av*- and *Gd*-MoFeP. The *dashed lines* in P^{OX} highlight bonds to the central sulfide S1 that are broken upon oxidation.

To determine if redox-dependent structural changes in the P-cluster are conserved, and thus likely to be important, we searched genomic databases and literature to find nitrogenases that showed sequence variations in residues that are proximal to the P-cluster, in particular positions equivalent to *Av*-βSer188. We limited our search to organisms with fully sequenced genomes and only considered bacterial species that could be readily cultured in a laboratory. As described later, we concluded that the diazotroph *Gluconacetobacter diazotrophicus* (*Gd*) represented a promising model organism for studying nitrogenase.

1.5 Redox-Dependent Structural Changes in the P-Cluster of *G. diazotrophicus* MoFe Protein

Most of our structural and mechanistic understanding on enzymatic nitrogen fixation has been obtained through the study of nitrogenases from a small number of diazotrophic bacteria: *A. vinelandii*, *Klebsiella pneumoniae* (*Kp*), and *Clostridium pasteurianum* (*Cp*). *A. vinelandii* and *K. pneumoniae* are closely related γ-proteobacteria and express group 1 nitrogenases (Howard, Kechris, Rees, & Glazer, 2013; Raymond, Siefert, Staples, & Blankenship, 2004) that are structurally and functionally very similar to one another (Duyvis, Wassink, & Haaker, 1996, 1998; Einsle et al., 2002; Lowe & Thorneley, 1984; Mayer, Lawson, Gormal, Roe, & Smith, 1999; Thorneley & Lowe, 1983; Wilson et al., 2001). *Cp*-nitrogenase is structurally slightly dissimilar in the α-subunit (Zhang, Morrison, Kaiser, & Rees, 2015) and is classified as a group 2 nitrogenase (Howard et al., 2013; Raymond et al., 2004). Nevertheless, these three structurally well-characterized nitrogenases feature the same cofactor architecture and there is little sequence or structural variability in functionally critical regions of the enzyme. To gain information on the conformational gating mechanism, we decided to investigate a nitrogenase with sequence differences compared to nitrogenase from the aforementioned species, namely, *Gd*-nitrogenase.

G. diazotrophicus harbors a group 1 nitrogenase like *Av*-nitrogenase (Howard et al., 2013). *Gd*- and *Av*-nitrogenases share 60% sequence identity. However, *Gd*-nitrogenase exhibits key amino acid sequence differences surrounding the P-cluster. *Gd*-MoFeP features an Ala (*Gd*-βAla187) at the equivalent position to *Av*-βSer188 (Table 1) and a Tyr residue at the position equivalent to *Av*-βPhe99 (*Gd*-βTyr98). Interestingly, an analysis of nitrogenase sequences indicates that in the great majority of type 1 nitrogenases, the aforementioned amino acid pairs Ser/Phe and Ala/Tyr are covariant (Howard et al., 2013): if MoFeP has a Ser in the first position, it features a Phe in the second. If an organism has an Ala in the first position, it has a Tyr in the second.

To investigate whether this two-position covariance has a functional significance and if *Gd*-MoFeP P-cluster can undergo similar redox-dependent structural changes as its *A. vinelandii* counterpart, we expressed and isolated *Gd*-MoFeP protein and solved its crystal structure in P^N and P^{ox} states at 1.8 and 2.6 Å resolution, respectively (Owens, Katz, Carter, Oswald, & Tezcan,

Table 1 Sequence Alignment of *Av*-, *Kp*-, *Cp*-, *Gd*-MoFeP Illustrating Sequence Differences Surrounding the P-Cluster

	***Av*-β185–190**	***Av*-β95–100**
A. vinelandii	HTP*S*FV	CVAY*F*R
K. pneumoniae	HTP*S*FI	CVAY*F*R
C. pasteurianum	NTP*S*YV	CCCY*H*R
G. diazotrophicus	HTP*A*FV	CVAY*Y*R

2016). As expected from their high sequence homologies, the overall architecture of *Gd*-MoFeP is very similar to that of *Av*-MoFeP with an RMSD of 0.7 Å over all αC atoms (Fig. 1B). The coordination and the outer-sphere environments of FeMoco are identical to one another in both species. Furthermore, in the P^N state, *Av*-P-cluster and *Gd*-P-cluster are fully superimposable (Fig. 4).

In contrast, the P^{ox} form of the *Gd*-MoFeP P-cluster displays striking differences from its *Av*-MoFeP counterpart, as shown in Fig. 4 and Table 2. Analogous to structural changes observed in *Av*-MoFeP, Fe5 becomes ligated to the backbone N of *Gd*-αCys104 and dissociates from the central S1 upon P-cluster oxidation. However, Ser, which coordinates Fe6 in *Av*-, *Kp*-, and *Cp*-MoFeP, is replaced by Ala in *Gd*-MoFeP. Instead, Fe8 now coordinates the O atom of the *Gd*-βTyr98 side chain. Tyr-ligated Fe8 also fully dissociates from S1 like the Ser-coordinated Fe6 in other species. Thus, while $2e^-$ oxidation of the P-clusters in both *Gd*- and *Av*-MoFeP yields an opened 4Fe–3S unit with two O- and N-based ligands, the mechanism by which this is achieved is distinct in these species.

While ligation of a Ser or Tyr to an FeS cluster is rare (Nicolet, Rohac, Martin, & Fontecilla-Camps, 2013), the role of an O-based ligand as a redox switch for an FeS-cluster can be readily explained. Coordination by the singly anionic tyrosinate (Tyr^-) or serinate (Ser^-) side chains would decrease the reduction potential of the P-cluster to induce ET to FeMoco, much like a cysteinate (Cys^-) ligand. However, unlike the borderline soft Cys^- ligand, the coordination of the hard Tyr^- and Ser^- ligands to an FeS-cluster would be highly dependent on whether the Fe center that they coordinate to is in the +3 (hard, favored) or +2 (borderline soft, disfavored) oxidation state. In other words, while Cys^- is a good permanent ligand that can stably anchor an Fe–S cluster in all oxidation states, Tyr^- and Ser^- are conditionally labile ligands that are well suited for reversible redox switching.

Table 2 Changes in Coordination Distances of the P-Cluster Upon Oxidation ($P^N \rightarrow P^{OX}$) in *A. vinelandii* and *G. diazotrophicus*

	d (S1–Fe6/Fe8) (Å)	*d* ($O^{Ser/Tyr}$–Fe6/Fe8) (Å)	*d* (N^{Cys}–Fe5) (Å)
Av-MoFeP	2.5 → 4.0	3.4 → 1.9	3.4 → 2.1
Cp-MoFeP	2.4 → 3.6	3.2 → 2.0	3.3 → 2.3
Kp-MoFeP	2.4 → 3.8	3.6 → 2.2	3.4 → 2.2
Gd-MoFeP	2.5 → 3.4	3.6 → 2.1	3.4 → 2.6

Our findings strongly suggest that redox-dependent structural changes at P-cluster are conserved and therefore likely relevant for nitrogenase function. Furthermore, the fact that P-cluster can cycle between the P^N and P^{OX} state through the engagement of conserved structural features hints at the possibility that it may be capable of transferring two electrons to FeMoco during catalytic turnover. Such $2e^-$ transfer steps between P-cluster and FeMoco may help minimize wasteful H^+ reduction during turnover by increasing the electron flux to FeMoco (Katz et al., 2016; Owens et al., 2016). Nevertheless, the functional importance of the P^{OX} state has yet to be experimentally established, as only P^N and P^{+1} have been detected under catalytic turnover conditions (Igarashi & Seefeldt, 2003; Lanzilotta, Christiansen, Dean, & Seefeldt, 1998). Likewise, the mechanism by which FeP–MoFeP interactions can induce conformational changes in the P-cluster at a distance of >15 Å and modulate its coordination-dependent redox equilibria also remains an exciting topic for future studies.

2. MATERIALS AND METHODS

The methods presented herein describe how to efficiently culture *G. diazotrophicus* in large scale for expression of *Gd*-nitrogenase component proteins. These procedures are based on protocols first introduced by Fisher and Newton (2005), and then adopted and modified by us (Owens et al., 2016). For more detailed information on factors influencing *G. diazotrophicus* growth, readers are directed to Reis and Döbereiner (1998), Flores-Encarnacion et al. (1999), and Tejera, Ortega, Rodes, and Lluch (2004). We furthermore present methods on the crystallographic and spectroscopic (EPR) characterization of *Gd*-MoFeP.

2.1 *G. diazotrophicus* as a Model Nitrogen-Fixing Organism

G. diazotrophicus is a gram-negative α-proteobacterium that was first isolated from sugarcane in Brazil (Cavalcante & Dobereiner, 1988). It is an endophyte that establishes symbiosis with its host by infecting the host plant's roots where it colonizes intercellular spaces (Saravanan, Madhaiyan, Osborne, Thangaraju, & Sa, 2008). Unlike the other nitrogenase model organisms *A. vinelandii*, *K. pneumoniae*, and *C. pasteurianum*, *G. diazotrophicus* is an agriculturally important species due to its plant growth enhancing properties which have been documented in several crops, including sugar cane, coffee, and rice (Pedraza, 2008; Saravanan et al., 2008; Sevilla, Burris,

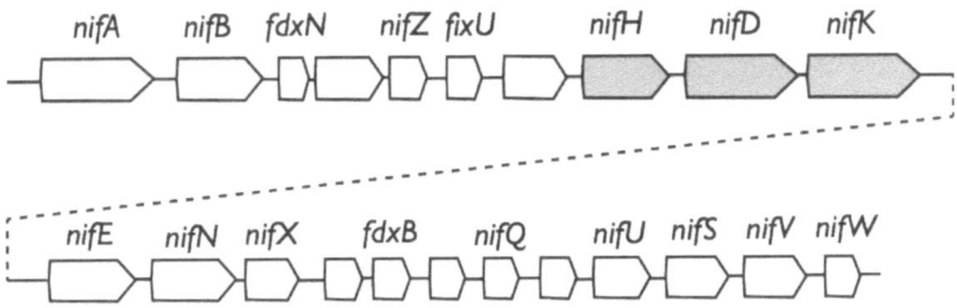

Fig. 5 *Nif* operon of *G. diazotrophicus*. *NifH* represents FeP, *nifD*, the α-subunit of MoFeP and *nifK* the β-subunit. For clarity, *nif* genes downstream of nifW (*fixA—modD*) have been omitted.

Gunapala, & Kennedy, 2001). The economic potential of *G. diazotrophicus* is highlighted by the fact that it is often included in "biofertilizer" products, which contain a variety of plant growth promoting, beneficial bacteria.

The entire genome *G. diazotrophicus* has been sequenced and it indicates that the organization of its nitrogen fixing genes resembles that of other proteobacteria such as *A. vinelandii* (Bertalan et al., 2009; Giongo, Tyler, Zipperer, & Triplett, 2010; Lee, Reth, Meletzus, Sevilla, & Kennedy, 2000). *Gd*-nitrogenase is part of the *nif* operon, which contains the genes for FeP, both MoFeP subunits, and several accessory proteins (Fig. 5).

2.2 Equipment and Reagents

2.2.1 Major Equipment and Instruments Needed

Anaerobic chamber (e.g., Coy glovebag)
Crystal trays and crystallography tools
Liquid nitrogen
Fast protein liquid chromatography (FPLC) workstation
Gas chromatograph with flame ionization detector (FID)
Gastight syringes (multiple sizes from 0.025 to 1 mL)
Large capacity shaker (e.g., Amerex 767, New Brunswick Excella E25)
Large capacity Erlenmeyer (6 L) or Fernbach (2.8 L) flasks
Large round bottom flasks (1.5–3 L) to hold buffers with adaptors to connect to Schlenk line and FPLC
Microfluidizer
Pear-shaped flasks (5, 10, 200, and 500 mL) for loading protein onto FPLC
Schlenk line
Stoppered 10 or 14-mL glass serum vials (Wheaton)

2.2.2 Reagents Required for the Growth of G. diazotrophicus

Metal salts

$CaCl_2 \cdot 2H_2O$
$FeSO_4 \cdot 7H_2O$
$MgSO_4 \cdot 7H_2O$
$Na_2MoO_4 \cdot 2H_2O$

Sugar

Glucose
Sucrose

Buffer

K_2HPO_4
KH_2PO_4

Nitrogen source

$(NH_4)_2SO_4$
NH_4Cl

C2 media components

Peptone
NaCl
Yeast extract
Agar (for solid media only)

Indicators

Bromophenol blue
Bromothymol blue

2.2.3 Reagents Required for Nitrogenase Purification, Enzymatic Assays, and Crystallization

Nitrogenase purification

NaCl
Tris base
Sodium dithionite (DT)

Enzymatic assays

ATP (as Na_2ATP)
Creatine phosphate
Creatine phosphatase
Glacial acetic acid to quench reaction
$MgCl_2$

Concentration measurement

Urea
2,2-Bipyridine

Crystallography

2-Methyl-2,4-pentanediol (MPD)
Indigo sulfonate (indigo carmine, IDS)
Sodium cacodylate
Spermine
Zwittergent (a zwitterionic detergent)

2.3 Growth Media for *G. diazotrophicus* Cultures

G. diazotrophicus is grown in LGI media for nitrogenase expression. To maintain cell stocks, LGI, LGIP, or C2 media can be used (Figs. 6 and 7).

(1) LGI contains 10% glucose 0.14 m*M* $CaCl_2$, 0.81 m*M* $MgSO_4$, 0.036 m*M* $FeSO_4$, 0.008 m*M* $NaMo_2O_4$, 0.5 m*M* $(NH_4)_2SO_4$ or 1 m*M* NH_4Cl, 10 m*M* KPO_4 (pH 6.0).

(2) LGIP can be used as an alternative to LGI for cell culture maintenance. LGIP uses the same recipe as LGI with the exception that sucrose is used instead of glucose. We do not recommend using LGIP for nitrogenase expression since the amount of nitrogenase produced is higher when using LGI, an observation that can be attributed to the higher respiration rate of *G. diazotrophicus* when glucose is used as the carbon source (Tejera et al., 2004).

(3) To make solid media, agar is added to a concentration of 12 g/L.

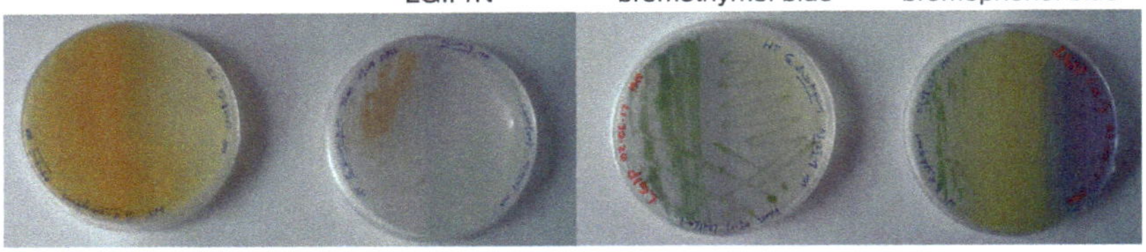

Fig. 6 *G. diazotrophicus* grown on different types of solid media. *G. diazotrophicus* grows to highest density on rich media (C2) compared to minimal, NH_4^+-free media (LGIP and LGI). Both bromothymol blue and bromophenol blue can be used to better visualize bacteria. Bromothymol blue is taken up by the bacteria, where it is *green* at neutral pH. This provides contrast to media, where bromothymol blue is almost clear under acidic conditions. Bromophenol is *blue-green* when taken up by *G. diazotrophicus*. As *G. diazotrophicus* acidifies the surrounding media during growth, bromophenol blue in the media changes color from *blue* to *yellow*. This color change is useful for visualizing the extent of bacterial growth. When grown to full density, the LGI plate containing bromophenol blue should be almost completely *yellow*.

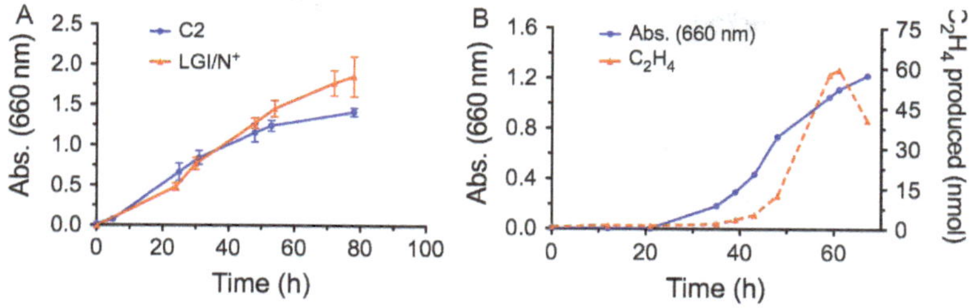

Fig. 7 (A) Growth curves for *G. diazotrophicus* grown in rich C2 and LGI/N$^+$ (10 m*M* NH_4^+) media. (B) Growth curves for *G. diazotrophicus* grown under N_2-fixing conditions. The growth medium contains 1 m*M* NH_4^+. Nitrogenase expression is monitored by measuring reduction of the alternative substrate C_2H_2 to C_2H_4. Nitrogenase expression starts in mid-exponential phase ($\geq$40 h) when NH_4^+ originally present in the medium is depleted. It is important to continuously monitor C_2H_4 production since MoFeP expression decreases slightly once intracellular NH_4^+ levels reach homostasis (Hamilton et al., 2011). Note that the different durations of the lag phases in (A) and (B) are due to the fact that different amounts of starter cultures were used for inoculation.

(4) We observed that while growing *G. diazotrophicus* in absence of any fixed nitrogen is possible on solid media (Fig. 6), doing so in liquid media results in impractically long lag phases. Therefore, 0.5 m*M* $(NH_4)_2SO_4$ or 1 m*M* NH_4Cl is added to liquid media (Fig. 7). If high-level nitrogenase expression is not the primary goal, more NH_4^+ (up to a concentration of 10 m*M*) can be added to increase the bacterial growth rate (Fig. 7).

(5) C2 media is a rich media, made by mixing 10 g peptone, 15 g glucose, 5 g NaCl, and 5 g yeast extract per liter and adjusting the pH of the media to 6.0 with acetic acid.

For the improved visualization of cell colonies on solid media, pH indicators (bromothymol blue or bromophenol blue) can be added to the media. During growth, *G. diazotrophicus* secretes organic acids, which cause the pH of growth media to decrease from 6.0 to 2.5–3.5 (Fig. 6) and a concomitant change in indicator color. Per liter of solid media, 5 mL of indicator is added from a 0.5% solution dissolved in 0.5 N KOH.

2.4 Freezer Stocks of *G. diazotrophicus*

(1) Starting with a single colony from C2 or an LGI agar plate, start a 100-mL C2 or LGI/N$^+$ (10 m*M* NH_4^+) culture in a 500-mL Erlenmeyer flask. Cultures are grown in an orbital shaker, with a speed of 200 rpm at 30°C.

(2) To make a freezer stock, grow *G. diazotrophicus* to late exponential phase such that OD_{660nm} is ca. 1.0.

(3) Spin down the cells, at 2000 × *g* for 15 min.
(4) Resuspend the cells gently in 1/20th of the original volume in LGI supplemented with 20 mg/L yeast extract and 20% (v/v) glycerol.
(5) Freeze the cells in liquid N_2 and store at −80°C.

2.5 Nitrogenase Expression in *G. diazotrophicus*

2.5.1 General Growth Considerations

To date, no heterologous expression system exists to produce nitrogenase in high yield. Although it is possible to express both FeP and MoFeP in *Escherichia coli* (Buren et al., 2017) and MoFeP in yeast (Wang et al., 2013), expression levels are very low and not suitable for large scale biochemical and structural investigations. Therefore, nitrogenase is produced in the native host organism under the control of its native promoter. To induce nitrogenase expression in *G. diazotrophicus*, cells are cultured in LGI media containing 1 m*M* NH_4^+, which is added to allow *G. diazotrophicus* to grow rapidly until mid-exponential phase, where the bacteria run out of fixed nitrogen. After NH_4^+ is depleted, *G. diazotrophicus* starts transcribing the *nif* operon and reducing N_2 (Fig. 7) (Hamilton et al., 2011).

To obtain sufficient quantities of protein required for protein crystallography, we found that growth of multiple medium-sized batches of *G. diazotrophicus* cultures (10 − 12 L) in 6 L Erlenmeyer or 2.8 L Fernbach flasks works quite well. While it was reported that large-volume fermenters can be used for nitrogenase expression in *G. diazotrophicus* (Fisher & Newton, 2005), we found that the protein yields from flask-grown cultures are superior to those of grown in fermenters (60 L, in our case).

2.5.2 Factors Affecting Nitrogenase Expression in G. diazotrophicus

The levels of nitrogenase expression in *G. diazotrophicus* are affected by aeration, the type of carbon source, and the concentration of sugar (carbon source) in the medium (Flores-Encarnacion et al., 1999; Reis & Döbereiner, 1998). To maximize the yield of *Gd*-nitrogenase, bacteria are grown in LGI media containing 10% (w/v) glucose. *G. diazotrophicus* is an obligate aerobe, and high aeration rates are necessary to meet the bacterium's respiratory needs under N_2-fixing conditions (Flores-Encarnacion et al., 1999). Optimal aeration conditions for nitrogenase expression in *G. diazotrophicus* entail agitation at 200 rpm in an orbital shaker, with Erlenmeyer flasks containing 0.4 L of media per liter flask volume, or Fernbach flasks filled with 0.55 L of media per liter flask volume. We found that lower or higher rates of aeration negatively affects nitrogenase yields.

(1) Starting with a freezer stock, prepare a 200-mL LGI starter culture and grow the culture until the cells reach late exponential or early stationary phase. The cell density at this point should be between 1.5 and 2 at 660 nm.
(2) To start large-scale growth, inoculate fresh LGI with 20 mL/L of starter culture.
(3) Monitor cell growth by measuring OD_{660nm} and measure nitrogenase expression by observing C_2H_2 reduction (Fig. 7 and Section 2.5.3).
(4) Nitrogenase expression is first observed after 2–3 days when OD_{660nm} ranges between 1 and 1.5.
(5) Peak nitrogenase activity is approximately 40–80 nmol C_2H_4 per 1 mL of cell culture (Fig. 7).
(6) After detecting high levels of C_2H_4, grow *G. diazotrophicus* for an additional 2–4 h. During this period, monitor C_2H_2 reduction every 1 h.
(7) Harvest *G. diazotrophicus* by centrifugation at 2000 × *g* for 15 min. Typical *G. diazotrophicus* yield is 2.5 g per liter of LGI.
(8) The cell pellet may be stored indefinitely at −80°C.

2.5.3 Monitoring Nitrogenase Activity in Whole Cells

(1) Place 1 mL of cells in a stoppered serum vial and add C_2H_2 to a pressure of 0.072 atm. This step is done by transferring C_2H_2 from a previously degassed round bottom flask filled with C_2H_2 to the stoppered vial filled using a gastight syringe.
(2) Place the cells in a shaking water bath set to 30°C, and incubate them for 10–15 min.
(3) *G. diazotrophicus* expressing nitrogenase reduces C_2H_2 to C_2H_4. C_2H_2 and C_2H_4 can be separated by GC on an alumina column and detected using an FID detector.

2.6 Buffers and Solutions for *G. diazotrophicus* Lysis and Nitrogenase Purification

Protein purification	
Resupension, lysis, and wash buffer (Buffer R)	50 m*M* Tris (pH 8), 100 m*M* NaCl, 5 m*M* DT
Elution buffer (Buffer E)	50 m*M* Tris (pH 8), 500 m*M* NaCl, 5 m*M* DT
Gel-filtration buffer (Buffer GF)	50 m*M* Tris (pH 8), 500 m*M* NaCl, 5 m*M* DT

Continued

Activity assay	
Buffer A	50 m*M* Tris (pH 8), 5 m*M* $MgCl_2$, 5 m*M* ATP, 30 m*M* creatine phosphate, 0.125 mg/mL creatine phosphatase
Buffer B	0.5 *M* Tris base
Buffer C	50 m*M* Tris (pH 8), 500 m*M* NaCl
Protein concentration measurement	
Urea	8 *M*
Bipyridine	100 m*M* in 50% glacial acetic acid

2.7 Isolation and Purification of *G. diazotrophicus* Nitrogenase

2.7.1 General Procedures for Working With Nitrogenase

Nitrogenase component proteins are highly air sensitive. *Gd*-nitrogenase is manipulated using anaerobic methods on a Schlenk line and in an anaerobic chamber at O_2 levels below 5 ppm. All buffers and equipment are degassed and, unless specifically indicated, DT is added to a concentration of 5 m*M* to maintain a reducing environment.

2.7.2 Lysis

(1) Thaw *G. diazotrophicus* on ice.

(2) Resuspend *G. diazotrophicus* in chilled Buffer R. Add 3 mL of buffer per gram of cells. It is important to keep all solutions cool during lysis to avoid heat denaturation of *Gd*-FeP. Prior to lysis, add DT to a final concentration of 5 m*M* to the resuspended cell pellet.

Because neither osmotic lysis methods nor sonication efficiently lyse *G. diazotrophicus* cells, we prefer using a microfluidizer. The exact protocol of cell lysis will be dependent on the specific hardware. We use an LM10 instrument fitted with an F12Y interaction chamber (Microfluidics Corporation) to lyse *G. diazotrophicus* at 16,000–18,000 psi. Typically, three passes are needed to fully lyse the entire cell pellet. To avoid oxygen damage to nitrogenase, lysis is carried out under a blanket of Ar.

(3) Place the cell lysate in an airtight centrifuge bottle equipped with a rubber septum and degas the solution for a total of 20 min on a Schlenk line by repeatedly exchanging the headspace with Ar.

(4) Clarify the cell lysate by centrifugation at 17,500 $\times g$ for 1 h. During centrifugation, the supernatant may darken slightly.
(5) After clarification, transfer the supernatant to a pear-shaped flask in an anaerobic chamber (Fig. 8A).

2.7.3 Purification

(1) Load the clarified lysate onto a DEAE column preequilibrated with anaerobic buffer R. Use ca. 7.5 mL of DEAE resin per gram of cell paste.
(2) Wash the column with fresh Buffer E until the flow through is clear.
(3) Elute *Gd*-MoFeP and *Gd*-FeP using a 100–500 m*M* linear NaCl gradient. *Gd*-MoFeP and *Gd*-FeP coelute between 250 and 300 m*M* NaCl (Fig. 8B).
(4) Determine the presence and purity of *Gd*-MoFeP and *Gd*-FeP by SDS-PAGE. *Gd*-MoFeP and *Gd*-FeP mostly coelute and there is no benefit to separating *Gd*-MoFeP and *Gd*-FeP fractions at this stage.
(5) Pool *Gd*-FeP/*Gd*-MoFeP fractions, concentrate the resulting solution to 5–10 mL in an anaerobic chamber, and load it onto a Superdex 200 gel filtration column equilibrated with anaerobic Buffer GF. The gel filtration column removes remaining impurities from *Gd*-MoFeP and *Gd*-FeP and separates the proteins from each other (Fig. 8C). The major impurity is a ferredoxin (ca. 8 kDa), which elutes after both *Gd*-nitrogenase components.
(6) Use SDS-PAGE to identify proteins contained in each collected fraction.
(7) Typical *Gd*-MoFeP yields are 2–4 mg protein per liter of cell culture (0.08–0.2 mg *Gd*-MoFeP per gram of cell pellet).
(8) Pool and concentrate the respective fractions of *Gd*-FeP and *Gd*-MoFeP and run an SDS-PAGE to evaluate the purity of the respective proteins (Fig. 8D).
(9) Measure the concentration of the respective proteins using the Bradford assay (Bradford, 1976) and by measuring the protein's respective Fe content. For the later assay, mix 10–20 μL of protein with 780–790 μL of 8 *M* urea and 100 μL of 0.1 *M* bipyridine. Let the sample incubate for 5 min and determine the Fe concentration using $\varepsilon_{522\text{nm}} = 8650\,M^{-1}\,\text{cm}^{-1}$.
(10) Purified proteins can be stored in liquid nitrogen indefinitely.

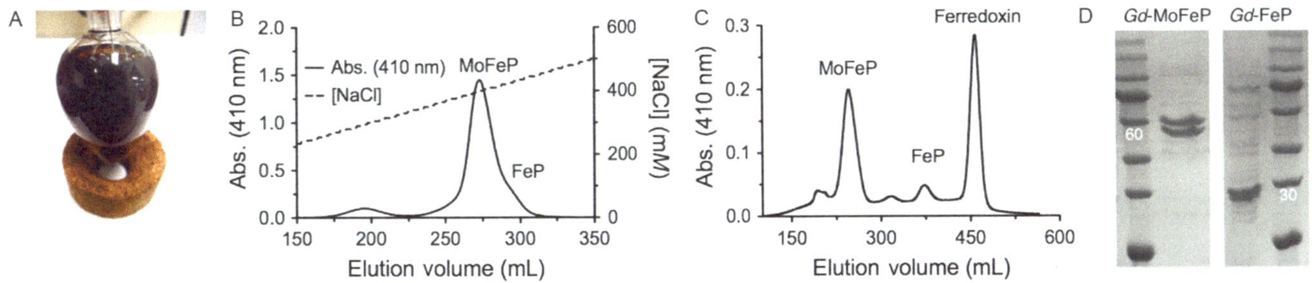

Fig. 8 (A) Clarified cell lysate collected in a pear-shaped flask. The *brown color* arises from the FeS clusters of MoFeP and FeP. (B) DEAE and (C) gel filtration chromatograms of *Gd*-nitrogenase. Note that in the ion exchange step, MoFeP and FeP coelute. We do not recommend trying to separate the nitrogenase components in this step, since they are readily separated by gel filtration chromatography at a later stage. (D) SDS-PAGE of purified *Gd*-MoFeP and *Gd*-FeP with relevant molecular weight marker bands indicated. The double band for *Gd*-MoFeP arises from the α and β subunits.

2.8 Enzymatic Activity Assays for *Gd*-Nitrogenase

The activity of *Gd*-nitrogenase is measured according to previously described protocols for C_2H_2 reduction (Fisher & Newton, 2005; Owens et al., 2015, 2016). Experiments are conducted under an Ar atmosphere in stoppered serum vials containing a buffered activity assay solution of 50 m*M* Tris (pH 8), 60 m*M* NaCl, 5 m*M* ATP, 5 m*M* $MgCl_2$, 30 m*M* creatine phosphate, 0.125 mg/mL creatine phosphokinase, and 13 m*M* DT.

(1) Mix the appropriate amounts of Buffer A and Buffer C in serum vials to make the activity assay solution. The total volume of a typical reaction in our experiments is typically ~1.1 mL. The amount of Buffer C that needs to be added to Buffer A will vary depending on the volume of *Gd*-MoFeP and *Gd*-FeP that is added in steps 5 and 6.

(2) Degas 2 mL of Buffer B in a stoppered serum vial and use it to anaerobically dissolve DT to a final concentration of 0.5 *M*.

(3) Degas the serum vials containing the activity assay solution and add DT to a final concentration of 13 m*M* using a gastight syringe.

(4) Add C_2H_2 to each serum vial to a final pressure of 0.072 atm using a gastight syringe.

(5) Add *Gd*-MoFeP, suspended in buffer C, to final concentration of 0.2 μ*M* to each serum vial using a gastight syringe.

(6) To initiate the reaction, anaerobically add increasing amounts of *Gd*-FeP, suspended in buffer C, to final concentrations ranging between 0 and 12 μ*M* using a gastight syringe.

(7) Incubate the reaction for 10 min at 30°C in a shaking water bath.

(8) Terminate the reaction by adding 0.3 mL of glacial acetic acid.

(9) C_2H_2 and C_2H_4 can be separated by GC using an alumina column and detected using an FID.

The specific C_2H_2 reduction activity of *Gd*-MoFeP ranges between 1000 and 1500 nmol mg^{-1} min^{-1}.

2.9 Electron Paramagnetic Resonance Spectroscopy of *G. diazotrophicus* MoFe Protein

Electron paramagnetic resonance (EPR) spectroscopy is a powerful method to monitor and characterize different redox states of the P-cluster and FeMoco. In the perpendicular collection mode, DT-reduced MoFeP displays signals at $g \sim 4.3$, 3.6, and 2.0, arising from the resting-state ($S=3/2$) FeMoco (termed M^N) (Igarashi & Seefeldt, 2003). Under these conditions, the all-ferrous P-cluster (i.e., P^N) is EPR-silent in both parallel and perpendicular modes. In the $2e^-$ oxidized P^{OX} state, the P-cluster features a diamagnetic $S \geq 3$ non-Kramers doublet with a signal at large *g*-values ($g=12$ in *Av*-MoFeP, $g=16$ in *Gd*-MoFeP) (Igarashi & Seefeldt, 2003), observable in parallel collection mode. For more information on EPR spectroscopy of nitrogenase, the reader is directed to reviews by Igarashi and Seefeldt (2003) and Danyal, Yang, and Seefeldt (2011).

(1) Set up EPR experiments using protein concentrations of ≥ 50 μ*M* in an anaerobic chamber.

(2) To generate P^N, add DT to a concentration of 5 m*M* to *Gd*-MoFeP and incubate for ≥ 5 min. Place the protein in an EPR tube and seal the tube.

(3) For P^{OX} samples, first remove any DT by running MoFeP over a desalting column (e.g., 10DG from Bio-Rad) prior to adding IDS. Then, add IDS to *Gd*-MoFeP a final concentration of 0.42 m*M*.

(4) Flash freeze EPR tubes containing P^N and P^{OX} samples in liquid N_2.

(5) All EPR experiments are carried out at 10 K.

(6) Collect perpendicular-mode spectra on an X-band EPR spectrometer using the following settings: microwave frequency, 9.64 GHz; modulation frequency, 100.00 Hz; modulation amplitude, 10.02; microwave power, 6.4 mW; attenuation, 15.0 dB; time constant, 81.92 ms; conversion time, 40.96 ms.

DT-reduced *Gd*-MoFeP features a signal for FeMoco in the M^N state with peaks at $g=4.3$, 3.6, and 2.0 (Fig. 9).

(7) Collect parallel-mode spectra on an X-band EPR spectrometer using the following settings: microwave frequency, 9.36 GHz; modulation frequency, 100.00 Hz; modulation amplitude, 10.02; microwave power, 6.4 mW; attenuation, 15.0 dB; time constant, 81.92 ms; conversion time, 40.96 ms.

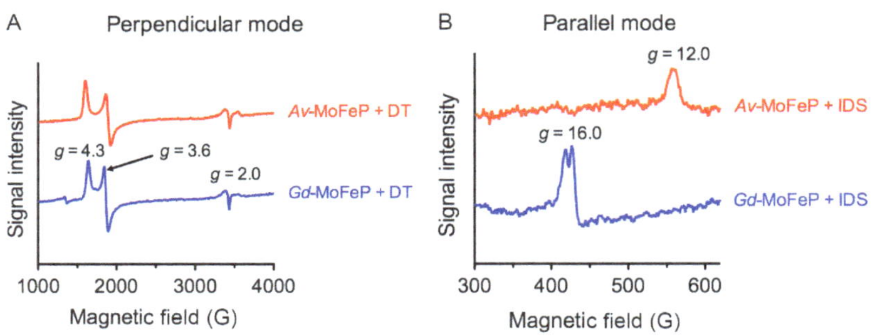

Fig. 9 EPR spectra of *Gd*-MoFeP compared to *Av*-MoFeP. The experimental conditions are described in the main text.

IDS-oxidized *Gd*-MoFeP features a signal at $g = 16.0$ in the parallel collection mode, indicative of $2e^-$ oxidized P-cluster with $S \geq 3$ (Fig. 9). In perpendicular mode, the signal of DT-reduced and IDS-oxidized *Gd*-MoFeP is identical since no change in the FeMoco oxidation state occurs (Owens et al., 2016).

(8) To obtain high signal-to-noise spectra, collect at least 20 scans per sample.

Since EPR spectroscopy is a nondestructive technique, protein can be removed from the EPR tube and used for subsequent experiments. Furthermore, redox-dependent changes in the P-cluster are reversible. MoFeP in the P^{OX} state can be reduced by addition of excess DT, and MoFeP in the P^{N} state can be oxidized by addition of excess IDS.

2.10 Crystallography of *G. diazotrophicus* MoFe Protein

2.10.1 Crystallization

Several successful methods and conditions for MoFeP crystallization have been developed. For a thorough review on nitrogenase crystallography, we direct the reader to a review by Roth and Tezcan (2011). To crystallize *Gd*-MoFeP, we chose to set up crystallization trials in a 24-well sitting drop format. The general format of the tray uses a solution of sodium 0.1 *M* cacodylate (pH 6.5), 40%–50% MPD, and 0.2–0.6 *M* NaCl.

(1) Place a stereomicroscope in the anaerobic chamber to be able to view crystals without removing the crystal trays from the chamber.

(2) Crystallization of DT-reduced *Gd*-nitrogenase: Grow crystals anaerobically using the sitting drop method with a *Gd*-MoFeP concentration of ca. 170 μ*M* (40 mg/mL). The well solution contains 45% MPD, 300 m*M* NaCl, 100 m*M* sodium cacodylate (pH 6.5), 1 m*M* spermine, 0.1% Zwittergent, and 5 m*M* DT. The drop consists of 2 μL *Gd*-MoFeP in a buffered solution of 25 m*M* Tris (pH 8), 100 m*M* NaCl, and 2 μL of well solution.

(3) Crystallization of IDS-oxidized *Gd*-nitrogenase: Remove DT from MoFeP by running the protein over a desalting column (e.g., 10DG column by Bio-Rad) or by buffer exchanging the protein at least 3 × using a spin concentrator (e.g., Millipore spin-concentrator with 10 or 20 kDa molecular weight cutoff).

To obtain oxidized Gd-MoFeP crystals, 2 μL of *Gd*-MoFeP (170 μ*M*) is mixed with 2 μL of well solution containing of 50% MPD, 300 m*M* NaCl, 100 m*M* sodium cacodylate (pH 6.5), 1 m*M* spermine, and 0.43 m*M* IDS. IDS is added from a 42 m*M* stock, which is spun at 12,000 × *g* prior to use to remove undissolved IDS.

For both reduced and oxidized *Gd*-MoFeP, the best diffracting crystals were obtained with 45%–50% MPD, however, we recommend using a broader range of MPD concentrations (40%–55% in initial screens. While the inclusion of NaCl led to the highest quality crystals, we also observed diffraction-quality crystals when NaCl was replaced with KCl. Similarly, high-quality crystals were found both in presence and absence of Zwittergent and spermine. To find best diffracting crystals, we recommend adjusting the concentration of spermine between 0 and 1 m*M* in a fine screen.

The crystal morphology is identical in both redox conditions. *Gd*-MoFeP crystals appear as rectangular rods with dimensions varying between approximately 0.04 and 0.4 mm. Crystals start to appear within 24 h, with additional crystals forming for up to 1 week.

(4) *Gd*-MoFeP crystals are harvested aerobically and flash frozen in liquid N_2. At 40%–50%, MPD itself acts as a glassing agent so no additional cryoprotectants are needed. To prevent oxygen damage during crystal harvesting, unseal only one well at a time, and complete harvesting within 5 min per tray. To further reduce oxygen damage and to keep the crystals anaerobic, harvesting can be carried out under a stream of inert gas.

2.10.2 Data Collection, Data Processing, and Structural Refinement

Diffraction experiments are carried out at a synchrotron or a home source, where the experimental strategy needs to be optimized for the respective hardware. Multiple free and commercially available indexing and integrating options are available today, including XDS (Kabsch, 2010), HKL2000 (Otwinowski & Minor, 1997), and iMosflm (Battye, Kontogiannis, Johnson, Powell, & Leslie, 2011). Aimless (Evans & Murshudov, 2013) can be used for scaling and merging, and Ctruncate (Evans & Murshudov, 2013) to convert structure factors to amplitudes.

2.10.3 Structural Refinement

The structure of *Gd*-MoFeP can be solved by molecular replacement, using the structure of *Av*-MoFeP as the search model. We recommend removing metal clusters from the search model. Phasing can be carried out using freely available software, including PhaserMR (Adams et al., 2010) and Molrep (Winn et al., 2011). The structure can be refined automatically using refmac

(Murshudov, Vagin, & Dodson, 1997), phenix.refine (Adams et al., 2010), and manually in Coot (Emsley, Lohkamp, Scott, & Cowtan, 2010).

To build the metal clusters, start with manual refinement.

(1) Open the coordinates of the P-cluster and FeMoco from an existing MoFeP structure and the coordinates of *Gd*-MoFeP in Coot (Emsley & Cowtan, 2004) and move the clusters into the corresponding regions of electron density, simultaneously using the $2F_o–F_c$ and $F_o–F_c$ maps. Then, merge the coordinates of the clusters with those of the polypeptide.

(2) After initial cluster placement, move the individual cluster atoms manually until they fit the electron density map.

Geometry constraints are necessary for automated refinement of the FeMoco and P-cluster. The respective constraint file contains coordinate and geometry information for each metal cluster. The files include values for each bond length, bond angle, and torsion (dihedral) angle in the respective cluster, as well as the maximum allowed deviation for each of the values. For both P-cluster and FeMoco, the maximum allowed deviation is 0.020 Å for bond distances, 3 degrees for bond angles, and 20 degrees for torsion angles. Making a new constraint file is necessary whenever the structure that needs to be refined differs substantially from one used to build the original constraint file. Since the structures of FeMoco and P^N in *Gd*-MoFeP closely resemble those in *Av*-MoFeP, existing constraint files originally generated for *Av*-MoFeP can be used. In this case, automated refinement will accurately place the atoms of the metal clusters into the electron density map and little additional manual refinement is needed.

The structure of P^{OX} in *Gd*-FeMoco, however, features large structural changes involving Fe6 and Fe8 that differ from *Av*-MoFeP P^{OX}. Therefore, oxidized *Gd*-MoFeP requires a new constraint file for P^{OX}. To make a new constraint file, the protein structure is first opened in Coot and the P-cluster is manually rebuilt by moving Fe and S ions until all cluster atoms fit the $2F_c–F_c$ electron density map. After fully placing the cluster, distances, bond angles, and torsion angles between cluster atoms are measured in Coot or in Pymol (Schrödinger). The values for respective distances and angles are input into a text editor according to cif format guidelines (Hall, Allen, & Brown, 1991) to create a new constraint file. Since the main difference between the structures of *Gd*-P^{OX} and *Gd*-P^N is the location of Fe5 and Fe8, we recommend using the constraint file for P^N as a template and modifying only those parameters that differ between P^{OX} and P^N. The text file can be saved as a ".cif" file that is recognized as a constraint file by refmac and phenix.refine.

3. CONCLUSION

Here, we have summarized recent studies on the ATP- and oxidation-state dependent structural dynamics of nitrogenase and the insights gained in these studies. Our investigations were greatly enriched by the examination of MoFeP from a new model organism, *G. diazotrophicus*. To afford easy access to similar investigations, we have described here methods for the isolation and characterization of *Gd*-MoFeP. We hope that the relatively straightforward expression, purification, and crystallization protocols for *Gd*-MoFeP will encourage researchers to include this variant in their mechanistic investigations on biological nitrogen fixation, and rekindle the lost interest in the study of nitrogenases from different diazotrophic bacteria (Brendt, Lowe, & Yates, 1977; Emerich & Burris, 1978a, 1978b; Ludden, Okon, & Burris, 1978; Song, Hartman, & Burris, 1985). Thanks to advances in bioinformatics, researchers have conducted detailed sequence analyses of known nitrogenases and uncovered amino acid variances in mechanistically critical regions of the protein, including selenocysteine ligation to P-cluster and amino acid variability in the outer coordination sphere of FeMoco (Howard et al., 2013). We anticipate that the growing pool of genome sequences from diazotrophic bacteria will greatly augment the biochemical and structural investigation of the nitrogenase mechanism and possibly lead to the discovery of nitrogenases with novel chemistries.

ACKNOWLEDGMENTS

This work was supported by NIH (GM099813 to F.A.T.). F.A.T. also acknowledges the Frasch Foundation for further support. C.P.O. gratefully acknowledges support by USDA NIFA (Grant 2015-67012-22895). The authors also thank Michael Medina for his contributions and valuable discussions.

REFERENCES

Adams, P. D., Afonine, P. V., Bunkóczi, G., Chen, V. B., Davis, I. W., Echols, N., et al. (2010). PHENIX: A comprehensive Python-based system for macromolecular structure solution. *Acta Crystallographica. Section D, Biological Crystallography*, *66*, 213–221.

Battye, T. G., Kontogiannis, L., Johnson, O., Powell, H. R., & Leslie, A. G. (2011). iMOSFLM: A new graphical interface for diffraction-image processing with MOSFLM. *Acta Crystallographica. Section D, Biological Crystallography*, *67*, 271–281.

Bertalan, M., Albano, R., de Pádua, V., Rouws, L., Rojas, C., Hemerly, A., et al. (2009). Complete genome sequence of the sugarcane nitrogen-fixing endophyte *Gluconacetobacter diazotrophicus* Pal5. *BMC Genomics*, *10*, 450.

Bradford, M. M. (1976). A rapid and sensitive method for the quantitation of microgram quantities of protein utilizing the principle of protein-dye binding. *Analytical Biochemistry, 72*, 248–254.

Brendt, H., Lowe, D. J., & Yates, M. G. (1977). The nitrogen-fixing system of *Corynebacterium autotrophicum. European Journal of Biochemistry, 86*, 133–142.

Buren, S., Young, E. M., Sweeny, E. A., Lopez-Torrejon, G., Veldhuizen, M., Voigt, C. A., et al. (2017). Formation of nitrogenase NifDK tetramers in the mitochondria of Saccharomyces cerevisiae. *ACS Synthetic Biology, 6*, 1043–1055.

Cavalcante, V. A., & Dobereiner, J. (1988). A new acid-tolerant nitrogen-fixing bacterium associated with sugarcane. *Plant and Soil, 108*, 22–31.

Chan, J. M., Christiansen, J., Dean, D. R., & Seefeldt, L. C. (1999). Spectroscopic evidence for changes in the redox state of the nitrogenase P-cluster during turnover. *Biochemistry, 38*, 5779–5785.

Chiu, H., Peters, J. W., Lanzilotta, W. N., Ryle, M. J., Seefeldt, L. C., Howard, J. B., et al. (2001). MgATP-bound and nucleotide-free structures of a nitrogenase protein complex between the Leu 127 Delta-Fe-protein and the MoFe-protein. *Biochemistry, 40*, 641–650.

Danyal, K., Dean, D. R., Hoffman, B. M., & Seefeldt, L. C. (2011). Electron transfer within nitrogenase: Evidence for a deficit-spending mechanism. *Biochemistry, 50*, 9255–9263.

Danyal, K., Mayweather, D., Dean, D. R., Seefeldt, L. C., & Hoffman, B. M. (2010). Conformational gating of electron transfer from the nitrogenase Fe protein to MoFe protein. *Journal of the American Chemical Society, 132*, 6894–6895.

Danyal, K., Yang, Z. Y., & Seefeldt, L. C. (2011). Electron paramagnetic resonance spectroscopy. *Methods in Molecular Biology, 766*, 191–205.

Dos Santos, P. C., Fang, Z., Mason, S. W., Setubal, J. C., & Dixon, R. (2012). Distribution of nitrogen fixation and nitrogenase-like sequences amongst microbial genomes. *BMC Genomics, 13*, 162.

Duyvis, M. G., Wassink, H., & Haaker, H. (1996). Pre-steady-state kinetics of nitrogenase from *Azotobacter vinelandii*. Evidence for an ATP-induced conformational change of the nitrogenase complex as part of the reaction mechanism. *The Journal of Biological Chemistry, 271*, 29,632–29,636.

Duyvis, M. G., Wassink, H., & Haaker, H. (1998). Nitrogenase of *Azotobacter vinelandii*: Kinetic analysis of the Fe protein redox cycle. *Biochemistry, 37*, 17,345–17,354.

Einsle, O., Tezcan, F. A., Andrade, S. L., Schmid, B., Yoshida, M., Howard, J. B., et al. (2002). Nitrogenase MoFe-protein at 1.16 Å resolution: A central ligand in the FeMo-cofactor. *Science, 297*, 1696–1700.

Emerich, D. W., & Burris, R. H. (1978a). Nitrogenase from Bacillus polymyxa. Purification and properties of the component proteins. *Biochimica et Biophysica Acta, 536*, 172–183.

Emerich, D. W., & Burris, R. H. (1978b). Preparation of nitrogenase. *Methods in Enzymology, 53*, 314–329.

Emsley, P., & Cowtan, K. (2004). Coot: Model-building tools for molecular graphics. *Acta Crystallographica. Section D, Biological Crystallography, 60*, 2126–2132.

Emsley, P., Lohkamp, B., Scott, W. G., & Cowtan, K. (2010). Features and development of Coot. *Acta Crystallographica. Section D, Biological Crystallography, 66*, 486–501.

Enemark, E. J., & Joshua-Tor, L. (2008). On helicases and other motor proteins. *Current Opinion in Structural Biology, 18*, 243–257.

Evans, P. R., & Murshudov, G. N. (2013). How good are my data and what is the resolution? *Acta Crystallographica. Section D, Biological Crystallography, 69*, 1204–1214.

Fisher, K., & Newton, W. E. (2005). Nitrogenase proteins from *Gluconacetobacter diazotrophicus*, a sugarcane-colonizing bacterium. *Biochimica et Biophysica Acta, 1750*, 154–165.

Flores-Encarnacion, M., Contreras-Zentella, M., Soto-Urzua, L., Aguilar, G. R., Baca, B. E., & Escamilla, J. E. (1999). The respiratory system and diazotrophic activity of Acetobacter diazotrophicus PAL5. *Journal of Bacteriology, 181*, 6987–6995.

Giongo, A., Tyler, H. L., Zipperer, U. N., & Triplett, E. W. (2010). Two genome sequences of the same bacterial strain, *Gluconacetobacter diazotrophicus* PAl 5, suggest a new standard in genome sequence submission. *Standards in Genomic Sciences, 2*, 309–317.

Hall, S. R., Allen, F. H., & Brown, I. D. (1991). The crystallographic information file (CIF): A new standard archive file for crystallography. *Acta Crystallographica, A47*, 655–685.

Hamilton, T. L., Jacobson, M., Ludwig, M., Boyd, E. S., Bryant, D. A., Dean, D. R., et al. (2011). Differential accumulation of nif structural gene mRNA in *Azotobacter vinelandii*. *Journal of Bacteriology, 193*, 4534–4536.

Hanson, P. I., & Whiteheart, S. W. (2005). AAA+ proteins: Have engine, will work. *Nature Reviews. Molecular Cell Biology, 6*, 519–529.

Hirsch, P. R., & Mauchline, T. H. (2015). The importance of the microbial N cycle in soil for crop plant nutrition. *Advances in Applied Microbiology, 93*, 45–71.

Hoffman, B. M., Lukoyanov, D., Yang, Z.-Y., Dean, D. R., & Seefeldt, L. C. (2014). Mechanism of nitrogen fixation by nitrogenase: The next stage. *Chemical Reviews, 114*, 4041–4062.

Howard, J. B., Kechris, K. J., Rees, D. C., & Glazer, A. N. (2013). Multiple amino acid sequence alignment nitrogenase component 1: Insights into phylogenetics and structure-function relationships. *PLoS One, 8*, e72751.

Igarashi, R. Y., & Seefeldt, L. C. (2003). Nitrogen fixation: The mechanism of the Mo-dependent nitrogenase. *Critical Reviews in Biochemistry and Molecular Biology, 38*, 351–384.

Kabsch, W. (2010). Integration, scaling, space-group assignment and post-refinement. *Acta Crystallographica. Section D, Biological Crystallography, 66*, 133–144.

Katz, F. E. H., Owens, C. P., & Tezcan, F. A. (2016). Electron transfer reactions in biological nitrogen fixation. *Israel Journal of Chemistry, 56*, 682–692.

Lanzilotta, W. N., Christiansen, J., Dean, D. R., & Seefeldt, L. C. (1998). Evidence for coupled electron and proton transfer in the [8Fe-7S] cluster of nitrogenase. *Biochemistry, 37*, 11,376–11,384.

Lee, S., Reth, A., Meletzus, D., Sevilla, M., & Kennedy, C. (2000). Characterization of a major cluster of nif, fix, and associated genes in a sugarcane endophyte, Acetobacter diazotrophicus. *Journal of Bacteriology, 182*, 7088–7091.

Lowe, D. J., & Thorneley, R. (1984). The mechanism of *Klebsiella pneumoniae* nitrogenase action. Pre-steady-state kinetics of H2 formation. *The Biochemical Journal, 224*, 877–886.

Ludden, P. W., Okon, Y., & Burris, R. H. (1978). The nitrogenase system of *Spirillum lipoferum*. *The Biochemical Journal, 173*, 1001–1003.

Mayer, S. M., Lawson, D. M., Gormal, C. A., Roe, S. M., & Smith, B. E. (1999). New insights into structure-function relationships in nitrogenase: A 1.6 Å resolution X-ray crystallographic study of *Klebsiella pneumoniae* MoFe-protein. *Journal of Molecular Biology, 292*, 871–891.

Murshudov, G. N., Vagin, A. A., & Dodson, E. J. (1997). Refinement of macromolecular structures by the maximum-likelihood method. *Acta Crystallographica. Section D, Biological Crystallography, 53*, 240–255.

Nicolet, Y., Rohac, R., Martin, L., & Fontecilla-Camps, J. C. (2013). X-ray snapshots of possible intermediates in the time course of synthesis and degradation of protein-bound Fe4S4 clusters. *Proceedings of the National Academy of Sciences of the United States of America, 110*, 7188–7192.

Oldroyd, G. E. D., & Dixon, R. (2014). Biotechnological solutions to the nitrogen problem. *Current Opinion in Biotechnology, 26*, 19–24.

Otwinowski, Z., & Minor, W. (1997). Processing of X-ray diffraction data collected in oscillation mode. *Methods in Enzymology*, *276*, 307–326.

Owens, C. P., Katz, F. E. H., Carter, C. H., Luca, M. A., & Tezcan, F. A. (2015). Evidence for functionally relevant encounter complexes in nitrogenase catalysis. *Journal of the American Chemical Society*, *137*, 12,704–12,712.

Owens, C. P., Katz, F. E., Carter, C. H., Oswald, V. F., & Tezcan, F. A. (2016). Tyrosine-coordinated P-cluster in G. diazotrophicus nitrogenase: Evidence for the importance of O-based ligands in conformationally gated electron transfer. *Journal of the American Chemical Society*, *138*, 10,124–7.

Pedraza, R. O. (2008). Recent advances in nitrogen-fixing acetic acid bacteria. *International Journal of Food Microbiology*, *125*, 25–35.

Peters, J. W., Stowell, M. H., Soltis, S. M., Finnegan, M. G., Johnson, M. K., & Rees, D. C. (1997). Redox-dependent structural changes in the nitrogenase P-cluster. *Biochemistry*, *36*, 1181–1187.

Rappas, M., Bose, D., & Zhang, X. (2007). Bacterial enhancer-binding proteins: Unlocking sigma54-dependent gene transcription. *Current Opinion in Structural Biology*, *17*, 110–116.

Raymond, J., Siefert, J. L., Staples, C. R., & Blankenship, R. E. (2004). The natural history of nitrogen fixation. *Molecular Biology and Evolution*, *21*, 541–554.

Rees, D. C., Akif Tezcan, F., Haynes, C. A., Walton, M. Y., Andrade, S., Einsle, O., et al. (2005). Structural basis of biological nitrogen fixation. *Philosophical Transactions of the Royal Society A: Mathematical, Physical and Engineering Sciences*, *363*, 971–984.

Reis, V. M., & Döbereiner, J. (1998). Effect of high sugar concentration on nitrogenase activity of Acetobacter diazotrophicus. *Archives of Microbiology*, *171*, 13–18.

Rogers, C., & Oldroyd, G. E. (2014). Synthetic biology approaches to engineering the nitrogen symbiosis in cereals. *Journal of Experimental Botany*, *65*, 1939–1946.

Roth, L. E., Nguyen, J. C., & Tezcan, F. A. (2010). ATP- and iron–protein-independent activation of nitrogenase catalysis by light. *Journal of the American Chemical Society*, *132*, 13,672–13,674.

Roth, L. E., & Tezcan, F. A. (2011). X-ray crystallography. *Methods in Molecular Biology (Clifton, N.J.)*, *766*, 147–164.

Roth, L. E., & Tezcan, F. A. (2012). ATP-uncoupled, six-electron photoreduction of hydrogen cyanide to methane by the molybdenum-iron protein. *Journal of the American Chemical Society*, *134*, 8416–8419.

Saravanan, V. S., Madhaiyan, M., Osborne, J., Thangaraju, M., & Sa, T. M. (2008). Ecological occurrence of *Gluconacetobacter diazotrophicus* and nitrogen-fixing Acetobacteraceae members: Their possible role in plant growth promotion. *Microbial Ecology*, *55*, 130–140.

Schindelin, H., Kisker, C., Schlessman, J. L., Howard, J. B., & Rees, D. C. (1997). Structure of ADP x AIF4(-)-stabilized nitrogenase complex and its implications for signal transduction. *Nature*, *387*, 370–376.

Seefeldt, L. C., Hoffman, B. M., & Dean, D. R. (2009). Mechanism of Mo-dependent nitrogenase. *Annual Review of Biochemistry*, *78*, 701–722.

Sevilla, M., Burris, R. H., Gunapala, N., & Kennedy, C. (2001). Comparison of benefit to sugarcane plant growth and 15N2 incorporation following inoculation of sterile plants with Acetobacter diazotrophicus wild-type and Nif- mutants strains. *Molecular Plant-Microbe Interactions: MPMI*, *14*, 358–366.

Song, S., Hartman, A., & Burris, R. H. (1985). Purification and properties of nitrogenase from *Azospirillum amazonense*. *Journal of Bacteriology*, *164*, 1271–1277.

Spatzal, T., Perez, K. A., Einsle, O., Howard, J. B., & Rees, D. C. (2014). Ligand binding to the FeMo-cofactor: Structures of CO-bound and reactivated nitrogenase. *Science*, *345*, 1620–1623.

Tejera, N. A., Ortega, E., Rodes, R., & Lluch, C. (2004). Influence of carbon and nitrogen sources on growth, nitrogenase activity, and carbon metabolism of *Gluconacetobacter diazotrophicus*. *Canadian Journal of Microbiology*, *50*, 745–750.
Tezcan, F. A., Kaiser, J. T., Howard, J. B., & Rees, D. C. (2015). Structural evidence for asymmetrical nucleotide interactions in nitrogenase. *Journal of the American Chemical Society*, *137*, 146–149.
Tezcan, F. A., Kaiser, J. T., Mustafi, D., Walton, M. Y., Howard, J. B., & Rees, D. C. (2005). Nitrogenase complexes: Multiple docking sites for a nucleotide switch protein. *Science (New York, N.Y.)*, *309*, 1377–1380.
Thorneley, R. N., & Lowe, D. J. (1983). Nitrogenase of *Klebsiella pneumoniae*. Kinetics of the dissociation of oxidized iron protein from molybdenum-iron protein: Identification of the rate-limiting step for substrate reduction. *Biochemical Journal*, *215*, 393–403.
Ubbink, M. (2009). The courtship of proteins: Understanding the encounter complex. *FEBS Letters*, *583*, 1060–1066.
Wang, L., Zhang, L., Liu, Z., Zhao, D., Liu, X., Zhang, B., et al. (2013). A minimal nitrogen fixation gene cluster from Paenibacillus sp. WLY78 enables expression of active nitrogenase in Escherichia coli. *PLoS Genetics*, *9*, e1003865.
Willing, A. H., Georgiadis, M. M., Rees, D. C., & Howard, J. B. (1989). Cross-linking of nitrogenase components. Structure and activity of the covalent complex. *The Journal of Biological Chemistry*, *264*, 8499–8503.
Willing, A., & Howard, J. B. (1990). Cross-linking site in *Azotobacter vinelandii* complex. *The Journal of Biological Chemistry*, *265*, 6596–6599.
Wilson, P. E., Nyborg, A. C., & Watt, G. D. (2001). Duplication and extension of the Thorneley and Lowe kinetic model for *Klebsiella pneumoniae* nitrogenase catalysis using a mathematica software platform. *Biophysical Chemistry*, *91*, 281–304.
Winn, M. D., Ballard, C. C., Cowtan, K. D., Dodson, E. J., Emsley, P., Evans, P. R., et al. (2011). Overview of the CCP4 suite and current developments. *Acta Crystallographica. Section D, Biological Crystallography*, *67*, 235–242.
Zhang, L. M., Morrison, C. N., Kaiser, J. T., & Rees, D. C. (2015). Nitrogenase MoFe protein from *Clostridium pasteurianum* at 1.08 A resolution: Comparison with the *Azotobacter vinelandii* MoFe protein. *Acta Crystallographica. Section D, Biological Crystallography*, *71*, 274–282.

CHAPTER THIRTEEN

Protein Film Electrochemistry of Iron–Sulfur Enzymes

Fraser A. Armstrong[1], Rhiannon M. Evans, Clare F. Megarity
Inorganic Chemistry Laboratory, Oxford, United Kingdom
[1]Corresponding author: e-mail address: fraser.armstrong@chem.ox.ac.uk

Contents

Abstract

A suite of dynamic electrochemical techniques known as protein film electrochemistry (PFE) offers important insight into the roles of active sites in enzymes, including properties of electron-transfer centers (individually or collectively), rates and dependences of catalytic electron transport, and binding and dissociation of inhibitors. In this chapter, we explain how PFE is used to investigate the properties of FeS clusters—centers lacking distinctive or convenient spectroscopic signatures that are often very sensitive to O_2. We see that PFE allows simultaneous detection and control of the reactions of individual FeS clusters, and measurement of their relaying efficiency in long-range electron transfer.

1. INTRODUCTION

The use of dynamic electrochemical techniques to study reversible redox processes in proteins began about 40 years ago with two papers reporting the diffusion-controlled cyclic voltammetry (CV) of cytochrome *c* in solution, and grew from that time (Armstrong & Wilson, 2000). A major advance occurred 10–15 years later with the discovery that meaningful information could be obtained about FeS clusters in a small protein

Methods in Enzymology, Volume 599
ISSN 0076-6879
https://doi.org/10.1016/bs.mie.2017.11.001

that is tightly adsorbed on an electrode, making it possible to measure reduction potentials and reactivities on minuscule quantities of protein (Butt et al., 1991). Confinement of a protein to the electrode surface brings many advantages—as well as superb sample economy, much sharper current signals are observed, and complex, seemingly chaotic behavior can be controlled. Early investigations provided detailed quantitative insight into rates of interconversion between different FeS cluster types, rates of ligand binding, and rates and energetics of coupled proton transfer. Attention soon turned to larger proteins, and it was noted quickly that despite their size and low coverage on an electrode, the high catalytic activity of many enzymes gave rise to large currents that displayed highly informative dependences on electrode potential (Léger et al., 2003). Not only were many enzymes identified to behave as reversible electrocatalysts when attached to an electrode (Armstrong & Hirst, 2011), but some displayed unusual rises and drops in activity at particular characteristic potentials, a sign that internal redox transitions were altering the performance of the enzyme. Using chronoamperometry (CA), the rates of these interconversions could be investigated and related to structural changes that were studied independently by spectroscopic techniques. Such investigations included the reaction of O_2 with hydrogenases, the characterization of site-directed variants of different proteins, altered both at the active site and at individual FeS clusters of the relay, and even the assembly of an active site from its components (Megarity et al., 2016). Most recently, AC modulation applied through electrochemical impedance spectroscopy (EIS) has allowed measurement of the "catalytic exchange rate" of hydrogenases—the frequency at which the cycle can turn back and forth at the equilibrium potential and a fundamental measure of kinetic capability (Pandey, Islam, Happe, & Armstrong, 2017).

2. TECHNICAL ASPECTS

The practical and theoretical backgrounds to executing and interpreting electrochemical experiments are well known, and protein film electrochemistry (PFE) makes good use of the standard material available in textbooks (Bard & Faulkner, 2001). The central requirement is an electrode (the working electrode, WE) upon which is immobilized the protein under investigation. The WE, which may be attached to a motor shaft that allows it to be rotated at high speed (typically up to 5000 rpm), is placed in a sealed electrochemical cell, a general example of which is described in Fig. 1. The cell solution contains buffer, supporting electrolyte, and reagents with which

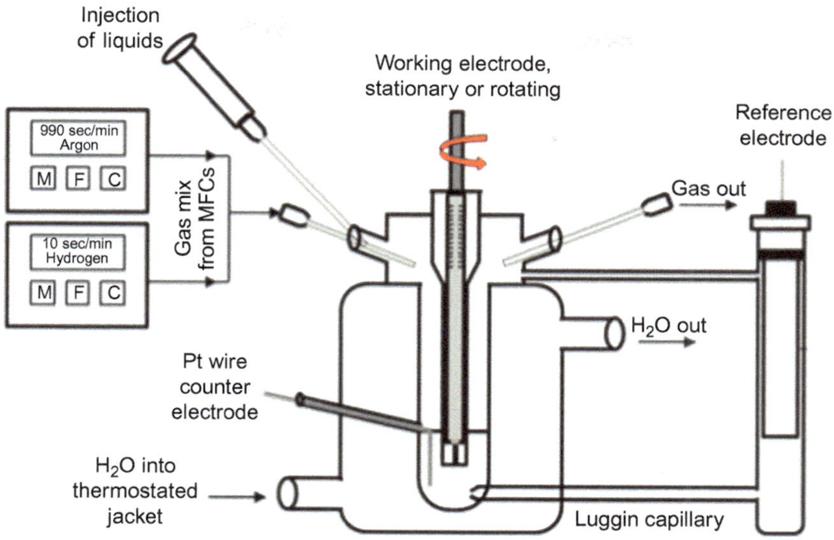

Fig. 1 Schematic representation of a typical electrochemical cell setup. All ports are sealed with gas-tight septa. A three-electrode system is represented here, comprising a working electrode (connected to a motorized spindle for optional rotation or stationary measurements), a reference electrode (e.g., Ag/AgCl or saturated calomel SCE) housed in a side arm containing electrolyte solution and connected to the electrochemical cell via a Luggin capillary, and a counter electrode (Pt wire). The use of a circulating water bath allows for temperature control via the water jacket. The use of Mass Flow Controllers (MFC, Sierra Instruments) allows precise gas mixtures at a constant flow rate to be delivered to the cell headspace. Introduction of liquids (gas-purged buffers, inhibitors, substrates, buffer exchanges, etc.) to the cell is carried out via a needle/syringe.

the protein may react (such as substrates and inhibitors), but, importantly, all the protein is confined to the electrode surface. The potential imposed on the redox active sites in the protein molecules is controlled relative to a reference electrode (RE) linked to the main compartment by a high-impedance Luggin capillary junction. The RE potential is usually scaled to correspond to the standard hydrogen electrode (SHE). The current is measured using a third, "counter" electrode (CE) which is usually a platinum wire. The entire cell is usually surrounded by a Faraday cage, to minimize noise, and placed inside an anaerobic glove box, which may also serve as the Faraday cage. Experiments are controlled from an electrochemical workstation that features a potentiostat able to measure the current passing at a particular voltage, and a voltage ramp generator (generally using digital steps, although a smooth analogue ramp is used for CV at high scan rates).

The first special requirement of PFE is preparation of an electroactive protein film on the WE. Although we should not refer to this aspect as a "black art," some proteins form much more stable films than others, and a few recommendations can be provided. A typical film contains $<5\,\mathrm{pmol\,cm^{-2}}$ of protein, so the method is incredibly economical, allowing hundreds of experiments to be carried out on a small sample of protein. The best electrodes appear to have a rough surface, not surprisingly since the overall protein–electrode binding depends on numerous weak interactions. The atoms at the electrode surface should also have "satisfied" valences; thus, benign functionalities such as oxides or carboxylates are good options, whereas bare metals, which promote denaturation, should be avoided. Pyrolytic graphite polished on its edge surface (PGE) and rich in C–O functionalities is a good basis: this electrode can be modified with carbon nanotubes and equipped with sites for covalent coupling to the protein surface (Alonso-Lomillo et al., 2007; Krishnan & Armstrong, 2012). Conducting metal oxides such as indium tin oxide are also useful electrode materials. Polyionic molecules such as polymyxin (a polycation) are often necessary to stabilize protein binding at the electrode. Another consideration is that an electrode for studying FeS enzymes must have a wide potential window, mainly so that very negative electrode potentials can be applied.

Some relevant equations which will be used in this chapter are explained in Box 1. Dynamic electrochemistry is distinguished from redox potentiometry (which is far more familiar to biochemists) because current, proportional to the net rate of electron transfer, is measured as a function of electrode potential and/or time. In CV, the electrode potential is scanned back and forth across a region of interest and the current that passes is measured simultaneously. The faradaic current due to electron transfer appears above a nonfaradaic current that is due to the charging and discharging of the electrode–solution interface (capacitance). The faradaic current is governed by the thermodynamics, expressed in the Nernst equation (Box 1, Eq. 2), and by kinetics, typically expressed as the Butler–Volmer equation (Box 1, Eq. 5). In chronoamperometry, the current is measured as a function of time at a constant potential following a perturbation such as introduction of a reactant or a potential step. Use of a rotating disc electrode helps ensure that the reactions occurring at the immobilized protein are not limited by mass transfer of reactants to the electrode surface. A diffusion-controlled current varies as the square root of rotation rate (ω) up to a limiting value: care is therefore taken to ensure ω is sufficiently high that the current is at its maximum value.

BOX 1 Useful Equations in Protein Film Electrochemistry

$$\text{Ox} + ne^- \rightleftharpoons \text{Red} \tag{1}$$

Generic half-cell reaction that is the basis of electrochemistry

$$E_{\text{appl}} = E^0 + 2.3\frac{RT}{nF}\log\left(\frac{[\text{Ox}]}{[\text{Red}]}\right) \tag{2}$$

A version of the Nernst equation useful for redox centers on an electrode: E_{appl} is the potential applied at the WE, E^0 is the standard potential under the conditions of the experiment, R is the gas constant, T is the absolute temperature, n is the number of electrons (as in Eq. 1), F is the Faraday constant, and [Ox] and [Red] are the surface concentrations of oxidized and reduced sites.

$$i_p = \frac{n^2F^2A\Gamma}{4RT}\upsilon \tag{3}$$

Equation describing the magnitude of the peak current i_p observed during a voltammetric scan, arising from a redox center attached to the electrode: A is the electrode surface area, Γ is the electroactive coverage, n is the number of electrons transferred in a single reaction, υ is the scan rate (V s^{-1})

$$i = knFA\Gamma \tag{4}$$

Equation linking current i and rate k, useful for converting catalytic current to turnover rate.

$$i = i_0[\exp(0.5nF\eta/RT) - \exp(-0.5nF\eta/RT)] \tag{5}$$

The Butler–Volmer equation describing the exponential increase in current (*ET* rate) with driving force (overpotential) η.

$$\frac{k_{\text{red}}}{k_{\text{ox}}} = \exp\left[\frac{-2F}{RT}(E_{\text{CS}} - E_{\text{substrate}})\right] \tag{6}$$

An equation expressing an enzyme's catalytic bias: the relative rates of reduction and oxidation depend on the difference between the reduction potential of the electrochemical control center E_{CS} and that of the substrate.

3. EXAMPLES IN WHICH SPECIFIC FeS CLUSTERS IN PROTEINS ARE STUDIED

A surface-confined redox couple that is fully reversible and uncomplicated by slow kinetics or coupling to a slower reaction is observed as a pair of peaks, corresponding to reduction and oxidation, termed the "signal." The oxidation and reduction peaks are virtually mirror images of each other, lying above the relatively flat nonfaradaic current background that is due to double-layer capacitance. For a reaction studied at 25°C, where n is the number of electrons transferred in a single step, the width at half height (mV) varies as $90/n$, and the height varies as n^2. These relationships mean that a cooperative two-electron transfer reaction shows up as a particularly prominent and narrow pair of peaks. Provided the scan rate is sufficiently slow, the peak positions coincide, but as the scan rate is increased kinetic factors take effect and the peak positions begin to separate.

As an example, Fig. 2 shows the voltammetry obtained for small ferredoxins, each containing a [3Fe–4S] cluster and a [4Fe–4S] cluster. With these small proteins (<10 kDa) the electroactive coverage is generally high

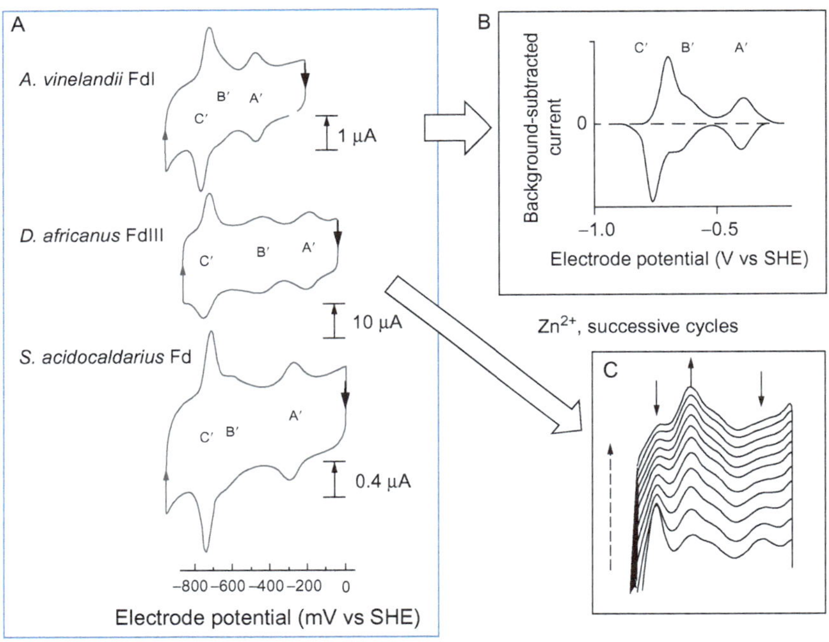

Fig. 2 (A) Cyclic voltammograms for three small FeS proteins, each containing one [3Fe–4S] cluster and one [4Fe–4S] cluster, showing the signals observed for the redox couples associated with each center: A′ = $[3Fe–4S]^{+/0}$, B′ = $[4Fe–4S]^{2+/+}$, C′ = $[3Fe–4S]^{0/2-}$. (B) The voltammogram for FdI from *Azotobacter vinelandii*, with capacitance background subtracted by curve fitting. (C) Voltammograms (successive cycles, only oxidative direction shown) for a film of 7Fe form of ferredoxin III from *Desulfovibrio africanus*, after Zn^{2+} is added to the cell solution. The [3Fe–4S] cluster converts rapidly into a [Zn3Fe–4S] cluster.

enough to observe clear and sharp signals due to electron transfers in and out of each center. The average of the oxidation and reduction peak potentials corresponds to the reduction potential of the site, there usually being only a small difference from that measured for the protein free in solution. After baseline correction by removing the capacitance background (Fig. 2B), the areas under each oxidation or reduction peak (current multiplied by time, derived from the scan rate) are used to determine the charge passed for each signal. The stoichiometries of the different sites in the protein are thus determined or confirmed, along with the presence of centers that transfer two electrons cooperatively. These ferredoxins exhibit one signal due to the [4Fe–4S] cluster (2+/+) and two signals due to the [3Fe–4S] cluster (+/0 and 0/2−). The signal at the most negative potential is sharp and narrow: the $[3Fe–4S]^{0/2-}$ couple is a cooperative, proton-coupled two-electron transfer, and it is observed in nearly all examples we have investigated (Duff, Breton, Butt, Armstrong, & Thomson, 1996). Despite the (0, 2−) redox couple being a characteristic of the [3Fe–4S] cluster in PFE experiments, the $[3Fe–4S]^{2-}$ state, which is an all-ferrous cluster, has only been poorly characterized in free solution, owing to its strongly reducing nature. The [3Fe–4S] cluster in ferredoxin III from *Desulfovibrio africanus* (*Da*) is easily converted into a [4Fe–4S] cluster. One of the early successes of PFE was to enable vacillatory FeS clusters to be tamed and their interconversions brought under control (Butt et al., 1991). Fig. 2C shows the changes in voltammetry that occur over several cycles after the electrode loaded with a film of *Da* FdIII is placed in a solution containing Zn^{2+} ions. The two signals due to the [3Fe–4S] cluster are replaced by a new signal due to $[Zn3Fe–4S]^{2+/+}$. Analogous investigations led to the generation and characterization of [M3Fe–4S] clusters containing a wide range of metal ions M, including Cu and Tl (Butt, Fawcett, Breton, Thomson, & Armstrong, 1997). The reaction is reversed and the exogenous metal ion released, by placing the electrode in a solution containing a complexing ligand such as EDTA and scanning to high potential.

Increasing the scan rate reveals the kinetics of electron transfer and coupled reactions through representations known as "trumpet plots." The oxidation and reduction peaks separate in a way that is conveniently represented by plotting peak potentials against log(scan rate). The data are analyzed using a procedure introduced by Laviron (1979) for an immobilized redox couple, using the Butler–Volmer model of electrochemical kinetics in which rate increases exponentially with overpotential (see Box 1, Eq. 5). Alternatively, Marcus theory can be used, which increases the complexity but is more appropriate for dealing with redox centers having a low reorganization energy (Hirst & Armstrong, 1998). Trumpet plots are very instructive for analyzing and interpreting data for coupled reactions such as proton transfer. Fig. 3A shows a reaction scheme in which a [3Fe–4S] cluster is protonated upon reduction from the 1+ to the 0 oxidation level—a reaction that occurs in ferredoxin I from *Azotobacter vinelandii* (Fig. 2A). Proton transfer to the buried [3Fe–4S] cluster occurs via the mobile carboxylate side chain of aspartate-15, and is greatly retarded in the D15N variant. The trumpet plot in Fig. 3B displays the data obtained up to a scan rate of $100\,V\,s^{-1}$. At high pH, only an electron is transferred and the peaks separate quite symmetrically with increasing scan rate, giving a standard (exchange) electrochemical rate constant $>200\,s^{-1}$. In contrast, at pH 5.5, which is below the pK of the $[3Fe–4S]^0$ cluster, the trumpet plot looks very different: at low scan rate, the shift in average peak position to a more positive value reflects the thermodynamics of the coupled proton–electron transfer; then, as the scan rate is increased the oxidation peak

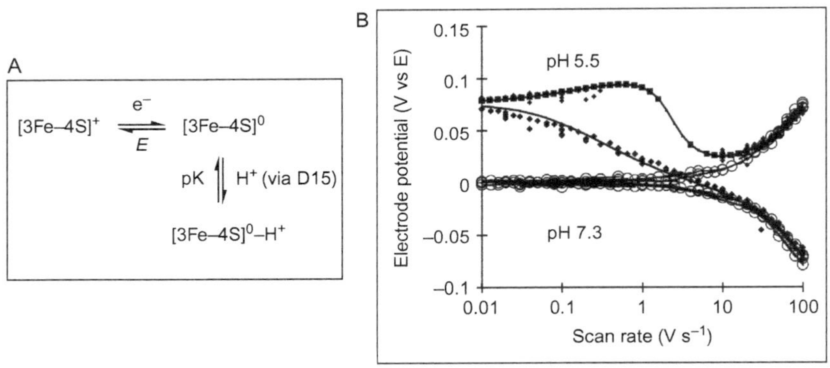

Fig. 3 (A) Investigation of a proton-coupled electron transfer reaction occurring at the [3Fe–4S] cluster in a variant (D15N) of ferredoxin I from *Azotobacter vinelandii*. In this protein, the replacement of aspartate-15 by asparagine slows down proton transfer to reveal three distinct time domains of the reaction in each direction. (B) Trumpet plots used to depict and analyze the data, which were obtained for single cycles starting from the high potential limit. Peak positions for oxidative and reductive directions are plotted against log (scan rate). At pH 7.3 (pH ≫ pK, *open circles*) no proton transfer occurs and the plot is essentially symmetrical: the intersection with the ordinate gives the reduction potential for the pure electron transfer and the peaks separate quite symmetrically as the scan rate is increased. At pH 5.5 (*closed circles*) the peaks at low scan rate give the proton-equilibrated reduction potential; at intermediate scan rate, the oxidation peak disappears as proton transfer off the cluster is too slow (thus blocking the following electron transfer); at highest scan rate, the peaks revert to the positions obtained at pH ≫ pK because electron transfer is much faster than proton transfer and the lifetime of the [3Fe–4S] cluster is too short to allow protonation to occur.

disappears because the system is trapped in the $[3Fe–4S]^0$-H^+ state. At much higher scan rates, the $[3Fe–4S]^0$ cluster does not have time to protonate before it is reoxidized, and the data points rejoin the lines obtained at high pH, showing that the pure electron transfer rate is not affected by pH. The data are relevant to an atomic-level understanding of the mechanism of proton pumps (Chen et al., 2000; Hirst et al., 1998).

4. STEADY-STATE CATALYTIC ELECTRON FLOW THROUGH FeS CLUSTER RELAYS

For larger proteins, it is much more difficult to observe the discrete "nonturnover" signals from individual FeS centers that we have mentioned above, mainly because the electroactive coverage is usually too low. For enzymes, however, it is still possible to observe the involvement of FeS centers in catalysis because the current may be greatly amplified. In contrast to normal kinetic studies, PFE excels when investigating enzymes with very high turnover rates: since current is directly proportional to catalytic turnover rate, any change in activity appears as a change in current, the higher the turnover rate the greater the signal strength. Instead of equilibrium reduction potentials, the voltammograms reveal characteristic steady-state potentials that reflect changes in the rate of catalytic electron flow. Despite their giant nature, many enzymes function as reversible electrocatalysts when attached to an electrode (Armstrong & Hirst, 2011). In this context "reversible" means that when both oxidized and reduced forms of the substrate are present in solution, the current due to catalytic electron flow cuts cleanly through the zero-current line and only a minimal overpotential is required to observe a net current in either direction. This situation may be contrasted with almost all chemical and industrial electrocatalysts, where a large overpotential is required to initiate and drive a reaction, a notable exemption being $2H^+/H_2$ interconversion at platinum. In most cases, electrons are transferred between the electrode and the catalytic site through an internal ET relay system, and PFE provides interesting insight into the properties that must have been crucial factors during evolution.

Hydrogenases are an important class of enzymes that use FeS clusters as internal electron relays and, in the case of [FeFe]-hydrogenases, as part of the active site at which H_2 is produced or oxidized at very high rates (Lubitz, Ogata, Rüdiger, & Reijerse, 2014). Fig. 4A compares the results obtained for two [NiFe]-hydrogenases, the membrane-extrinsic subunits of membrane-bound enzymes from *Escherichia coli*, termed hydrogenase-1

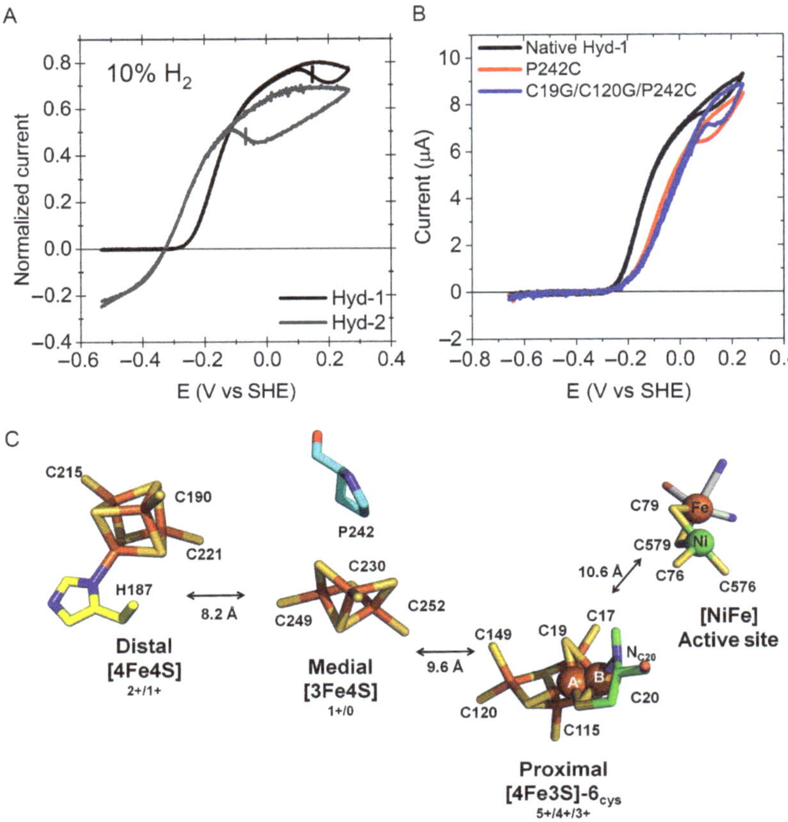

Fig. 4 (A) Catalytic cyclic voltammograms of two hydrogenases, Hyd-1 and Hyd-2, from *E. coli*, adsorbed on a PGE electrode at pH 6, 10% H_2, 90% N_2. The CVs show that Hyd-1 has no H_2 production (H^+ reduction) activity under these conditions, whereas Hyd-2 is a reversible electrocatalyst. The potential at which activation occurs (E_{switch}) is shown as vertical bars. (B) The electron-transfer pathway in Hyd-1. Key residues that have been substituted in the FeS variants shown in (C) are highlighted. The reduction potentials for the two proximal cluster transitions (with associated conformational change of the Fe, labeled "A" and "B," and deprotonated backbone N-atom of cys20 upon formation of a localized 3+ state on the Fe, as the second electron is discharged) and the medial cluster +/0 transition are given in mV vs SHE, as determined by EPR spectroscopic potentiometric titrations of the C19G/C120G and P242C variants (Roessler, Evans, Davies, Harmer, & Armstrong, 2012). The reduction potential of the distal [4Fe–4S] cluster was estimated by model analysis (Hexter et al., 2012). (C) A comparison of H_2 oxidation by native Hyd-1 to variants in which the medial (P242C) and proximal (C19G/C120G/P242C) FeS clusters have been modified by site-directed mutagenesis. The onset potentials are the same, consistent with the distal [4Fe–4S] cluster being unaffected by mutation, but the more shallow climb in current signifies that the "conductivity" of the FeS relay is decreased—more driving force being required to draw off the electrons from H_2 oxidation at the [NiFe]-active site.

(Hyd-1) and hydrogenase-2 (Hyd-2) (Lukey et al., 2010). In both cases, the active site is a [NiFe]-cofactor, and electrons are transferred to and from the protein surface by a series of three FeS clusters, as illustrated in the lower panel (B) in the case of Hyd-1. The voltammograms, measured under 10% H_2 at pH 6.0, reveal important differences between Hyd-1 and Hyd-2. First, Hyd-2 catalyzes H_2 production (proton reduction), whereas Hyd-1 does not: the potential at which the catalytic current crosses the zero-current axis is the formal equilibrium potential for the $2H^+/H_2$ couple under the experimental conditions. Under these conditions Hyd-2 is a reversible electrocatalyst, whereas Hyd-1 has an increased overpotential requirement for H_2 oxidation. Second, both enzymes undergo reversible inactivation to a Ni^{III}-OH species (known as Ni–B, effectively a resting state of the enzyme) at high potential, and the potential (E_{switch}) at which reductive reactivation occurs is more positive for Hyd-1 than Hyd-2. This interconversion, where inactivation is much slower than activation, appears as hysteresis on the cyclic voltammogram. A third factor not evident from these voltammograms is that Hyd-1 is an O_2-tolerant hydrogenase that can operate in the presence of O_2, whereas Hyd-2 undergoes serious inactivation by O_2 that is not easily reversed (oxygenation of cysteine-S coordinated to the Ni atom (Volbeda et al., 2015)). A major reason for the O_2 tolerance of Hyd-1 is the presence of an unusual FeS cluster in the proximal position (closest to the active site): instead of a standard [4Fe–4S] cluster, Hyd-1 has a [4Fe–3S]-6cys cluster that can undergo two successive electron transfers, the additional step being a proton-coupled electron transfer in which one Fe becomes coordinated by a deprotonated peptide-N (Fig. 4C) (Goris et al., 2011; Pandelia et al., 2011; Shomura, Yoon, Nishihara, & Higuchi, 2011; Volbeda et al., 2012). In addition, the reduction potentials of the FeS clusters in Hyd-1 appear to be more positive than those in Hyd-2. Having a proximal cluster able to discharge two electrons, along with a relay system comprising centers with relatively high potentials, helps the enzyme to have enough electrons available for immediate transfer back to the active site when O_2 attacks, thus converting it rapidly to water and avoiding destructive oxygen intermediates (Cracknell, Wait, Lenz, Friedrich, & Armstrong, 2009; Evans et al., 2013). The Ni–B species that results is easily reactivated, as mentioned above.

There are two major reasons why Hyd-2 acts as a reversible electrocatalyst, whereas Hyd-1 does not. One reason is that H_2 is a potent inhibitor of Hyd-1, and the 10% H_2 used in the experiment shown is sufficient to suppress H_2 evolution. The second reason relates to the reduction potentials of the FeS clusters. One part of a model that we proposed to account for reversibility in electron-transport enzymes predicts that the catalytic bias (the tendency to operate preferentially in one direction) is related (see Box 1, Eq. 6) to the difference between the reduction potential of the reaction being catalyzed ($E_{substrate}$) and the reduction potential E_{CS} of the "electrochemical control center" of the enzyme (Hexter, Esterle, & Armstrong, 2012, 2014). The latter is usually the center at which electrons enter or leave the enzyme, which for Hyd-1 and Hyd-2 is the distal [4Fe–4S] cluster. The reason why Hyd-1 does not produce H_2 at neutral pH (even under H_2-free conditions) is that the reduction potential of the distal cluster is too positive relative to $E_{substrate}$ ($E_{H+/H2}$): instead, E_{CS} correlates with the onset of H_2 oxidation. Lowering the pH raises $E_{H+/H2}$ much more than E_{CS} (FeS centers normally having little pH dependence) and the bias gap closes up (Murphy, Sargent, & Armstrong, 2014), so Hyd-1 becomes a proficient H_2 producer at pH 3.

As expected, properties of the FeS clusters of the relay also affect the rate of intramolecular electron transfer from the active site to the distal cluster. This aspect is revealed by examining the voltammetric waveshapes that result when mutations have been carried out to alter the FeS clusters, avoiding either the active site or the distal cluster that is the electron/entry exit site. The results shown in Fig. 4B reveal the decrease in H_2 oxidation efficiency that results when one or two of the FeS centers are altered by site-directed mutagenesis—a medial [3Fe–4S]-to-[4Fe–4S] cluster change (P242C) alone or combined with the proximal [4Fe–3S]-6cys cluster mutated to a conventional [4Fe–4S] cluster (C19G/C120G/P242C), either of which also causes a loss in O_2 tolerance (Evans et al., 2013). The native enzyme shows a strong increase in current above the onset potential, whereas the H_2 oxidation current for the FeS cluster variants responds much more sluggishly to the increase in driving force. We note that there is an interesting, albeit distant, resemblance to a quantity known as the "fill factor" (FF) that is used to judge the efficiency of a photovoltaic device for solar–electricity conversion. An optimal solar cell has a high FF, meaning that a large current is obtained with minimal voltage expenditure. We are unsure as yet as to what advantages were lost upon mutation, optimized reduction potentials or reorganization energies, but the result is fully in accord with evolutionary expectations.

The [FeFe]-hydrogenases use a complex cofactor known as the "H-cluster" to catalyze $2H^+/H_2$ interconversion and they are generally

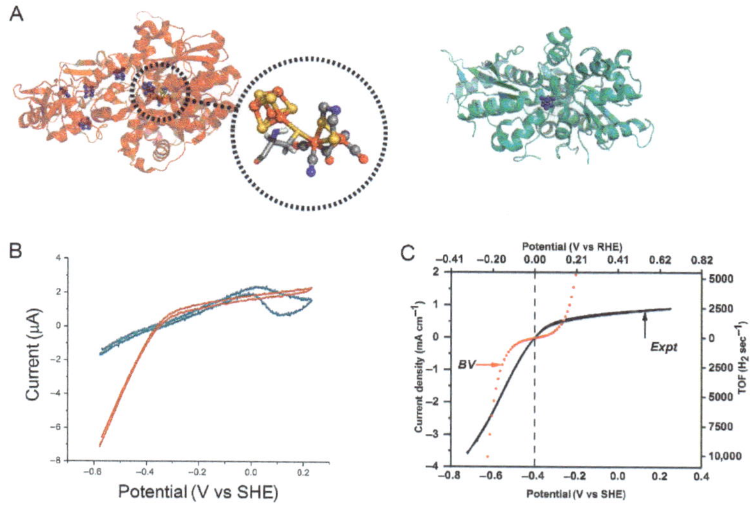

Fig. 5 (A) Structures of two [FeFe]-hydrogenase, Hydrogenase-1 (*Cp*1) from *Clostridium pasteurianum* (lilac, pdb 4XDC) and Hydrogenase A1 (*Cr*HydA1) from *Chlamydomonas reinhardtii* (apoenzyme, *cyan*, pdb 3LX4). Whereas *Cp*1 has a relay system comprising three [4Fe–4S] clusters and an additional distal [2Fe–2S] cluster (shown as *navy stick* and *spheres*), *Cr*HydA1 lacks such a relay. Also shown is the H-cluster, comprised of a [2Fe] complex containing diatomic ligands CO and CN^- and bridging azadithiolate and a [4Fe–4S] cluster. (B) Typical cyclic voltammograms showing the catalytic bias for both *Cp*1 (*maroon*) and *Cr*HydA1 (*teal*). Experiment conditions: 5°C; 50 m*M* sodium phosphate, 0.1 *M* NaCl, pH 6; 100% H_2; 1000 rpm; scan rate $30\,mV\,s^{-1}$. (C) Voltammogram for *Cp*1 scaled after measuring the exchange rate constant, calculating the electroactive coverage, and converting observed current into current density (*left ordinate*) and catalytic turnover frequency, TOF (*right ordinate*). The potential axis at the *top* has also been adjusted to the reversible hydrogen electrode, using the intersection (zero-current potential).

considered to be more active than [NiFe]-hydrogenases. As shown in Fig. 5, the H-cluster consists of two components—(1) an unusual [2Fe] complex, containing diatomic ligands CO, CN^-, and a bridging azadithiolate, which is the site of formation or oxidation of H_2, and (2) a [4Fe–4S] cluster, coordinated by four cysteine thiolates, which can store one electron during catalysis (Fig. 5A, magnified inset). The two components are connected by one of the cysteine thiolate-S atoms which also provides the only covalent link between the [2Fe] complex and the protein. In Fig. 5, we show the structures of two [FeFe]-hydrogenases (Fig. 5A), along with their "signature" electrocatalytic voltammograms measured under 1 bar H_2 at pH 6.0

(Fig. 5B). The hydrogenase *Cp*1 from *Clostridium pasteurianum* has a four-FeS center relay to carry electrons between the H-cluster and the docking area for the cytochrome redox partner, whereas *Cr*HydA1 from the photosynthetic green alga *Chlamydomonas reinhardtii* contains only the H-cluster, electron supply coming directly from a [2Fe–2S] ferredoxin that must bind and dissociate twice for each turnover. The voltammograms overlaid in Fig. 5B suggest that although *Cp*1 is excellent at reducing protons, its oxidation of H_2 is markedly suppressed not far into the potential region that thermodynamically favors this direction. Even so, the steep cut across the zero-current axis shows that it is a highly reversible electrocatalyst (Armstrong & Hirst, 2011). In contrast, the catalytic action of *Cr*HydA1 is evenly balanced, although the broad inflection across the zero-current intersection region suggests that electron transfer is rather sluggish.

To explore electron transfer further, we applied a technique known as electrochemical impedance spectroscopy (EIS), which uses a small AC modulation and measures the complex impedance characteristics of catalytic electron transfer through the enzymes (Pandey et al., 2017). Although details lie outside the scope of this chapter, the experiments resolve two components of catalytic electron transport, the "DC" resistance component R, which is independent of frequency, and the capacitance component C, which is frequency dependent. The origins of R lie mainly in long-range ET through the FeS clusters, whereas the origins of C lie more in the rate of catalytic turnover that draws or delivers electrons. The results show (Fig. 5C) that the potential region appropriate for H_2 evolution by *Cp*1 is one in which the R value is particularly low, reinforcing the casual description that the FeS centers form a "wire," in this case one in which conductivity is highest close to the reduction potentials of these sites (there is evidence that the FeS clusters in *Cp*1 have quite negative potentials) (Adams, Eccleston, & Howard, 1989). It is instructive to consider how electron hopping occurs in such a relay—an interesting analogy being the mineral known as magnetite Fe_3O_4. Magnetite is an inverse spinel in which the octahedral sites contain equal numbers of Fe^{2+} and Fe^{3+}, and it conducts electricity because electrons hop between Fe^{2+} and Fe^{3+} with minimal reorganization energy (Franke, Rosenberg, Whall, & Jones, 1978). It would be expected that a biological electron-hopping relay might also have evolved to promote alternating vacancies, analogous to the $Fe^{2+} \rightarrow Fe^{3+}$. $Fe^{2+} \rightarrow Fe^{3+}$ relays set up naturally in magnetite. Optimized electron flow would thus occur at potentials close to the reduction potentials of the centers (if these are similar in value). Another significant outcome of the EIS experiments was determination of the "catalytic exchange" rate constant—the turnover frequency at zero driving force which cannot otherwise be determined directly. Exchange rate constants of 78 and 13 H_2 molecules s^{-1} were obtained for *Cp*1 and *Cr*HydA1, respectively, reinforcing and quantifying the remarkable activity of these enzymes, which have been compared to platinum catalysts. The electroactive coverage of hydrogenase is too low to be determined, so current cannot be correlated with turnover frequency. However, using Eq. (4), Box 1, the $2H^+/H_2$ exchange rate per active site is easily converted to exchange current and used to construct a profile of the rate–overpotential relationship for *Cp*1, which is shown alongside the current density dependence in Fig. 5C. The FeS relay is so efficient that the current in the region of the equilibrium potential far exceeds that predicted by the Butler–Volmer equation (Box 1).

5. A CASE STUDY IN CONTROL: THE FINAL STAGES OF H-CLUSTER ASSEMBLY IN HYDROGENASES

Much attention has been devoted to the mechanism of the multistep process by which the complex H-cluster is assembled and finally inserted into the protein host to give active [FeFe]-hydrogenase. In living organisms the [2Fe] subcomplex is assembled by two maturation proteins, HydE and HydG, and then transferred to the apoenzyme (in which the [4Fe–4S] cluster is already present) via a third protein, HydF (Berggren et al., 2013; Shepard et al., 2010). In 2013, it was discovered that a synthetic CO adduct of the [2Fe] subcluster $[Fe_2[\mu\text{-}(SCH_2)_2NH](CN)_2(CO)_4]^{2-}$ (abbreviated as $[2Fe]^{MIM}$) reacts directly with the apoenzyme, in the absence of the three maturation enzymes, to yield active holoenzyme (Esselborn et al., 2013). The final stages, represented in Fig. 6, suggested an interesting mechanistic study that would exploit three characteristics of PFE. First, completion of the assembly process, resulting in active enzyme, is observed immediately and directly through the catalytic current that builds up, the rate of change of current reflecting the rate of the activation process. Second, by immobilizing the apoenzyme on an electrode, it is possible to use extremely small amounts of cofactor and remove any remaining in solution after a certain period of time. Third, any steps in the reaction that might be sensitive to the oxidation level of either the [4Fe–4S] cluster or the [2Fe] subcomplex can be addressed by controlling the electrode potential.

The potential dependence of activation of apoenzyme by $[2Fe]^{MIM}$ was investigated using chronoamperometry experiments, the change in current

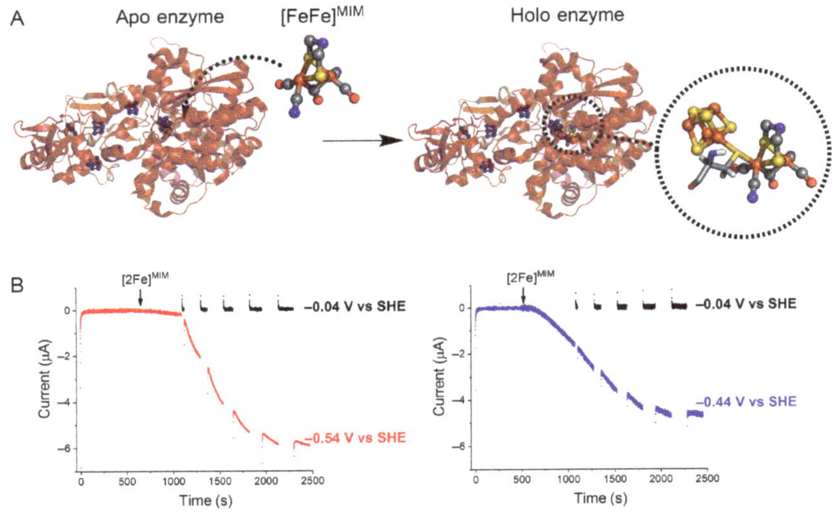

Fig. 6 (A) Cartoon diagram illustrating the activation of apoenzyme (in this instance, *Cp*1) by a synthetic CO adduct of the [2Fe] subcluster $[Fe_2[\mu\text{-}(SCH_2)_2NH](CN)_2(CO)_4]^{2-}$ (abbreviated to $[2Fe]^{MIM}$), represented as *sticks* and *spheres*. $[2Fe]^{MIM}$ reacts directly with the apoenzyme to yield active, holoenzyme in which the [2Fe] complex is connected covalently to a [4Fe4S] subcluster by a cysteine thiolate (μ-SCys) (see *magnified inset*). (B) Chronoamperometry experiments showing how activation of *Cp*1 adsorbed on a PGE electrode by $[2Fe]^{MIM}$ is affected by the electrode potential. *Left*: Electrode potential held at −0.54 V vs SHE (*red*) initially for ~1000 s during which $[2Fe]^{MIM}$ was injected to a final concentration of 600 n*M* at ~500 s. Potential then stepped to −0.04 V vs SHE (*black*) for five separate intervals of increasing lengths of time with steps back to −0.54 V in between to monitor H^+ reduction activity. *Right:* As for above with starting potential held at −0.44 V vs SHE. Experiment conditions: 100% Ar throughout; electrode rotated at 1000 rpm. (H_2 absent during the whole experiment; therefore, no H_2 oxidation current was observed during the steps to −0.04 V.)

relating directly to the rate of assembly of active H-cluster. Some results obtained for *Cp*1 are shown in Fig. 6. The experiment starts with adsorption of a film of apoenzyme on a PGE electrode, after which the electrode is held at a constant potential and a small amount of $[2Fe]^{MIM}$ is introduced to the solution. Simply cycling the electrode potential revealed that activity develops over the course of a few minutes, but a more detailed examination exploited the ability to hold the system at fixed potentials and exchange the cell solution. By holding the electrode potential at −0.54 V vs SHE, no activation was observed, but a brief step to −0.04 V and back to −0.54 V initiated the reaction, which occurs with a half-life of about 100 s. Only one step to high potential was required. Holding the potential instead at −0.44 V revealed that activation was already underway before the step to more positive potential. Importantly also, if $[2Fe]^{MIM}$ was introduced at −0.5 V and the cell solution was completely exchanged, the reaction was initiated as before upon stepping to −0.05 V, proving that an intermediate is formed in which the [2Fe] complex is strongly bound in the apoenzyme but the reaction cannot proceed until the reducing condition is relaxed. A summary of the proposed mechanism involving necessary intermediates P and Q is given in Scheme 1.

A plausible explanation for the ability to use the electrode potential to trap the intermediate P (as P_{red}) is that the cysteine-S bond between the already trapped 2Fe subsite and the [4Fe–4S] cluster cannot form if the [4Fe–4S] cluster is reduced, because the resulting H-cluster would be too electron rich (Megarity et al., 2016). The potential region close to −0.5 V does not correlate with a redox couple of $[2Fe]^{MIM}$ but is close to where the $[4Fe–4S]^{2+/+}$ transition is expected. The so-called 18-electron rule is invoked, as each Fe atom in $[Fe_2[\mu\text{-}(SCH_2)_2NH](CN)_2(CO)_4]^{2-}$

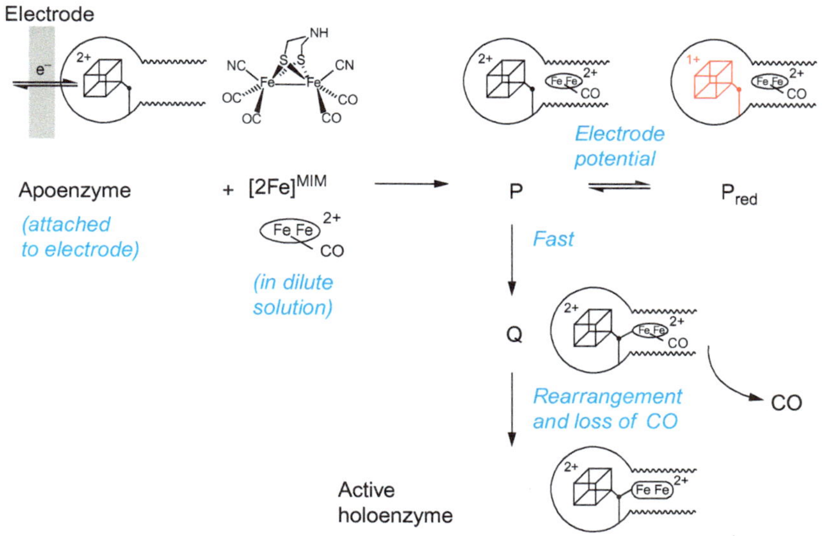

Scheme 1 Proposed mechanism for the order of steps involved in the formation of active H-cluster from the apoenzyme and the synthetic mimic $[2Fe]^{MIM}$. *Reproduced from Megarity, C. F., Esselborn, J., Hexter, S. V., Wittkamp, F., Apfel, U. P., Happe, T., et al. (2016). Electrochemical investigations of the mechanism of assembly of the active-site H-cluster of [FeFe]-hydrogenases.* Journal of the American Chemical Society, 138*(46), 15227–15233. doi:10.1021/jacs.6b09366 with permission of the American Chemical Society.*

formally has 17 valence electrons if we ignore the likelihood of a metal–metal bond. Formation of an additional μ-SCys bond to one Fe (which adds 2 electrons to its valence shell) can be tolerated if the [4Fe–4S] cluster is oxidized (2+) because 1 electron can be shunted back to the cluster, yielding 18 and 17 valence electrons for proximal and distal Fe atoms, respectively. When the [4Fe–4S] cluster is reduced (1+), such a shunt is not possible and an antibonding orbital on the proximal Fe would need to be occupied—the only other option being to transfer an electron to the distal Fe. But this transfer would result in a Fe(0) site, which is unlikely given the presence of cyanide and thiolate ligands that are good donors and therefore disfavor such an electron-rich metal site. Fusion of the 2Fe and [4Fe–4S] cluster to give intermediate Q is irreversible (accounting for why only a single step to a potential > -0.4 V is required to initiate assembly). Formation of the final active state requires dissociation of a CO ligand, and it is possible that this is the rate-determining step.

6. SUMMARY

The small selection of systems we have discussed should give a good idea of the interesting insight into FeS clusters that can be obtained using PFE. We started by describing experiments conducted on small electron-transfer proteins, for which the electroactive coverage is high enough to observe discrete peak-like signals. For larger protein molecules such as enzymes, the coverage is usually too low to see such signals; instead, we exploit the fact that high catalytic activity greatly amplifies the current, and the action of FeS clusters is now observable during turnover. The abilities to control an active site's reactivity through the electrode potential, to immobilize a tiny quantity of sample and subject it to different conditions, and to monitor reactions simultaneously together make PFE a very powerful tool for discovery and detailed investigation. In enzymes, PFE reveals interesting characteristics of electron flow along FeS relays and shows, as with a real wire, how the rate (current) is controlled by potential.

ACKNOWLEDGMENTS

Research highlighted in this work has been supported over numerous years by the UK Biotechnology and Biological Sciences Research Council. F.A.A. is a Royal Society Wolfson Research Merit Award holder.

REFERENCES

Adams, M. W. W., Eccleston, E., & Howard, J. B. (1989). Iron-sulfur clusters of hydrogenase-1 and hydrogenase-2 of Clostridium pasteurianum. *Proceedings of the National Academy of Sciences of the United States of America, 86*(13), 4932–4936. https://doi.org/10.1073/pnas.86.13.4932.

Alonso-Lomillo, M. A., Rüdiger, O., Maroto-Valiente, A., Velez, M., Rodriguez-Ramos, I., Munoz, F. J., et al. (2007). Hydrogenase-coated carbon nanotubes for efficient H_2 oxidation. *Nano Letters*, 7(6), 1603–1608. https://doi.org/10.1021/nl070519u.

Armstrong, F. A., & Hirst, J. (2011). Reversibility and efficiency in electrocatalytic energy conversion and lessons from enzymes. *Proceedings of the National Academy of Sciences of the United States of America, 108*(34), 14049–14054. https://doi.org/10.1073/pnas.1103697108.

Armstrong, F. A., & Wilson, G. S. (2000). Recent developments in faradaic bioelectrochemistry. *Electrochimica Acta, 45*(15–16), 2623–2645. https://doi.org/10.1016/s0013-4686(00)00342-x.

Bard, A. J., & Faulkner, L. R. (2001). *Electrochemical methods: Fundamentals and applications*. Wiley.

Berggren, G., Adamska, A., Lambertz, C., Simmons, T. R., Esselborn, J., Atta, M., et al. (2013). Biomimetic assembly and activation of [FeFe]-hydrogenases. *Nature, 499*(7456), 66–69. https://doi.org/10.1038/nature12239.

Butt, J. N., Armstrong, F. A., Breton, J., George, S. J., Thomson, A. J., & Hatchikian, E. C. (1991). Investigation of metal-ion uptake reactivities of [3Fe-4S] clusters in proteins: Voltammetry of coadsorbed ferredoxin-aminocyclitol films at graphite electrodes and spectroscopic identification of transformed clusters. *Journal of the American Chemical Society, 113*(17), 6663–6670. https://doi.org/10.1021/ja00017a045.

Butt, J. N., Fawcett, S. E. J., Breton, J., Thomson, A. J., & Armstrong, F. A. (1997). Electrochemical potential and pH dependences of 3Fe-4S <-> M3Fe-4S cluster transformations (M = Fe, Zn, Co, and Cd) in ferredoxin III from Desulfovibrio africanus and detection of a cluster with M = Pb. *Journal of the American Chemical Society, 119*(41), 9729–9737. https://doi.org/10.1021/ja971403a.

Chen, K. S., Hirst, J., Camba, R., Bonagura, C. A., Stout, C. D., Burgess, B. K., et al. (2000). Atomically defined mechanism for proton transfer to a buried redox centre in a protein. *Nature, 405*(6788), 814–817. https://doi.org/10.1038/35015610.

Cracknell, J. A., Wait, A. F., Lenz, O., Friedrich, B., & Armstrong, F. A. (2009). A kinetic and thermodynamic understanding of O_2 tolerance in [NiFe]-hydrogenases. *Proceedings of the National Academy of Sciences of the United States of America, 106*, 20681–20686. https://doi.org/10.1073/pnas.0905959106.

Duff, J. L. C., Breton, J. L. J., Butt, J. N., Armstrong, F. A., & Thomson, A. J. (1996). Novel redox chemistry of 3Fe-4S clusters: Electrochemical characterization of the all-Fe(II) form of the 3Fe-4S cluster generated reversibly in various proteins and its spectroscopic investigation in Sulfolobus acidocaldarius ferredoxin. *Journal of the American Chemical Society, 118*(36), 8593–8603. https://doi.org/10.1021/ja961465l.

Esselborn, J., Lambertz, C., Adamska-Venkatesh, A., Simmons, T., Berggren, G., Nothl, J., et al. (2013). Spontaneous activation of [FeFe]-hydrogenases by an inorganic 2Fe active site mimic. *Nature Chemical Biology, 9*(10), 607–609. https://doi.org/10.1038/nchembio.1311.

Evans, R. M., Parkin, A., Roessler, M. M., Murphy, B. J., Adamson, H., Lukey, M. J., et al. (2013). Principles of sustained enzymatic hydrogen oxidation in the presence of oxygen—The crucial influence of high potential Fe-S clusters in the electron relay of [NiFe]-hydrogenases. *Journal of the American Chemical Society, 135*(7), 2694–2707. https://doi.org/10.1021/ja311055d.

Franke, H., Rosenberg, M., Whall, T. E., & Jones, M. R. B. (1978). Low-temperature Mössbauer spectra, electron transfer and spin density distribution in substituted maganetite. *Journal of Magnetism and Magnetic Materials*, 7(1–4), 223–229. https://doi.org/10.1016/0304-8853(78)90189-0.

Goris, T., Wait, A. F., Saggu, M., Fritsch, J., Heidary, N., Stein, M., et al. (2011). A uniques iron-sulfur cluster is crucial for oxygen tolerance of a [NiFe]-hydrogenase (vol 7, pg, 310, 2011). *Nature Chemical Biology*, 7(9), 648.

Hexter, S. V., Esterle, T. F., & Armstrong, F. A. (2014). A unified model for surface electrocatalysis based on observations with enzymes. *Physical Chemistry Chemical Physics*, *16*(24), 11822–11833. https://doi.org/10.1039/c3cp55230f.

Hexter, S. V., Grey, F., Happe, T., Climent, V., & Armstrong, F. A. (2012). Electrocatalytic mechanism of reversible hydrogen cycling by enzymes and distinctions between the major classes of hydrogenases. *Proceedings of the National Academy of Sciences of the United States of America*, *109*(29), 11516–11521. https://doi.org/10.1073/pnas.1204770109.

Hirst, J., & Armstrong, F. A. (1998). Fast-scan cyclic voltammetry of protein films on pyrolytic graphite edge electrodes: Characteristics of electron exchange. *Analytical Chemistry*, *70*(23), 5062–5071. https://doi.org/10.1021/ac980557l.

Hirst, J., Duff, J. L. C., Jameson, G. N. L., Kemper, M. A., Burgess, B. K., & Armstrong, F. A. (1998). Kinetics and mechanism of redox-coupled, long-range proton transfer in an iron-sulfur protein. Investigation by fast-scan protein-film voltammetry. *Journal of the American Chemical Society*, *120*(28), 7085–7094. https://doi.org/10.1021/ja980380c.

Krishnan, S., & Armstrong, F. A. (2012). Order-of-magnitude enhancement of an enzymatic hydrogen-air fuel cell based on pyrenyl carbon nanostructures. *Chemical Science*, *3*(4), 1015–1023. https://doi.org/10.1039/c2sc01103d.

Laviron, E. (1979). General expression of the linear potential sweep voltammogram in the case of diffusionless electrochemical systems. *Journal of Electroanalytical Chemistry*, *101*(1), 19–28. https://doi.org/10.1016/s0022-0728(79)80075-3.

Léger, C., Elliott, S. J., Hoke, K. R., Jeuken, L. J. C., Jones, A. K., & Armstrong, F. A. (2003). Enzyme electrokinetics: Using protein film voltammetry to investigate redox enzymes and their mechanisms. *Biochemistry*, *42*(29), 8653–8662. https://doi.org/10.1021/bi034789c.

Lubitz, W., Ogata, H., Rüdiger, O., & Reijerse, E. (2014). Hydrogenases. *Chemical Reviews*, *114*(8), 4081–4148. https://doi.org/10.1021/cr4005814.

Lukey, M. J., Parkin, A., Roessler, M. M., Murphy, B. J., Harmer, J., Palmer, T., et al. (2010). How Escherichia coli is equipped to oxidize hydrogen under different redox conditions. *Journal of Biological Chemistry*, *285*(6), 3928–3938. https://doi.org/10.1074/jbc.M109.067751.

Megarity, C. F., Esselborn, J., Hexter, S. V., Wittkamp, F., Apfel, U. P., Happe, T., et al. (2016). Electrochemical investigations of the mechanism of assembly of the active-site H-cluster of [FeFe]-hydrogenases. *Journal of the American Chemical Society*, *138*(46), 15227–15233. https://doi.org/10.1021/jacs.6b09366.

Murphy, B. J., Sargent, F., & Armstrong, F. A. (2014). Transforming an oxygen-tolerant NiFe uptake hydrogenase into a proficient, reversible hydrogen producer. *Energy & Environmental Science*, 7(4), 1426–1433. https://doi.org/10.1039/c3ee43652g.

Pandelia, M.-E., Nitschke, W., Infossi, P., Giudici-Orticoni, M.-T., Bill, E., & Lubitz, W. (2011). Characterization of a unique [FeS] cluster in the electron transfer chain of the oxygen tolerant [NiFe] hydrogenase from Aquifex aeolicus. *Proceedings of the National Academy of Sciences of the United States of America*, *108*(15), 6097–6102. https://doi.org/10.1073/pnas.1100610108.

Pandey, K., Islam, S. T. A., Happe, T., & Armstrong, F. A. (2017). Frequency and potential dependence of reversible electrocatalytic hydrogen interconversion by [FeFe]-hydrogenases. *Proceedings of the National Academy of Sciences of the United States of America*, *114*(15), 3843–3848. https://doi.org/10.1073/pnas.1619961114.

Roessler, M. M., Evans, R. M., Davies, R. A., Harmer, J., & Armstrong, F. A. (2012). EPR spectroscopic studies of the Fe-S clusters in the O-2-tolerant [NiFe]-hydrogenase Hyd-1 from Escherichia coli and characterization of the unique 4Fe-3S cluster by HYSCORE. *Journal of the American Chemical Society*, *134*(37), 15581–15594. https://doi.org/10.1021/ja307117y.

Shepard, E. M., McGlynn, S. E., Bueling, A. L., Grady-Smith, C. S., George, S. J., Winslow, M. A., et al. (2010). Synthesis of the 2Fe subcluster of the [FeFe]-hydrogenase H cluster on the HydF scaffold. *Proceedings of the National Academy of Sciences of the United States of America*, *107*(23), 10448–10453. https://doi.org/10.1073/pnas.1001937107.

Shomura, Y., Yoon, K.-S., Nishihara, H., & Higuchi, Y. (2011). Structural basis for a [lsqb] 4Fe-3S[rsqb] cluster in the oxygen-tolerant membrane-bound [lsqb]NiFe[rsqb]-hydrogenase. *Nature*, *479*(7372), 253–256.

Volbeda, A., Amara, P., Darnault, C., Mouesca, J. M., Parkin, A., Roessler, M. M., et al. (2012). X-ray crystallographic and computational studies of the O-2-tolerant [NiFe]-hydrogenase 1 from Escherichia coli. *Proceedings of the National Academy of Sciences of the United States of America*, *109*(14), 5305–5310. https://doi.org/10.1073/pnas.1119806109.

Volbeda, A., Martin, L., Barbier, E., Gutierrez-Sanz, O., De Lacey, A. L., Liebgott, P. P., et al. (2015). Crystallographic studies of [NiFe]-hydrogenase mutants: Towards consensus structures for the elusive unready oxidized states. *Journal of Biological Inorganic Chemistry*, *20*(1), 11–22. https://doi.org/10.1007/s00775-014-1203-9.

CHAPTER FOURTEEN

NRVS for Fe in Biology: Experiment and Basic Interpretation

Leland B. Gee[1,2], Hongxin Wang, Stephen P. Cramer
University of California, Davis, Davis, CA, United States
[1]Corresponding author: e-mail address: lbgee@stanford.edu

Contents

Abstract

For over 20 years, nuclear resonance vibrational spectroscopy (NRVS) has been used to study vibrational dynamics of iron-containing materials. With the only selection rule being iron motion, ^{57}Fe NRVS has become an excellent tool to study iron-containing enzymes. Over the past decade, considerable progress has been made in the study of complex metalloenzymes using NRVS. Iron cofactors in heme-containing globins; [2Fe2S], [3Fe4S], [4Fe4S] proteins; the [NiFe] and [FeFe] hydrogenases; and nitrogenases have been explored in a fashion not possible through traditional vibrational spectroscopy. In this chapter, we discuss the basics of NRVS, a strategy to perform NRVS, and a discussion of the application of NRVS on rubredoxin and [FeFe] hydrogenase.

1. INTRODUCTION

Nuclear resonance vibrational spectroscopy (NRVS) has become a staple technique in the study of Mössbauer-active isotopes since its inception

[2] *Current address*: Department of Chemistry, Stanford University, Stanford, CA, United States.

Methods in Enzymology, Volume 599
ISSN 0076-6879
https://doi.org/10.1016/bs.mie.2017.11.002

over two decades ago (Seto, Yoda, Kikuta, Zhang, & Ando, 1995; Sturhahn et al., 1995). The synchrotron-based technique has particularly enjoyed success with ^{57}Fe, stemming from favorable nuclear parameters that align with contemporary synchrotron technology. Originally, NRVS was used for materials sciences and enzyme model complexes, but in recent years the method has been expanded as a probe of metallocofactors in biological systems.

NRVS is uniquely suited for studying low-frequency Fe-based vibrational modes in biological systems and can solve questions of metal–ligand bond strength and ligand identity. Ultimately, the power of NRVS is derived directly from its specificity, selection, and sensitivity. Fe is ubiquitous in biological metallocofactors either as part of electron transfer or catalysis. Studying these cofactors with traditional vibrational spectroscopy, such as infrared (IR) or Raman spectroscopy, is impeded by the background created by the biomolecule proper or even fluorescence of the cofactor. As such, the extreme specificity of ^{57}Fe NRVS for only ^{57}Fe motion makes it the perfect tool to study vibrational dynamics of the critical Fe-containing components of biomolecules. Without contribution from the biological sample background, NRVS enjoys minimal spectroscopic noise; for example, modes of Fe hydrides within enzymes (Ogata, Kramer, et al., 2015; Reijerse et al., 2017) have been observed which are typically difficult to directly identify with other techniques (Kaesz & Saillant, 1972; Rosenberg, 1989). The lack of conventional selection rules allows for ^{57}Fe NRVS to reveal both IR- and Raman-active modes that contain ^{57}Fe motion—thus providing a complete picture for Fe vibration.

NRVS is related to, and descended from, Mössbauer spectroscopy, which relies on disconnecting the vibrational motions from a Mössbauer nuclear excitation to learn about the electronic structure of the atom—a recoilless excitation (Mössbauer, 1958a, 1958b, 1959). The logical antithesis of this idea was to realize and differentiate vibrational states that are created or annihilated upon nuclear excitation (Singwi & Sjölander, 1960; Visscher, 1960). Mössbauer spectroscopy is achieved by mounting a decaying parent isotope to a velocity transducer to sweep an energy range by Doppler effect. When the energy matches a nuclear transition of the daughter isotope in the sample, a nuclear resonance excitation occurs. However, the energy resonances in ^{57}Fe Mössbauer spectroscopy are typically on the order of neV and vibrational quanta observed through NRVS are on the order of meV. This implies a Doppler-shifted source would need to move with a velocity 6 orders of magnitude faster than a typical ^{57}Fe Mössbauer experiment in order to perform ^{57}Fe NRVS which is widely considered impractical. To access the ^{57}Fe vibrational modes, a different source with high flux, high monochromation, and high tuneability is necessary. Fortunately, beamlines at third-generation high-energy synchrotron X-ray light sources meet these requirements and make NRVS possible (Wang, Alp, Yoda, & Cramer, 2014).

2. MATERIALS

As described earlier, NRVS requires the use of a synchrotron radiation light source. Few synchrotron beamlines can accommodate the challenging NRVS requirements for flux, monochromation, and time-sensitive detection. As such, an NRVS experiment can only be performed by applying for beamtime at four synchrotron radiation light sources: The Advanced Photon Source (APS—http://www.aps.anl.gov), The Positron-Electron Tandem Ring Accelerator (PETRA-III—http://petra3.desy.de), The European Synchrotron Radiation Facility (ESRF—http://www.esrf.eu), and the Super Photon Ring—8 GeV (SPring-8—http://www.spring8.or.jp/en/).

Despite differences in synchrotron facilities, the principle setup of an NRVS experiment remains the same. Charged particles circulating at relativistic speeds within an ultrahigh vacuum (UHV) storage ring pass through a set of magnets called an undulator that generates a tangential beam of photons, called synchrotron radiation, that covers the energy of nuclear resonance (14.4 keV for ^{57}Fe). The "white light" beam ($\Delta E \approx 100$ eV) travels down a UHV transport channel, with optics to aperture and focus the beam, toward the target beamline and is initially monochromated by a high heat load monochromator ($\Delta E \approx 1$ eV). The incident light is then monochromated by a tuneable high-resolution monochromator ($\Delta E \approx 1$ meV) composed of Si or Ge crystals. Finally, the beam interacts with the sample within a cryostat or cryojet stream.

After excitation, ^{57}Fe within the sample will relax either through direct nuclear relaxation (photon energy $\approx$14.4 keV photon) or internal conversion followed by, primarily, Kα emission (photon energy $\approx$6.4 keV photon) with the internal conversion process favored 8:1. Both relaxation channels are relevant to NRVS and must be detected. The detection system is time resolved and designed to exclude early-time electronic scattering events while focusing on photons resulting from the ^{57}Fe relaxation (with a $\tau = 141$ ns). The sophisticated detection scheme is made possible by the nanosecond response avalanche photodiode (APD) detector analyzed by a constant fraction discriminator (CFD). Multiple "bunches," or groups, of particles within a storage ring are spaced throughout the period of the synchrotron such that photons arrive at the beamline in a periodic pulsed pattern (the parameters of which define a "bunch mode"). As such, $t=0$ must be defined to the CFD for each bunch—a process synchronized by the "bunch clock." The CFD then vetoes photon counts 10–20 ns around $t=0$—giving time for the APD to recover from the prompt (incident) pulse of photons and isolating the NRVS emission signal from electronic scattering. Collectively, the properly implemented setup typically allows for dark count rates of less than 5×10^{-3} per second.

Given the low natural abundance of ^{57}Fe (2.14%), ^{57}Fe NRVS requires that biological samples be enriched with ^{57}Fe. Generally for bioinorganic applications, enrichment can be achieved by dissolving elemental ^{57}Fe in *aqua regia* and adding to an Fe-free growth medium. Likewise, ^{57}Fe-enriched salts (e.g., $^{57}FeCl_3$ or $^{57}FeSO_4$) can provide an avenue for integration to the protein or enzyme of interest. To study specific Fe atoms or clusters out of many within a protein—care must be taken to integrate ^{57}Fe by targeting steps in the enzyme maturation or reintroducing the enriched cofactor of interest to an unenriched apoprotein.

Although some variants exist, a typical ^{57}Fe NRVS sample cell for proteins has dimensions of $10 \times 3 \times 1$ mm. These parameters take into account the dimensions of the synchrotron X-ray beam, the penetration depth of the beam, the solid angle projection from the sample to the APD, and the escape depth of the Fe emission following internal conversion. The cell is sealed liquid-tight with an adhesive tape and a port is available in the cell to inject sample solution, followed by subsequent freezing in liquid nitrogen. The cell and tape should be made from cryocompatible materials that do not interact with the sample nor the X-ray beam. A commonly used sample cell material is poly(methyl methacrylate) (often referred to by the eponym "Lucite") and a common tape material can be polyimide (eponymously "Kapton Tape"). When in doubt, materials can be checked for their X-ray transmission properties at http://henke.lbl.gov/optical_constants/filter2.html.

NRVS experiments for bioinorganic applications are often performed at cryogenic temperatures with the sample loaded onto a copper cold finger within a helium cryostat. Although, typically the nominal temperature reading in the cryostat may be near helium temperatures <10 K, the actual sample temperature can vary widely based on the sample properties and the solvents used to marry the sample cell to the cold finger. Due to favorable physical properties, particularly a melting point of 147 K, 1-propanol is considered one of the best mounting liquids (Wang, Yoda, Kamali, Zhou, & Cramer, 2012). The true sample temperature can be obtained by fitting the ratio of phonon annihilation events to creation events using the PHOENIX software and can be approximated by a Maxwell–Boltzmann distribution. To ensure the temperature can be approximated, a standard NRVS spectrum is usually collected slightly below the elastic resonance at $-180\ cm^{-1}$ (annihilation of phonons), through the elastic resonance, and at energies higher than the resonance into the region of interest (creation of phonons) (Fig. 1).

It is generally favorable to analyze the spectrum in terms of partial vibrational density of states (PVDOS) to isolate single phonon events and remove their energy dependence. Sturhahn created a program called PHOENIX that is the standard to convert raw NRVS data to PVDOS (among many other features not discussed here) (Sturhahn, 2000) using the Fourier-log method (Johnson & Spence, 1974). The program can be downloaded from: http://www.nrixs.com/downloads.html. Alternatively, data can be converted and analyzed through a web-based graphical user interface for PHOENIX through http://www.spectra.tools/.

The transformation into PVDOS, which is proportional to the normal mode composition factors for ^{57}Fe, allows for a straightforward prescription to estimate the PVDOS intensity through normal mode analysis using density functional theory or other in silico methods; the normal mode composition factors and normal mode energies can be extracted from diagonalization of the Hessian force matrix used in typical computational normal mode analysis (Petrenko, Sturhahn, & Neese, 2008).

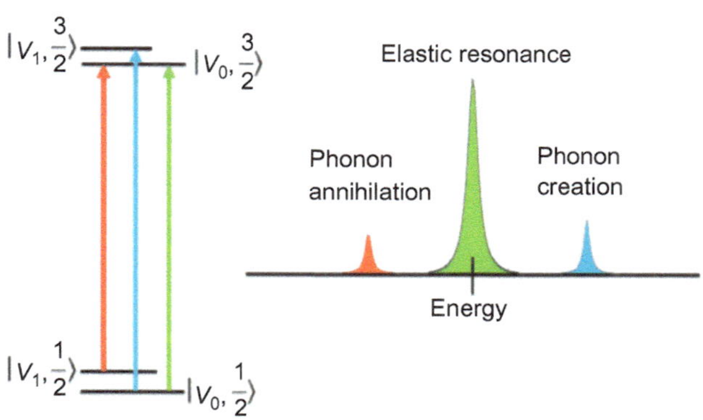

Fig. 1 A simplification of the NRVS processes. The *green line* and *peak shape* contain the recoilless Mössbauer transition that is beyond the resolving power of NRVS. The *red* and *blue peaks* correspond to phonon annihilation and creation, respectively—and the ratio of their intensity is temperature dependent.

2.1 Calibration

The monochromator crystals are very sensitive to the experimental environment specifically temperature. As such, the proportion between expected energy and true energy of monochromation can vary between experiments. It is ideal to calculate a linear "calibration factor" for the monochromator using a secondary standard that has been analyzed using a technique with a reliable energy domain. Typically, this can be handled by small ^{57}Fe compounds such as: $(Et_4N)^{57}FeCl_4$ or $[MgFe(CN)_6]^{2-}$.

For $(Et_4N)^{57}FeCl_4$, assuming an idealized T_d point group, two T_2 modes can be observed with maxima in the FT-IR-calibrated NRVS spectrum at 378 and 139 cm^{-1} (asymmetric Fe—Cl stretching modes and Fe—Cl bending modes, respectively)—as well as mixed optical and acoustic modes grouped as "lattice modes" below 60 cm^{-1} (Fig. 2) (Smith et al., 2005). Notably absent from the NRVS spectrum are the A_1 Fe—Cl symmetric stretching mode and the E, Fe—Cl symmetric bend as these modes do not contain Fe motion. A user need only collect an NRVS spectrum for $(Et_4N)^{57}FeCl_4$ and align the T_2 peaks to the previously calibrated positions with a linear factor and apply this factor to the energy axes of their subsequent spectra.

Interpretation of NRVS data generally involves fundamental vibrational analysis, particularly the practice of isotopic substitution, group theory, and empirical comparison to identify spectral features. A tool to assist in analysis, called Vibratz (Dowty, 1987), allows the user to fit a Urey–Bradley forcefield to their spectrum based on a set of internal coordinate force constants. The user may also define the point group for the molecule of interest

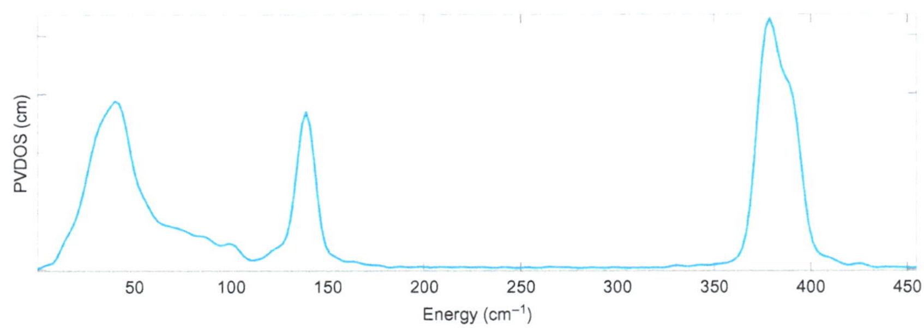

Fig. 2 The NRVS spectrum of $(et_4N)^{57}FeCl_4$. The low energy intensity is related to lattice phonons and the features at 378 and 139 cm^{-1} can be approximated as T_2 modes in T_d symmetry. The energies of these modes are used to calibrate NRVS spectra. Noticeably absent is the A_1 symmetric stretch as that mode does not involve ^{57}Fe motion.

to connect the irreducible representations of the point group to the features of the spectrum. This type of analysis may lead to new interpretations about the geometry of the molecule of interest when minimal structural data is available.

3. PROCEDURE

This procedural outline will describe a standard process of collecting and analyzing bioinorganic NRVS data using the spectra.tools website.

a. Obtain beamtime at one of the above listed synchrotron beamlines where NRVS is available.

b. Follow established protocols to produce the enzyme of interest in an Fe-free media.

c. Add ^{57}Fe as either a pure powder or salt to the media.

d. Follow established protocols to purify the enzyme of interest.

  i. A typical concentration goal is at least 1 mM of ^{57}Fe in the final NRVS sample, but this depends on the intensity of the signals of interest.

e. Obtain a three walled 10 × 3 × 1 mm cell made of Lucite with an injection port milled through the top.

f. Wrap the cell, liquid tight, in Kapton tape that is at most 1 mil in thickness.

  i. Take care to minimize and flatten the tape on the back of the cell—as this will interfere with thermal contact to a cryostat cold finger.

g. Inject the purified enriched enzyme into the cell.

  i. For air-sensitive proteins, the sample should be injected into the cell anaerobically then immediately frozen in liquid nitrogen.

h. If the synchrotron is not local, ship a dry, liquid nitrogen-cooled shipping dewar containing the sample(s) to the synchrotron. Then travel to the synchrotron.

i. Assuming the cryogenics are already established at the beamline, at a nominal temperature of 110–120 K vent the sample space vacuum with helium and remove the cold finger housing.

j. Apply 1-propanol to the cold finger surface. Generally, a thin layer is required to establish good thermal contact—too much will cause poor contact.

k. Attach the NRVS cell to the 1-propanol on the cold finger, with attention paid that the "sample side" is "up" or in position to receive the incident X-ray.

- **l.** Reattach the cold finger housing, turn off the venting gas, reestablish a vacuum in the sample space, and allow the cryostat to cool.
- **m.** At a nominal temperature of 10 K, begin data collection.
  - **i.** This step is beamline dependent, and the local beamline scientist should be able to assist.
  - **ii.** If specific regions of the spectrum are of interest, the beamline scientist should be able to perform specialized region of interest scans.
  - **iii.** As data workup later depends on the elastic region, even region of interest scans need to contain the elastic peak, but the acquisition time for the region of interest will be much higher.
- **n.** Once the data are satisfactory the sample may be switched following the steps above and after warming up to 147 K (e.g., for 1-propanol). If a thin even layer of 1-propanol was applied, most of it will come off with the sample cell—otherwise the surface may need to be cleaned slightly with a lab tissue.
- **o.** Depending on the beamline, there may be inhouse programs/methods to analyze data. Otherwise navigate a web browser to http://www.spectra.tools/.
- **p.** At the navigation bar toward the top, under the drop down menu for NRVS, click "NRVS Spectra Processing Tool."
- **q.** In the "Data Upload" window, select the data source type (supported currently: SPring-8, APS .mda files, and PETRA-III). Click "Choose Files" and navigate to the folder that contains the NRVS scan files—on most systems SHIFT + DRAG (or holding CTRL) can be used to highlight multiple files. Then select the submit button.
- **r.** A three-panel window will appear:
  - **i.** The left panel (Parameters) allows the user to alter different NRVS-related parameters. By default, the parameters listed are the most commonly changed, but advanced parameters can be altered by clicking "Advanced Submit."
  - **ii.** The central window displays the PVDOS spectrum—by popular demand it is displayed as a function of wavenumbers. The name of the final file processed is the title of the graph. The graph is zoomable and exportable to image formats. Error bars calculated by PHOENIX are shown and can be turned off by clicking "Error Bars." To obtain the complete data (including everything required by PHOENIX) the "Download" button in the left panel is used (Fig. 3).

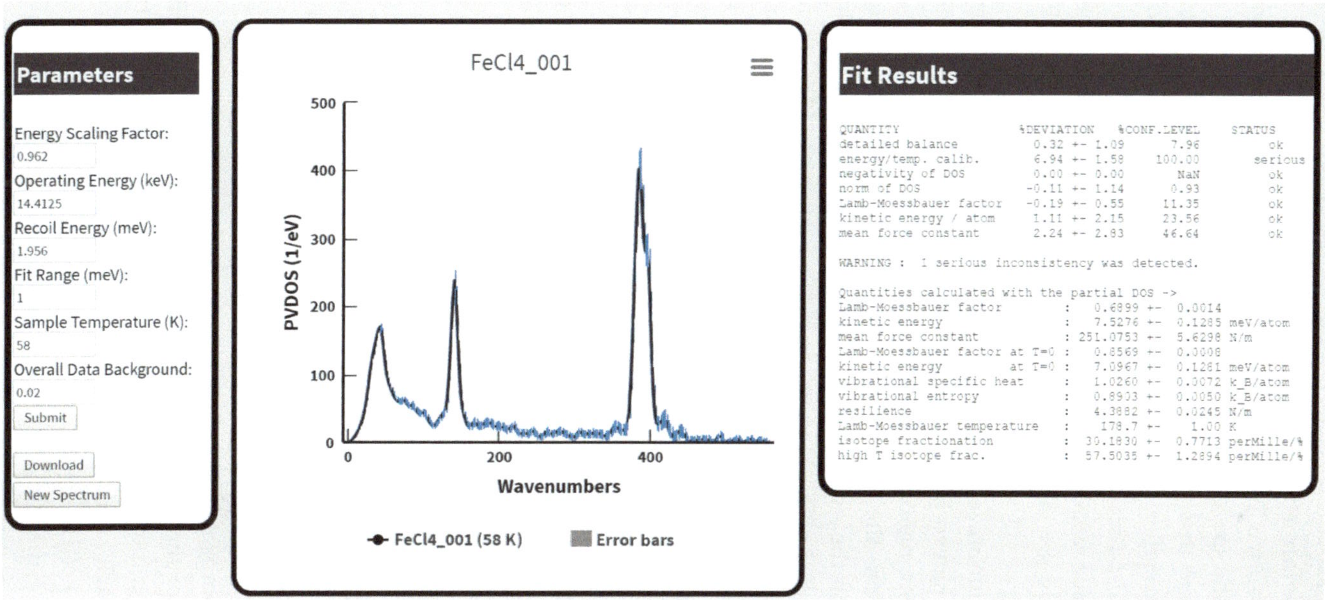

Fig. 3 The output window of the spectra.tools NRVS data processing. Here, parameters (on the *left*) can be adjusted to achieve desirable statistics (on the *right*). The graph in the center can be exported in image formats and the data downloaded by the "Download" button in the parameters window.

  iii. The right panel (Fit Results) indicates the results of the statistical analysis performed by PHOENIX.

s. Once the spectrum is loaded, adjust the parameters on the left-hand side and resubmit the spectrum for analysis.

  i. First, ensure that the "Energy Scaling Factor" is appropriate. As mentioned earlier, this factor is determined by collecting the NRVS spectrum for a secondary standard particularly $(Et_4N)^{57}FeCl_4$. The normal value ranges are from 0.92 to 1.10, are beamline dependent, and should be determined at least once during each NRVS beamtime.

  ii. Second, minimize the background contribution to the spectrum. This is achieved by increasing "Overall Data Background" until the signal in regions with no expected signal is minimized.

  iii. Third, alter the "Sample Temperature" until the "detailed balance" in the right hand "Fit Results" window is acceptable.

t. Ideally, once all the "Fit Results" indicate a status of "OK" then the spectral parameters are acceptable. Occasionally this is an unattainable goal, especially at a very high number of scans or if there were significant changes in sample temperature during the experiment. In this case, each scan should be carefully examined.

u. If everything is satisfactory click the "Download" button in the parameters window. A download of a compressed folder should begin. The file in the folder with the suffix *.dos will contain the final PVDOS.

4. APPLICATION

4.1 Rubredoxin

Rubredoxins are small electron transfer proteins. Rubredoxin from *Pyrococcus furiosus* is composed of 53 amino acid residues for nearly 6 kDa in mass. Electron transfer is handled by a single Fe redox center coordinated to four cysteinyl thiolates. The two redox states are formally Fe(II) and Fe(III). The simplicity and small size made rubredoxin an excellent early candidate for bioinorganic ^{57}Fe NRVS (Xiao et al., 2005).

The spectrum for the oxidized and reduced proteins is shown in Fig. 4. There are three prominent features near 70, 150, and 360 for the Fe(III) site in the oxidized sample, and the high frequency features shift to 303 cm^{-1} upon reduction to Fe(II).

Naively speaking, the initial assumption is that the three features are analogous to the T_d symmetry $FeCl_4$ modes discussed earlier (Fig. 2). However,

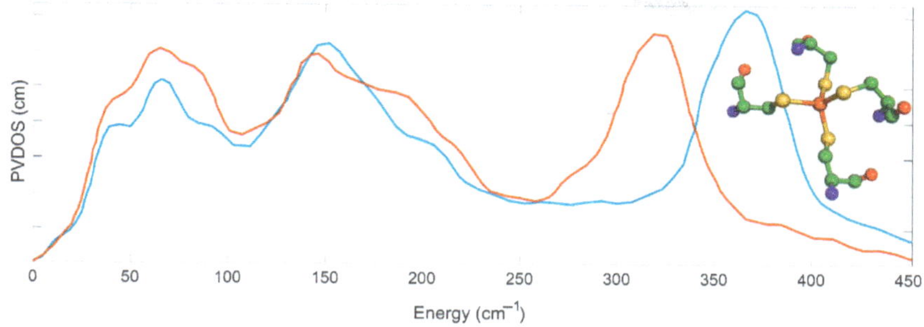

Fig. 4 The NRVS spectrum of oxidized (---) and reduced (---) rubredoxin in solution. *Inset*: structure of the Fe site from PDB: 1CAD. *Figure reproduced from Xiao, Y., Wang, H., George, S. J., Smith, M. C., Adams, M. W., & Jenney, F. E., Jr. (2005). Normal mode analysis of* Pyrococcus furiosus *rubredoxin via nuclear resonance vibrational spectroscopy (NRVS) and resonance Raman spectroscopy.* Journal of the American Chemical Society, 127*(42), 14,596–14,606. https://doi.org/10.1021/ja042960h.*

the broadening and obvious shouldering of the three distinct features in the spectrum (Fig. 4) indicate that interpretation under presumption of a perfect T_d symmetry, as with $(Et_4N)^{57}FeCl_4$, is not appropriate. To fit the internal coordinate force constants to the collected spectrum, it is effective to use a descent in symmetry ($T_d \rightarrow D_{2d} \rightarrow C_1$) while increasing the complexity of the Vibratz model ($Fe(SC)_4 \rightarrow Fe(SCC)_4 \rightarrow Fe(SCys)_4$). For the final model incorporating four cysteines, Xiao and coworkers used the geometries from the pertinent crystal structures (reduced PDB: 1CAD (Day et al., 1992) and oxidized PDB: 1BQ8 (Bau et al., 1998)). This iterative treatment allowed for a reasonable fit of the internal coordinate force constants. The results of modeling the NRVS spectra were a calculated 36% decrease in Fe—S bond force constant (1.24 vs 0.92 mdyne/Å) upon reduction of the protein.

The Vibratz program also describes the motion of the components of the fitted spectrum. For the oxidized protein model, the broad region from 345 to 375 cm^{-1} is composed of three strong components at 375, 358, 350 cm^{-1} and three weaker contributions at 365, 364, and 340 cm^{-1} that are principally described as Fe—S asymmetric stretching modes. The intensity in the region from 300 to 340 cm^{-1} arises from nearly symmetric Fe—S stretch modes—a consequence of the deviation from ideal symmetry. Due to much lower ^{57}Fe motion, the contributions in this so-called breathing region are approximately 10% that of the asymmetric stretch components. The area around 150 cm^{-1} is composed of nearly degenerate S–Fe–S bending modes. The region between 100 and 150 cm^{-1} are assigned as Fe–S–C bending modes with mixing of S-Fe-S modes. Finally, the low energy region below

100 cm^{-1} appears analogous to the "lattice modes" described in $^{57}FeCl_4$; however, they are more complex and involved significant dihedral, torsional, acoustic, and delocalized vibrational modes within the greater protein and can provide a measure on the degree of vibrational coupling between the cofactor and the protein matrix (Guo et al., 2012).

When the same analysis is performed for the reduced Fe center, the order of assignments still hold, but the stretching modes downshift by 18%. It is noteworthy that the symmetric mode centroids shift to approximately 270 cm^{-1} and start to mix with the bending modes leading to enhanced intensity throughout the 150–300 cm^{-1} region (Fig. 4).

These results confirmed the delocalization of the dynamic properties of the redox-active Fe site which at the time was debated (Hiroshi, Takeo, Kaori, Akio, & Tatsuhiko, 1991). Besides the broadness of the Fe—S stretch region in both redox states, the weaker Fe—S force constant in the reduced rubredoxin leads to even stronger coupling between the Fe site and the motion of the local protein environment, evidenced experimentally by increased intensity below 300 cm^{-1}. This work on rubredoxin laid the framework for a later NRVS and molecular dynamics study that revealed non-Rieske [2Fe2S] clusters can couple even more dramatically to the motion of their environment and possibly induce binding and release of redox partners (Xiao et al., 2008).

4.2 [FeFe] Hydrogenase

The [FeFe] hydrogenase is an enzyme that catalyzes the reversible evolution of H_2. It is an FeS cluster-containing enzyme with a catalytic [2Fe] cofactor supported by the electron transfer activity of one or more other [FeS] clusters of varying composition depending on the source organism. The conversion of protons to dihydrogen presumably involves at least one Fe—H intermediate state. Unlike the bridging hydride in the related [NiFe] hydrogenase intermediate states (Ogata, Kramer, et al., 2015; Ogata, Nishikawa, & Lubitz, 2015), the consensus was that the intermediate hydride is terminally bound to the cofactor (Mulder, Guo, Ratzloff, & King, 2017), and a direct vibrational observation was recently made for the Fe—H bond by NRVS (Reijerse et al., 2017).

To trap the hydride-bound intermediate in significant amounts, and to selectively label the relevant diiron site, Reijerse and coworkers inserted a synthetic ^{57}Fe-labeled [2Fe] site into the enzyme. Whereas the natural enzyme contains an azadithiolate bridge between the two irons, proton transport to the synthetic active site was disrupted by an oxadithiolate (ODT) bridge, inhibiting full catalysis and increasing the yield of the hydride-bound intermediate called "H_{hyd}."

The experimental NRVS spectrum for the [2Fe]-ODT site in *Chlamydomonas reinhardtii* [FeFe] hydrogenase in the H_{hyd} state is shown in Fig. 5. The vibrational modes from 0 to 400 cm^{-1} are predominantly Fe—S in nature. The modes from 400 to 600 cm^{-1} are Fe-CO/CN stretching and bending vibrations. However, the most noteworthy observations are at 727 and 670 cm^{-1} which involve Fe—H bending motions within and perpendicular to the Fe–Fe–H plane, respectively. The two hydride features disappear upon deuteration of the sample. Beside a similar Fe—S region, the deuterated enzyme features a general shuffling of the Fe-CO/CN region intensities as the Fe-CO/CN vibrational motions couple differently to the deuteride motion in comparison to the hydride. The deuterated bending modes shift to 625 and 564 cm^{-1} for the in-plane and perpendicular Fe–D bending modes, respectively.

These observations, coupled to concomitant work on the natural cofactor in a mutant [FeFe] hydrogenase (Mulder et al., 2017), provided an unambiguous determination of the Fe—H bonding moiety in the H_{hyd} state. This work pushes the limits of the technique as hydridic modes normally involve

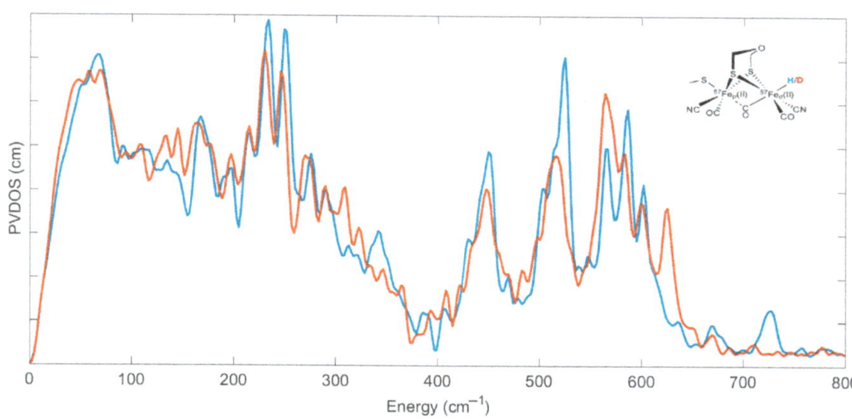

Fig. 5 The NRVS spectrum of the hydride (---) and deuteride (---) forms of the modified cofactor-inserted [FeFe] hydrogenase H_{hyd} state. *Inset*: schematic structure of the modified diiron cofactor. *Figure reproduced from Reijerse, E. J., Pham, C. C., Pelmenschikov, V., Gilbert-Wilson, R., Adamska-Venkatesh, A., & Siebel, J. F. (2017). Direct observation of an iron-bound terminal hydride in [FeFe]-hydrogenase by nuclear resonance vibrational spectroscopy.* Journal of the American Chemical Society, 139*(12), 4306–4309. https://doi.org/10.1021/jacs.7b00686.*

very little Fe motion. Consequently, another frontier for ^{57}Fe NRVS will be the observation of higher energy hydride/deuteride stretching vibrational modes in a biological system.

5. PERSPECTIVE

Here, we have discussed the NRVS spectral acquisition and analysis of bioinorganic FeS systems. Besides rubredoxin (Guo et al., 2012; Xiao et al., 2005); non-Rieske [2Fe2S] (Xiao et al., 2008); [3Fe4S] (Lauterbach et al., 2016); and [4Fe4S] (Mitra et al., 2011) proteins, it is important to note that ^{57}Fe NRVS has been expanded to study electron transfer and catalysis in larger enzyme metallocofactors such as (but not comprehensively) the NiFe site in [NiFe] hydrogenase (Kamali et al., 2013; Lauterbach et al., 2015; Ogata, Kramer, et al., 2015), the diiron site in [FeFe] hydrogenase (Gilbert-Wilson et al., 2015; Kuchenreuther et al., 2013), the P [8Fe7S]- and M [7Fe9SMo]-clusters of Mo-nitrogenase (George et al., 2012; Mitra et al., 2013; Scott et al., 2014; Xiao et al., 2006), heme systems (Leu et al., 2009; Sage et al., 2001), nonheme systems (Sutherlin et al., 2016; Wong et al., 2013), and nitric oxide-sensing [4Fe4S] clusters (Serrano et al., 2016). Further as the systems of interest have become more complicated, with multiple FeS clusters of differing composition, innovative strategies of site-specific ^{57}Fe labeling have been implemented that target maturation or a modified cofactor insertion (Gilbert-Wilson et al., 2015; Reijerse et al., 2017).

Since its inception over two decades ago, improvements in synchrotron flux, resolution, and sample preparation have allowed for the NRVS investigations of weaker features and more dilute biological samples. The technique is continuing to mature and is solving questions in ways that resonance Raman and infrared spectroscopies cannot. The continued advancement of synchrotron technologies and the development of X-ray free-electron lasers will enable NRVS to become a more prolific tool to solve complex questions in bioinorganic systems.

REFERENCES

Bau, R., Rees, D. C., Kurtz, D. M., Jr., Scott, R. A., Huang, H., Adams, M. W. W., et al. (1998). Crystal structure of rubredoxin from *Pyrococcus furiosus* at 0.95 Å resolution, and the structures of N-terminal methionine and formylmethionine variants of Pf Rd. Contributions of N-terminal interactions to thermostability. *JBIC Journal of Biological Inorganic Chemistry*, *3*(5), 484–493. https://doi.org/10.1007/s007750050258.

Day, M. W., Hsu, B. T., Joshua-Tor, L., Park, J. B., Zhou, Z. H., Adams, M. W., et al. (1992). X-ray crystal structures of the oxidized and reduced forms of the rubredoxin from the marine hyperthermophilic archaebacterium *Pyrococcus furiosus*. *Protein Science*, *1*(11), 1494–1507. https://doi.org/10.1002/pro.5560011111.

Dowty, E. (1987). Fully automated microcomputer calculation of vibrational spectra. *Physics and Chemistry of Minerals*, *14*(1), 67–79. https://doi.org/10.1007/bf00311150.

George, S. J., Barney, B. M., Mitra, D., Igarashi, R. Y., Guo, Y. S., Dean, D. R., et al. (2012). EXAFS and NRVS reveal a conformational distortion of the FeMo-cofactor in the MoFe nitrogenase propargyl alcohol complex. *Journal of Inorganic Biochemistry*, *112*, 85–92. https://doi.org/10.1016/j.jinorgbio.2012.02.004.

Gilbert-Wilson, R., Siebel, J. F., Adamska-Venkatesh, A., Pham, C. C., Reijerse, E., Wang, H., et al. (2015). Spectroscopic investigations of [FeFe] hydrogenase maturated with [57Fe2(adt)(CN)2(CO)4]2^{-}. *Journal of the American Chemical Society*, *137*(28), 8998–9005. https://doi.org/10.1021/jacs.5b03270.

Guo, Y., Brecht, E., Aznavour, K., Nix, J. C., Xiao, Y., Wang, H., et al. (2012). Nuclear resonance vibrational spectroscopy (NRVS) of rubredoxin and MoFe protein crystals. *Hyperfine Interactions*, *222*(S2), 77–90. https://doi.org/10.1007/s10751-012-0643-2.

Hiroshi, S., Takeo, I., Kaori, W., Akio, U., & Tatsuhiko, Y. (1991). Resonance Raman active vibrations of rubredoxin. Normal coordinate analysis of a 423-atom model. *Bulletin of the Chemical Society of Japan*, *64*(3), 829–836. https://doi.org/10.1246/bcsj.64.829.

Johnson, D. W., & Spence, J. C. H. (1974). Determination of the single-scattering probability distribution from plural-scattering data. *Journal of Physics D: Applied Physics*, 7, 771–780.

Kaesz, H. D., & Saillant, R. B. (1972). Hydride complexes of the transition metals. *Chemical Reviews*, *72*(3), 231–281. https://doi.org/10.1021/cr60277a003.

Kamali, S., Wang, H., Mitra, D., Ogata, H., Lubitz, W., Manor, B. C., et al. (2013). Observation of the Fe-CN and Fe-CO vibrations in the active site of [NiFe] hydrogenase by nuclear resonance vibrational spectroscopy. *Angewandte Chemie (International Ed. in English)*, *52*(2), 724–728. https://doi.org/10.1002/anie.201204616.

Kuchenreuther, J. M., Guo, Y. S., Wang, H. X., Myers, W. K., George, S. J., Boyke, C. A., et al. (2013). Nuclear resonance vibrational spectroscopy and electron paramagnetic resonance spectroscopy of Fe-57-enriched [FeFe] hydrogenase indicate stepwise assembly of the H-cluster. *Biochemistry*, *52*(5), 818–826. https://doi.org/10.1021/Bi301336r.

Lauterbach, L., Gee, L. B., Pelmenschikov, V., Jenney, F. E., Kamali, S., Yoda, Y., et al. (2016). Characterization of the [3Fe-4S]0/1 + cluster from the D14C variant of *Pyrococcus furiosus* ferredoxin via combined NRVS and DFT analyses. *Dalton Transactions*, *45*(17), 7215–7219. https://doi.org/10.1039/C5DT04760A.

Lauterbach, L., Wang, H., Horch, M., Gee, L. B., Yoda, Y., Tanaka, Y., et al. (2015). Nuclear resonance vibrational spectroscopy reveals the FeS cluster composition and active site vibrational properties of an O_2-tolerant NAD^+-reducing [NiFe] hydrogenase. *Chemical Science*, *6*(2), 1055–1060. https://doi.org/10.1039/c4sc02982h.

Leu, B. M., Ching, T. H., Zhao, J. Y., Sturhahn, W., Alp, E. E., & Sage, J. T. (2009). Vibrational dynamics of iron in cytochrome c. *Journal of Physical Chemistry B*, *113*(7), 2193–2200. https://doi.org/10.1021/Jp806574t.

Mitra, D., George, S. J., Guo, Y. S., Kamali, S., Keable, S., Peters, J. W., et al. (2013). Characterization of [4Fe-4S] cluster vibrations and structure in nitrogenase Fe protein at three oxidation levels via combined NRVS, EXAFS, and DFT analyses. *Journal of the American Chemical Society*, *135*(7), 2530–2543. https://doi.org/10.1021/Ja307027n.

Mitra, D., Pelmenschikov, V., Guo, Y., Case, D. A., Wang, H., Dong, W., et al. (2011). Dynamics of the [4Fe-4S] cluster in *Pyrococcus furiosus* D14C ferredoxin via nuclear resonance vibrational and resonance Raman spectroscopies, force field simulations, and

density functional theory calculations. *Biochemistry*, *50*(23), 5220–5235. https://doi.org/10.1021/bi200046p.
Mössbauer, R. (1958a). Kernresonanzabsorption von Gammastrahlung in Ir191. *Naturwissenschaften*, *45*(22), 538–539. https://doi.org/10.1007/bf00632050.
Mössbauer, R. (1958b). Kernresonanzfluoreszenz von Gammastrahlung in Ir191. *Zeitschrift für Physik*, *151*(2), 124–143. https://doi.org/10.1007/BF01344210.
Mössbauer, R. (1959). Kernresonanzabsorption von γ-Strahlung in Ir191. *Zeitschrift für Naturforschung A*, *14*, 211.
Mulder, D. W., Guo, Y., Ratzloff, M. W., & King, P. W. (2017). Identification of a catalytic iron-hydride at the H-cluster of [FeFe]-hydrogenase. *Journal of the American Chemical Society*, *139*(1), 83–86. https://doi.org/10.1021/jacs.6b11409.
Ogata, H., Kramer, T., Wang, H., Schilter, D., Pelmenschikov, V., van Gastel, M., et al. (2015). Hydride bridge in [NiFe]-hydrogenase observed by nuclear resonance vibrational spectroscopy. *Nature Communications*, *6*, 7890. https://doi.org/10.1038/ncomms8890.
Ogata, H., Nishikawa, K., & Lubitz, W. (2015). Hydrogens detected by subatomic resolution protein crystallography in a [NiFe] hydrogenase. *Nature*, *520*(7548), 571–574. https://doi.org/10.1038/nature14110.
Petrenko, T., Sturhahn, W., & Neese, F. (2008). First-principles calculation of nuclear resonance vibrational spectra. *Hyperfine Interactions*, *175*(1–3), 165–174. https://doi.org/10.1007/s10751-008-9600-5.
Reijerse, E. J., Pham, C. C., Pelmenschikov, V., Gilbert-Wilson, R., Adamska-Venkatesh, A., Siebela, J. F., et al. (2017). Direct observation of an iron-bound terminal hydride in [FeFe]-hydrogenase by nuclear resonance vibrational spectroscopy. *Journal of the American Chemical Society*, *139*(12), 4306–4309. https://doi.org/10.1021/jacs.7b00686.
Rosenberg, E. (1989). Kinetic deuterium isotope effects in transition metal hydride clusters. *Polyhedron*, *8*(4), 383–405. https://doi.org/10.1016/S0277-5387(00)80733-0.
Sage, J., Durbin, S., Sturhahn, W., Wharton, D., Champion, P., Hession, P., et al. (2001). Long-range reactive dynamics in myoglobin. *Physical Review Letters*, *86*(21), 4966–4969. https://doi.org/10.1103/PhysRevLett.86.4966.
Scott, A., Pelmenschikov, V., Guo, Y., Wang, H., Yan, L., George, S., et al. (2014). Structural characterization of CO-inhibited Mo-nitrogenase by combined application of nuclear resonance vibrational spectroscopy, extended X-ray absorption fine structure, and density functional theory: New insights into the effects of CO binding and the role of the interstitial atom. *Journal of the American Chemical Society*, *136*, 15942.
Serrano, P. N., Wang, H., Crack, J. C., Prior, C., Hutchings, M. I., Thomson, A. J., et al. (2016). Nitrosylation of nitric-oxide-sensing regulatory proteins containing [4Fe-4S] clusters gives rise to multiple iron–nitrosyl complexes. *Angewandte Chemie International Edition*, *55*(47), 14,575–14,579. https://doi.org/10.1002/anie.201607033.
Seto, M., Yoda, Y., Kikuta, S., Zhang, X., & Ando, M. (1995). Observation of nuclear resonant scattering accompanied by phonon excitation using synchrotron radiation. *Physical Review Letters*, *74*(19), 3828–3831. https://doi.org/10.1103/PhysRevLett.74.3828.
Singwi, K. S., & Sjölander, A. (1960). Resonance absorption of nuclear gamma rays and the dynamics of atomic motions. *Physical Review*, *120*(4), 1093–1102.
Smith, M. C., Xiao, Y., Wang, H., George, S. J., Coucouvanis, D., Koutmos, M., et al. (2005). Normal-mode analysis of FeCl4- and Fe2S2Cl42- via vibrational Mössbauer, resonance Raman, and FT-IR spectroscopies. *Inorganic Chemistry*, *44*(16), 5562–5570. https://doi.org/10.1021/ic0482584.
Sturhahn, W. (2000). CONUSS and PHOENIX: Evaluation of nuclear resonant scattering data. *Hyperfine Interactions*, *125*, 149–172.
Sturhahn, W., Toellner, T., Alp, E., Zhang, X., Ando, M., Yoda, Y., et al. (1995). Phonon density of states measured by inelastic nuclear resonant scattering. *Physical Review Letters*, *74*(19), 3832–3835. https://doi.org/10.1103/PhysRevLett.74.3832.

Sutherlin, K. D., Liu, L. V., Lee, Y.-M., Kwak, Y., Yoda, Y., Saito, M., et al. (2016). Nuclear resonance vibrational spectroscopic definition of peroxy intermediates in nonheme iron sites. *Journal of the American Chemical Society*, *138*(43), 14,294–14,302. https://doi.org/10.1021/jacs.6b07227.

Visscher, W. M. (1960). Study of lattice vibrations by resonance absorption of nuclear gamma rays. *Annals of Physics*, *9*(2), 194–210. https://doi.org/10.1016/0003-4916(60)90028-2.

Wang, H., Alp, E. E., Yoda, Y., & Cramer, S. P. (2014). A practical guide for nuclear resonance vibrational spectroscopy (NRVS) of biochemical samples and model compounds. In C. J. Fontecilla-Camps & Y. Nicolet (Eds.), *Metalloproteins: Methods and protocols* (pp. 125–137). Totowa, NJ: Humana Press.

Wang, H., Yoda, Y., Kamali, S., Zhou, Z. H., & Cramer, S. P. (2012). Real sample temperature: A critical issue in the experiments of nuclear resonant vibrational spectroscopy on biological samples. *Journal of Synchrotron Radiation*, *19*(Pt. 2), 257–263. https://doi.org/10.1107/S0909049512001380.

Wong, S. D., Srnec, M., Matthews, M. L., Liu, L. V., Kwak, Y., Park, K., et al. (2013). Elucidation of the Fe(IV)=O intermediate in the catalytic cycle of the halogenase SyrB2. *Nature*, *499*(7458), 320–323. https://doi.org/10.1038/nature12304.

Xiao, Y., Smith, M. C., Newton, W., Case, D. A., George, S., Wang, H., et al. (2006). How nitrogenase shakes—Initial information about P-cluster and FeMo-cofactor normal modes from nuclear resonance vibrational spectroscopy (NRVS). *Journal of the American Chemical Society*, *128*(23), 7608–7612. https://doi.org/10.1021/ja0603655.

Xiao, Y., Tan, M. L., Ichiye, T., Wang, H., Guo, Y., Smith, M. C., et al. (2008). Dynamics of Rhodobacter capsulatus [2FE-2S] ferredoxin VI and Aquifex aeolicus ferredoxin 5 via nuclear resonance vibrational spectroscopy (NRVS) and resonance Raman spectroscopy. *Biochemistry*, *47*(25), 6612–6627. https://doi.org/10.1021/bi701433m.

Xiao, Y., Wang, H., George, S. J., Smith, M. C., Adams, M. W., Jenney, F. E., Jr., et al. (2005). Normal mode analysis of *Pyrococcus furiosus* rubredoxin via nuclear resonance vibrational spectroscopy (NRVS) and resonance Raman spectroscopy. *Journal of the American Chemical Society*, *127*(42), 14596–14606. https://doi.org/10.1021/ja042960h.

CHAPTER FIFTEEN

Advanced X-ray Spectroscopic Methods for Studying Iron–Sulfur-Containing Proteins and Model Complexes

Serena DeBeer[1]
Max Planck Institute for Chemical Energy Conversion, Mülheim an der Ruhr, Germany
[1]Corresponding author: e-mail address: serena.debeer@cec.mpg.de

Contents

Abstract

In this chapter, a brief overview of X-ray spectroscopic methods that may be utilized to obtain insight into the geometric and electronic structure of iron–sulfur proteins is provided. These methods include conventional methods, such as metal and ligand K-edge X-ray absorption, as well as more advanced methods including nonresonant and resonant X-ray emission. In each section, the basic information content of the spectra is

Methods in Enzymology, Volume 599
ISSN 0076-6879
https://doi.org/10.1016/bs.mie.2017.09.008

highlighted and important experimental considerations are discussed. Throughout the chapter, recent applications to iron–sulfur-containing models and proteins are highlighted.

1. INTRODUCTION

Iron–sulfur proteins are near universal in nature, playing essential roles in a wide range of processes, including electron transfer, homeostatic regulation, and enzymatic catalysis (Johnson, Dean, Smith, & Johnson, 2005; Lill & Muhlenhoff, 2008). The active sites of iron–sulfur clusters range from simple mononuclear active sites, to more complex dimers and tetramers, and finally to the complex seven and eight iron clusters, found in the respective M- and P-clusters of nitrogenase (Hoffman, Lukoyanov, Dean, & Seefeldt, 2013; Rees & Howard, 2000). A longstanding goal within the bioinorganic community has been to understand how iron–sulfur proteins are optimized for their specific functions (Holm, Kennepohl, & Solomon, 1996; Lee & Holm, 2004; Lee, Lo, & Holm, 2014). Ultimately, one would like to understand the electronic and geometric structural changes, which occur upon redox events or substrate-binding events, and from this, glean insights into Nature's evolved complexity. As such, spectroscopic tools, which can provide detailed insights into the changes that occur in the active site geometric and electronic structure, are highly desirable. It is here that X-ray spectroscopy has long played an important role (Corbett et al., 2006; Cramer et al., 1978; George et al., 2012; George, Coyle, Hales, & Cramer, 1988; Glaser, Hedman, Hodgson, & Solomon, 2000; Kowalska & DeBeer, 2015; Kowalska, Hahn, et al., 2016; Lancaster et al., 2011; Lancaster, Hu, Bergmann, Ribbe, & DeBeer, 2013; Musgrave, Angove, Burgess, Hedman, & Hodgson, 1998; Musgrave, Liu, et al., 1998; Pollock, Tan, et al., 2014; Shulman, Eisenberger, & Kincaid, 1978; Shulman, Eisenberger, Teo, Kincaid, & Brown, 1978). Some of the earliest X-ray absorption spectroscopic studies on biological systems focused on the active site metrics of rubredoxins using extended X-ray absorption fine structure (EXAFS) (Shulman, Eisenberger, Teo, et al., 1978). In parallel, researchers had taken on the challenging task of trying to elucidate the active site metrics of the M-cluster in nitrogenase (Cramer et al., 1978). While the shear complexity of the M-cluster, which is now known to be a Fe_7MoS_9C cluster (Lancaster et al., 2011; Spatzal et al., 2011), prohibited an accurate description of the M-cluster three-dimensional structure, this early work nonetheless paved the way for numerous future studies of iron–sulfur clusters using X-ray-based spectroscopy. Today, X-ray spectroscopy continues to play an important role in our understanding of iron–sulfur active sites and the complexity of the applied experimental methods continue to evolve rapidly.

Herein, we provide an overview of the most important X-ray spectroscopic methods, which have been applied to both iron–sulfur proteins and to related model complexes. This chapter begins with a brief review of X-ray absorption spectroscopy (XAS), and its applications at the metal K- and ligand K-edge. We then move on to both nonresonant and resonant X-ray emission spectroscopy (XES) methods, which have recently seen increased applications in the hard X-ray regime. Finally soft X-ray methods, including 2p3d resonant inelastic X-ray scattering (2p3d RIXS) and Fe L-edge X-ray magnetic circular dichroism (XMCD), are briefly highlighted. Throughout the chapter important experimental considerations for each method are emphasized, with the goal of providing the basic information that a new user would require before applying these methods.

2. X-RAY ABSORPTION SPECTROSCOPY

Perhaps the best-known and most widely applied X-ray spectroscopic method is XAS (DeBeer, 2011; de Groot, 2001; Penner-Hahn, 1999). An XAS spectrum involves the measurement of the absorption coefficient of a given photoabsorbing atom as a function of energy (Fig. 1). An X-ray absorption "edge" occurs whenever a core electron has an absorbed an X-ray with an energy that is equal to or greater than its binding energy. When this occurs, a sharp discontinuity in the absorption spectrum is observed, which is most commonly referred to as an "edge." Beyond the edge is the so-called EXAFS region. This results from the photoelectron wave vector scattering off of neighboring atoms before returning to the photoabsorber. These scattering events result in a modulation of the absorption coefficient from which information about the distance, coordination number, and type of ligands surrounding the photoabsorber can be extracted. As EXAFS has been amply reviewed elsewhere, we will not discuss this spectral region further here (DeBeer, 2011; Penner-Hahn, 1999).

For purposes of the present discussion on iron–sulfur proteins, it is most useful to consider the so-called K-edges, which involve the excitation of 1s core electrons. The Fe K-edge occurs at ~7112 eV, while the S K-edge occurs at ~2472 eV. The very large differences in core-ionization energies

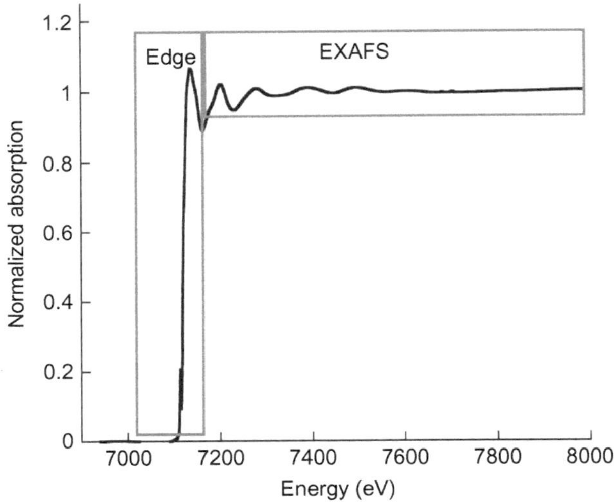

Fig. 1 An iron K-edge X-ray absorption spectrum. The XAS edge and EXAFS regions are indicated.

are precisely the reason that XAS is an element-selective technique. Importantly, however, XAS is not simply element selective, it is also sensitive to the changes in oxidation state, spin state, coordination environment, and site symmetry of the photoabsorber. In order to more clearly illustrate this point, the specific information that can be obtained at the Fe and S K-edges is discussed later.

2.1 Metal K-Edge XAS

Fig. 2 depicts the Fe K-edge of two dimeric iron–sulfur model complexes, the diferric complex $[L_2Fe(III)_2S_2]^{2-}$ and the diferrous complex $[L_2Fe(II)_2S_2]^{4-}$ (where L = bis(benzimidazolato) (Kowalska, Hahn, et al., 2016). Inspection of the figure clearly shows there are two primary spectral regions: (1) an intense edge feature with a first inflection point at ~7120 eV and (2) a weak preedge feature at ~7113 eV.

The edge results primarily from a dipole allowed 1s to 4p transition. However, additional "shakedown" transitions (due to possible ligand-to-metal charge transfer processes) may be superimposed on the rising edge (Kau, Spirasolomon, Pennerhahn, Hodgson, & Solomon, 1987). Hence, it is important that comparisons are made between metal sites within similar ligation environments when attempting to extract oxidation state information. In a simple picture, the rising edge inflection point primarily reflects the effective charge on the absorbing atom. Hence, ongoing from a diferrous to

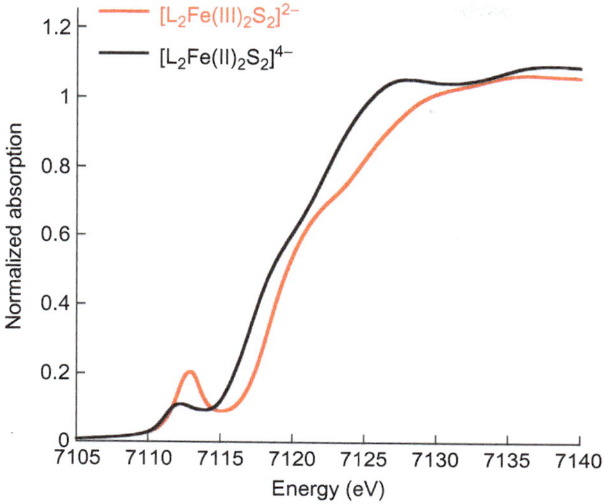

Fig. 2 Normalized Fe K-edge XAS spectra of $[L_2Fe(III)_2S_2]^{2-}$ and $[L_2Fe(II)_2S_2]^{4-}$ (where L = bis(benzimidazolato). See open access article at: http://pubs.acs.org/doi/abs/10.1021/acs.inorgchem.6b00295.

a diferric iron site, the increase in oxidation state results in the 1s ionization occurring at higher energies, with the general "rule of thumb" being an increase in the edge energy of 1–2 eV for each one unit increase in oxidation state. While this trend clearly holds for the model complexes shown here, it turns out that for iron–sulfur clusters the situation is in fact far more complex, this point will be discussed in further detail later.

Before doing so, however, it is important to note that similar energy trends are also observed in the weak 1s to 3d preedge region, with the preedge clearly appearing at higher energy for the diferric model than for the diferrous analogue. We note that 1s to 3d transitions are formally dipole forbidden, but can gain intensity due to 3d–4p mixing in the appropriate symmetry (DeBeer George, Brant, & Solomon, 2005; Westre et al., 1997). As iron–sulfur clusters support locally $\sim T_d$ coordination environments around the iron atom, symmetry dictates that the metal 4p and 3d orbitals are allowed to mix. Hence, iron–sulfur clusters are typically characterized by relatively intense preedge features relative to O_h iron sites (Kowalska, Hahn, et al., 2016; Musgrave, Angove, et al., 1998; Musgrave, Liu, et al., 1998). The increase in preedge intensity of the diferric complex, relative to the diferrous complex, largely reflects the increase iron d-hole character and the increased covalency that occurs upon oxidation of the iron.

At this point, we wish to briefly revisit the point made above about using Fe K-edges as a fingerprint for oxidation state. In a recent report by Kowalska et al., the mixed-valent analogue of the dimers shown in Fig. 2 was also examined (Kowalska, Hahn, et al., 2016). Surprisingly, it was shown that the edges of the mixed-valent and diferrous analogues overlaid, hence indicating that a simple correlation to oxidation state alone was not sufficient to explain the observed experimental trends. Rather, it was shown that the edges reflect not only oxidation state but also the extent of electron delocalization between the two iron sites. As such, ferro- vs antiferromagnetically coupled iron dimers will result in different rising edge behavior. While this serves as a cautionary note in terms of using edges as simple fingerprints of oxidation states (particularly for iron–sulfur clusters), it also emphasizes the pronounced sensitivity of the edge and the ability (when coupled to other spectroscopic methods) to determine even more detailed electronic structure information. Importantly, it has also been demonstrated that even relatively simple time-dependent density functional theory calculations can capture the changes in the rising edge for iron–sulfur dimers (Kowalska, Hahn, et al., 2016). Hence, by coupling theory with experiment, the ability to use XAS for a quantitative understanding of the complex electronic structure of iron–sulfur clusters is greatly enhanced. In our view, these recent developments also motivate the need to reevaluate previous Fe K-edge data reports on FeS proteins (Musgrave, Angove, et al., 1998; Musgrave, Liu, et al., 1998), in order to more quantitatively evaluate the detailed electronic structural changes.

2.2 Ligand K-Edge XAS

As noted earlier, in certain cases the information that can be extracted from the metal K-edge may be limited and supporting data from other methods is often highly desirable in order to arrive at a more cohesive picture of the electronic structure of highly covalent iron–sulfur clusters. In this context, strong complementary information can be obtained by directly probing the sulfur ligands (Glaser, Hedman, et al., 2000). Fig. 3 depicts the S K-edge XAS spectra for the same diferric and diferrous dimers discussed in Section 2.1 (Kowalska, Hahn, et al., 2016). As with the Fe K-edge data, the S K-edge data can also be described in terms of two general spectral regions: (1) a rising edge at ~2475 eV and (2) a preedge at ~2470 eV.

The edge reflects a sulfur 1s to 4p transition. One clearly sees that the S K-edge appears at higher energy for the diferric complex than for the diferrous complex. As in both cases the ligand is the same (S^{2-}), the shift

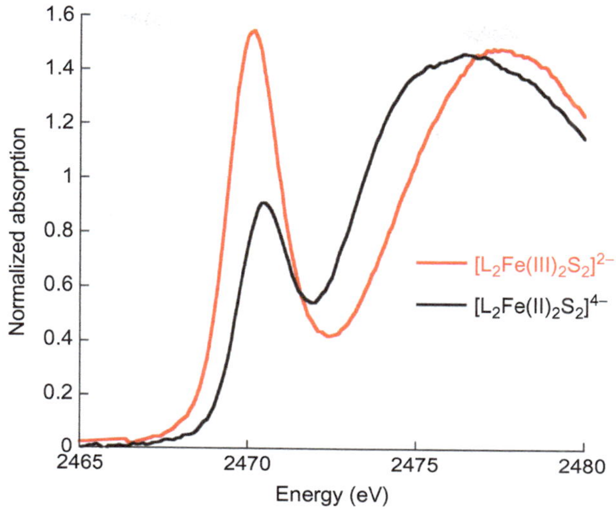

Fig. 3 Normalized S K-edge XAS spectra of $[L_2Fe(III)_2S_2]^{2-}$ and $[L_2Fe(II)_2S_2]^{4-}$ (where L = bis(benzimidazolato). See open access article at: http://pubs.acs.org/doi/abs/10.1021/acs.inorgchem.6b00295.

in the edge reflects a change in charge donation. As the Fe–S bond is much stronger and more covalent in the diferric case than the diferrous case, the S^{2-} ligand appears effectively "more oxidized" in the former.

In the case of an isolated S^{2-} sulfide ligand, one would observe only the edge feature, and no preedge, as the 3p shell is filled. However, when sulfide interacts within an open-shell transition metal such as iron, bonding and antibonding combinations of molecular orbitals form, which will be comprised of a mixture of Fe 3d character and S 3p character. It is transitions to the unoccupied antibonding molecular orbitals that give rise to the preedge feature. As such, the preedge correlates with extent of metal–ligand mixing and hence the covalency of metal–ligand bonds. The stronger and shorter a metal–ligand bond, the more intense the ligand preedge (Glaser, Hedman, et al., 2000). Hence, the reason that the preedge is more intense for the diferric complex than for the diferrous complex (Fig. 3). We also note that in the case of a sulfur K-edge, the preedge transitions are much more intense than what was observed at the Fe K-edge. This is due to the fact that the S 1s to 3p transition is dipole allowed (in contrast to the formally dipole forbidden 1s to 3d transitions at the metal K-edge).

Finally, in closing it is of interest to highlight the pioneering work of Solomon and coworkers in the area of S K-edge XAS (Dey et al., 2007; Glaser, Hedman, et al., 2000; Glaser et al., 1999, 2001; Glaser, Rose, et al., 2000;

Rose et al., 1999). In numerous detailed studies on iron–sulfur clusters, the ability to separate out the thiolate vs the sulfide contributions, as well as to determine hydrogen bonding and solvation contributions, has been demonstrated.

2.3 Metal and Ligand K-Edge XAS Experimental Considerations

In the previous sections, we have briefly highlighted the electronic structure information that one can obtain from metal and ligand K-edge studies on iron–sulfur complexes. Here, we briefly discuss some of the important experimental aspects that should be taken into account.

In general the measurement of XAS, particularly on dilute metalloproteins, requires a synchrotron to provide an intense source of X-rays. A beamline that is well optimized for biological X-ray absorption will generally have the following features:

(1) A wiggler or an undulator insertion device providing an intense source of incident X-ray ($\sim 10^{12}$ photons/s or higher).

(2) A scanning double crystal monochromator (typically Si(111), Si(220), or Si(311)) allowing for a 1 eV or better resolution for the incident X-ray beam.

(3) An energy-resolving detector, allowing for the rejection of unwanted scattering and/or fluorescent events.

(4) The ability to maintain the sample at low temperatures (typically 20 K or lower) or to flow the sample in order to reduce damage.

An advantage of XAS measurements is that samples can be measured in any form—frozen solutions, flowing solutions, lyophilized proteins, or single crystals. As a general rule of thumb, an optimal photoabsorber concentration is 1 m*M* or greater. However, more dilute samples certainly can be measured. This is often a question of the required signal to noise, as well as the available beamtime. Hence, in general, it is best to work with samples that are as concentrated as is reasonably possible. Depending on the size of the X-ray beam the sample size can be optimized in order to minimize the sample volume. Typical volume requirements for frozen solutions are on the order of 50–200 μL. For flowing solutions, the sample requirements will depend on flow rates and may often be prohibitive for biological samples that can be prepared only in limited quantities.

An additional consideration that must be taken into account for XAS measurements on metalloproteins (and on model complexes) is the propensity of the samples to suffer from X-ray beam-induced damage (van Schooneveld & DeBeer, 2015). Hence, the experimentalist must take special care to establish protocols for the maximum dose rates. For frozen solution samples, this may be done by performing successive XAS scans on the same sample spot and following the evolution of damage with time. For flowing solutions, this can be achieved by varying flow speeds in continuous flow setups.

Finally, we note that the lower energies of S K-edge XAS (relative to Fe) present additional challenges. The ~2.4 keV X-rays required for S K-edge XAS have a relatively short path length in air and hence the beamline setups are often enclosed in helium (or in some cases vacuum) atmosphere. This may present additional experimental challenges, particularly in terms of the sample environment.

3. X-RAY EMISSION SPECTROSCOPY

While XAS is a very well recognized and widely used method within the bioinorganic community, XES has only seen increased applications in the later part of the last decade (Beckwith et al., 2011; DeBeer & Bergmann, 2016; de Groot, 2001; Delgado-Jaime, DeBeer, & Bauer, 2013; Delgado-Jaime et al., 2011; Gallo & Glatzel, 2014; Gamblin & Urch, 2001; Glatzel & Bergmann, 2005; Lancaster et al., 2011; Lassalle-Kaiser et al., 2013; Lee, Petrenko, Bergmann, Neese, & DeBeer, 2010; Lu, Weng, & Liaw, 2014; Pollock & DeBeer, 2011; Pollock, Delgado-Jaime, Atanasov, Neese, & DeBeer, 2014; Pollock, Grubel, Holland, & DeBeer, 2013; Smolentsev et al., 2009). This was largely enabled by the construction of dedicated high-resolution XES spectrometers at synchrotron beamlines, which have allowed the more routine application of XES to be realized.

An XES spectrum results from the fluorescence decay, which results after the ionization of a core electron. For the purposes of the present section, we will focus only on the processes which follow after ionization of a 1s core electron. Fig. 4 depicts the primary XES spectral regions (top) and the corresponding transitions, which give rise to each X-ray emission line.

3.1 XES

After the ionization of a 1s core electron, the most likely event is that a 2p electron will refill the 1s core hole, producing the so-called Kα lines. The Kα lines are split into Kα_1 and Kα_2 lines due to 2p core hole spin orbit coupling—as such, the chemical information content of the lines is somewhat limited.

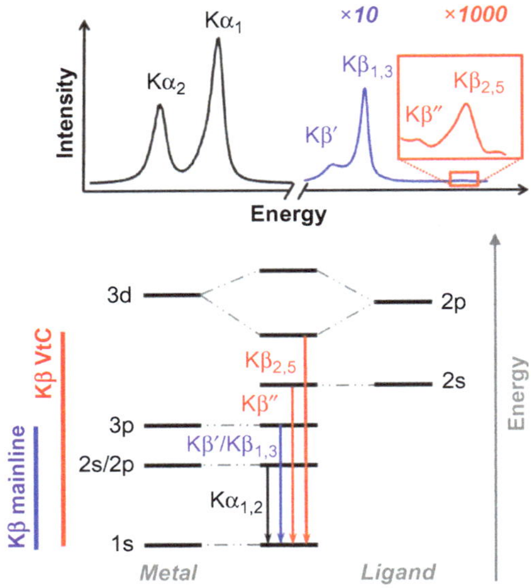

Fig. 4 The Kα (2p to 1s), Kβ (3p to 1s), and $K\beta_{2,5}/K\beta''$ (valence to 1s or valence to core, VtC) spectral regions (*top*) and their corresponding origins (*bottom*). *Adapted from Rees, J. A., Bjornsson, R., Kowalska, J. K., Lima, F. A., Schlesier, J., Sippel, D., et al. (2017). Comparative electronic structures of nitrogenase FeMoco and FeVco.* Dalton Transaction, *46, 2445–2455. Copyright 2016 John Wiley and Sons.*

3.2 Kβ XES

To higher energy are the Kβ lines, which result from a 3p electron refilling the 1s core hole. Though Kβ lines are almost an order of magnitude less intense than the Kα line, the chemical information content is much richer. This is due to the fact that 3p–3d exchange dominates the splitting of the Kβ lines (often referred to as the $K\beta_{1,3}$ and $K\beta'$ or collectively as the Kβ mainlines). Hence, for first row transition metals the Kβ mainlines serve as a sensitive probe of both spin state and oxidation state (DeBeer & Bergmann, 2016; Glatzel & Bergmann, 2005; Lee et al., 2010). The effect of spin state on Kβ XES spectra is illustrated in Fig. 5A and C. All of complexes shown are in the Fe^{3+} oxidation state; however, Fig. 5A shows high-spin ($S=5/2$) complexes, while Fig. 5C shows low spin ($S=1/2$ complexes) (Lee et al., 2010). One clearly observes that the high-spin Fe(III) complexes are characterized by two features split by ~15 eV, while in the low-spin complexes a well-resolved low-energy shoulder is no longer present. This is due to the decreased 3p–3d exchange contribution in the low-spin complexes, which results in decreased spectral splitting.

While these data suggest that Kβ mainline XES data should provide a simple fingerprint for spin and oxidation state, in fact, there are other

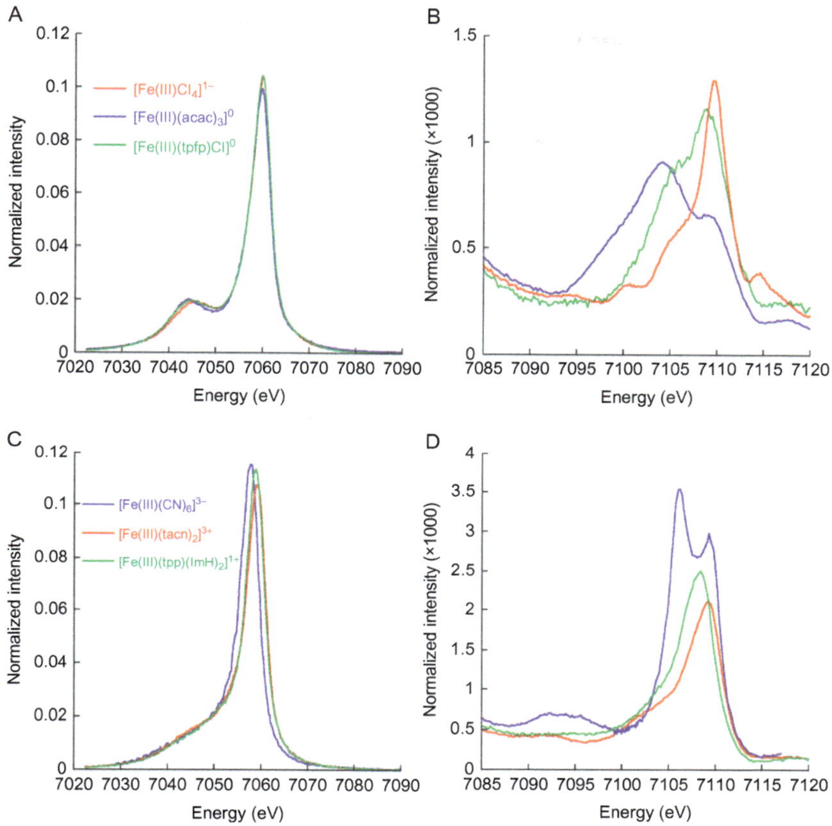

Fig. 5 Fe Kβ mainline XES spectrum for high-spin ferric complexes (A). VtC XES of the complexes shown in A (B). Fe Kβ mainline XES spectrum for low-spin ferric complexes (C). VtC XES of the complexes shown in C (D). *Adapted from Lee, N., Petrenko, T., Bergmann, U., Neese, F., & DeBeer, S. (2010). Probing valence orbital composition with iron K beta X-ray emission spectroscopy.* Journal of the American Chemical Society, 132, *9715–9727. Copyright 2010 American Chemical Society.*

competing factors. As covalency increases, the covalent dilution of 3d character onto the ligand results in a decrease in the effective 3p–3d exchange contribution to the spectral splitting (Pollock, Delgado-Jaime, et al., 2014). As a result, it is important to have additional spectroscopic data in order to understand the origin of the splitting. This is particularly true for highly covalent iron–sulfur clusters where diferric, mixed-valent, and diferrous dimers have been shown to have superimposable Kβ mainline XES spectra (Kowalska, Hahn, et al., 2016). This can be understood in terms of competing contributions of reduced d-count and decreased covalency upon reduction. Nonetheless in the complex M-clusters of the Mo- and

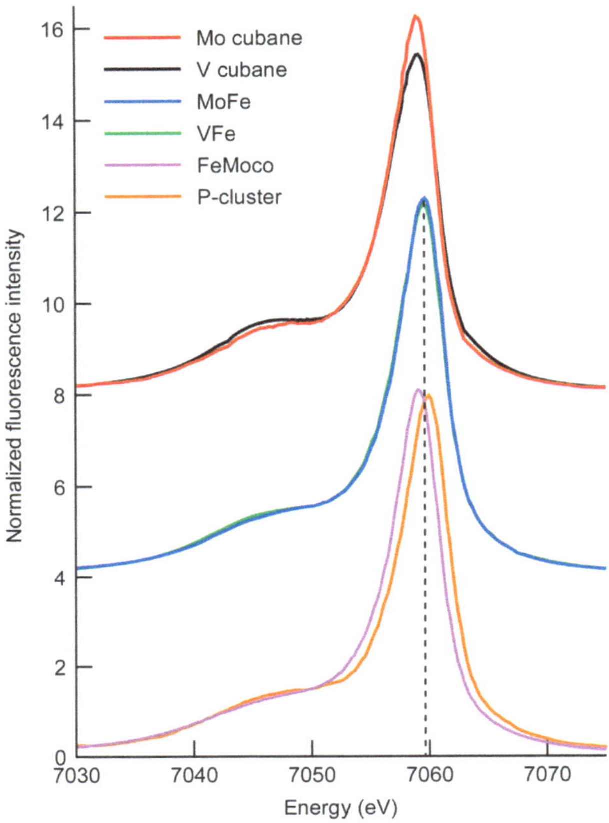

Fig. 6 Fe Kβ mainlines of the synthetic Mo cubane ((Et$_4$N)[(Tp)MoFe$_3$S$_4$Cl$_3$]) and V cubane ((Me$_4$N)[VFe$_3$S$_4$Cl$_3$](DMF)$_3$) clusters, MoFe, VFe, FeMoco, and the P-cluster of nitrogenase. *Reproduced from Rees, J. A., Bjornsson, R., Kowalska, J. K., Lima, F. A., Schlesier, J., Sippel, D., et al. (2017). Comparative electronic structures of nitrogenase FeMoco and FeVco.* Dalton Transaction, 46, *2445–2455, with permission from the Royal Society of Chemistry.*

V-dependent nitrogenases, it has been shown that the contribution of the central carbide is still significant enough to modulate the Kβ mainline splitting relative to other iron–sulfur clusters (Fig. 6; Rees et al., 2017).

3.3 Valence-to-Core XES

To highest energy are the valence-to-core (VtC) XES lines, which are also known as the K$\beta_{2,5}$ and Kβ'' features (DeBeer & Bergmann, 2016; Glatzel & Bergmann, 2005; Lee et al., 2010). This spectral region results from electrons in filled ligand orbitals refilling the 1s core hole on the metal. While the probability of a VtC emission event occurring is much lower (~100 time weaker than the Kβ mainlines), the information content is much richer. This is due to the fact that the VtC XES is intrinsically sensitive to the ionization energy of the ligands coordinated to the photoabsorber. Hence, while in Fig. 5A the Kβ mainlines for the ferric complexes are all very similar, one notes that the corresponding VtC XES spectra (Fig. 5B) vary greatly. This is a direct reflection of the differences in ligands. Early applications of VtC XES in bioinorganic chemistry have included its use to establish the presence of bridging oxo groups in the Mn_4CaO_5 active site of the oxygen-evolving complex in photosystem II (Pushkar et al., 2010) and to establish the presence of a central carbide in M-cluster of both the Mo- (Lancaster et al., 2011) and V-dependent (Rees, Bjornsson, et al., 2015) nitrogenases. In addition, VtC XES has been shown to be sensitive to the extent of bond activation (for bond N_2, O_2, and NO), as well as the protonation state of ligands (Kupper, Rees, Dechert, DeBeer, & Meyer, 2016; Lassalle-Kaiser et al., 2013; Lu et al., 2014; Pollock et al., 2013; Rees, Martin-Diaconescu, Kovacs, & DeBeer, 2015). Finally, VtC XES has also been utilized to understand the changes in active site structure in M-cluster biosynthesis (Lancaster et al., 2013).

3.4 XES Experimental Considerations

Having highlighted the utility of XES in the preceding sections, we now turn to some of the important experimental considerations (DeBeer & Bergmann, 2016; Glatzel & Bergmann, 2005; Kowalska, Lima, Pollock, Rees, & DeBeer, 2016). As with XAS on dilute proteins, XES on dilute proteins also generally requires an intense synchrotron beamline as the source of incident X-rays. Some of the general requirements for an ideal beamline for XES on biological samples are detailed later:

(1) A wiggler or an undulator insertion device providing an intense source of incident X-ray (~10^{12} Ph/s or higher).

(2) A double crystal monochromator or a multilayer mirror for ~1–100 eV band pass of incident X-rays. Note, that in contrast to XAS, XES does not require a narrow band incident beam. High resolution is only needed if one wishes to perform resonant XES (RXES) (as discussed in Section 4).

(3) A high-resolution crystal analyzer array, as described in detail in Anklamm et al. (2014), DeBeer and Bergmann (2016), Glatzel and Bergmann (2005), and Kowalska, Lima, et al. (2016). The setup may employ either a Rowland or a von Hamos geometry. As a detector, either a diode- or an energy-resolving detector may be used in a Rowland geometry. For dilute samples, an energy-resolving detector is desirable. In a von Hamos geometry either a line detector or a CCD is needed. The advantage of the von Hamos geometry is that it is intrinsically dispersive (all energies are measured at the same time), and hence it can readily be used for time-resolved measurements.

The sample requirements for XES are essential identical to those described for XAS in Section 3.4; therefore, we do not repeat those details here. We only wish to emphasize that for the weak VtC XES features, even higher sample concentrations (several m*M* or higher) are generally desirable.

Similar to XAS, it is highly desirable to have some means to assess sample damage. As the Kβ mainline may not be sufficiently sensitive to assess damage, the ability to also measure relatively short (~10 s or less) XAS scans to assess the maximum X-ray dose rate is often very helpful. Hence while a ~1-eV resolution incident beam in by no means required for a nonresonant XES measurement, it is often desirable for the assessment of damage.

4. RXES/RIXS MEASUREMENTS IN THE HARD X-RAY REGIME

In order to obtain even deeper electronic structural insights into iron–sulfur active sites, one can combine XAS and XES in a so-called RXES or resonant inelastic X-ray scattering (RIXS) measurements (DeBeer, 2015; DeBeer & Bergmann, 2016; de Groot, 2001; Glatzel & Bergmann, 2005). In such a measurement, one can measure XAS together with Kα, Kβ, or VtC XES. These are then referred to as 1s2p, 1s3p, and 1s-VtC RIXS, respectively. Here, we briefly highlight the information content of each type of RIXS measurement.

4.1 1s2p RIXS

In a 1s2p RIXS experiment, one measures the metal K-edge in one direction and the 2p–1s XES in the other direction (Fig. 7; Lundberg et al., 2013). If one considers a ferric complex with a $1s^2 2p^6 3d^5$ ground state, then after a 1s to 3d transition, one arrives at a $1s^1 2p^6 3d^6$ intermediate state. After 2p–1s emission, the final state in $1s^2 2p^5 3d^6$ or the equivalent of an "L-edge like" ground state. L-edges involve direct dipole allowed 2p to 3d transitions and thus can provide rich electronic structural information about an iron active site. However, as an Fe L-edge occurs at ~700 eV (in contrast to over

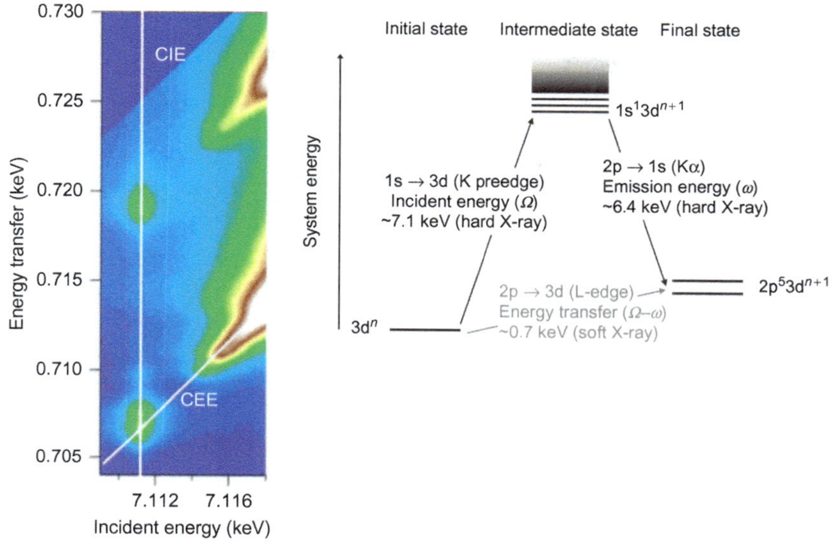

Fig. 7 1s2p RIXS for an iron model complex (*left*), showing the constant incident energy (CIE) and constant emission energy (CEE) directions. Schematic for the states involved in 1s2p RIXS. *Adapted from Lundberg, M., Kroll, T., DeBeer, S., Bergmann, U., Wilson, S. A., Glatzel, P., et al. (2013). Metal-ligand covalency of iron complexes from high-resolution resonant inelastic X-ray scattering.* Journal of the American Chemical Society, 135, *17121–17134. Copyright 2013 American Chemical Society.*

7100 eV for the Fe K-edge) ultrahigh vacuum conditions are typically required for the experiment. The 1s2p RIXS experiment circumvents this problem, allowing for soft X-ray information to be obtained with a hard X-ray probe. A disadvantage, however, is that collecting an entire RIXS plane is quite time consuming and also requires relatively large quantities of sample.

An alternative measurement that can be performed is a so-called high-energy resolution fluorescence-detected (HERFD) XAS experiment (Hamalainen, Siddons, Hastings, & Berman, 1991). A HERFD XAS spectrum basically involves measuring the XAS at a single XES energy. In the case of Kα-detected HERFD, the major advantage is improved spectral resolution. Fig. 8 shows a comparison of a typical total fluorescence yield XAS spectrum as compared to a HERFD XAS spectrum at the Mo K-edge (Lima et al., 2013). One immediately notes that the HERFD spectra are more richly featured, which thus allows for more quantitative information to be extracted. We refer the reader to the original literature and more recent review articles for a detailed discussion of the origins of HERFD spectral

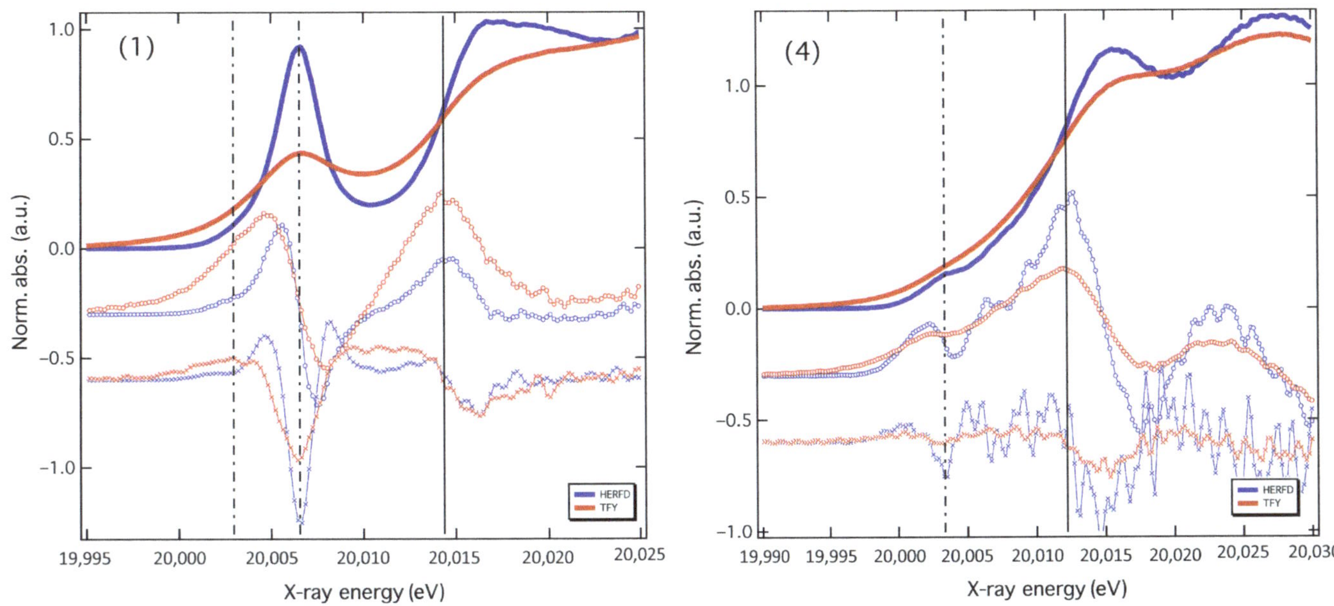

Fig. 8 A comparison of HERFD (*blue*) vs total fluorescence yield XAS at the Mo K-edge for two different model complexes. In each graph the *top* plots are the XAS, the *middle* plots are the first derivatives, and the *bottom* plots are the second derivatives. *Reproduced from Hamalainen, K., Siddons, D. P., Hastings, J. B., & Berman, L. E. (1991). Elimination of the inner-shell lifetime broadening in X-ray-absorption spectroscopy.* Physical Review Letters, 67*, 2850–2853, with permission from the PCCP Owner Societies.*

sharpening (Hamalainen et al., 1991; Kowalska, Lima, et al., 2016; Lima et al., 2013). We note that recent Fe Kα-detected HERFD XAS of V- and Mo-dependent nitrogenases has been important in identifying metal-to-metal charge transfer features in the nitrogenase M-clusters (Rees et al., 2017). Most notably, these studies have shown a significant Mo–Fe bonding contribution in FeMoco relative to FeVco.

4.2 1s3p RIXS

In a 1s–3p RIXS experiment, the final state which one arrives at is $1s^2 2p^5 3d^{n+1}$ configuration or the equivalent of an M-edge. To date, the M-edge-like energy transfer axis in 1s3p RIXS experiments has to be explored in detail. The primary use of 1s3p RIXS has been to obtain oxidation-state-dependent or spin-selective XAS (Glatzel & Bergmann, 2005; Lambertz et al., 2014; Leidel et al., 2012; Rees et al., 2017). In such an experiment, one takes advantage of the 3p–3d exchange contribution to the Kβ mainlines in order to select for XAS spectra from specific final state multiplets. This approach has been used to separate the high-spin FeS sites in hydrogenase from the low-spin carbonyl/cyanide coordinated H-cluster (Lambertz et al., 2014). In addition, Kβ-detected XAS has been used to examine spin-selective excitation channels in nitrogenase and related model complexes (Rees et al., 2017).

4.3 1s-VtC RIXS

Finally one can combine the weak VtC XES region with XAS in a 1s-VtC RIXS experiment. As the VtC region is strongly ligand dependent, such data can in principle be used to obtain "ligand-selective" XAS. To our knowledge, only one ligand-selective XAS study has been reported to date. Using this approach Hall et al. were able to show large spectral changes occur when XAS data are collected from different VtC XES (Hall et al., 2014). The ability to interpret these spectra theoretically, however, remains a significant challenge. These efforts are ongoing in our laboratory. In addition, applications to iron–sulfur models and nitrogenase are underway.

4.4 Experimental Considerations

The experimental considerations for hard X-ray RIXS experiments are essentially identical to those discussed for nonresonant XES in Section 3.4. The primary difference is that a ~1 eV or better resolution is required for the incident X-ray beam. This thus limits these experiments (for a practical purposes) to high-intensity undulator beamlines. This is particularly true when working with dilute metalloproteins. As it is generally desirable to have a focused beam for these experiments, the issues of radiation damage become even more acute. Finally, when one desires full RIXS planes the data collection time scales up tremendously. Hence, in our view, such RIXS experiments should only be performed when the desired information cannot be accessed by conventional XAS or XES approaches.

5. RIXS AND XMCD MEASUREMENTS IN THE SOFT X-RAY REGIME

Up to this point, we have focused on hard X-ray spectroscopies, which can be readily applied to biological samples. We note, however, that in many cases even more detailed electronic structural information can be obtained by performing measurements in the soft X-ray regime (Cramer, 2003; de Groot, 2001; Van Kuiken, Hahn, Maganas, & DeBeer, 2016; Wang, Bryant, LeGros, Wang, & Cramer, 2012; Wang, deGroot, & Cramer, 1996; Wang, Patil, Ralston, Bryant, & Cramer, 2001). Examples of this include 2p3d RIXS (Hahn et al., 2017; Van Kuiken et al., 2016) and Fe L-edge XMCD (Kowalska et al., 2017), which we briefly describe here.

In a 2p3d RIXS measurement, one begins at a $2p^63d^n$ ground state, arrives at $2p^53d^{n+1}$ intermediate state, and subsequent decay results in a $2p^63d^{n\prime}$ final state. The difference between the initial $2p^63d^n$ ground state and the $2p^63d^{n\prime}$ final state is the equivalent of a d-to-d transition. However, it has been arrived at via two dipole allowed processes. Further due to 2p spin orbit coupling in the intermediate state, formally spin-forbidden transitions may gain intensity. Such experiments, thus in principle, allow one to map out the low-lying excited states in an element-specific fashion. Recently, we have made some of the first applications of 2p3d RIXS to molecular systems (Hahn et al., 2017; Van Kuiken et al., 2016). While protein studies have not yet been reported, these efforts are underway in our laboratory.

Finally, we wish to briefly highlight the information content of Fe L-edge XMCD. In an XMCD measurement one examines the Fe L-edge with left and right circularly polarized X-rays. In a simple picture, the left circularly polarized X-rays dominantly excite the beta electrons, while the right circularly polarized X-ray dominantly excites the alpha electrons. As such, there will be intense L-edge absorption for a high-spin Fe on the left circularly polarized excitation channel (as there are five unoccupied beta holes) and very weak absorption on the right circularly polarized excitation channel. This approach has been recently applied to a series of iron–sulfur models and nitrogenase cluster analogues (Kowalska et al., 2017) and shows great promise for the extraction of subtle electronic structural details that were not readily observable by other methods.

6. SUMMARY AND OUTLOOK

Herein, we have provided a brief overview of both XAS and XES methods and their utility in obtaining electronic and geometric structural information, with a focus on iron–sulfur clusters. More advance RXES or RIXS methods have also been reviewed. In all cases, the experimental requirements are briefly discussed. Finally, emerging applications in the soft X-ray regime are highlighted. We note that the number of beamlines for high-resolution XES is growing rapidly, and while at the moment many of the applications discussed herein remain in the realm of the specialist, it is our view that increased availability of beamtime will open up these methods to a much broader community. Insights into the complex electronic structure of iron–sulfur proteins will likely greatly benefit from these ongoing experimental developments. Finally, we emphasize that ongoing efforts in theory developments will be important for obtaining a full understanding of complex X-ray spectroscopic data. These developments are ongoing in our laboratories in a collaborative effort with the Neese group, as well as in other research groups (Guo et al., 2016; Maganas, DeBeer, & Neese, 2014; Maganas, Kristiansen, et al., 2014; Maganas et al., 2013; Maganas, Roemelt, et al., 2014; Roemelt, Maganas, DeBeer, & Neese, 2013).

ACKNOWLEDGMENTS

I would like to thank all of my collaborators and coworkers, who contributed to the work referenced within this chapter. The Max Planck Society is acknowledged for funding.

REFERENCES

Anklamm, L., Schlesiger, C., Malzer, W., Grotzsch, D., Neitzel, M., & Kanngiesser, B. (2014). A novel von Hamos spectrometer for efficient X-ray emission spectroscopy in the laboratory. *The Review of Scientific Instruments*, *85*, 053110.

Beckwith, M. A., Roemelt, M., Collomb, M.-N. l., DuBoc, C., Weng, T.-C., Bergmann, U., et al. (2011). Manganese Kβ X-ray emission spectroscopy as a probe of metal–ligand interactions. *Inorganic Chemistry*, *50*, 8397–8409.

Corbett, M. C., Hu, Y. L., Fay, A. W., Ribbe, M. W., Hedman, B., & Hodgson, K. O. (2006). Structural insights into a protein-bound iron-molybdenum cofactor precursor.

Proceedings of the National Academy of Sciences of the United States of America, 103, 1238–1243.

Cramer, S. P. (2003). X-ray magnetic circular dichroism: A primer for chemists. *ACS Symposium Series, 858*, 358–402.

Cramer, S. P., Gillum, W. O., Hodgson, K. O., Mortenson, L. E., Stiefel, E. I., Chisnell, J. R., et al. (1978). Molybdenum site of nitrogenase. 2. Comparative-study of Mo-Fe proteins and iron-molybdenum cofactor by X-ray absorption spectroscopy. *Journal of the American Chemical Society, 100*, 3814–3819.

DeBeer, S. (2011). X-ray absorption spectroscopy. *Methods in Molecular Biology, 766*, 165–176.

DeBeer, S. (2015). Biological catalysts. In J. Sa (Ed.), *High-resolution XAS/XES* (pp. 153–168). New York: CRC Press.

DeBeer, S., & Bergmann, U. (2016). X-ray emission spectroscopic techniques in bioinorganic applications. In R. A. Scott (Ed.), *Encyclopedia of inorganic and bioinorganic chemistry*. Chichester: John Wiley, https://doi.org/10.1002/9781119951438.eibc2158.

DeBeer George, S., Brant, P., & Solomon, E. I. (2005). Metal and ligand K-edge XAS of organotitanium complexes: Metal 4p and 3d contributions to pre-edge intensity and their contributions to bonding. *Journal of the American Chemical Society, 127*, 667–674.

de Groot, F. (2001). High resolution X-ray emission and X-ray absorption spectroscopy. *Chemical Reviews, 101*, 1779–1808.

Delgado-Jaime, M. U., DeBeer, S., & Bauer, M. (2013). Valence-to-core X-ray emission spectroscopy of iron–carbonyl complexes: Implications for the examination of catalytic intermediates. *Chemistry—A European Journal, 19*, 15888–15897.

Delgado-Jaime, M. U., Dible, B. R., Chiang, K. P., Brennessel, W. W., Bergmann, U., Holland, P. L., et al. (2011). Identification of a single light atom within a multinuclear metal cluster using valence-to-core X-ray emission spectroscopy. *Inorganic Chemistry, 50*, 10709–10717.

Dey, A., Francis, E. J., Adams, M. W. W., Babini, E., Takahashi, Y., Fukuyama, K., et al. (2007). Solvent tuning of electrochemical potentials in the active sites of HiPIP versus ferredoxin. *Science, 318*, 1464–1468.

Gallo, E., & Glatzel, P. (2014). Valence to core X-ray emission spectroscopy. *Advanced Materials, 26*, 7730–7746.

Gamblin, S. D., & Urch, D. S. (2001). Metal K beta X-ray emission spectra of first row transition metal compounds. *Journal of Electron Spectroscopy and Related Phenomena, 113*, 179–192.

George, S. J., Barney, B. M., Mitra, D., Igarashi, R. Y., Guo, Y. S., Dean, D. R., et al. (2012). EXAFS and NRVS reveal a conformational distortion of the FeMo-cofactor in the MoFe nitrogenase propargyl alcohol complex. *Journal of Inorganic Biochemistry, 112*, 85–92.

George, G. N., Coyle, C. L., Hales, B. J., & Cramer, S. P. (1988). X-ray absorption of Azotobacter-Vinelandii vanadium nitrogenase. *Journal of the American Chemical Society, 110*, 4057–4059.

Glaser, T., Hedman, B., Hodgson, K. O., & Solomon, E. I. (2000). Ligand K-edge X-ray absorption spectroscopy: A direct probe of ligand-metal covalency. *Accounts of Chemical Research, 33*, 859–868.

Glaser, T., Rose, K., Anxolabehere-Mallart, E., Shadle, S. E., Frank, P., Hedman, B., et al. (1999). The electronic structures of iron-sulfur clusters in models and proteins studied by SK-edge XAS. *Journal of Inorganic Biochemistry, 74*, 142.

Glaser, T., Rose, K., Shadle, S. E., Hedman, B., Hodgson, K. O., & Solomon, E. I. (2000). Delocalization vs. localization in tetrameric and dimeric iron-sulfur clusters studied by SK-edge xas. *Abstracts of Papers of the American Chemical Society, 219*, U782.

Glaser, T., Rose, K., Shadle, S. E., Hedman, B., Hodgson, K. O., & Solomon, E. I. (2001). SK-edge X-ray absorption studies of tetranuclear iron-sulfur clusters: Mu-sulfide bonding and its contribution to electron delocalization. *Journal of the American Chemical Society, 123*, 442–454.

Glatzel, P., & Bergmann, U. (2005). High resolution 1s core hole X-ray spectroscopy in 3d transition metal complexes—electronic and structural information. *Coordination Chemistry Reviews, 249*, 65–95.

Guo, M. Y., Kallman, E., Sorensen, L. K., Delcey, M. G., Pinjari, R. V., & Lundberg, M. (2016). Molecular orbital simulations of metal 1s2p resonant inelastic X-ray scattering. *The Journal of Physical Chemistry. A, 120*, 5848–5855.

Hahn, A. W., Van Kuiken, B. E., al Samarai, M., Atanasov, M., Weyhermuller, T., Cui, Y.-T., et al. (2017). Measurement of the ligand field spectra of ferrous and ferric iron chlorides using 2p3d RIXS. *Inorganic Chemistry, 56*, 8203–8211.

Hall, E. R., Pollock, C. J., Bendix, J., Collins, T. J., Glatzel, P., & DeBeer, S. (2014). Valence-to-core-detected X-ray absorption spectroscopy: Targeting ligand selectivity. *Journal of the American Chemical Society, 136*, 10076–10084.

Hamalainen, K., Siddons, D. P., Hastings, J. B., & Berman, L. E. (1991). Elimination of the inner-shell lifetime broadening in X-ray-absorption spectroscopy. *Physical Review Letters, 67*, 2850–2853.

Hoffman, B. M., Lukoyanov, D., Dean, D. R., & Seefeldt, L. C. (2013). Nitrogenase: A draft mechanism. *Accounts of Chemical Research, 46*, 587–595.

Holm, R. H., Kennepohl, P., & Solomon, E. I. (1996). Structural and functional aspects of metal sites in biology. *Chemical Reviews, 96*, 2239–2314.

Johnson, D. C., Dean, D. R., Smith, A. D., & Johnson, M. K. (2005). Structure, function, and formation of biological iron-sulfur clusters. *Annual Review of Biochemistry, 74*, 247–281.

Kau, L. S., Spirasolomon, D. J., Pennerhahn, J. E., Hodgson, K. O., & Solomon, E. I. (1987). X-ray absorption-edge determination of the oxidation-state and coordination-number of copper—Application to the type-3 site in Rhus-Vernicifera Laccase and its reaction with oxygen. *Journal of the American Chemical Society, 109*, 6433–6442.

Kowalska, J., & DeBeer, S. (2015). The role of X-ray spectroscopy in understanding the geometric and electronic structure of nitrogenase. *Biochimica et Biophysica Acta-Molecular Cell Research, 1853*, 1406–1415.

Kowalska, J. K., Hahn, A. W., Albers, A., Schiewer, C. E., Bjornsson, R., Lima, F. A., et al. (2016). X-ray absorption and emission spectroscopic studies of [L2Fe2S2](n) model complexes: Implications for the experimental evaluation of redox states in iron-sulfur clusters. *Inorganic Chemistry, 55*, 4485–4497.

Kowalska, J. K., Lima, F. A., Pollock, C. J., Rees, J. A., & DeBeer, S. (2016). A practical guide to high-resolution X-ray spectroscopic measurements and their applications in bioinorganic chemistry. *Israel Journal of Chemistry, 56*, 803–815.

Kowalska, J. K., Nayyar, B., Rees, J. A., Schiewer, C. E., Lee, S. C., Kovacs, J. A., et al. (2017). Iron L2,3-edge X-ray absorption and magnetic circular dichroism studies of molecular iron complexes with relevance to the FeMoco and FeVco active sites of nitrogenase. *Inorganic Chemistry, 56*, 8147–8158.

Kupper, C., Rees, J. A., Dechert, S., DeBeer, S., & Meyer, F. (2016). Complete series of {FeNO}(8), {FeNO}(7), and {FeNO}(6) complexes stabilized by a tetracarbene macrocycle. *Journal of the American Chemical Society, 138*, 7888–7898.

Lambertz, C., Chernev, P., Klingan, K., Leidel, N., Sigfridsson, K. G. V., Happe, T., et al. (2014). Electronic and molecular structures of the active-site H-cluster in [FeFe]-hydrogenase determined by site-selective X-ray spectroscopy and quantum chemical calculations. *Chemical Science, 5*, 1187–1203.

Lancaster, K. M., Hu, Y. L., Bergmann, U., Ribbe, M. W., & DeBeer, S. (2013). X-ray spectroscopic observation of an interstitial carbide in NifEN-bound FeMoco precursor. *Journal of the American Chemical Society*, *135*, 610–612.

Lancaster, K. M., Roemelt, M., Ettenhuber, P., Hu, Y., Ribbe, M. W., Neese, F., et al. (2011). X-ray emission spectroscopy evidences a central carbon in the nitrogenase iron-molybdenum cofactor. *Science*, *334*, 974–977.

Lassalle-Kaiser, B., Boron, T. T., Krewald, V., Kern, J., Beckwith, M. A., Delgado-Jaime, M. U., et al. (2013). Experimental and computational X-ray emission spectroscopy as a direct probe of protonation states in oxo-bridged MnIV dimers relevant to redox-active metalloproteins. *Inorganic Chemistry*, *52*, 12915–12922.

Lee, S. C., & Holm, R. H. (2004). The clusters of nitrogenase: Synthetic methodology in the construction of weak-field clusters. *Chemical Reviews*, *104*, 1135–1157.

Lee, S. C., Lo, W., & Holm, R. H. (2014). Developments in the biomimetic chemistry of cubane-type and higher nuclearity iron-sulfur clusters. *Chemical Reviews*, *114*, 3579–3600.

Lee, N., Petrenko, T., Bergmann, U., Neese, F., & DeBeer, S. (2010). Probing valence orbital composition with iron K beta X-ray emission spectroscopy. *Journal of the American Chemical Society*, *132*, 9715–9727.

Leidel, N., Chernev, P., Havelius, K. G. V., Schwartz, L., Ott, S., & Haumann, M. (2012). Electronic structure of an [FeFe] hydrogenase model complex in solution revealed by X-ray absorption spectroscopy using narrow-band emission detection. *Journal of the American Chemical Society*, *134*, 14142–14157.

Lill, R., & Muhlenhoff, U. (2008). Maturation of iron-sulfur proteins in eukaryotes: Mechanisms, connected processes, and diseases. *Annual Review of Biochemistry*, *77*, 669–700.

Lima, F. A., Bjornsson, R., Weyhermuller, T., Chandrasekaran, P., Glatzel, P., Neese, F., et al. (2013). High-resolution molybdenum K-edge X-ray absorption spectroscopy analyzed with time-dependent density functional theory. *Physical Chemistry Chemical Physics*, *15*, 20911–20920.

Lu, T. T., Weng, T. C., & Liaw, W. F. (2014). X-ray emission spectroscopy: A spectroscopic measure for the determination of NO oxidation states in Fe-NO complexes. *Angewandte Chemie (International ed. in English)*, *53*, 11562–11566.

Lundberg, M., Kroll, T., DeBeer, S., Bergmann, U., Wilson, S. A., Glatzel, P., et al. (2013). Metal-ligand covalency of iron complexes from high-resolution resonant inelastic X-ray scattering. *Journal of the American Chemical Society*, *135*, 17121–17134.

Maganas, D., DeBeer, S., & Neese, F. (2014). Restricted open-shell configuration interaction cluster calculations of the L-edge X-ray absorption study of TiO_2 and CaF_2 solids. *Inorganic Chemistry*, *53*, 6374–6385.

Maganas, D., Kristiansen, P., Duda, L. C., Knop-Gericke, A., DeBeer, S., Schlogl, R., et al. (2014). Combined experimental and ab initio multireference configuration interaction study of the resonant inelastic X-ray scattering spectrum of CO_2. *Journal of Physical Chemistry C*, *118*, 20163–20175.

Maganas, D., Roemelt, M., Havecker, M., Trunschke, A., Knop-Gericke, A., Schlogl, R., et al. (2013). First principles calculations of the structure and V L-edge X-ray absorption spectra of V_2O_5 using local pair natural orbital coupled cluster theory and spin-orbit coupled configuration interaction approaches. *Physical Chemistry Chemical Physics*, *15*, 7260–7276.

Maganas, D., Roemelt, M., Weyhermuller, T., Blume, R., Havecker, M., Knop-Gericke, A., et al. (2014). L-edge X-ray absorption study of mononuclear vanadium complexes and spectral predictions using a restricted open shell configuration interaction ansatz. *Physical Chemistry Chemical Physics*, *16*, 264–276.

Musgrave, K. B., Angove, H. C., Burgess, B. K., Hedman, B., & Hodgson, K. O. (1998). All-ferrous titanium(III) citrate reduced Fe protein of nitrogenase: An XAS study of electronic and metrical structure. *Journal of the American Chemical Society*, *120*, 5325–5326.

Musgrave, K. B., Liu, H. I., Ma, L., Burgess, B. K., Watt, G., Hedman, B., et al. (1998). EXAFS studies on the P(N) and P(OX) states of the P-clusters in nitrogenase. *Journal of Biological Inorganic Chemistry, 3*, 344–352.

Penner-Hahn, J. E. (1999). X-ray absorption spectroscopy in coordination chemistry. *Coordination Chemistry Reviews, 190*, 1101–1123.

Pollock, C. J., & DeBeer, S. (2011). Valence-to-core X-ray emission spectroscopy: A sensitive probe of the nature of a bound ligand. *Journal of the American Chemical Society, 133*, 5594–5601.

Pollock, C. J., Delgado-Jaime, M. U., Atanasov, M., Neese, F., & DeBeer, S. (2014). K beta mainline X-ray emission spectroscopy as an experimental probe of metal-ligand covalency. *Journal of the American Chemical Society, 136*, 9453–9463.

Pollock, C. J., Grubel, K., Holland, P. L., & DeBeer, S. (2013). Experimentally quantifying small-molecule bond activation using valence-to-core X-ray emission spectroscopy. *Journal of the American Chemical Society, 135*, 11803–11808.

Pollock, C. J., Tan, L. L., Zhang, W., Lancaster, K. M., Lee, S. C., & DeBeer, S. (2014). Light-atom influences on the electronic structures of iron sulfur clusters. *Inorganic Chemistry, 53*, 2591–2597.

Pushkar, Y., Long, X., Glatzel, P., Brudvig, G. W., Dismukes, G. C., Collins, T. J., et al. (2010). Direct detection of oxygen ligation to the Mn4Ca cluster of photosystem II by XES. *Angewandte Chemie, International Edition, 49*, 800–803.

Rees, J. A., Bjornsson, R., Kowalska, J. K., Lima, F. A., Schlesier, J., Sippel, D., et al. (2017). Comparative electronic structures of nitrogenase FeMoco and FeVco. *Dalton Transactions, 46*, 2445–2455.

Rees, J. A., Bjornsson, R., Schlesier, J., Sippel, D., Einsle, O., & DeBeer, S. (2015). The Fe-V cofactor of vanadium nitrogenase contains an interstitial carbon atom. *Angewandte Chemie (International ed. in English), 54*, 13249–13252.

Rees, D. C., & Howard, J. B. (2000). Nitrogenase: Standing at the crossroads. *Current Opinion in Chemical Biology, 4*, 559–566.

Rees, J. A., Martin-Diaconescu, V., Kovacs, J. A., & DeBeer, S. (2015). X-ray absorption and emission study of dioxygen activation by a small-molecule manganese complex. *Inorganic Chemistry, 54*, 6410–6422.

Roemelt, M., Maganas, D., DeBeer, S., & Neese, F. (2013). A combined DFT and restricted open-shell configuration interaction method including spin-orbit coupling: Application to transition metal L-edge X-ray absorption spectroscopy. *The Journal of Chemical Physics, 138*, 204101.

Rose, K., Shadle, S. E., Glaser, T., de Vries, S., Cherepanov, A., Canters, G. W., et al. (1999). Investigation of the electronic structure of 2Fe-2S model complexes and the Rieske protein using ligand K-edge X-ray absorption spectroscopy. *Journal of the American Chemical Society, 121*, 2353–2363.

Shulman, R. G., Eisenberger, P., & Kincaid, B. M. (1978). X-ray absorption spectroscopy of biological molecules. *Annual Review of Biophysics and Bioengineering, 7*, 559–578.

Shulman, R. G., Eisenberger, P., Teo, B. K., Kincaid, B. M., & Brown, G. S. (1978). Fluorescence X-ray absorption studies of rubredoxin and its model compounds. *Journal of Molecular Biology, 124*, 305–321.

Smolentsev, G., Soldatov, A. V., Messinger, J., Merz, K., Weyhermuller, T., Bergmann, U., et al. (2009). X-ray emission spectroscopy to study ligand valence orbitals in Mn coordination complexes. *Journal of the American Chemical Society, 131*, 13161–13167.

Spatzal, T., Aksoyoglu, M., Zhang, L. M., Andrade, S. L. A., Schleicher, E., Weber, S., et al. (2011). Evidence for interstitial carbon in nitrogenase FeMo cofactor. *Science, 334*, 940.

Van Kuiken, B. E., Hahn, A. W., Maganas, D., & DeBeer, S. (2016). Measuring spin-allowed and spin-forbidden d-d excitations in vanadium complexes with 2p3d resonant inelastic X-ray scattering. *Inorganic Chemistry, 55*, 11497–11501.

van Schooneveld, M. M., & DeBeer, S. (2015). A close look at dose: Toward L-edge XAS spectral uniformity, dose quantification and prediction of metal ion photoreduction. *Journal of Electron Spectroscopy and Related Phenomena, 198*, 31–56.
Wang, H. X., Bryant, C., LeGros, M., Wang, X., & Cramer, S. P. (2012). Fluorescence-detected X-ray magnetic circular dichroism of well-defined Mn-II and Ni-II doped in MgO crystals: Credential evaluation for measurements on biological samples. *The Journal of Physical Chemistry. A, 116*, 10082–10089.
Wang, X., deGroot, F., & Cramer, S. P. (1996). X-ray magnetic circular dichroism spectra and distortions at Fe^{2+} L(2,3) edges. *Journal of Electron Spectroscopy and Related Phenomena, 78*, 337–340.
Wang, H., Patil, D. S., Ralston, C. Y., Bryant, C., & Cramer, S. P. (2001). L-edge X-ray magnetic circular dichroism of Ni enzymes: Direct probe of Ni spin states. *Journal of Electron Spectroscopy and Related Phenomena, 114*, 865–871.
Westre, T. E., Kennepohl, P., DeWitt, J. G., Hedman, B., Hodgson, K. O., & Solomon, E. I. (1997). A multiplet analysis of Fe K-edge 1s→3d pre-edge features of iron complexes. *Journal of the American Chemical Society, 119*, 6297–6314.

AUTHOR INDEX

Note: Page numbers followed by "*f*" indicate figures, and "*t*" indicate tables.

A

C

D

G

H

I

J

K

M

N

Q

R

S

T

U

V

W

X

Y

Z

SUBJECT INDEX

Note: Page numbers followed by "*f*" indicate figures, "*t*" indicate tables, and "*s*" indicate schemes.

D

E

H

I

K

L

M

N

O

P

Kolibri 360 GmbH
Ferdinand-Jühlke-Straße 7
99095 Erfurt, Deutschland
produktsicherheit@kolibri360.de